SCIENCE

A CLOSER LOOK

Mc Graw Hill Macmillan/McGraw-Hill

Program Authors

Dr. Jay K. Hackett
Professor Emeritus of Earth Sciences
University of Northern Colorado
Greeley, CO

Dr. Richard H. Moyer
Professor of Science Education and Natural Sciences
University of Michigan-Dearborn
Dearborn, MI

Dr. JoAnne Vasquez
Elementary Science Education Consultant
NSTA Past President
Member, National Science Board and NASA Education Board

Mulugheta Teferi, M.A.
Principal, Gateway Middle School
Center of Math, Science, and Technology
St. Louis Public Schools
St. Louis, MO

Dinah Zike, M.Ed.
Dinah Might Adventures LP
San Antonio, TX

Kathryn LeRoy, M.S.
Chief Officer
Curriculum Services
Duval County Public Schools, FL

Dr. Dorothy J. T. Terman
Science Curriculum Development Consultant
Former K–12 Science and Mathematics Coordinator
Irvine Unified School District, CA
Irvine, CA

Dr. Gerald F. Wheeler
Executive Director
National Science Teachers Association

Bank Street College of Education
New York, NY

Contributing Authors

Dr. Sally Ride
Sally Ride Science
San Diego, CA

Lucille Villegas Barrera, M.Ed.
Elementary Science Supervisor
Houston Independent School District
Houston, TX

American Museum of Natural History
New York, NY

Contributing Writer

Ellen C. Grace, M.S.
Consultant
Albuquerque, NM

RFB&D learning through listening Students with print disabilities may be eligible to obtain an accessible, audio version of the pupil edition of this textbook. Please call Recording for the Blind & Dyslexic at 1-800-221-4792 for complete information.

The McGraw-Hill Companies

Macmillan/McGraw-Hill

Send all inquiries to:
Macmillan/McGraw-Hill
8787 Orion Place
Columbus, OH 43240-4027

FOLDABLES® is a registered trademark of The McGraw-Hill Companies, Inc.

ISBN: 978-0-02-288007-1
MHID: 0-02-288007-0

Printed in the United States of America.

2 3 4 5 6 7 8 9 10 WDQ/LEH 15 14 13 12 11 10

Content Consultants

Editorial Advisory Board

The American Museum of Natural History in New York City is one of the world's preeminent scientific, educational, and cultural institutions, with a global mission to explore and interpret human cultures and the natural world through scientific research, education, and exhibitions. Each year the Museum welcomes around 4 million visitors, including 500,000 schoolchildren in organized field trips. It provides professional development activities for thousands of teachers; hundreds of public programs that serve audiences ranging from preschoolers to seniors; and an array of learning and teaching resources for use in homes, schools, and community-based settings. Visit www.amnh.org for online resources.

Be a Scientist

AMERICAN MUSEUM OF NATURAL HISTORY

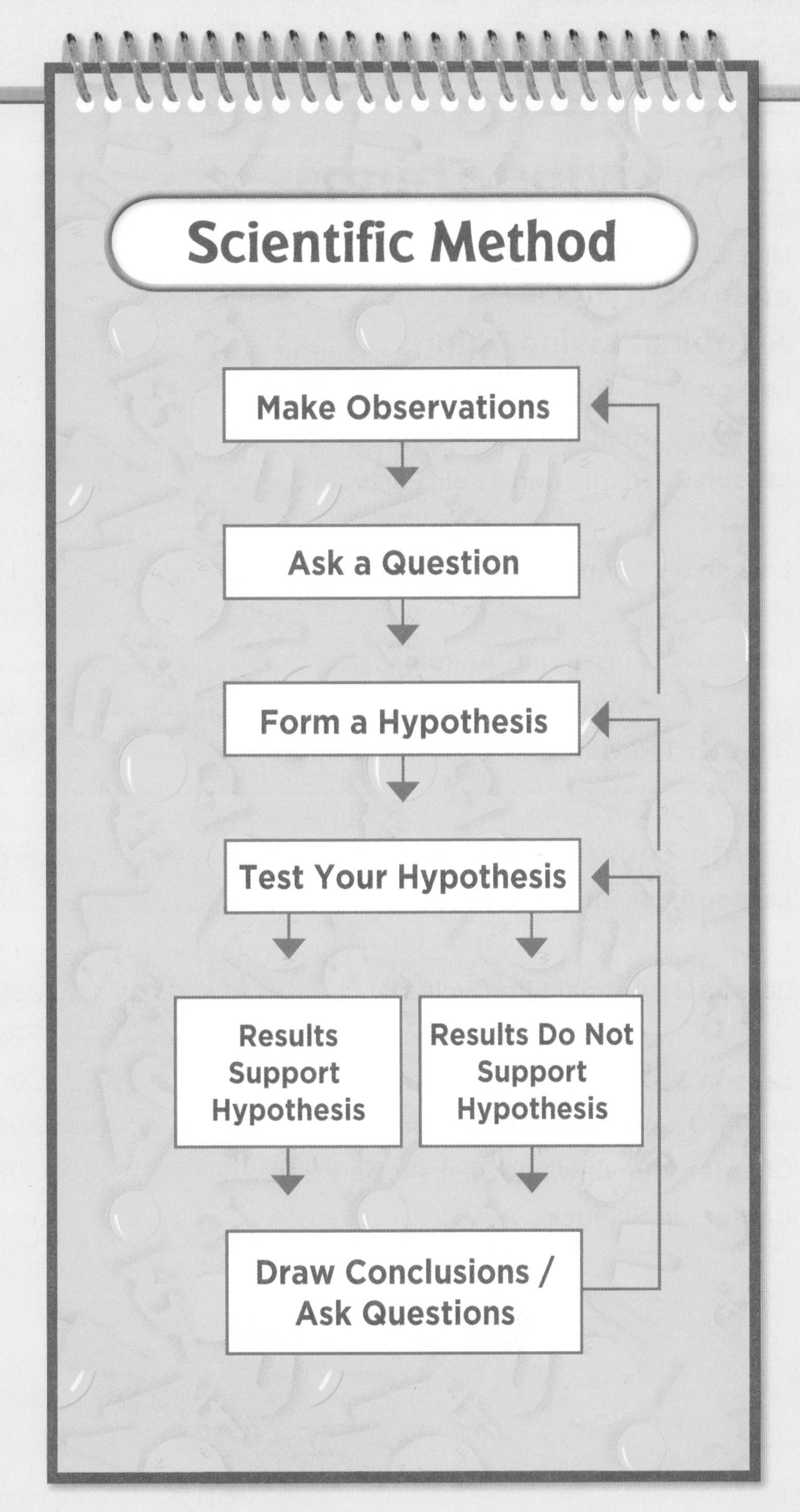
Scientific Method
Make Observations
Ask a Question
Form a Hypothesis
Test Your Hypothesis
Results Support Hypothesis
Results Do Not Support Hypothesis
Draw Conclusions / Ask Questions

Life Science

UNIT A Living Things

UNIT B Ecosystems

Earth and Space Science

UNIT C Earth and Its Resources

UNIT D Weather and Space

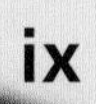

Physical Science

UNIT E Matter

UNIT F Forces and Energy

Online Resources

Animations

Science in Motion Animations online at www.macmillanmh.com

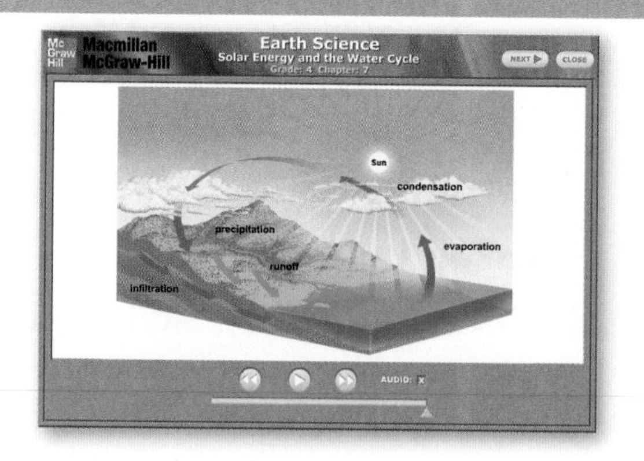

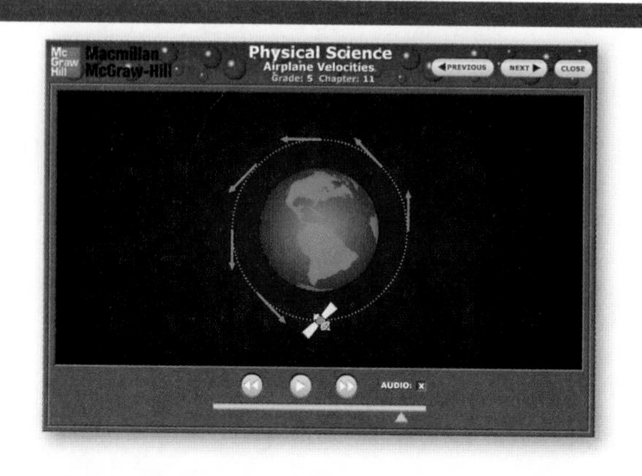

Additional Student Resources

Visit www.macmillanmh.com for additional student resources.

Online Student Edition

See the book online.

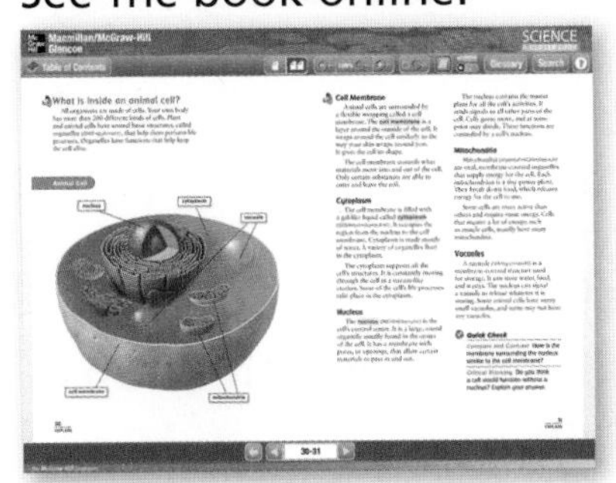

e-Journal

Discover and write about science.

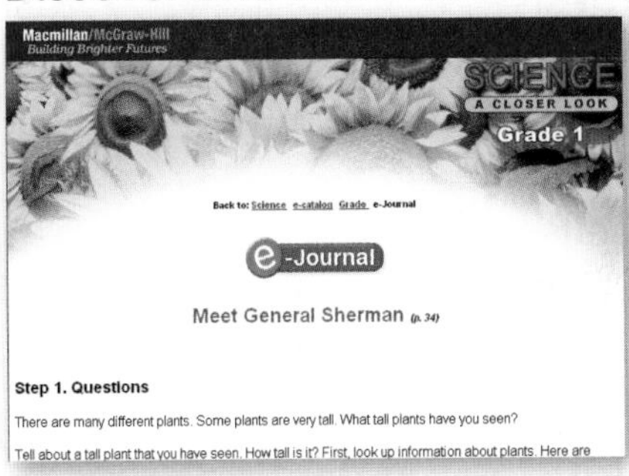

Vocabulary Games

Test your vocabulary knowledge.

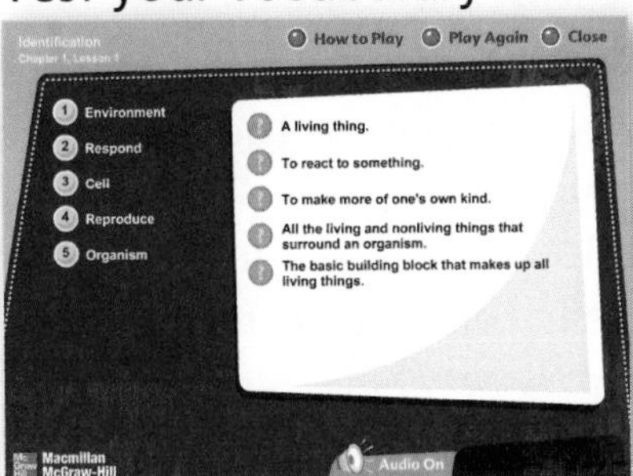

e-Glossary

Hear terms and review definitions.

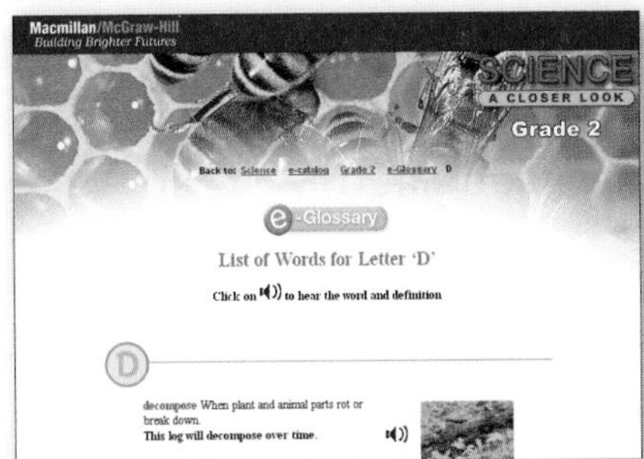

e-Career

Learn about science careers.

e-Review

Watch an animated summary and take a quiz for each lesson.

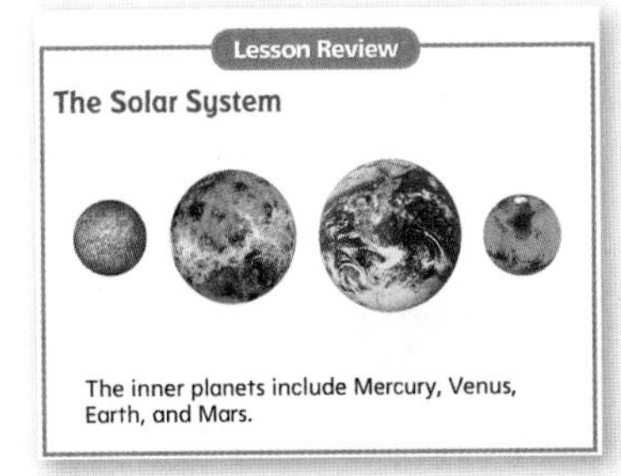

Other Online Resources

View images from NASA or see real-world connections at the American Museum of Natural History online.

Activities and Investigations

Life Science

CHAPTER 1

CHAPTER 2

CHAPTER 3

CHAPTER 4

snake

Sun

hawk

Activities and Investigations

Earth and Space Science

CHAPTER 5

Explore Activities

Quick Labs

Inquiry Skills and Investigations

CHAPTER 6

Explore Activities

Quick Labs

Inquiry Skills and Investigations

CHAPTER 7

Explore Activities

Quick Labs

Inquiry Skills and Investigations

CHAPTER 8

Explore Activities

Quick Labs

Inquiry Skills and Investigations

Activities and Investigations

Physical Science

Be a Scientist

giant Madagascan chameleon

Be a Scientist

The Scientific Method

Look and Wonder

Madagascar is a tropical island off the coast of Africa. It is home to plants and animals found nowhere else on Earth! What would it be like to live on a tropical island? What kinds of things might you find there?

Explore

What do you know about animals that live in Madagascar?

- How do you look for animals in their natural habitat?
- What kinds of animals would you find in a forest?
- What does an animal need to live in a forest?
- How do scientists find answers to these questions?

Chris Raxworthy

Paule Razafimahatratra

Meet two scientists who are curious about the natural world and everything that lives in it. Chris Raxworthy and Paule Razafimahatratra study animals that live in Madagascar. They work at the American Museum of Natural History in New York City and at the University of Antananarivo in Madagascar.

What do scientists do?

Chris and Paule want to find out about the many amazing animals that live in Madagascar. Much of the island has never been explored by scientists. New plants and animals are discovered all the time.

The scientific method is a process that scientists use to investigate the world around them. It helps them answer questions about the natural world.

Right now, Chris and Paule are studying a lizard called a giant Madagascan chameleon. Chris has observed these chameleons in dry forests. He wants to know where else in Madagascar the chameleons live.

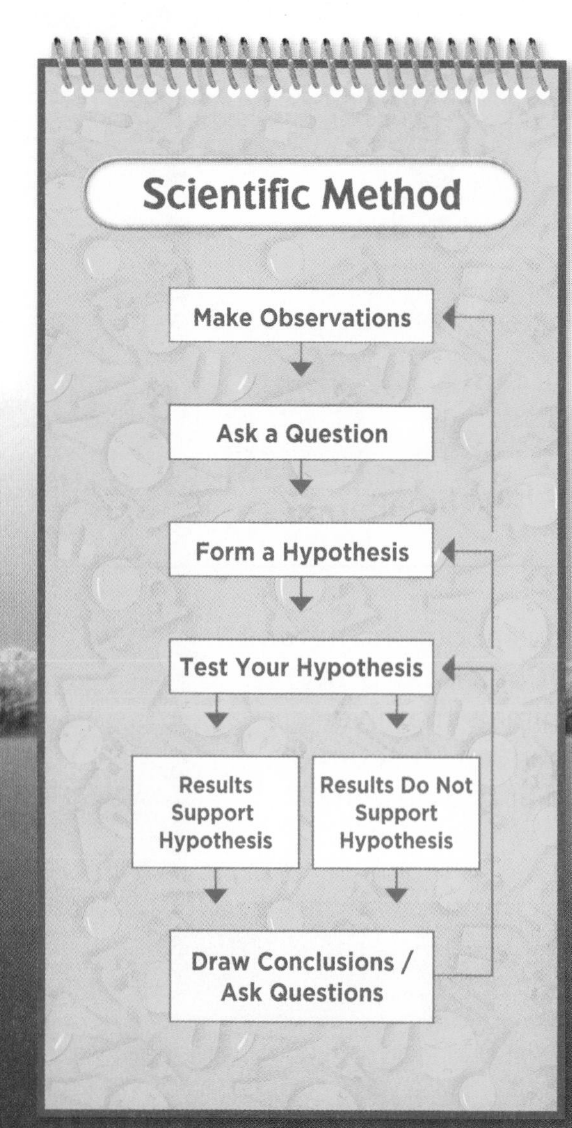

All scientists use the scientific method. However, they might not use all the steps, or they might do the steps in a different order.

Chris knows that variables such as temperature and rainfall affect where animals live. A variable is something that can change.

Chris uses this information to form a hypothesis. A hypothesis is a statement that can be tested to answer a question.

Here is Chris's hypothesis. If a place has temperatures between 10 and 40 degrees Celsius and between 50 and 150 centimeters of rainfall every year, then giant Madagascan chameleons could live there.

Form a Hypothesis

1. Ask lots of "why" questions.
2. Look for connections between important variables.
3. Suggest possible explanations for those connections.

▶ **Make sure the explanations can be tested.**

giant Madagascan chameleon ▲

How do scientists test a hypothesis?

The giant Madagascan chameleon is about as long as a banana. It is hard to find in the dense forest, though, because it hides. People in Madagascar say you can never find a chameleon when you are looking for one!

Where should Chris and Paule look for chameleons? In order to find out, they study their data about temperature and rainfall. Data is information. They put this data into a computer and make a map. The computer colors yellow all the areas that are likely to have chameleons. Those areas have similar temperatures and rainfall to places where chameleons have been found before. Chris predicts that if they go to those areas, they will find giant Madagascan chameleons.

observed
predicted

▲ The purple dots on this map show where giant Madagascan chameleons have been seen before. The yellow areas show where Chris and Paule think the chameleons live.

◀ Chris uses his headlamp to find chameleons at night when they sleep.

Chris and Paule choose new places to look for chameleons. They choose places that are in the yellow areas on the map. They collect data in these places to test their hypothesis. They use procedures that other scientists can repeat. That way other scientists can check Chris's and Paule's results.

"We wear headlamps and search at night, when the chameleons are sleeping and are easier to find," Chris explains. "We look up in the branches for pale-colored comma shapes." Every time they find a chameleon, Chris and Paule make careful notes and take photographs. They record the exact date, time, and place in their field journals.

Test Your Hypothesis

1. Think about the different kinds of data that could be used to test the hypothesis.
2. Choose the best method to collect this data.
   - **perform an experiment** (in the lab)
   - **observe the natural world** (in the field)
   - **make and use a model** (on a computer)
3. Then plan a procedure and gather data.

▶ **Make sure that the procedure can be repeated.**

Chris and Paule record data when they find giant Madagascan chameleons and other lizards.

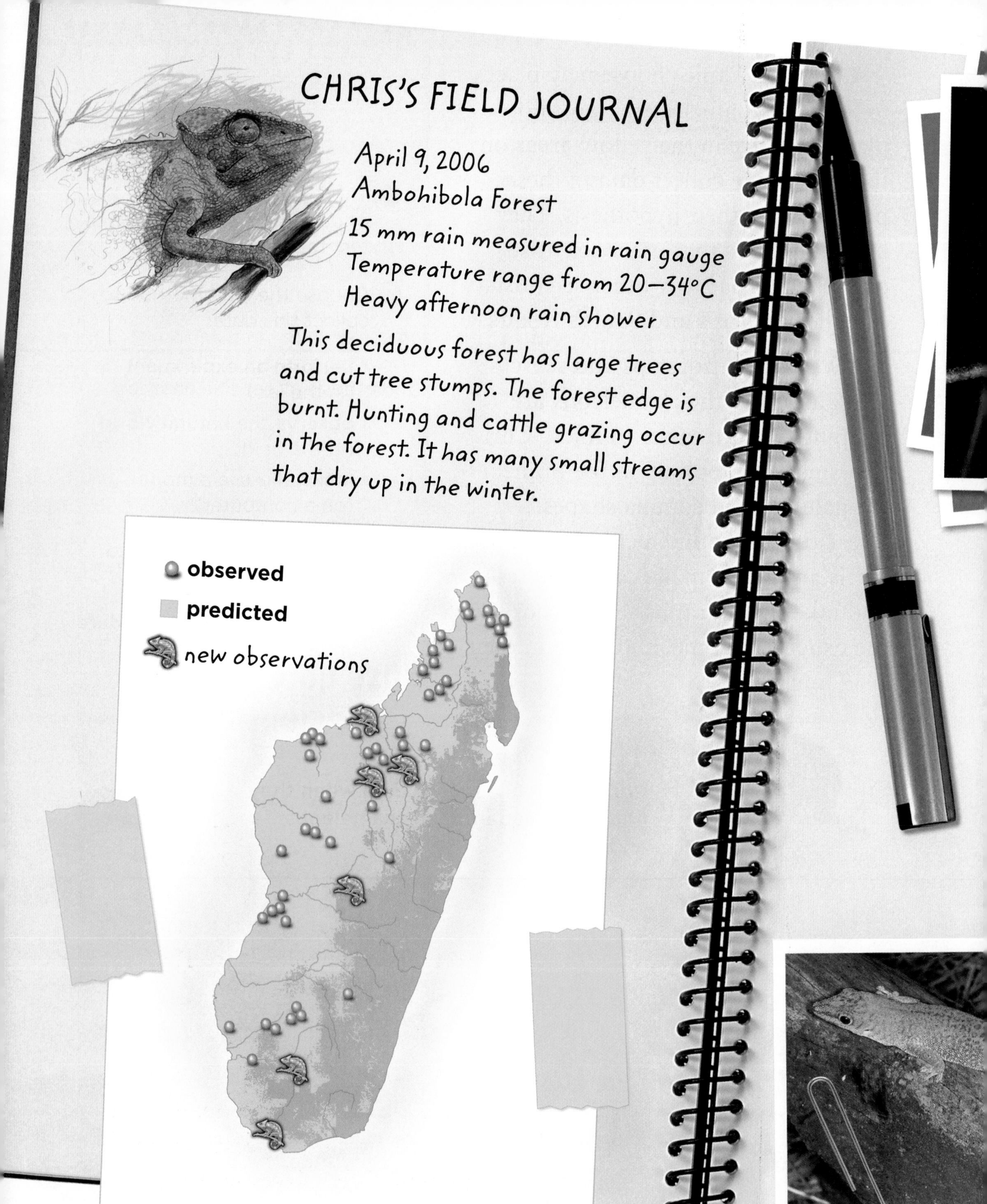

CHRIS'S FIELD JOURNAL
April 9, 2006
Ambohibola Forest
15 mm rain measured in rain gauge
Temperature range from 20–34°C
Heavy afternoon rain shower
This deciduous forest has large trees and cut tree stumps. The forest edge is burnt. Hunting and cattle grazing occur in the forest. It has many small streams that dry up in the winter.
observed
predicted
new observations

giant Madagascan chameleon (Furcifer oustaleti)
Found at 10:45 A.M. in grassland with scattered trees. Laid 17 eggs 14 x 8 mm.

Madagascan day gecko (Phelsuma madagascariensis) on a tree trunk at 11:30 A.M., in a small clump of trees growing by a small stream.

Analyze the Data

1. Organize the data as a chart, table, graph, diagram, map, or group of pictures.
2. Look for patterns in the data. These patterns can show how important variables in the hypothesis affect one another.

▶ **Make sure to check the data by comparing it to data from other sources.**

How do scientists analyze data?

Part of testing a hypothesis is looking for patterns in the data that has been collected. Chris and Paule study the information from all of the locations they visited. They mark the six places on the map where they found a giant Madagascan chameleon. Then they look for patterns in their data.

They observe that the chameleons they found were in the yellow area on the map. They talk about the temperatures and rainfall in the places where they found the chameleons.

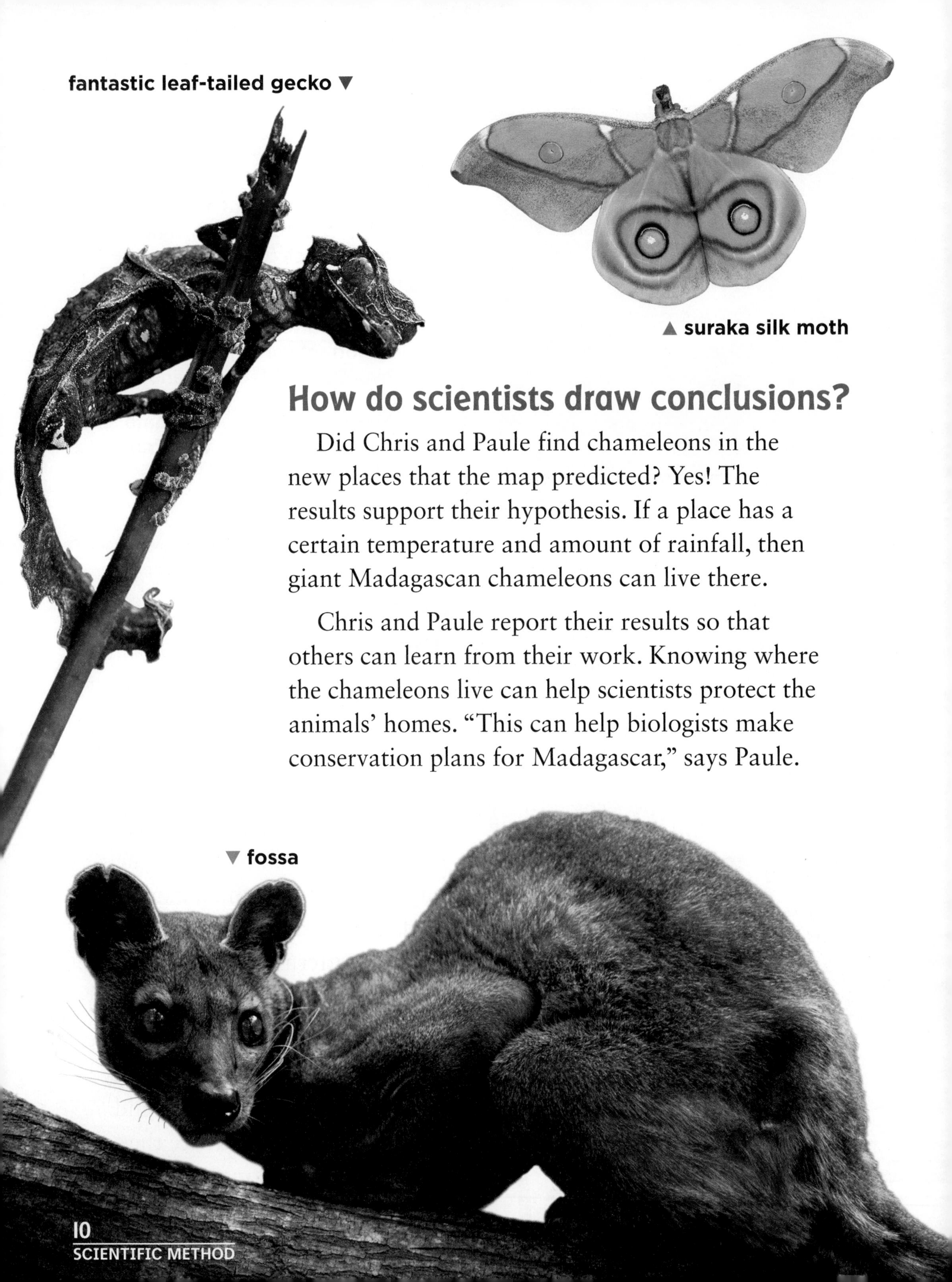

fantastic leaf-tailed gecko ▼

▲ suraka silk moth

How do scientists draw conclusions?

Did Chris and Paule find chameleons in the new places that the map predicted? Yes! The results support their hypothesis. If a place has a certain temperature and amount of rainfall, then giant Madagascan chameleons can live there.

Chris and Paule report their results so that others can learn from their work. Knowing where the chameleons live can help scientists protect the animals' homes. "This can help biologists make conservation plans for Madagascar," says Paule.

▼ fossa

Malagasy tree frog ▲

Draw Conclusions

1. Decide if the data clearly support or do not support the hypothesis.
2. If the results are not clear, rethink how the hypothesis was tested and make a new plan.
3. Write down the results to share with others.

▶ **Make sure to ask questions.**

Chris and Paule's results lead them to new questions. What other variables affect where giant Madagascan chameleons live? The animals shown on this page all live in Madagascar. Could scientists search for these living things in the same way? Which places on the island are home to the greatest number of plants and animals? New questions can lead to a new hypothesis and to learning new things. Learning more about the animals that live in Madagascar will help protect them.

ring-tailed lemur ▶

Think, Talk, and Write

1. **Why is the scientific method useful to scientists?**
2. **What other questions about animals can you think of? Choose one and form a hypothesis that can be tested.**

▼ red millipede

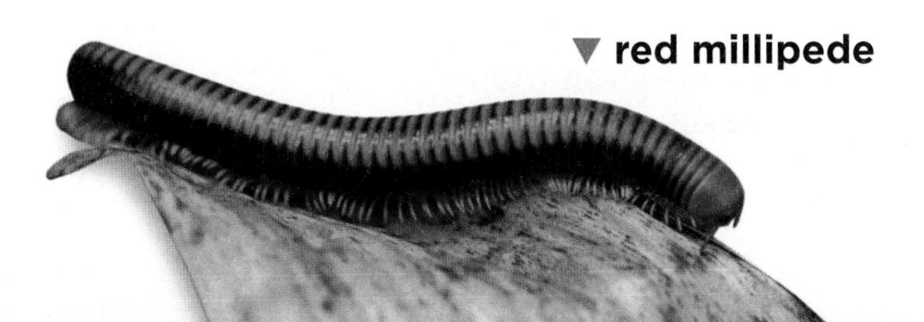

Focus on Skills

Scientists use many skills as they work through the scientific method. Skills help them gather information and answer questions they have about the world around us. Here are some skills they use.

Observe Use your senses to learn about an object or event.

Form a Hypothesis Make a statement that can be tested to answer a question.

Communicate Share information with others.

Classify Place things with similar properties into groups.

Use Numbers Order, count, add, subtract, multiply, and divide to explain data.

Make a Model Make something to represent an object or event.

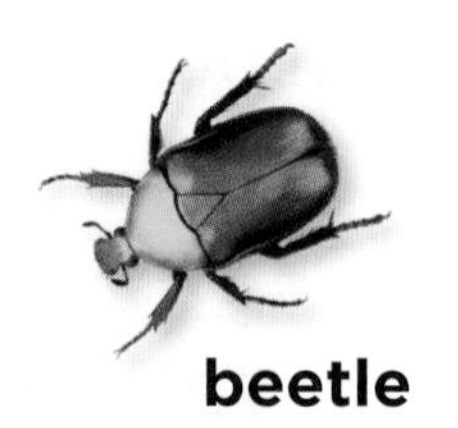

animal	what I observed

▲ **Observe the animals on these pages. Then make a chart to communicate your observations.**

Use Variables Identify things that can control or change the outcome of an experiment.

Interpret Data Use information that has been gathered to answer questions or solve a problem.

Measure Find the size, distance, time, volume, area, mass, weight, or temperature of an object or event.

Predict State possible results of an event or experiment.

Infer Form an idea from facts or observations.

Experiment Perform a test to support or disprove a hypothesis.

Inquiry Skill Builder

In each chapter of this book, you will find an Inquiry Skill Builder. These features will help you build the skills you need to become a great scientist.

snail

Animal Young

Animal	Average Number of Young
beetle	75
sea star	2,000,000
lizard	14
hedgehog	4
gazelle	1

▲ **Use this chart to infer how an animal's size affects how many young it has at a time.**

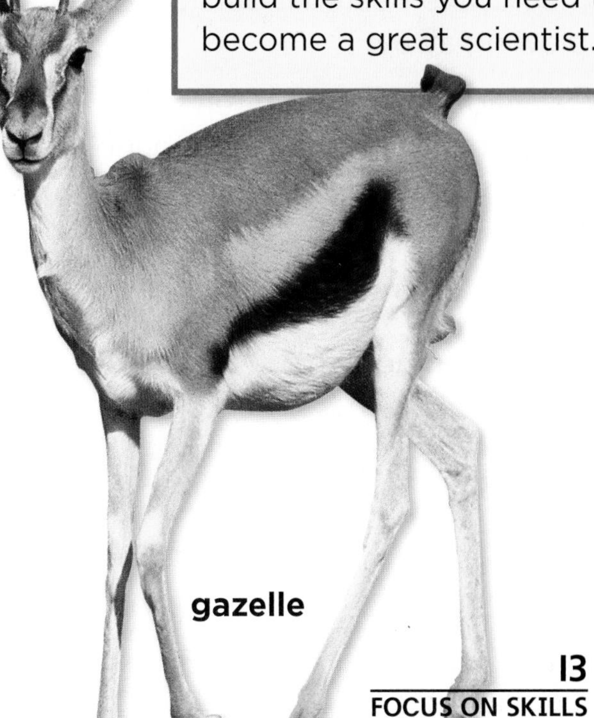

Focus on Skills

Science and Technology: The Design Process

Have you ever invented something? You have used the design process! The **design process** is a series of steps for solving problems.

School Picnic this Friday!

Sign up to bring:

Fruit

Sandwiches

Drinks

▶ Learn It

The **design process** is like a process you probably use already. First you identify a problem. Then you identify several possible solutions.

Sometimes the best idea can not be designed. It can be too expensive or the materials are not available. Other times the design can have harmful results. Finally, not all great ideas work the way they are supposed to. That is why they need to be tested.

▶ Try It

Kay made a sign to tell students about the school picnic. She posted it outside. But the sign kept blowing away!

Help Kay design three solutions to solve her problem. Illustrate them and include labels. When you have permission from your teacher, test your design.

▶ Apply It

Explain how the **design process** has improved your life. How has it improved your family's life? How has it improved your community?

A Creative Design Strategy

Identify a Problem

Fredo tried to change his bicycle tire. One of the bolts fell through a sewer grate! Now he can not put the wheel back on.

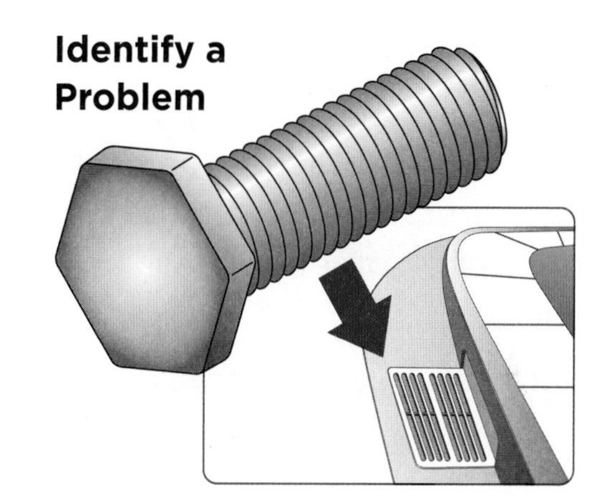

Identify Possible Solutions

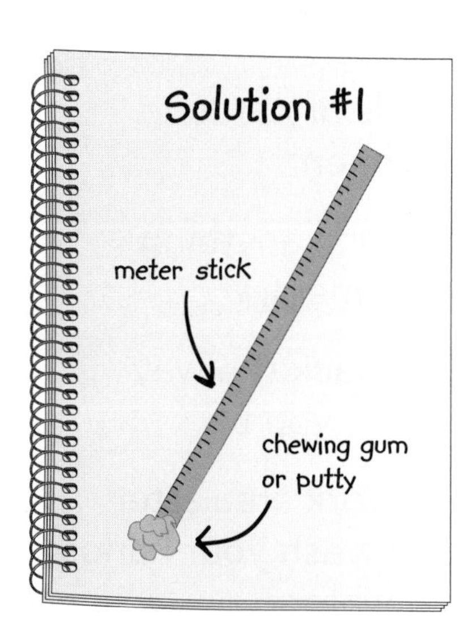

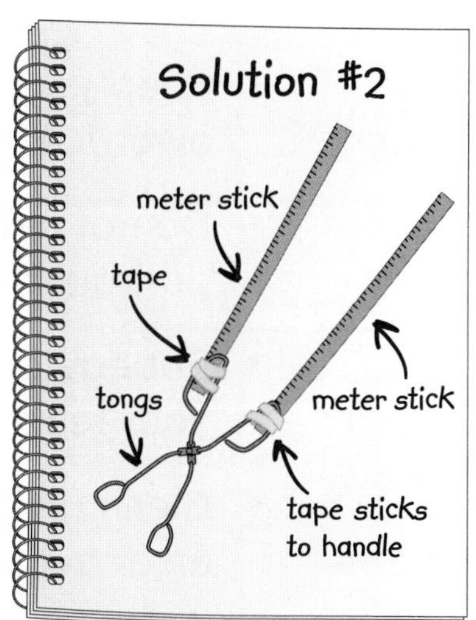

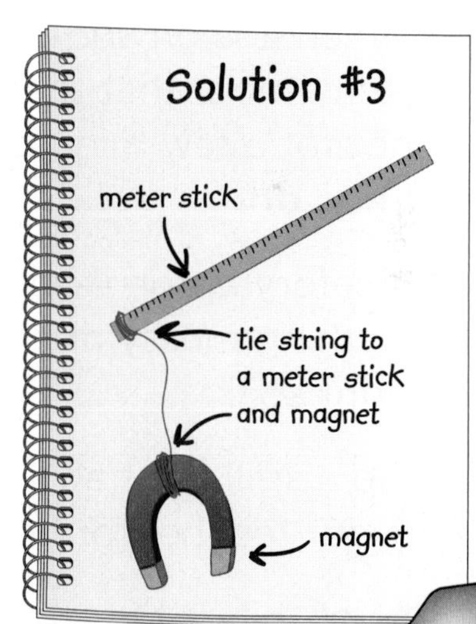

Design a Solution

Look at the solutions above. Fredo did not have a magnet. He did not have tongs. However, he did have chewing gum. He used a flat stick in place of a meter stick. Solution number one worked!

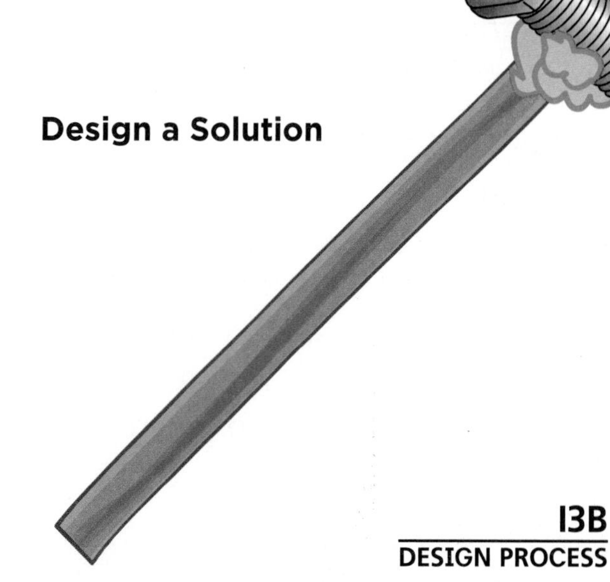

Safety Tips

In the Classroom

- Read all of the directions. Make sure you understand them. When you see "△ **Be Careful,**" follow the safety rules.
- Listen to your teacher for special safety directions. If you do not understand something, ask for help.
- Wash your hands with soap and water before an activity.
- Be careful around a hot plate. Know when it is on and when it is off. Remember that the plate stays hot for a few minutes after it is turned off.
- Wear a safety apron if you work with anything messy or anything that might spill.
- Clean up a spill right away, or ask your teacher for help.
- Dispose of things the way your teacher tells you to.
- Tell your teacher if something breaks. If glass breaks, do not clean it up yourself.
- Wear safety goggles when your teacher tells you to wear them. Wear them when working with anything that can fly into your eyes or when working with liquids.
- Keep your hair and clothes away from open flames. Tie back long hair, and roll up long sleeves.
- Keep your hands dry around electrical equipment.
- Do not eat or drink anything during an experiment.
- Put equipment back the way your teacher tells you to.
- Clean up your work area after an activity, and wash your hands with soap and water.

In the Field

- Go with a trusted adult—such as your teacher, or a parent or guardian.
- Do not touch animals or plants without an adult's approval. The animal might bite. The plant might be poison ivy or another dangerous plant.

Responsibility

Treat living things, the environment, and one another with respect.

Living Things

Touching a frog will not give you warts.

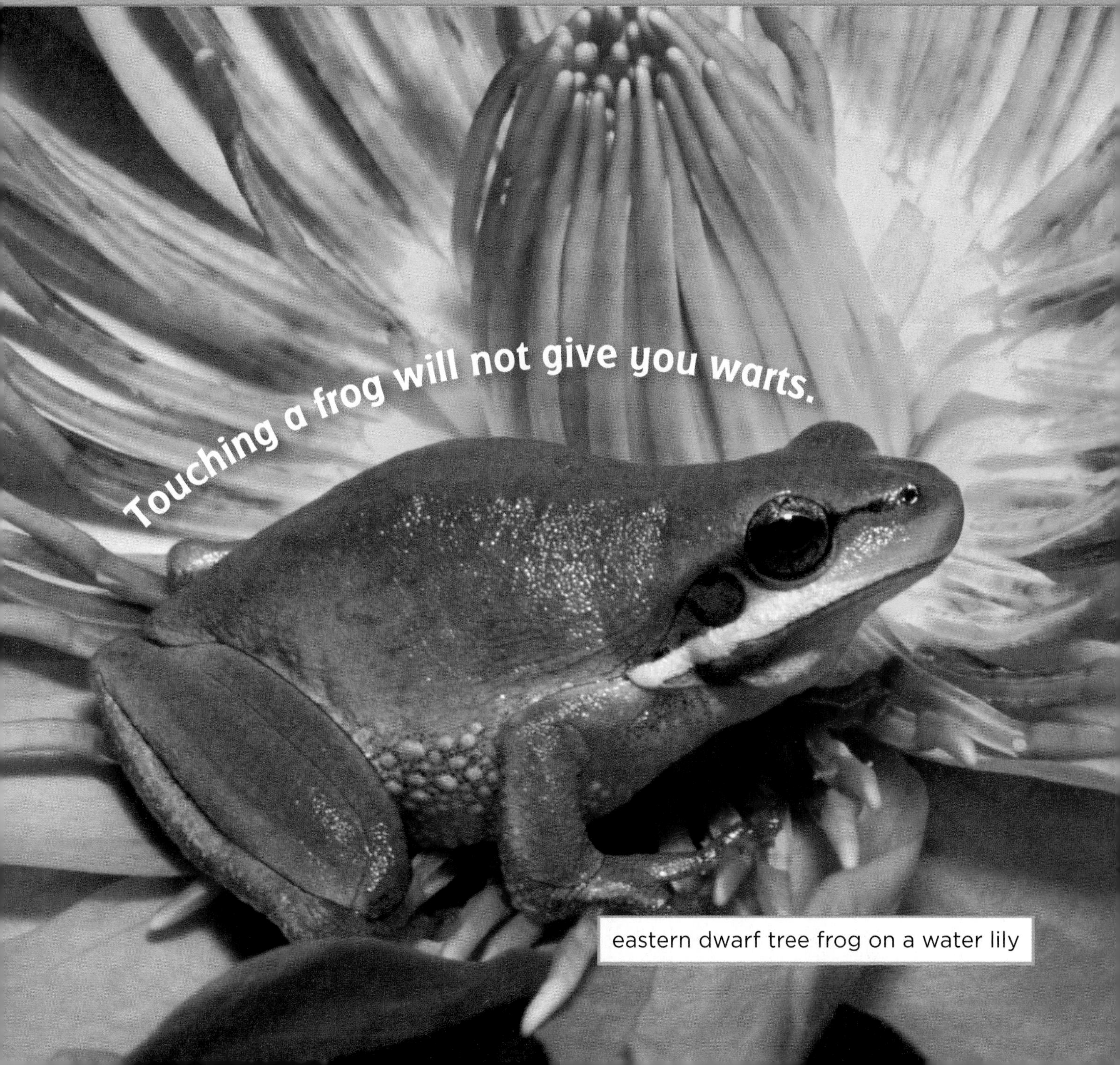

eastern dwarf tree frog on a water lily

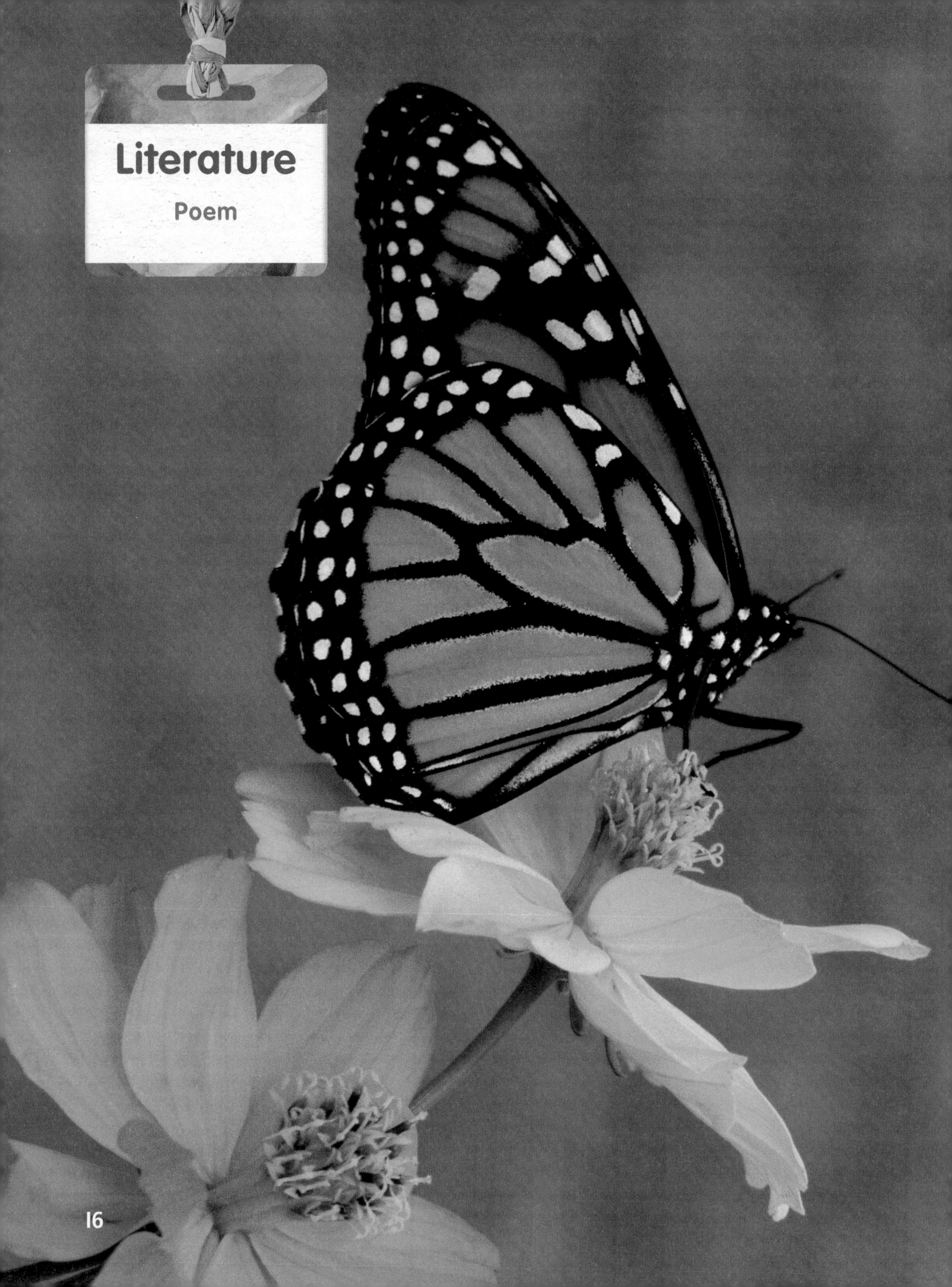
Literature
Poem

Monarch Butterfly

by Marilyn Singer

Wait I can wait
For the fullness of wings
For the lift   For the flight
Wait I can wait
A moment less
A moment more
I have waited much longer before
For the taste of the flower
For the feel   For the sight
Wait I can wait
For the prize of the skies
For the gift of the air
Almost finished
Almost there
Almost ready
to rise

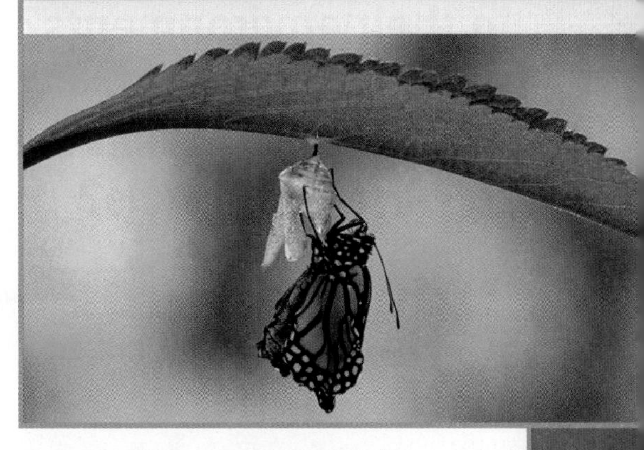

After about two weeks, a monarch butterfly comes out of its protective covering.

Write About It

Response to Literature This poem describes a caterpillar changing into a butterfly. All living things change as they grow. Write a poem about how you have changed as you have grown. Write about some exciting things you are waiting for.

Write about it online at www.macmillanmh.com

CHAPTER 1

A Look at Living Things

How do living things get what they need to live and grow?

Essential Questions

Lesson 1
How are all living things alike?

Lesson 2
How do plant structures compare?

Lesson 3
What helps animals survive in their environments?

Lesson 4
Which features can we use to classify animals?

A gecko hatches from an egg.

Big Idea Vocabulary

organism a living thing (p. 22)

environment all the living and nonliving things that surround an organism (p. 24)

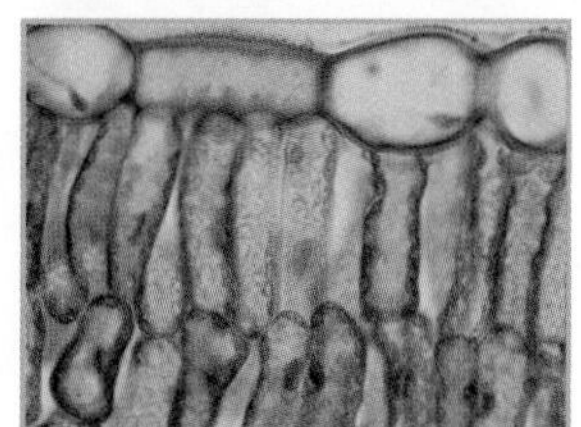

cell the basic building block that makes up all living things (p. 26)

photosynthesis the process through which plants make food (p. 36)

vertebrate an animal with a backbone (p. 54)

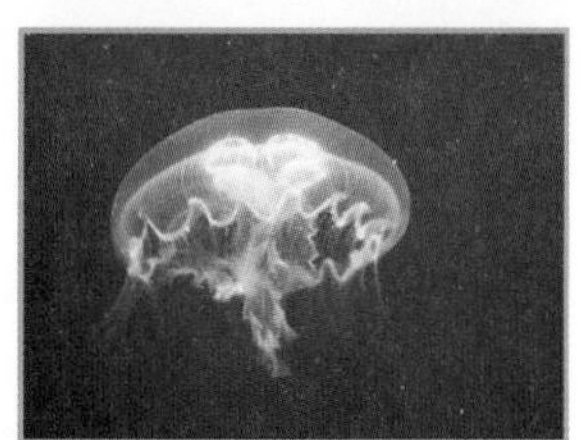

invertebrate an animal that does not have a backbone (p. 55)

Visit www.macmillanmh.com for online resources.

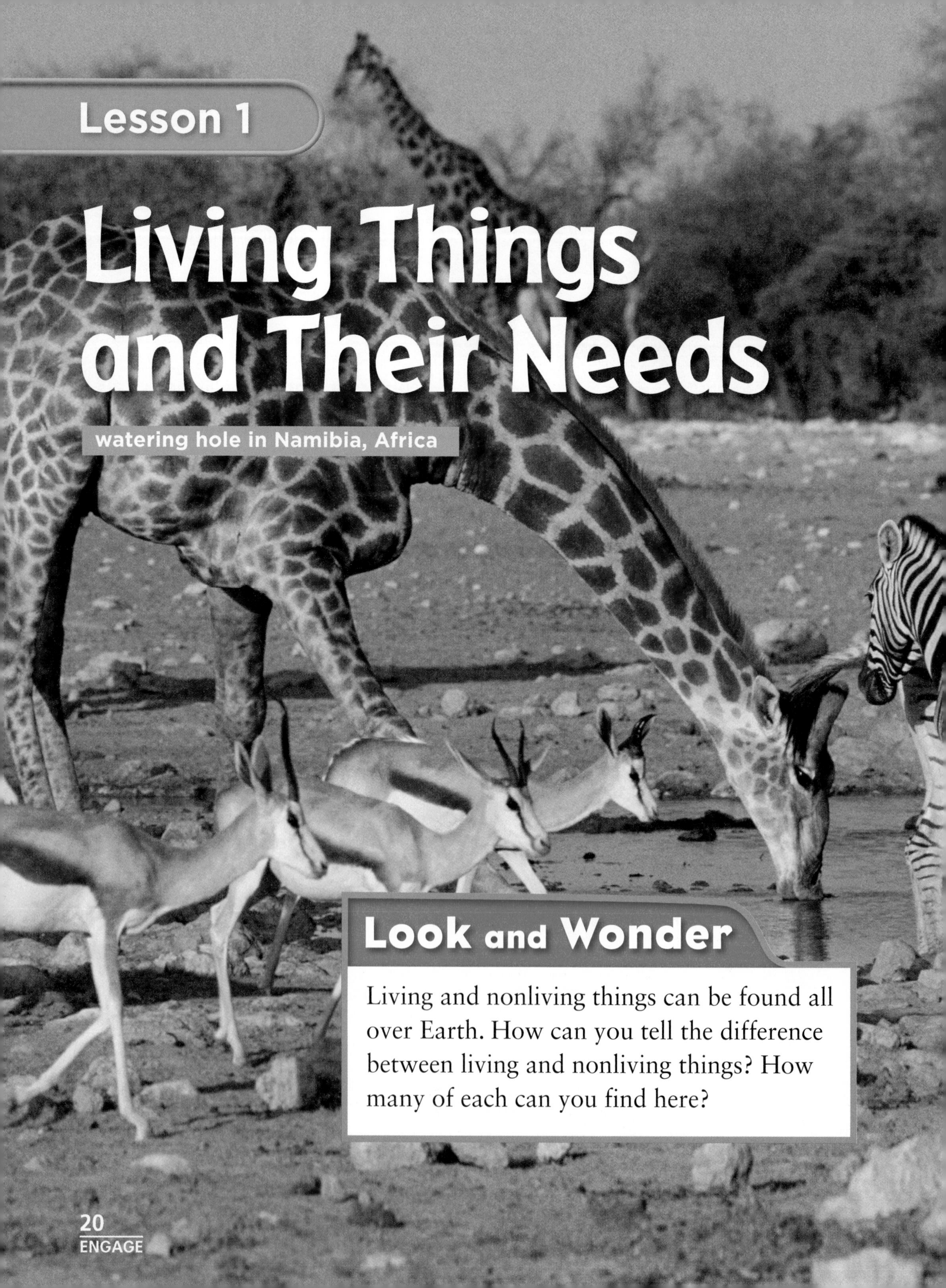

Lesson 1

Living Things and Their Needs

watering hole in Namibia, Africa

Look and Wonder

Living and nonliving things can be found all over Earth. How can you tell the difference between living and nonliving things? How many of each can you find here?

Explore

Inquiry Activity

How do living and nonliving things differ?

Purpose

Find out some characteristics of living and nonliving things.

Procedure

1. **Predict** How are living things alike? How are nonliving things alike?
2. Make a table. Label the columns *Living Things* and *Nonliving Things*.
3. Place four pieces of string outside on the ground so they form a square.
4. **Observe** Find living things in the square area. List them in your table. Tell how you know they are living. Do the same with nonliving things that you find.

Draw Conclusions

5. **Interpret Data** Which characteristics do the living things share? Which do the nonliving things share?
6. Trade tables with a partner. Do the things in your partner's table share the same characteristics as yours?
7. **Infer** How are living things different from nonliving things?

Explore More

Experiment Does the amount of sunlight affect how many living things are in an area? How could you test this?

Materials

four 1-meter pieces of string

Step 2

Living Things	Nonliving Things

Step 4

Read and Learn

Essential Question

How are all living things alike?

Vocabulary

organism, p. 22

respond, p. 22

reproduce, p. 23

environment, p. 24

cell, p. 26

Reading Skill 

Main Idea and Details

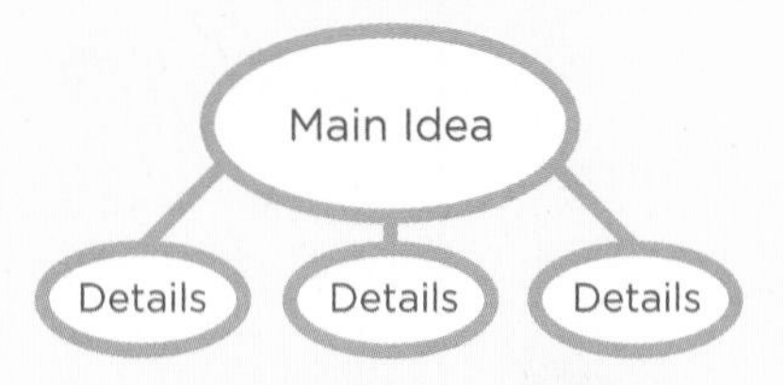

Technology LOG ON

e-Glossary and e-Review online at www.macmillanmh.com

What are living things?

What plants or animals can you find outside? Plants and animals are living things. What are some characteristics that all living things share?

Living Things Grow

Living things are called **organisms** (OR•guh•nih•zumz). All organisms use energy to grow. To *grow* means to change with age. A young sunflower plant is small and green. Over time, it grows taller and its stem grows harder. Eventually, a flower forms. A young bird grows into an adult. It gets bigger and less fuzzy.

Living Things Respond

Living things **respond** (rih•SPOND), or react, to the world around them. When a plant is in shade, it responds by bending toward sunlight. When a bird sees a cat and senses danger, it may fly high into the trees. When a day gets hot, a mouse may go underground to keep cool.

Living Things Grow

Read a Photo

How will the small gulls change as they grow?

Clue: Young organisms grow to be more similar to their parents.

◀ When skinks reproduce, the female lays eggs. New skinks hatch from the eggs.

Living Things Reproduce

Living things reproduce (ree•pruh•DEWS). To **reproduce** means to make more of one's own kind. An apple tree reproduces by making apple seeds. The seeds can grow into new apple trees. A turtle reproduces by laying eggs. Young turtles hatch from the eggs.

Nonliving Things

Nonliving things are all around you. Rocks, soil, and water are nonliving things that come from nature. Cars and roads are nonliving things made by people. Nonliving things are different from living things. They do not use energy to grow, respond, or reproduce.

Quick Check

Main Idea and Details What are some characteristics of living things?

Critical Thinking Is a toy a living thing? How can you tell?

When the weather gets cooler in autumn, this tree responds by losing its leaves. ▶

What do living things need?

Living things have needs. They need food, water, and space. Many also need gases found in air or water. A living thing will die if its needs are not met.

Living things get everything they need to survive from their environment (en•VI•run•munt). An **environment** is all the living and nonliving things that surround an organism.

Food

Living things need energy to live and grow. They get energy from food. Animals get food by eating other organisms. Plants make their own food using energy from sunlight.

▲ A caterpillar gets the energy it needs to grow by eating leaves.

Water

Did you know that more than half of your body is water? All living things are full of water. They use the water in their bodies to break down food and get rid of waste. They use it to transport food throughout their bodies. Living things need a regular supply of water to stay healthy.

These plants soak up water from wet soil in their environment.

Gases

Animals need oxygen (AHK•sih•jun) to survive. *Oxygen* is a gas found in air and water. Every time you breathe, you take in oxygen from the air. Fish, clams, and most other sea animals get oxygen from the water around them.

Plants need oxygen and a gas called *carbon dioxide* (KAR•bun di•AHK•side). Plants use energy from sunlight to change carbon dioxide and water into food.

▲ Some water animals, such as this manatee, must come to the surface to take in oxygen from the air.

Space

Organisms need space, or room. Plants need space to grow and to get water and sunlight. Animals need space to move and find food. Different organisms need different amounts of space. Whales swim for miles in oceans. Goldfish can live in tiny ponds.

Quick Check

Main Idea and Details What are some things that all organisms need to survive?

Critical Thinking What might happen to an animal in a crowded environment?

▲ Foxes hunt in forests and fields. Small dens help them stay safe.

Quick Lab

Observe Cells

1. **Observe** Look at a piece of onion. Then observe it using a hand lens. What do you see?
2. **Communicate** Draw how the onion looks when viewed with a hand lens.
3. **Observe** Look at a slide of an onion under a microscope. What do you see? Is there any space between the cells?

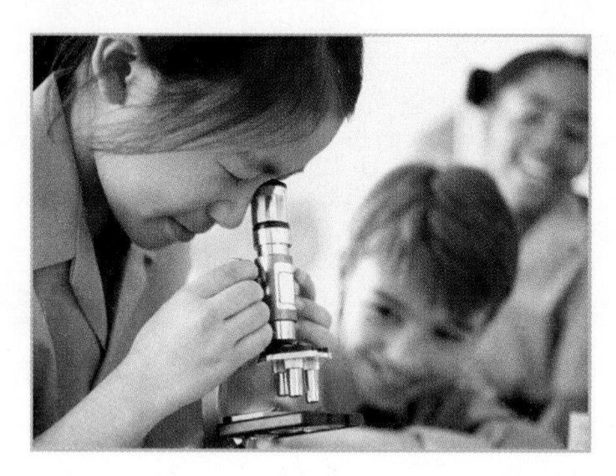

4. **Communicate** Draw how the onion looks when viewed with a microscope. Then compare your two drawings.
5. **Infer** How small are cells? What tool do you need to observe cells?

What are living things made of?

How are you like a brick building? The building is made of many small bricks. You are made of many small parts called cells. **Cells** are the building blocks of life. All organisms are made up of one or more cells.

Cells are too small to see with only your eyes. Cells are so small that it takes millions to make one little ant. You need a tool called a *microscope* (MI•kruh•skohp) to observe cells. A microscope makes tiny things look larger.

Some organisms are made of a single cell. Organisms called bacteria are an example. They live in soil and water. Some live on our skin and in our bodies!

Quick Check

Main Idea and Details What are cells?

Critical Thinking What do you think cells need to survive?

◀ These cells from a lilac leaf were magnified with a microscope.

Lesson Review

Visual Summary

Living things grow, respond, and reproduce.

Living things **need** food, water, gases from air or water, and space to live.

Living things are made of **cells.**

Make a FOLDABLES® Study Guide

Make a three-tab book. Use it to summarize what you learned about living things and their needs.

Think, Talk, and Write

1. **Vocabulary** What is an environment?

2. **Main Idea and Details** What do living things need to survive?

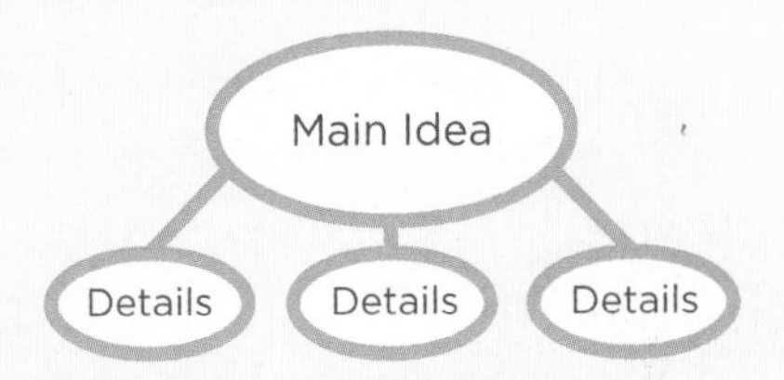

3. **Critical Thinking** Suppose you wanted to grow plants in your backyard. What would you do?

4. **Test Prep** **People need all of the following to survive except**
   A air.
   B water.
   C cars.
   D space.

5. **Essential Question** How are all living things alike?

Writing Link

Write a Story
Suppose you were a bird. What would you need to live? How would your life be different? Research to learn more about birds. Write a story about life as a bird.

Health Link

Food Pyramid
People need the right balance of foods to stay healthy. The Food Guide Pyramid shows this balance. Do research to find out about MyPyramid. Look on page R26 of this book.

LOG ON e-Review Summaries and quizzes online at www.macmillanmh.com

EATING AWAY AT POLLUTION

You can not see microorganisms, but they are all around you. Microorganisms are tiny living things. You need a microscope to see them. Many are made of just one cell.

Some microorganisms are harmful. They can make animals and plants sick. Others are helpful. They can eat things that are harmful to plants and animals. Some can even help clean up Earth's water, land, and air. Scientists use these tiny organisms to eat pollution.

Connect to

at www.macmillanmh.com

Some microorganisms eat oil. When oil spills on water or soil, the tiny creatures eat the oil. The waste they leave behind is safe for the environment. Other microorganisms can help keep air clean. Factories and power plants often produce a lot of smoke. Microorganisms can eat dangerous chemicals in the smoke that would pollute the air.

Classify

When you classify,

- you compare things to learn how they are alike and different;
- you put things into groups based on their characteristics.

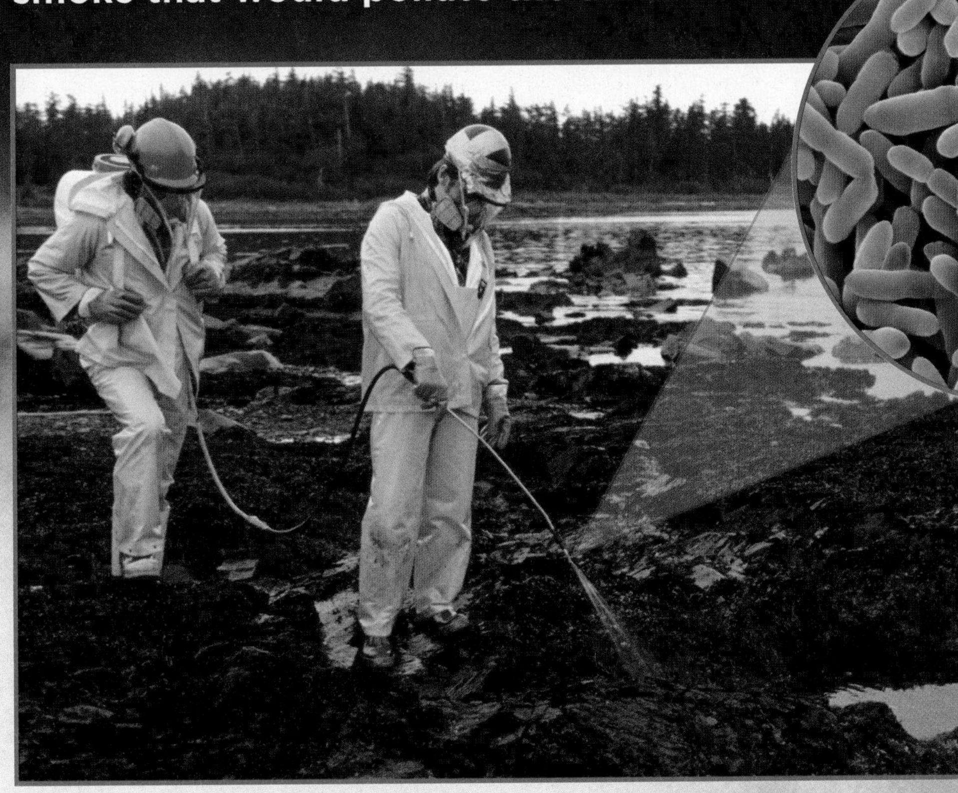

Workers spray microorganisms onto an oil spill in Alaska. The round photo shows the microorganisms as seen through a microscope.

Write About It

Classify This article explains that some microorganisms are harmful and others are helpful. This is a way to classify them. Read the article again with a partner. Look for another way to classify microorganisms. Then write about it.

-Journal Write about it online at www.macmillanmh.com

Lesson 2

Plants and Their Parts

Look and Wonder

Some plants catch flies. Some smell like rotting meat. Some can grow three feet in one day. Plants come in many shapes and sizes. How are all plants alike?

Explore

Inquiry Activity

How are plants alike?

Purpose

Find out about some characteristics of plants.

Procedure

1. **Observe** Study each plant carefully. Which plants have leaves? How do their leaves compare? Describe them using words and pictures.
2. **Infer** Which part of each plant grows underground? How is this part the same on each plant? How is it different? Record your observations with words and pictures.
3. **Observe** Look carefully at each plant again. What other parts does each plant have? How are these parts the same? How are they different? Record your observations.

Draw Conclusions

4. **Infer** Which parts do most plants have?
5. How are plants alike?

Explore More

Experiment Can different-looking plants survive under the same conditions? How could you find out? Make a plan and try it.

Materials

Step 1

Read and Learn

Essential Question

How do plant structures compare?

Vocabulary

structure, p. 33

root, p. 34

nutrient, p. 34

stem, p. 35

leaf, p. 36

photosynthesis, p. 36

Reading Skill

Summarize

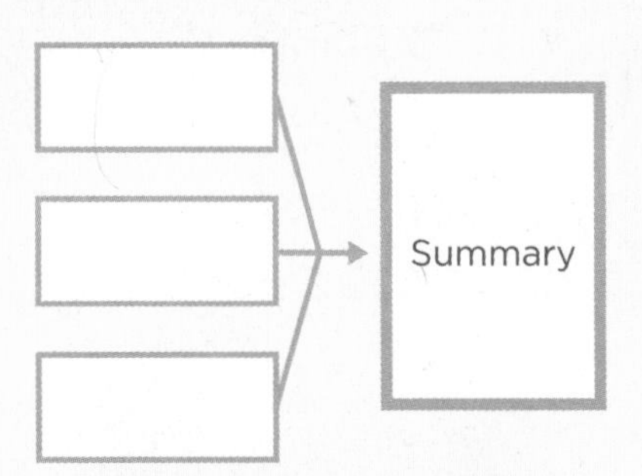

Technology

e-Glossary, e-Review, and animations online at www.macmillanmh.com

What are plants?

From tall trees to tiny wildflowers, plants come in different shapes and sizes. What do plants have in common? One characteristic that all plants share is that they can make their own food. Plants do not eat other organisms to get energy like animals do. Instead they use energy from the Sun to make food.

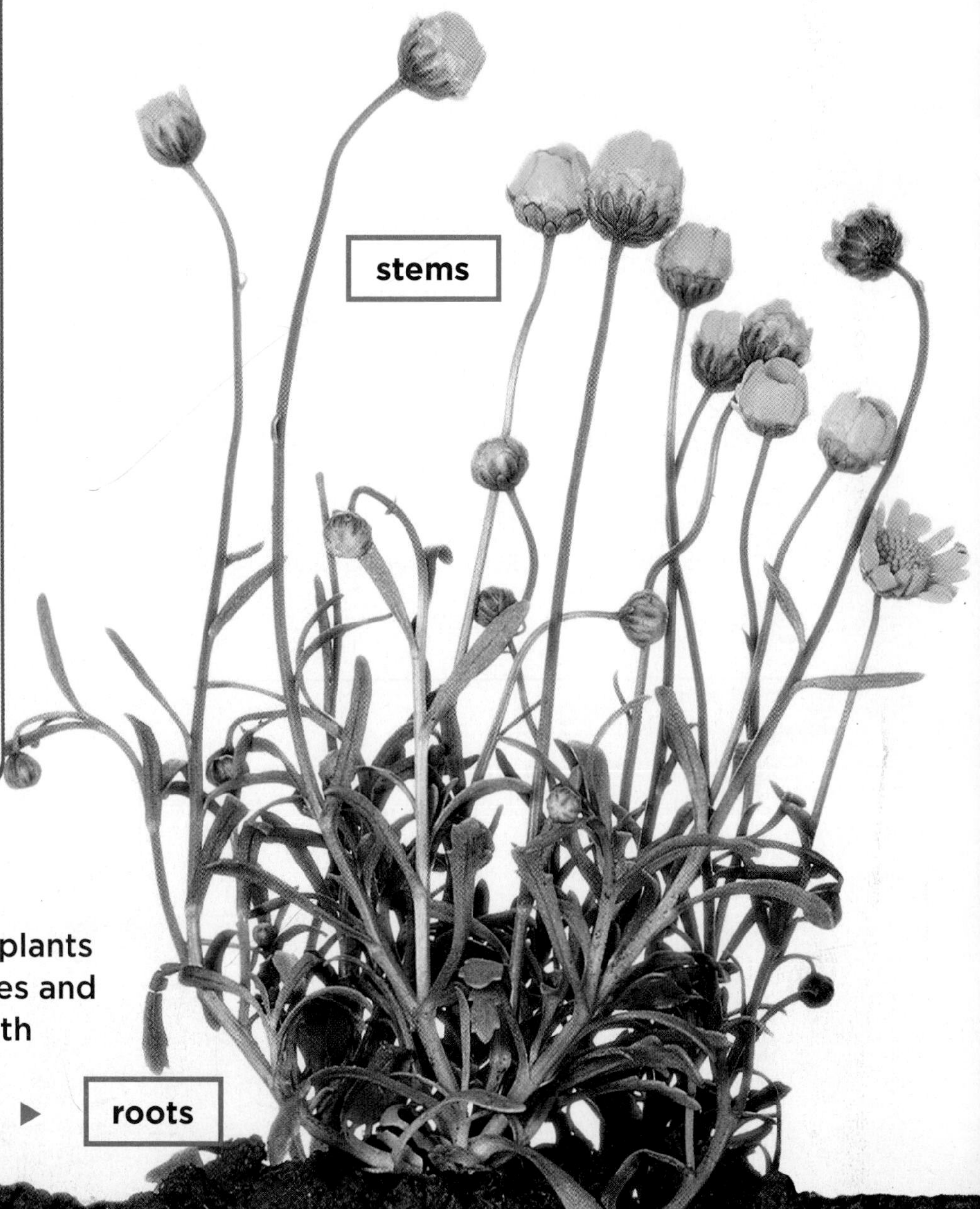

These two kinds of plants have different shapes and colors. However, both have roots, stems, leaves, and flowers. ▶

Most plants share other characteristics too. Most plants have parts that are green. Most grow in the ground. Plants can not move around like animals can.

Many plants have the same basic **structures** (STRUK•churz), or parts. Most plants have roots, stems, and leaves. These structures help plants get what they need to survive. Some plants have flowers and fruits. Some have cones. These structures help plants reproduce.

Quick Check

Summarize What three structures do most plants have?

Critical Thinking Most plants do not have structures for eating. What might be a reason for this?

How do roots and stems help plants?

Most plants have roots and stems. These structures help plants get what they need to survive.

◀ A beet has one main root.

Roots

Plants need water. They take in water with their roots. **Roots** are structures that take in water and hold a plant in place. Some plants, such as carrots and radishes, have one thick root called a taproot. Others have a web of thinner roots. Roots may grow deep in the soil to find water far underground. They may spread wide to take in water from a large area.

Roots also take in nutrients (NEW•tree•unts). **Nutrients** are substances that help living things grow and stay healthy. They are part of soil. When roots take in water, they also take in nutrients.

This plant has many thin roots. ▶

Some roots have the job of storing food for a plant. These roots can be good for people too. Carrots, radishes, and sweet potatoes are roots that you eat.

A palm tree's leaves blow in the wind, but the roots cling tightly to the sand.

Stems

A **stem** is a structure that holds up a plant. Leaves attach to it. The stem holds leaves upright so they can get sunlight. The stem also carries water, nutrients, and food throughout a plant. Water and nutrients flow up from the roots through tubes in the stem. Food flows from the leaves through other tubes.

Not all stems are alike. Stems can be soft and green like tulip stems. They can be hard and woody like tree trunks. A trunk is a tree's stem.

Quick Check

Summarize How do stems help a plant meet its needs?

Critical Thinking What happens to a plant with diseased roots?

Quick Lab

Observe Stems

1. Get a stalk of celery with leaves on it. Carefully cut two inches off the bottom.
   △ **Be Careful.**
2. Half fill a plastic jar with water. Then add five drops of food coloring to the water. Mix the water with a spoon.
3. **Observe** Place the celery into the jar. Observe the celery stalk a few times throughout the day. What do you notice?
4. **Communicate** How has your celery stalk changed? Draw a picture. Write a description.
5. **Infer** What do stems do?

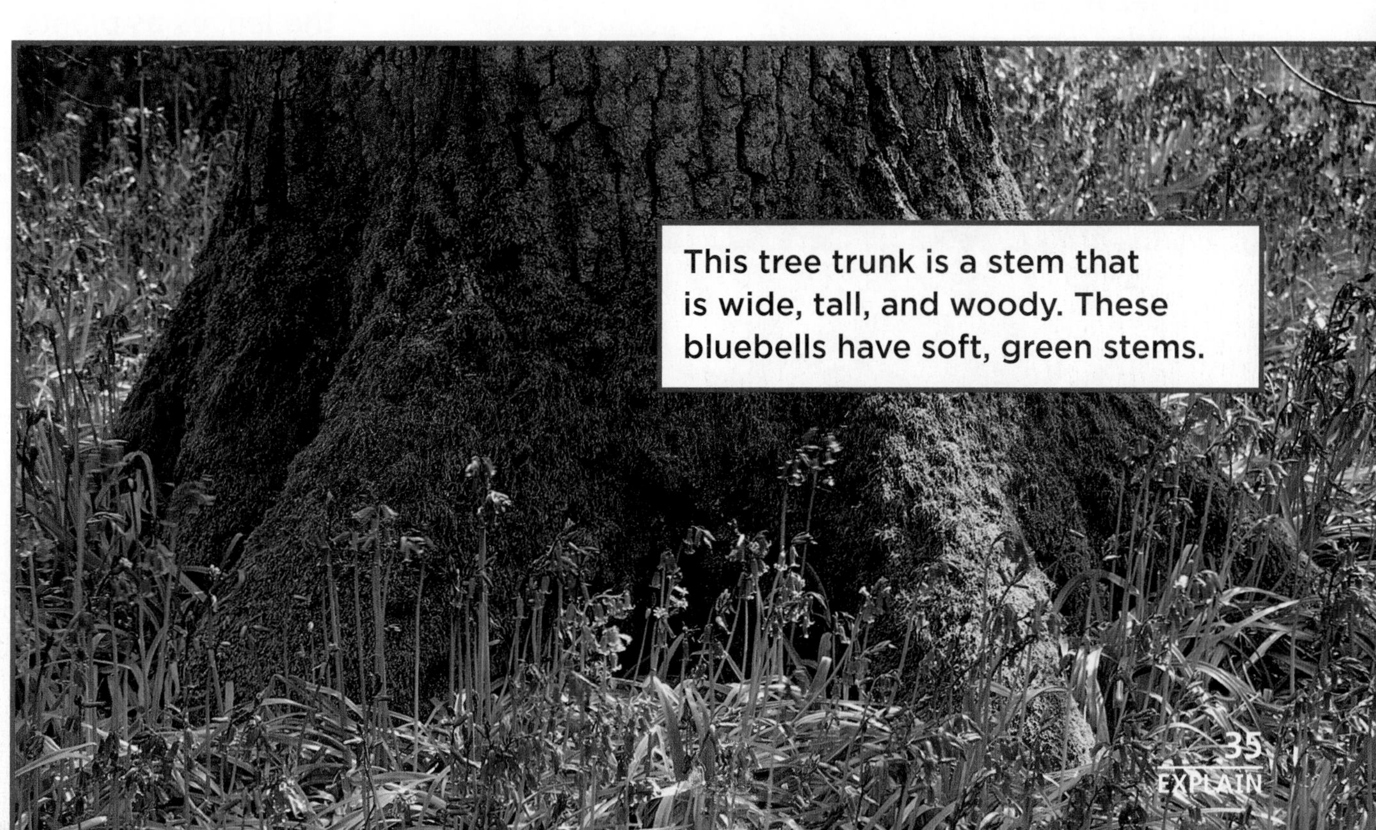

This tree trunk is a stem that is wide, tall, and woody. These bluebells have soft, green stems.

Why are leaves important?

Leaves come in many shapes and sizes. Leaves on a pine tree are like short needles. Leaves on a maple tree are wide and flat. Whatever their shape and size, leaves do an important job for a plant. A **leaf** is the structure where a plant makes food.

maple leaf

pine needles

Plants make food in a process called **photosynthesis** (foh•toh•SIN•thuh•sus). During photosynthesis plants use energy from the Sun to change carbon dioxide and water into sugars. Sugars are food for a plant. Sugars give plants the energy they need to grow.

Photosynthesis

Sunlight soaks into leaves and provides energy.

Oxygen flows from the leaves as plants make food.

Food made inside leaves travels to the rest of the plant.

Carbon dioxide flows into holes in leaves.

Water and nutrients flow from the roots to the leaves.

Read a Diagram

Which gas does the plant give off when it makes food?

Clue: Find the arrow that points away from the plant.

Science in Motion Learn more at www.macmillanmh.com

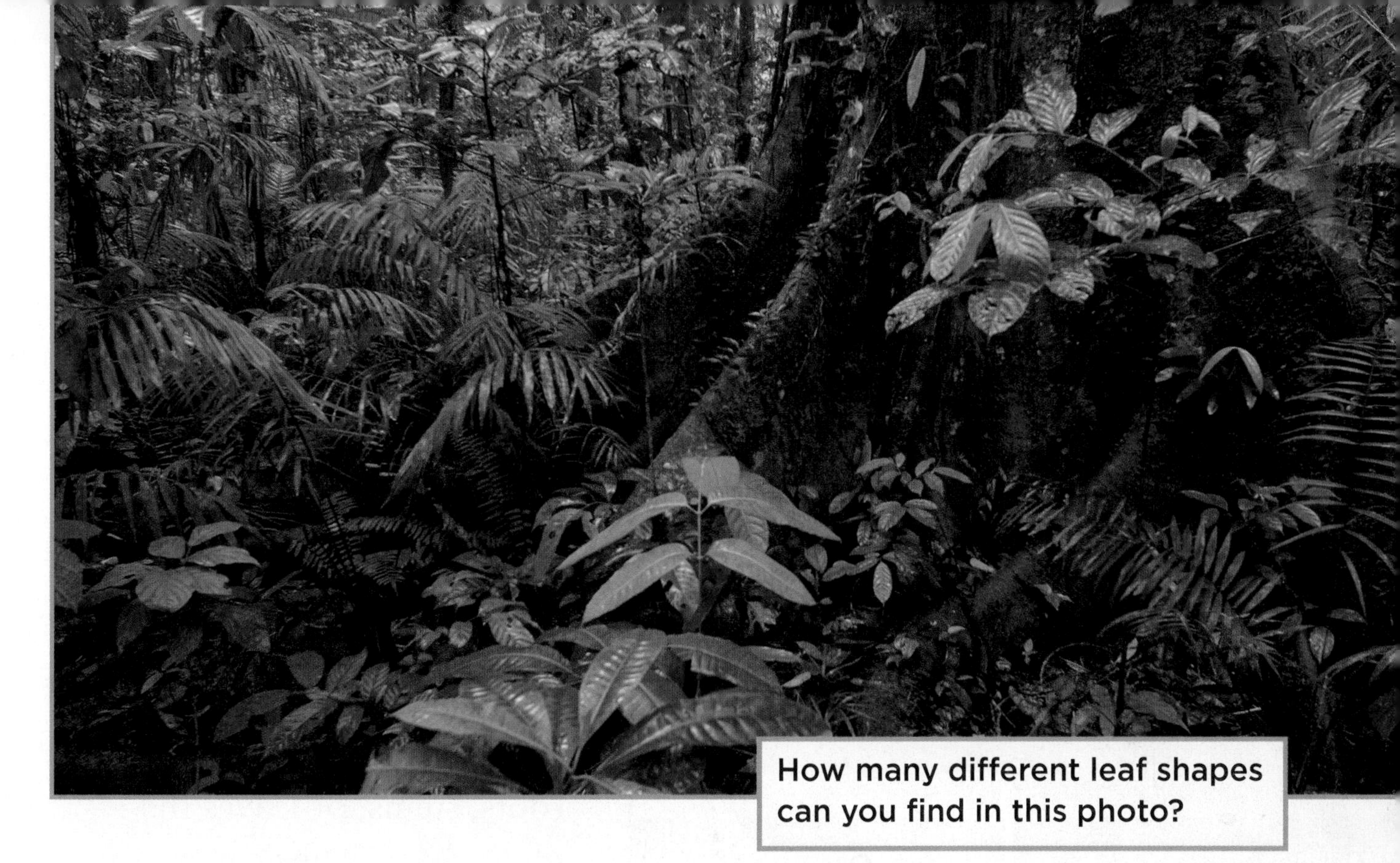
How many different leaf shapes can you find in this photo?

Leaves take in the carbon dioxide and sunlight that plants need for photosynthesis. Carbon dioxide enters through tiny holes on the underside of a leaf. Leaves trap energy from sunlight with chlorophyll (KLOR•uh•fil). *Chlorophyll* is a substance inside a plant's cells. Chlorophyll gives leaves their green color.

During photosynthesis plants also give off oxygen. People and animals need oxygen to live. You inhale the oxygen made by plants with each breath you take.

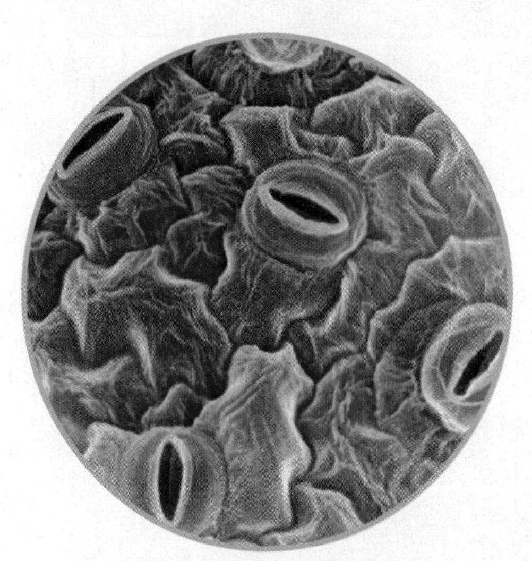
▲ This photo shows the tiny holes in leaves as seen through a microscope.

Quick Check

Summarize How do leaves help plants survive?

Critical Thinking How would air change if there were fewer plants?

FACT Plants need oxygen.

Classifying Plants	
Flowering	**Nonflowering**
cherry tree	ginkgo tree
prickly pear cactus	yew tree
squash plant	juniper tree

How can you classify plants?

There are thousands of different kinds of plants on Earth. Scientists put all these plants into groups to study and learn about them. That way, they can study a whole group of plants at once.

Scientists often group plants by their structures. They can group plants by the types of roots, stems, or leaves the plants have. They can group plants by whether the plants make flowers. There are several examples of flowering and nonflowering plants in the chart on this page. How would you group plants?

Quick Check

Summarize What are two ways to group plants?

Critical Thinking Why do you think it is useful for scientists to group plants?

Lesson Review

Visual Summary

	Plants have **structures** to help them survive. Most plants have roots, stems, and leaves.
	**Roots** take in water and hold a plant in place. **Stems** hold up a plant so it can get sunlight.
	Leaves are where food is made in a plant. Plants use sunlight, carbon dioxide, and water to make food.

Make a FOLDABLES® Study Guide

Make a three-tab book. Use it to summarize what you learned about plants and their parts.

Think, Talk, and Write

1. **Vocabulary** What is photosynthesis?
2. **Summarize** How do a plant's parts help it survive?

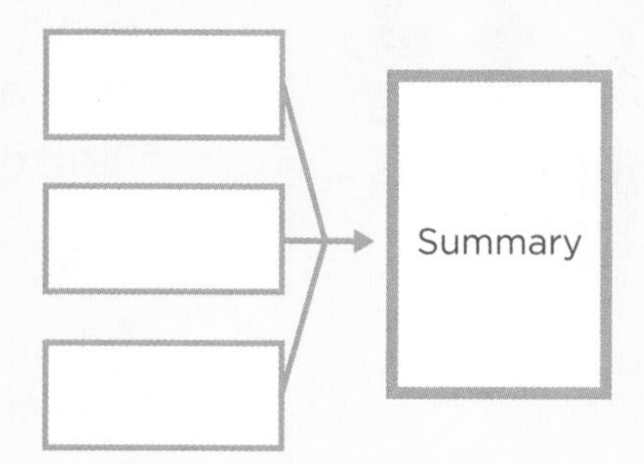

3. **Critical Thinking** How are plants different from animals?
4. **Test Prep** What is the main job of roots?
   - **A** They make the plant green.
   - **B** They take in water and nutrients.
   - **C** They produce seeds.
   - **D** They take in sunlight.
5. **Essential Question** How do plant structures compare?

Writing Link

Writing That Explains
Suppose you are taking care of a plant. Explain how you would make sure the plant gets what it needs.

Math Link

Plant Groups
Find leaves from ten types of plants near your home. Put the leaves into groups based on size. How many are in each group? Now group them another way. Show your groups in a chart.

LOG ON e-Review Summaries and quizzes online at www.macmillanmh.com

Be a Scientist

Materials

4 identical plants

measuring cup and water

ruler

Structured Inquiry

What do plants need to survive?

Form a Hypothesis

Do plants need light to grow? Do they need water? Write a hypothesis. Start with "If plants do not get light and water, then..."

Test Your Hypothesis

Light and Water	Light and No Water
Water and No Light	No Light and No Water

1. Label four identical plants as shown.
2. **Observe** How do the plants look? How tall are they? Measure them and record your observations in a chart. Use words and pictures.

Step 2

3. Put the plants labeled *No Light* in a dark place, such as in a closet. Put the plants labeled *Light* in a sunny place, such as on a windowsill.
4. **Predict** What do you think will happen to each plant? Record your predictions.

5 **Observe** Look at the plants on a regular basis. Water each plant labeled *Water* with 200 milliliters of water. Measure how tall the plants grow. Record your observations in your chart using words and pictures.

Draw Conclusions

6 **Interpret Data** Which plant grew the most after two weeks? Which plant looks the healthiest? Use your chart to help you.

7 What do plants need to survive?

Guided Inquiry

What else do land plants need to survive?

Form a Hypothesis

Do plants need air? Do they need soil? Write a hypothesis about one of these.

Test Your Hypothesis

Design an experiment to test your hypothesis. Decide which of the materials below you will use. Write the steps you will follow.

- two identical plants
- petroleum jelly
- measuring cup
- water
- soil

Draw Conclusions

Did your results support your hypothesis? Why or why not? Share your results with your classmates.

Open Inquiry

What other questions do you have about plants and their needs or structures? Talk with your classmates about questions you have. Choose one question to investigate. How might you answer this question? Make sure your experiment tests only one variable at a time.

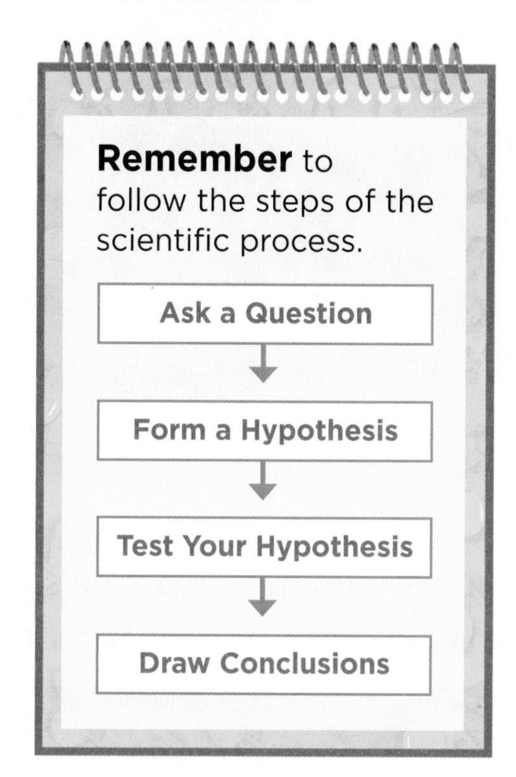

Lesson 3

Animals and Their Parts

Look and Wonder

This koala bear's strong arms and teeth help it get food. Do all animals use the same structures to get what they need?

Explore

Inquiry Activity

How do an animal's structures help it meet its needs?

Purpose

Observe a snail to learn about its structures.

Procedure

1. **Observe** Look at the snail. What parts does it have? Do you see legs or eyes?

   △ **Be Careful.** Handle animals with care.

2. Draw the snail. Label all the parts you can.
3. **Predict** Which parts help the snail move? Which parts help it get food or stay safe?
4. **Experiment** Gently touch the snail with a cotton swab. Observe the snail's actions for a few minutes. Record what you see.
5. **Experiment** Place a wet paper towel in the container. Record the snail's actions. Now repeat this step using a lettuce leaf.

Draw Conclusions

6. **Communicate** On your drawing, circle the parts that the snail used to move and to eat (if it ate). Describe how it responded to its environment.
7. **Infer** Think about other animals you have seen, such as hamsters, birds, and fish. Do they have the same parts as the snail? What parts do they use to meet their needs?

Explore More

Experiment Does the snail respond to light and dark? Make a plan and find out.

Materials

snail

clear plastic container

cotton swab

paper towel

water

lettuce leaf

Step 2

Step 4

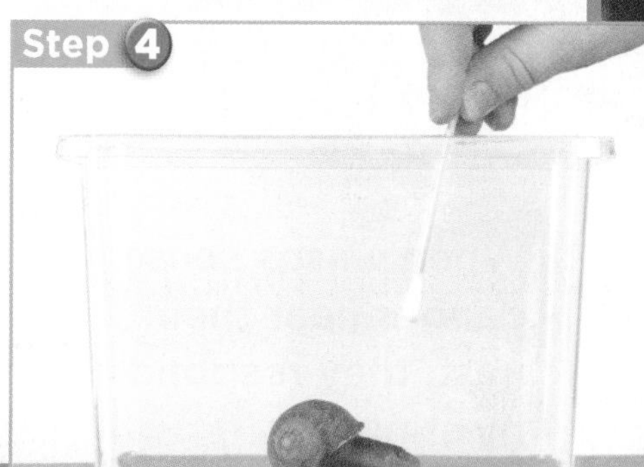

Read and Learn

Essential Question
What helps animals survive in their environments?

Vocabulary
lung, p. 47
gills, p. 47
shelter, p. 48

Reading Skill
Compare and Contrast

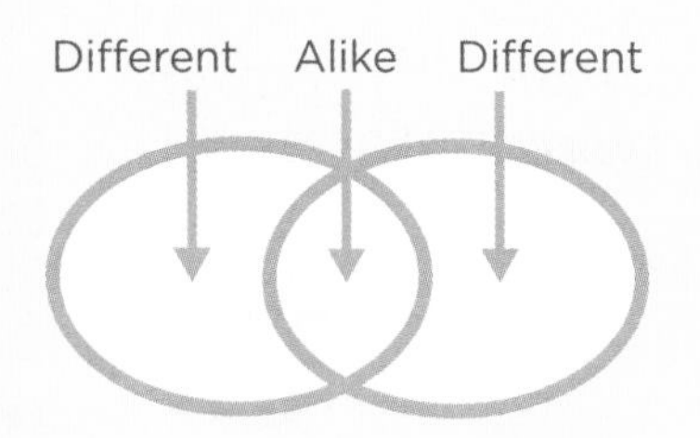

Technology
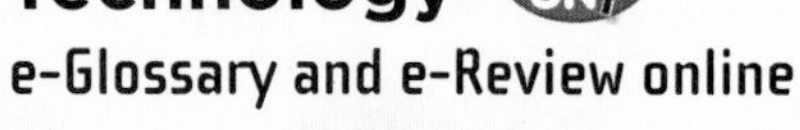

e-Glossary and e-Review online at www.macmillanmh.com

A hummingbird beats its wings so fast that it hovers like a helicopter.

What are animals?

What do snails, elephants, and even tiny ants have in common? They are all part of a group of living things called *animals*. What are animals? How are they different from plants?

Animals have certain traits in common. Most animals can move. Birds can fly. Foxes can run and jump. Sharks swim. Unlike plants, animals can not make their own food. They must eat other organisms to get energy and nutrients.

Animals respond to their environments in more noticeable ways than plants. They use their senses to get information. A wolf may growl when it sees, hears, or smells another wolf near its young. A snake may lie in sunlight when it feels cold. A cat may look for food when it is hungry.

Animals have certain kinds of structures, or parts, that help them get what they need. Legs, fins, wings, and tails are some animal structures.

▲ When wasps sense danger near their nest, they respond by stinging.

How Animals Move

Animals move to find food and water. They move to escape danger. Animals may use feet, legs, tails, wings, or other structures to move.

Animals such as wolves, cheetahs, and house cats have strong legs for running and jumping. Big, rough paws help them balance.

Some animals do fine without legs. Snails make a trail of slime to slide on. They use muscles on their underside to push themselves forward. Snakes use their whole bodies to slither forward. Birds fly and glide through the air with wings.

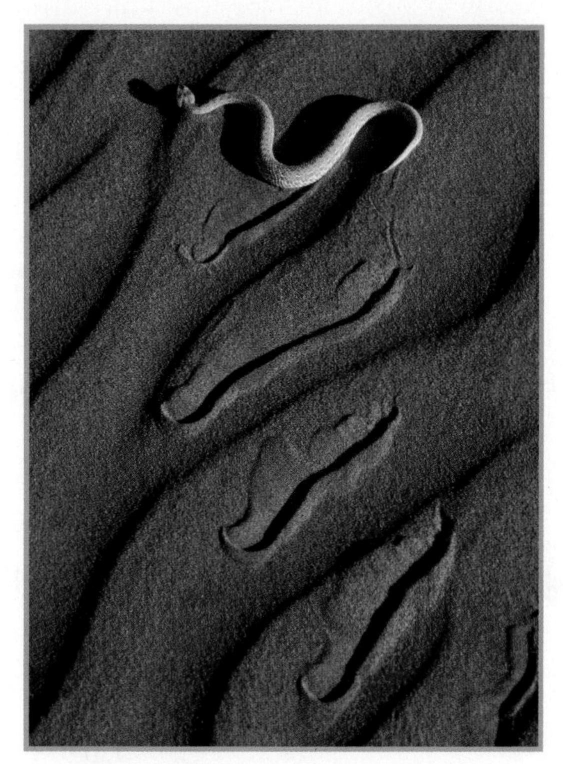

▲ This snake's slithering has left a trail on the sand.

Quick Check

Compare and Contrast What are some different ways that animals move?

Critical Thinking How can you tell that a cat is an animal?

To jump, this wolf pushes off the ground with its strong back legs.

How do animals get what they need?

Animals need water, food, and oxygen. They have structures that help them get these things.

▲ This squirrel uses its paws and sharp teeth to eat an acorn.

Getting Water and Food

Some animals have long tongues for lapping water. Birds scoop up water in their beaks. Elephants pick up water in their trunks and pour it into their mouths.

The same structures help animals get food. Lions scrape meat from bones with their rough tongues. Birds grab worms or seeds with their beaks. Elephants use their trunks to pull plants to their mouths.

Lions and many other animals have long, sharp front teeth. These are good for biting. Many animals have flat, back teeth for chewing. Strong jaws help some animals bite and chew.

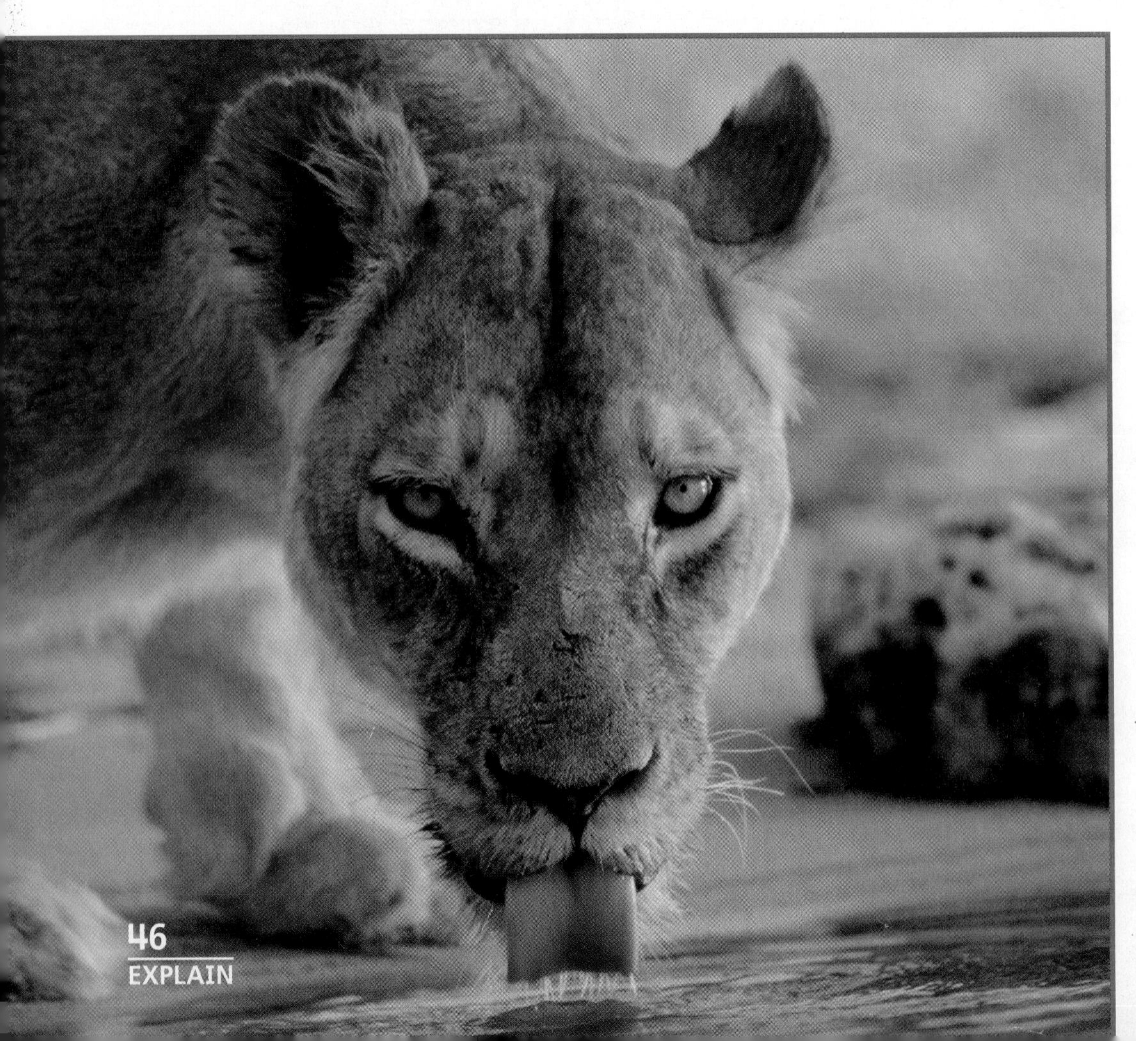

◀ A big, strong tongue helps this lion lap up water.

Getting Oxygen

Animals breathe to get oxygen. Many animals breathe with lungs. **Lungs** are structures that take in oxygen from the air. Fish get oxygen using gills. **Gills** are structures that take in oxygen from the water.

Some animals can breathe without lungs or gills. Worms and salamanders, for example, take in oxygen through their skin.

Quick Check

Compare and Contrast How are lungs like gills? How are they different?

Critical Thinking Which structures do you use to eat food?

Quick Lab

Observe Animal Structures

1. Find photos of dolphins in a magazine or an encyclopedia.
2. **Infer** Look at the dolphins' structures. How does a dolphin use its tail? How does it use its blowhole? Which structures help it get food?
3. **Communicate** Make a data table to show how each structure helps a dolphin meet its needs.

Breathing and Moving

Breathing Water enters the fish's mouth and exits through the gills. As water flows out, the gills take in oxygen from the water.

Moving A fish moves forward by waving its muscular tail. Fins help the fish steer toward food or away from danger.

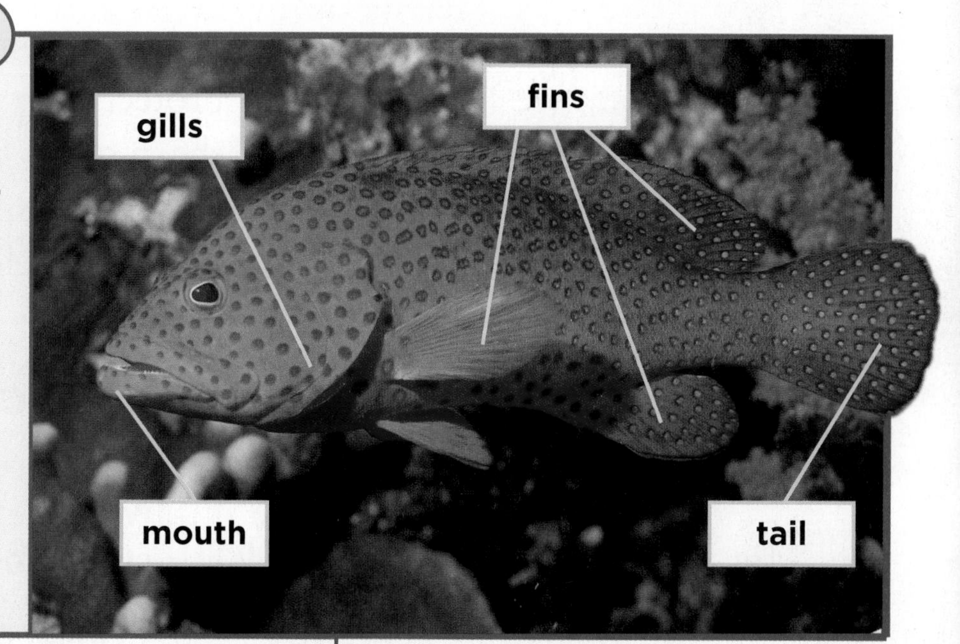

Read a Photo

Which two structures help fish get oxygen?

Clue: Labels and captions give information.

How do animals stay safe?

Animals need a way to stay safe in their environments. They must protect themselves from bad weather or from other animals. Some animals stay safe by finding a safe place, or **shelter**. Other animals have structures that help protect them.

▲ A bird builds a nest to keep its young safe.

Some animals find shelter in the ground. Groundhogs dig holes in the soil with their paws. Lizards flatten their bodies and crawl under rocks.

Other animals use trees or other plants for shelters. Birds build nests as shelters for their young. They use their beaks and feet to gather materials and build their nests.

Some animals have structures that protect their bodies. A porcupine's sharp quills help keep away other animals. A snail's hard shell protects it. Fur can shield animals from the cold.

▲ Young kangaroos stay safe in their mother's pouch.

Quick Check

Compare and Contrast Describe two different ways that animals stay safe.

Critical Thinking Think of some animals you know. How do they stay safe?

Lesson Review

Visual Summary

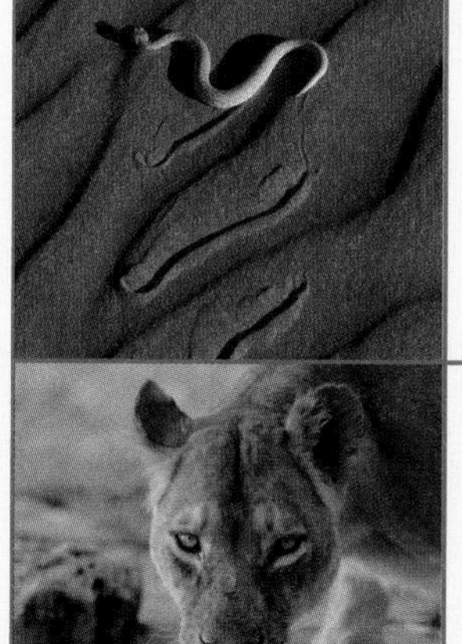

All animals have some characteristics in common. Most have structures that help them **move.**

Animals have structures that help them get **food, water,** and **oxygen.**

Some animals find shelter in which to **stay safe.** Others have structures that keep them safe.

Make a FOLDABLES® Study Guide

Make a layered-look book. Use it to summarize what you learned about animals and their structures.

Think, Talk, and Write

1. **Vocabulary** What is a shelter?

2. **Compare and Contrast** How are an animal's needs like a plant's needs? How are they different?

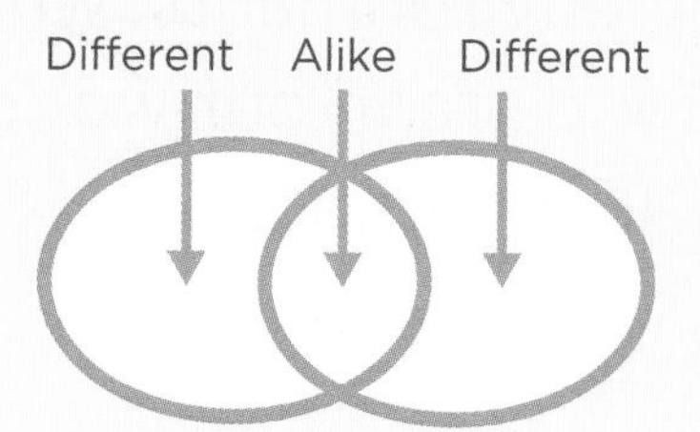

3. **Critical Thinking** How might long legs help a bird that lives in a pond environment?

4. **Test Prep** **Animals use all the structures below to get oxygen <u>except</u>**

   A lungs.
   B gills.
   C eyes.
   D skin.

5. **Essential Question** What helps animals survive in their environments?

Writing Link

Writing That Compares

Choose two animals. How are these animals alike? How are they different? Write an essay that compares the animals and how they meet their needs.

Social Studies Link

Do Research

People need shelter to stay safe from danger and harsh weather. Research homes from around the world. How do homes from different places compare?

LOG ON e-Review Summaries and quizzes online at www.macmillanmh.com

Focus on Skills

Inquiry Skill: Classify

Earth is a big place. Millions of living things find homes in many different environments around our planet. With so many living things and so many environments, what can scientists do to understand life in our world? One thing they do is **classify** living things.

▶ Learn It

When you **classify,** you put things into groups that are alike. Classifying is a useful tool for organizing and analyzing things. It is easier to study a few groups of things that are alike than millions of individual things.

▶ Try It

Scientists **classify** plants. They **classify** animals too. Can you?

1. To start, observe the animals shown on the next page. Look for things they have in common.
2. Now think of a rule. What characteristic can you use to group the animals? Let's try wings. Which animals have wings? Which animals do not? Make a table to show your groups.

Wings	No Wings

I have wings!

rhea

▶ Apply It

Classify these animals using your own rule.

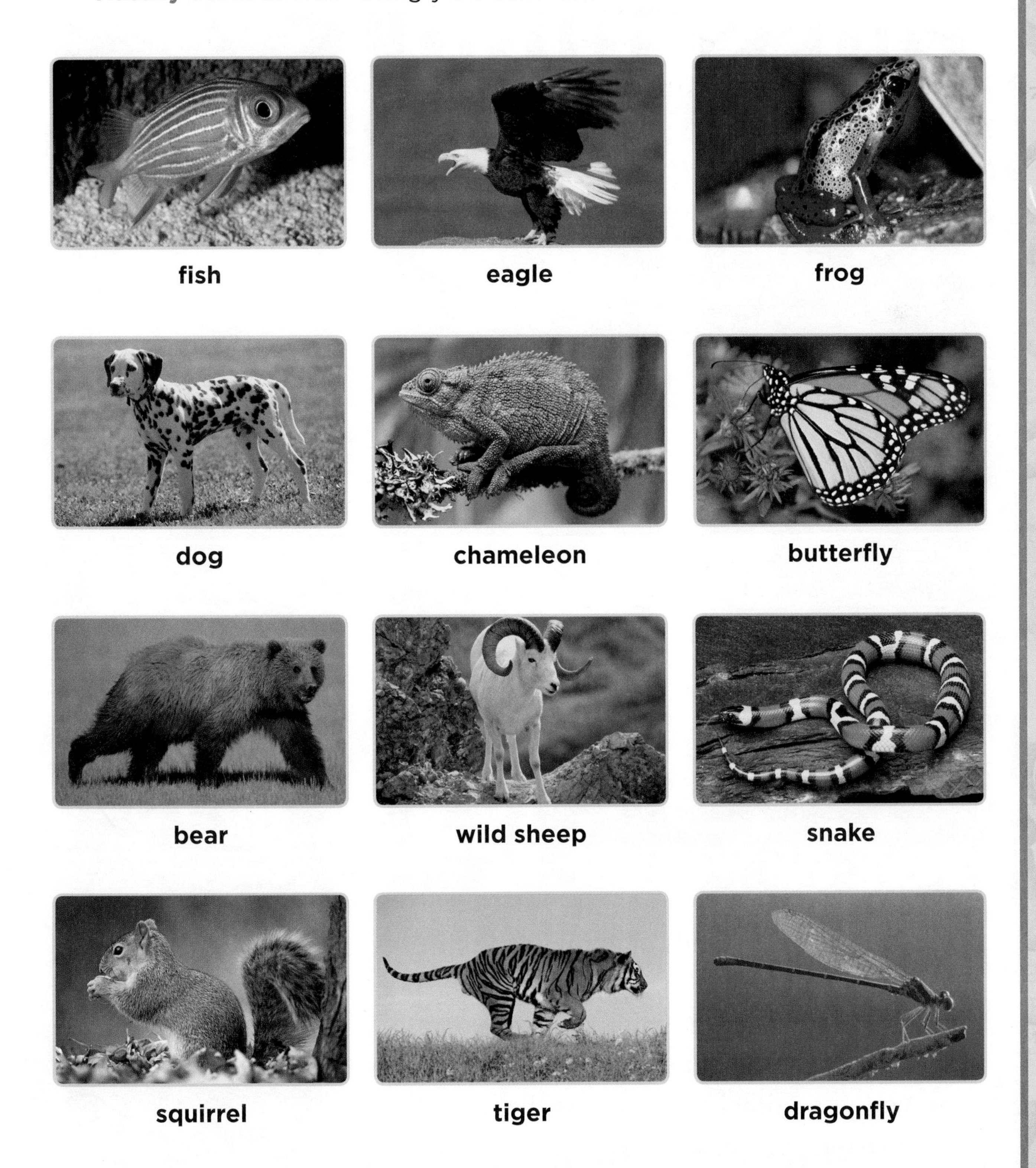

Lesson 4

Classifying Animals

Look and Wonder

Can you find two kinds of animals in this photo? The orange clown fish are easy to find. The green stuff they are hiding in is also an animal! There are many kinds of animals. How can you put animals into groups that are alike?

Explore

Inquiry Activity

How can you classify animals?

Purpose

Classify animals to form groups that have similar characteristics.

Procedure

1. **Observe** Look at each animal. What structures does each animal have? Does each animal have legs? If so, how many? Does each animal have a distinct head and body?
2. **Communicate** Make a chart like the one shown. Use words and pictures to describe characteristics of each animal.
3. **Classify** Put the animals into groups that are alike. Use the information in your chart to help you. Is there more than one way to group the animals?

Draw Conclusions

4. **Interpret Data** Which two animals are most similar to each other?
5. **Communicate** What rule did you use to classify the animals? Why did you classify the animals the way you did?

Explore More

Classify What other animals fit into your groups? Add animals to each of your groups. Research any animals you are not sure of.

Materials

- 4 plastic containers
- hand lens
- worm
- beetle
- snail
- ant

Step 1

Step 2

Animal structure	beetle	snail	worm	ant
legs	6			
antennae	2			
head				
mouth				
eyes				

Read and Learn

Essential Question
Which features can we use to classify animals?

Vocabulary
- **vertebrate,** p. 54
- **invertebrate,** p. 55
- **exoskeleton,** p. 57
- **bird,** p. 58
- **reptile,** p. 58
- **amphibian,** p. 59
- **fish,** p. 59
- **mammal,** p. 60

Reading Skill
Classify

Technology 

e-Glossary and e-Review online at www.macmillanmh.com

How can you classify animals?

What does an animal look like? There is no one answer. Tigers, ants, bluebirds, and sharks are all animals. They all move and respond to their environment. They all reproduce and have the same basic needs. Yet they are all very different from each other. Classifying animals to form smaller groups makes it easier for scientists to study them. One way scientists classify animals is by their structures.

One structure that is useful for classifying animals is a backbone. A backbone is made of many small bones running down the center of an animal's back. Animals with backbones are called **vertebrates** (VUR•tuh•brayts). Tigers, dogs, eagles, and goldfish are all examples of vertebrates.

backbone

A raccoon is a vertebrate. Its backbone helps hold up its body.

Read a Diagram

Where does a raccoon's backbone begin and end?

Clue: The raccoon's backbone is drawn on top of the photo.

Animals without backbones are called **invertebrates** (in•VUR•tuh•brayts). Most of the animals on Earth are invertebrates. Invertebrates lack more than backbones. They have no bones inside their bodies at all! Insects, spiders, worms, and jellies are common invertebrates.

Quick Check

Classify What characteristic separates vertebrates from invertebrates?

Critical Thinking How do bones help vertebrates?

Quick Lab

Model a Backbone

1. **Observe** Look at the photo of the raccoon. What does its backbone look like?
2. **Make a Model** Use clay and pipe cleaners to make a model of a backbone. Design your model so it can bend from side to side and forward and backward.

3. **Experiment** How can your model move? Can you move one bone without moving all the others?
4. **Infer** If a backbone were one solid bone, could it move in the same ways?

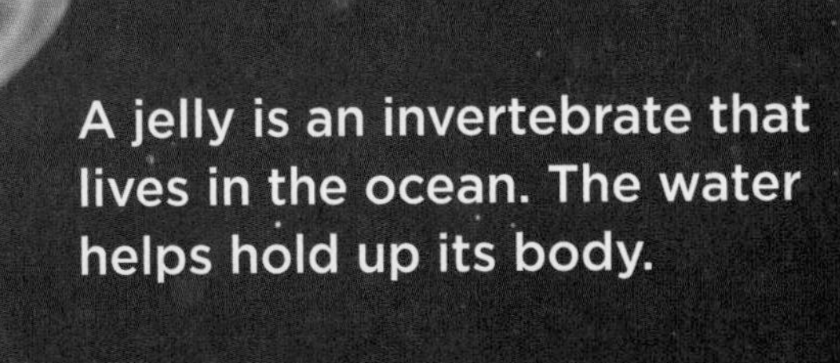

A jelly is an invertebrate that lives in the ocean. The water helps hold up its body.

What are some invertebrates?

Invertebrates can be found all over Earth. They live on land and in water. Most are small, like insects. A few, such as the giant squid, can grow as long as a school bus! The photos below show some common invertebrate groups.

sponges

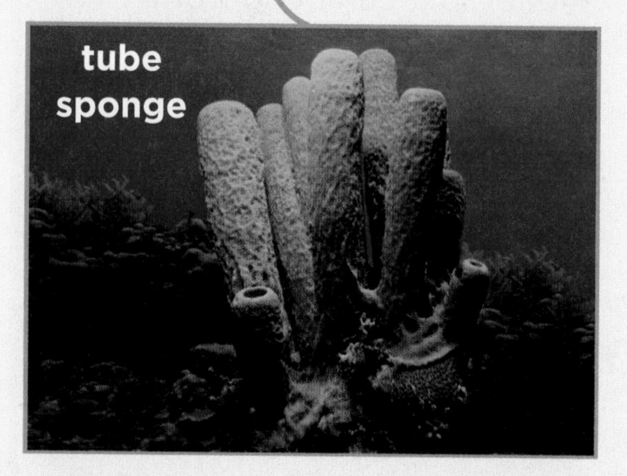

These simple animals have holes in their bodies. They pull water and floating food into the holes.

worms

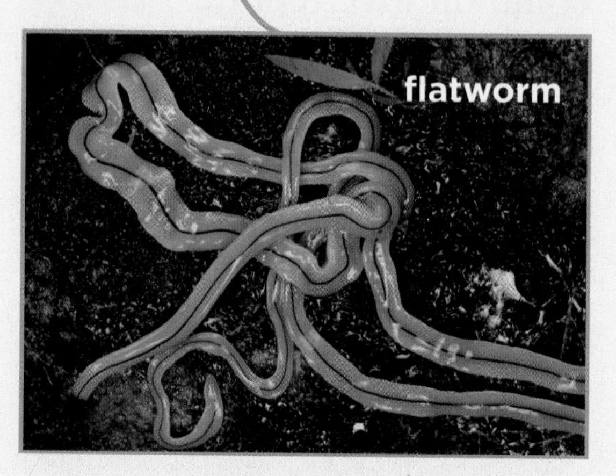

Worms have no skeleton, inside or out. There are more than one million types of worms.

sea stars and urchins

Sea stars and sea urchins have shells inside their bodies. They eat through tubes on their feet.

jellies

These invertebrates have no bones, brains, or eyes. Their tentacles can sting their prey.

Invertebrates have no bones. However, they have other structures that hold up and protect their bodies. Many have a thin, hard covering, for example. This outer covering is an **exoskeleton** (ek•so•SKEH•luh•tun).

Quick Check

Classify Name one invertebrate that lives in water and one that lives on land.

Critical Thinking Is an octopus an invertebrate? How can you tell?

arthropods

northern lobster

Arthropods make up the biggest group of invertebrates. Animals in this group have thin exoskeletons and legs that bend in many places. Insects, spiders, and lobsters are some arthropods.

beetle

mollusks

squid

This group of invertebrates has soft bodies. A few have hard shells. Most push their bodies along with a muscle called a *foot*. Clams, snails, and octopuses are mollusks.

snail

▲ Penguins are one of the few birds that can not fly.

What are some vertebrates?

Are all vertebrates alike? Compare these four types and see what you think.

Birds

A **bird** is a kind of animal with a beak, feathers, two wings, and two legs. Birds are built to fly. Birds breathe air with lungs. They reproduce by laying eggs. Most birds feed their young until the young can find food on their own.

Reptiles

Crocodiles, turtles, and snakes are reptiles. **Reptiles** (REP•tilez) are vertebrates with scaly skin. Tough scales help protect them. Some reptiles live on land and some live in water. All breathe through lungs. Most reproduce by laying eggs, but some give birth to live young.

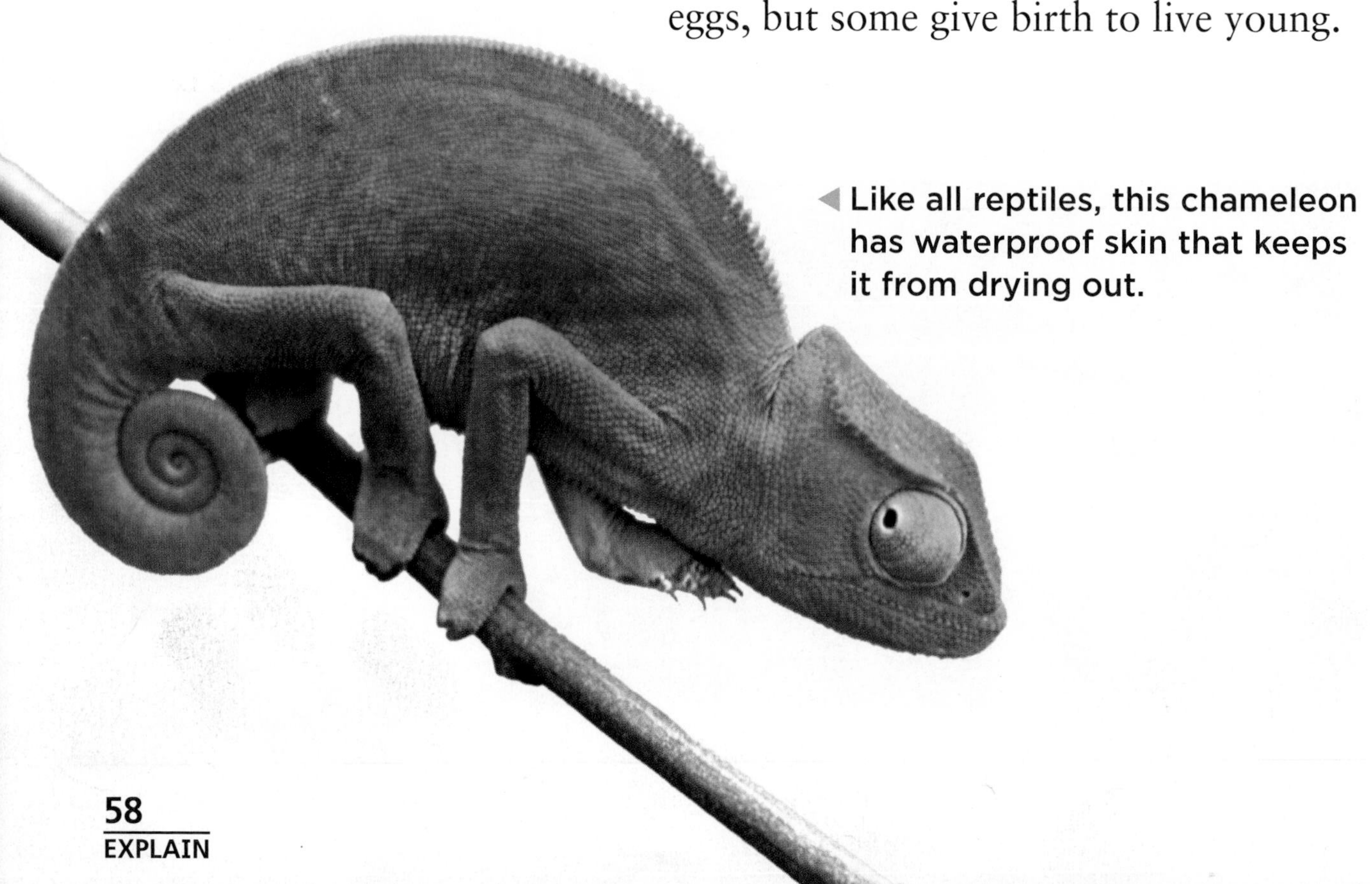

◀ Like all reptiles, this chameleon has waterproof skin that keeps it from drying out.

Amphibians

Some animals spend part of their lives in water and part on land. They are called **amphibians** (am•FIH•bee•unz). Frogs, toads, and salamanders are amphibians.

Most amphibians start out as an egg floating in water. When they hatch, they look like fish. They breathe through gills. As they get older, they grow legs and lungs and begin to live on land.

▲ Adult amphibians, like this frog, breathe through lungs or their skin.

Fish

Fish are vertebrates that spend their whole lives in water. Fish breathe oxygen using gills. They reproduce by laying eggs. Most are covered in scales and a slimy coating.

Quick Check

Classify What kind of vertebrate is a frog?

Critical Thinking Do you think turtles breathe with lungs or gills? Why?

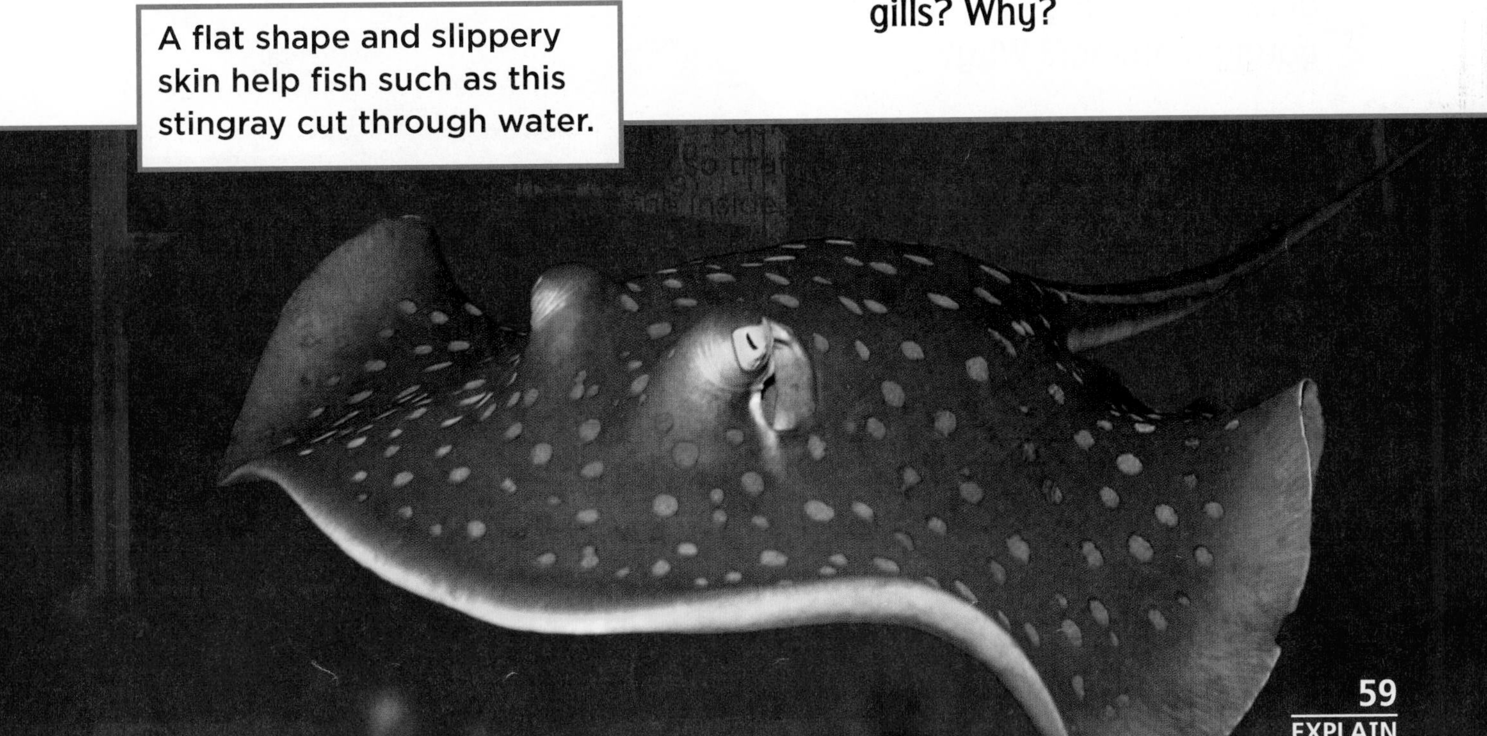

A flat shape and slippery skin help fish such as this stingray cut through water.

What are mammals?

The last type of vertebrate includes mice, dogs, and elephants. It includes people too! **Mammals** (MA•mulz) are vertebrates with hair or fur. Most mammals do not hatch from eggs. They are born live. Female mammals make milk to feed their young. They care for the young until the young can find food on their own.

Mammals are covered with hair or fur. Mammals such as cats and bears have thick fur. Others, such as elephants and people, have thinner hair.

Mammals breathe with lungs. Dolphins and whales are mammals that live in water. They poke their heads out of the water to breathe.

▲ A mammal's first food is milk from its mother.

Quick Check

Classify Which characteristics help you know that a bear is a mammal?

Critical Thinking Which could survive better alone—a young reptile or a young mammal? Why?

When whales swim fast, they jump out of the water to take a breath.

FACT Whales and dolphins are not fish. They are mammals.

Lesson Review

Visual Summary

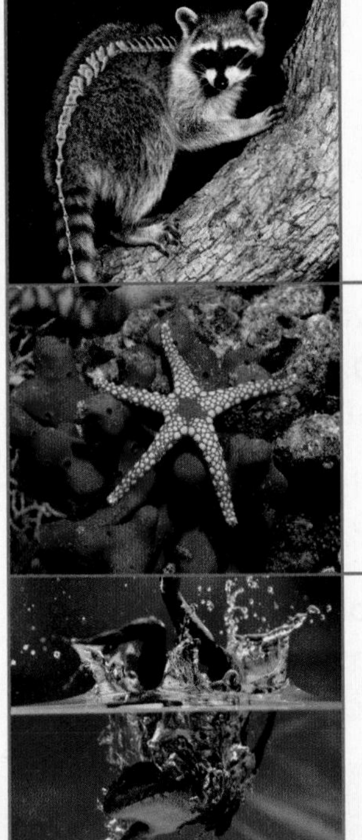

Animals are classified according to their structures and characteristics.

An **invertebrate** is an animal without a backbone. Insects, spiders, and lobsters are invertebrates.

Fish, birds, reptiles, amphibians, and mammals are **vertebrates,** animals with backbones.

Make a FOLDABLES® Study Guide

Make a shutter fold. Use it to summarize what you learned about classifying animals.

Think, Talk, and Write

1. **Vocabulary** What is an exoskeleton?
2. **Classify** What kind of animal is a zebra? How do you know?

3. **Critical Thinking** How do you think your bones affect your shape and the way you move?
4. **Test Prep** **All reptiles are animals that have**
   - A backbones and gills.
   - B lungs and legs.
   - C backbones and lungs.
   - D backbones and fins.
5. **Essential Question** Which features can we use to classify animals?

Math Link

Make a Graph

Research the number of bones different animals have in their bodies. Then put this information into a bar graph.

Art Link

Make an Animal Picture

Suppose you are a scientist who discovers a new creature. Draw a picture of the new creature. Label structures that could help classify it.

DESERT BIRDS

Roadrunners are birds that live in southwestern deserts. Roadrunners run fast on their strong feet. They are black and white with long white-tipped tails. They hunt lizards, snakes, and insects during the day.

Elf owls live in southwestern deserts too. They are the smallest owls. Unlike roadrunners, elf owls are active only at night. They have yellow eyes and very short tails. Their eyesight is excellent. They eat insects, lizards, and mice.

Descriptive Writing

A good description

- includes words that tell how something looks, sounds, smells, tastes, and/or feels;
- uses details to create a picture for the readers;
- can use words that compare and contrast, such as *like, similar,* and *different.*

roadrunner

elf owl

Write About It

Descriptive Writing Choose two animals. Learn more about them. Then write a paragraph that describes how the animals are alike and different.

e-Journal Write about it online at www.macmillanmh.com

ANIMAL LINEUP

Do all snakes look alike? Do all lizards? No! There are many kinds of snakes and lizards. Each kind is a little different from the others. Each kind is called a *species*. The table below lists the number of known species for four groups of reptiles.

Order Numbers

- ▶ To order numbers from greatest to least, first find the numbers with the most digits.
- ▶ Identify the place value of the digits. Compare the numbers with the highest place value to find out which number is larger.
- ▶ Then repeat this with the remaining numbers.

Kinds of Reptiles	Number of Known Species in 2005
crocodiles	23
lizards	4,765
snakes	2,978
turtles	307

The banded boa constrictor, at left, and the spotted bush snake, above, are two species of snakes.

Solve It

List these reptiles in order from greatest to least number of species. Which kinds of reptiles have more than 1,000 species? Which kind of reptile has the fewest species?

CHAPTER 1 Review

Visual Summary

Lesson 1 All living things have certain characteristics and needs in common.

Lesson 2 Most plants have roots, stems, and leaves. Each plant part does a special job to keep a plant alive.

Lesson 3 Different kinds of animals have different structures that help them get what they need.

Lesson 4 Animals can be classified based on their structures and characteristics.

Make a FOLDABLES® Study Guide

Glue your lesson study guides to a large sheet of paper as shown. Use your study guide to review what you have learned in this chapter.

Vocabulary

DOK I

Fill each blank with the best term from the list.

cells, p. 26	**organism,** p. 22
environment, p. 24	**photosynthesis,** p. 36
invertebrate, p. 55	**reproduce,** p. 23
mammal, p. 60	**shelter,** p. 48
nutrient, p. 34	**vertebrate,** p. 54

1. Animals often seek a safe place, or ______, to protect themselves.
2. Each living thing is an ______.
3. An animal with a backbone is called a ______.
4. Living things ______ to make more of their own kind.
5. A vertebrate that is born live is a ______.
6. A substance that helps living things grow and stay healthy is a ______.
7. Plants make their own food using the process of ______.
8. Living things are made of one or more tiny ______.
9. An animal without a backbone is called an ______.
10. All the living and nonliving things that surround an organism are part of an ______.

LOG ON e-Glossary Words and definitions online at www.macmillanmh.com

Skills and Concepts

DOK 2–3

Answer each of the following.

11. Main Idea and Details What makes living things different from nonliving things?

12. Descriptive Writing Describe the structures that different animals use to breathe.

13. Classify Group the following animals as vertebrates or invertebrates: butterfly, cow, snail, goldfish, owl, spider.

14. Critical Thinking What might happen to a plant if someone picked most of its leaves?

15. Critical Thinking On a cold day, you might see icicles that have grown from the roof of a building. Does this mean icicles are living? Explain your answer.

16. Classify Many animals move by using their legs, fins, or wings. List two animals for each type of movement.

17. Explain how each labeled part helps a plant survive.

18. Which nonliving parts of an environment do all living things need to survive?

A rocks and fire

B water and space

C soil and carbon dioxide

D fire and food

19. True or False *All animals breathe oxygen using either gills or lungs.* Is this statement true or false? Explain.

20. How do living things get what they need to live and grow?

Performance Assessment

DOK 2

It's Alive!

- Make a list of all the living things you find around your home.
- Make a chart to classify each as a plant or an animal. What kind of plant or animal was it?

Living Thing	Plant or Animal
Cat	Animal-mammal
Flower	Plant
Goldfish	Animal-fish
Tree	Plant
Mosquito	Animal-insect

- What kind of living thing was most common? Why do you think that was the case?

Test Preparation

1 Look at the picture below.

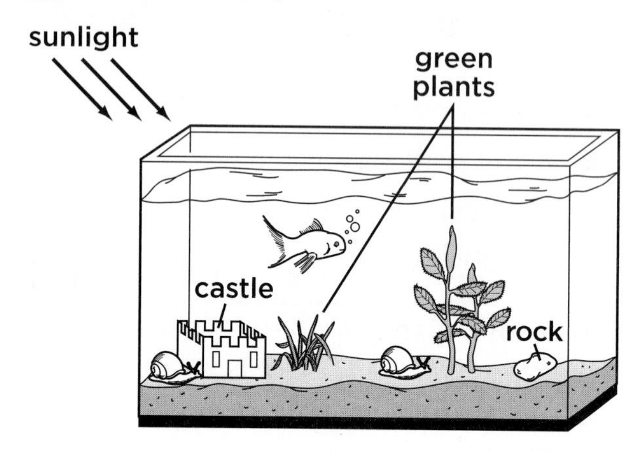

Which of these is a living thing?

A castle

B plant

C rock

D sunlight

DOK I

2 Which list best describes the needs of an animal?

A shelter, soil, water, wind

B shelter, water, food, oxygen

C food, oxygen, soil, rain

D food, water, carbon dioxide, shelter

DOK I

3 Which structure helps fish live in water?

A fins

B lungs

C hair

D blubber

DOK I

4 Look at the picture below.

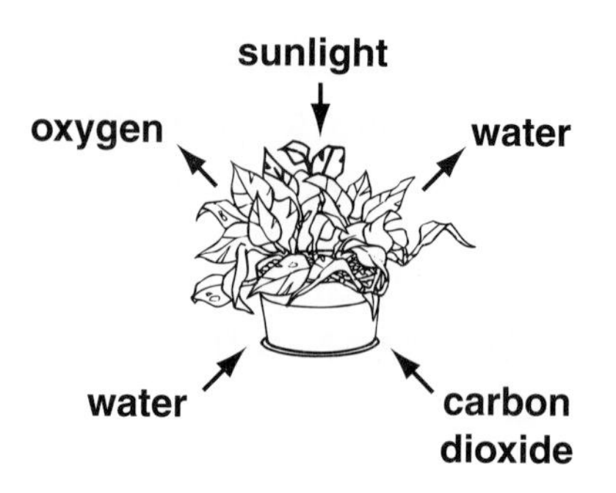

What does the picture show?

A breathing

B eating

C photosynthesis

D reproduction

DOK 2

5 Look at the picture of the bird.

Which structure helps the bird obtain food?

A wings

B feathers

C claws

D beak

DOK I

6 Which of the following is a vertebrate?

A sea star

B snail

C snake

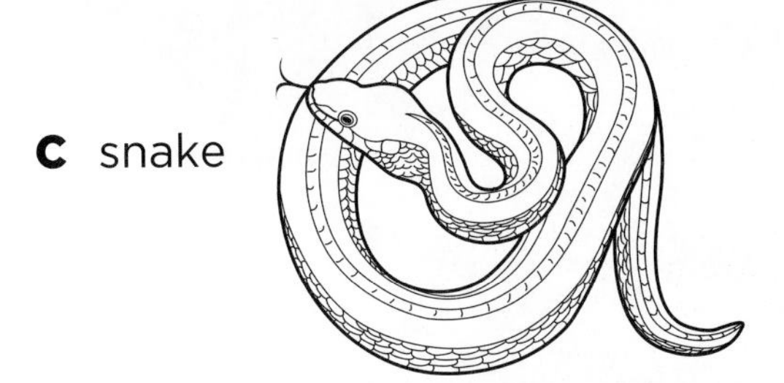

D spider

DOK I

7 What do birds and mammals have in common?

A They have exoskeletons.

B They have backbones.

C They lay eggs.

D They have fur.

DOK I

8 The diagram below shows four plant parts.

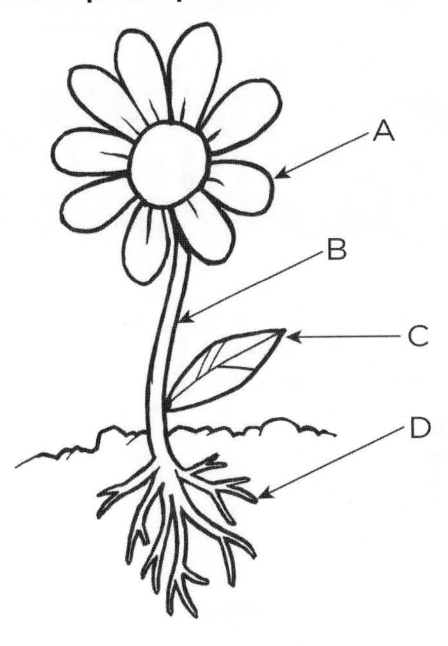

Make a table like the one shown below.

Plant Part	Function

Complete your table. Identify each plant part from the diagram above. Then explain the function of each plant part.

DOK 2

Check Your Understanding

Question	Review	Question	Review
1	pp. 22–25	5	p. 46
2	pp. 22–25, 44–48	6	pp. 54–58
3	p. 47	7	pp. 58–60
4	pp. 36–37	8	pp. 32–37

CHAPTER 2

Living Things Grow and Change

The Big Idea

How do living things change?

Essential Questions

Lesson 1

How do plants grow and reproduce?

Lesson 2

How do animals grow and reproduce?

Lesson 3

How do organisms get their features?

swan with young

Big Idea Vocabulary

seed a structure that can grow into a new plant (p. 70)

pollination when pollen moves from the male part of a plant to an egg, after which a seed can form (p. 73)

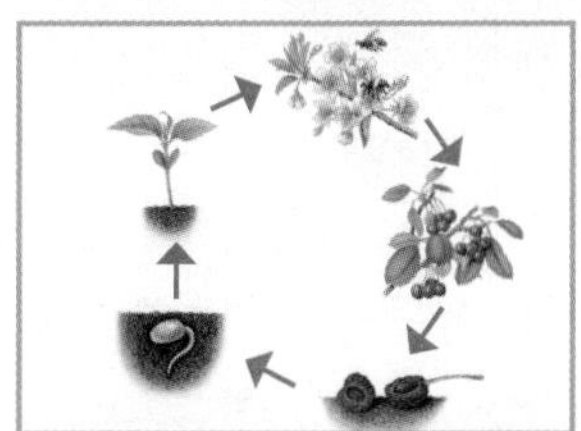

life cycle how an organism grows and reproduces (p. 74)

metamorphosis a series of changes through which an organism's body changes form (p. 83)

egg an animal structure that protects and feeds some very young animals such as birds (p. 83)

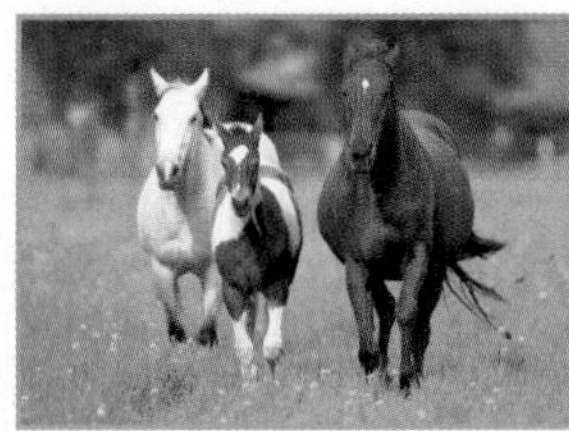

inherited trait a characteristic that is passed from parents to offspring (p. 92)

Visit www.macmillanmh.com for online resources.

Lesson 1

Plant Life Cycles

Look and Wonder

Did you know that when you blow on a dandelion, you are helping seeds spread? New plants grow from those seeds. Where do seeds come from? How do seeds grow into plants?

Explore

Inquiry Activity

What does a seed need to grow?

Form a Hypothesis

Do seeds need water to grow? Form a hypothesis. Write your answer in the form "If seeds do not get water, then..."

Test Your Hypothesis

1. **Observe** Look at the seeds with a hand lens. Draw what you see.
2. **Use Variables** Fold each paper towel into quarters. Then put two tablespoons of water onto one towel. Put the wet towel into a plastic bag. Label the bag *Water*. Put the dry towel into a bag. Label this bag *No Water*.
3. Place three seeds into each bag. Seal the bags and place them in a warm spot.
4. **Observe** Look at the seeds every day for one week. Record what you see with pictures and words. If the paper towel in the *Water* bag feels dry, add two tablespoons of water.

Draw Conclusions

5. **Interpret Data** Which seeds changed? How did they change?
6. **Infer** Why do you think the seeds changed?
7. Did your results support your hypothesis?

Explore More

Experiment What would happen if you wet the paper towel with something other than water? Experiment to find out.

Step 1

Step 3

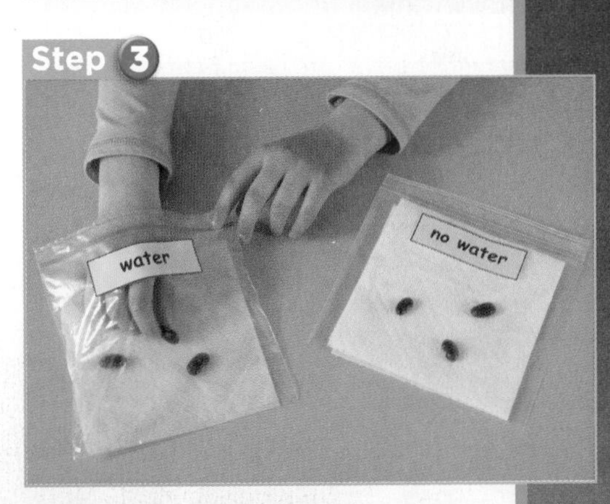

Read and Learn

Essential Question
How do plants grow and reproduce?

Vocabulary
- **seed,** p. 70
- **embryo,** p. 70
- **flower,** p. 72
- **pollination,** p. 73
- **fruit,** p. 73
- **life cycle,** p. 74
- **cone,** p. 75

Reading Skill
Sequence

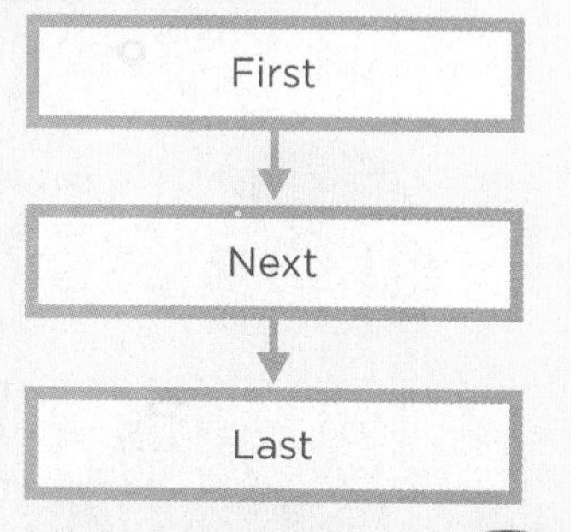

Technology
e-Glossary, e-Review, and animations online at www.macmillanmh.com

How do plants grow?

Did you know that when you eat corn, peas, or nuts, you are eating seeds? Seeds come in all shapes and sizes. Some are big like lima beans. Some are tiny like poppy seeds. Big or small, all seeds have the same function.

A **seed** is a structure that can grow into a new plant. It holds a young plant that is ready to grow. This young plant is called an **embryo** (EM•bree•oh). A seed has parts that help an embryo survive. It holds stored food that the embryo uses to grow. It has a tough covering that protects the embryo.

When a seed is planted in the soil, it can *germinate*, or begin to grow. A seed needs water, nutrients, and the right temperature to germinate. It can wait to grow for months or even years until conditions are right.

From Seed to Plant

1. A seed is planted in soil.

2. The seed germinates. Roots start growing down into the soil.

When a seed begins to germinate, it soaks up water. That makes it swell up and break through its covering. The embryo grows out of the seed. It grows into a small plant called a *seedling*. A seedling can grow into an adult plant.

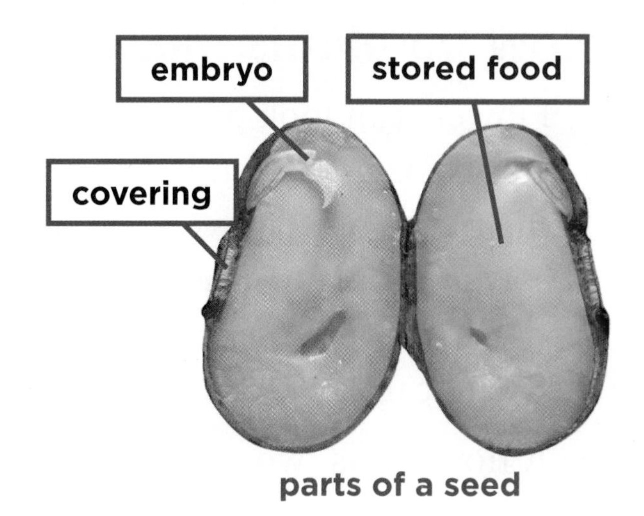

parts of a seed

Quick Check

Sequence What happens to a seed after it germinates?

Critical Thinking What might happen to a seed if it does not get enough water?

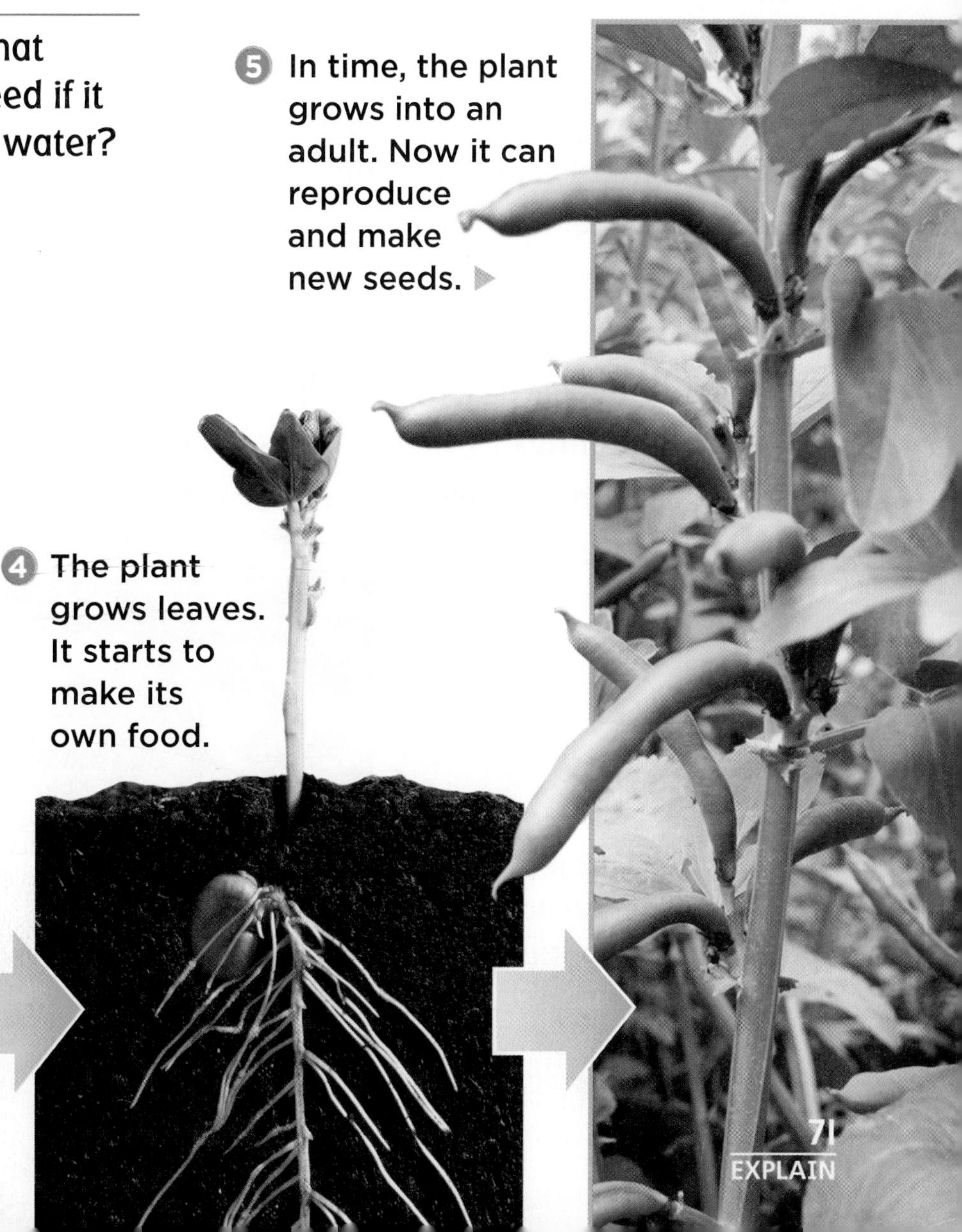

5 In time, the plant grows into an adult. Now it can reproduce and make new seeds.

3 The roots grow longer, and a stem pushes up out of the ground.

4 The plant grows leaves. It starts to make its own food.

How do plants make seeds?

Flowers can look pretty and smell sweet. They also do an important job. Many plants need flowers to reproduce. A **flower** is a plant structure that makes seeds. Plants that use flowers to make seeds are called *flowering plants*.

A flower has two parts that help it make seeds—a male part and a female part. The male part makes a powder called *pollen*. The female part makes tiny eggs. When pollen and an egg come together, a seed can form.

How does pollen get to an egg? Wind can blow pollen from one flower to another. Animals such as hummingbirds, bees, and bats can carry pollen too. Some animals are attracted to a flower's smell or bright colors. They drink a sweet liquid, called *nectar,* from the flower. Sticky pollen clings to their bodies. Then, they carry the pollen to another flower.

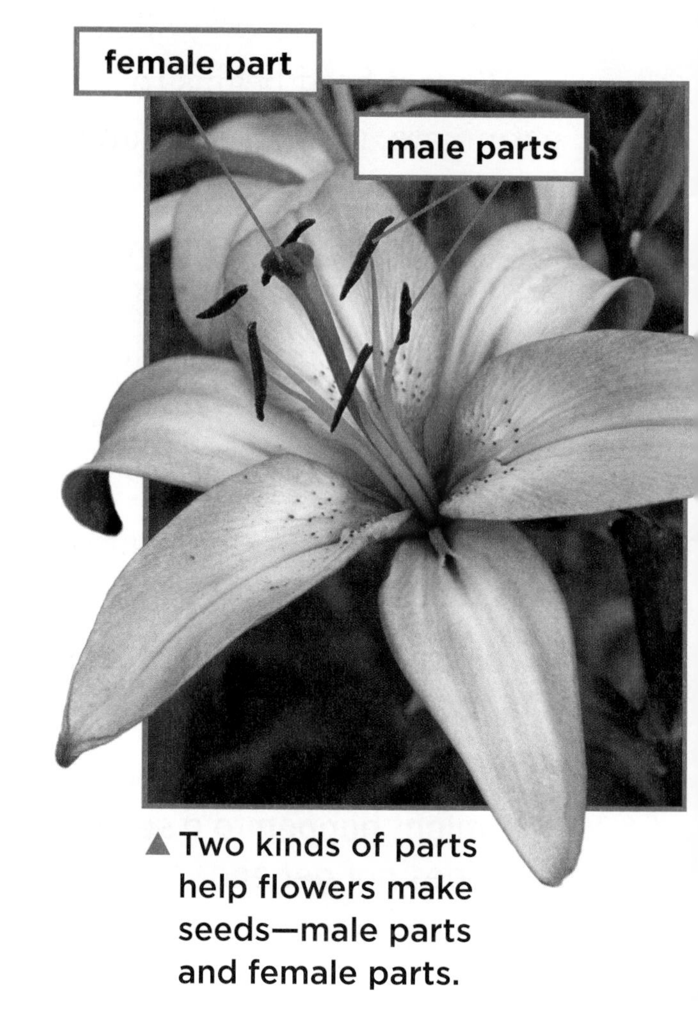

▲ Two kinds of parts help flowers make seeds—male parts and female parts.

When a bee drinks nectar from a flower, yellow pollen sticks to its body. ▶

The movement of pollen from the male part of a flower to the female part is called **pollination** (pah•luh•NAY•shun). After pollination, seeds can develop. In flowering plants, fruit forms around the seeds. **Fruit** is a structure that holds seeds.

How Seeds Travel

Before a seed can germinate, it must find its way to the soil. How does it get there? Some seeds, such as a fuzzy dandelion's, are made to blow in the breeze. Other seeds fall to the ground inside of ripe fruit. The fruit rots and spills its seeds.

Animals can help too. Seeds, such as acorns, can be buried by squirrels. Prickly seeds can stick to an animal's fur and be carried to a new place. When an animal eats fruit, the seeds can pass through the animal's body. They are left on the ground in the animal's waste.

Quick Lab

Fruits and Seeds

strawberry

1. **Observe** Look at the fruit from three different plants. Compare their shapes and sizes.

peach

2. △ **Be careful!** Cut open the fruit. How do their parts compare? Do they all have a peel or skin? Do they all have seeds?

kiwi

3. **Observe** Look at the seeds from each fruit. Compare the location of the seeds in each fruit.
4. **Infer** What do all fruit have in common? How might fruit help seeds survive and grow?

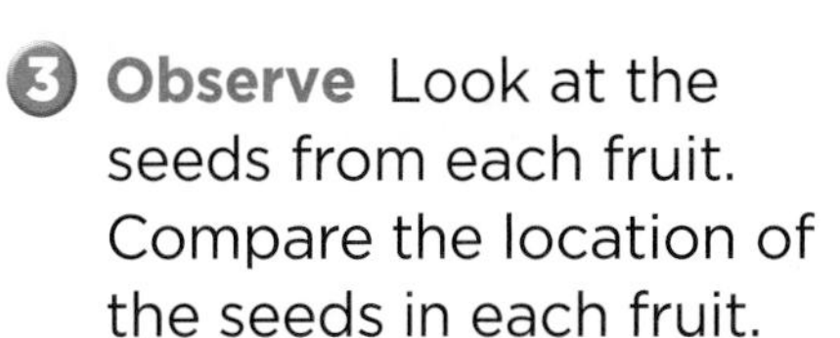

◀ Berry seeds will pass through this ermine's body and into the soil, where they can grow.

Quick Check

Sequence How does a seed form?

Critical Thinking How do bright, sweet-smelling flowers help plants?

FACT Tomatoes have seeds, so they are fruit.

What is a plant's life cycle?

How a plant germinates, grows, and reproduces is the plant's **life cycle**. Plants grow and reproduce in different ways. For example, some plants have flowers, and others have cones.

In time, adult plants die. They *decompose,* or break down, and become part of the soil. This adds nutrients to the soil that help other plants grow.

Flowering Plants

Most plants are flowering plants. Flowering plants grow from seeds into adult plants. As adults, they reproduce and make new seeds using flowers.

cherry tree

Life Cycle of a Cherry Tree

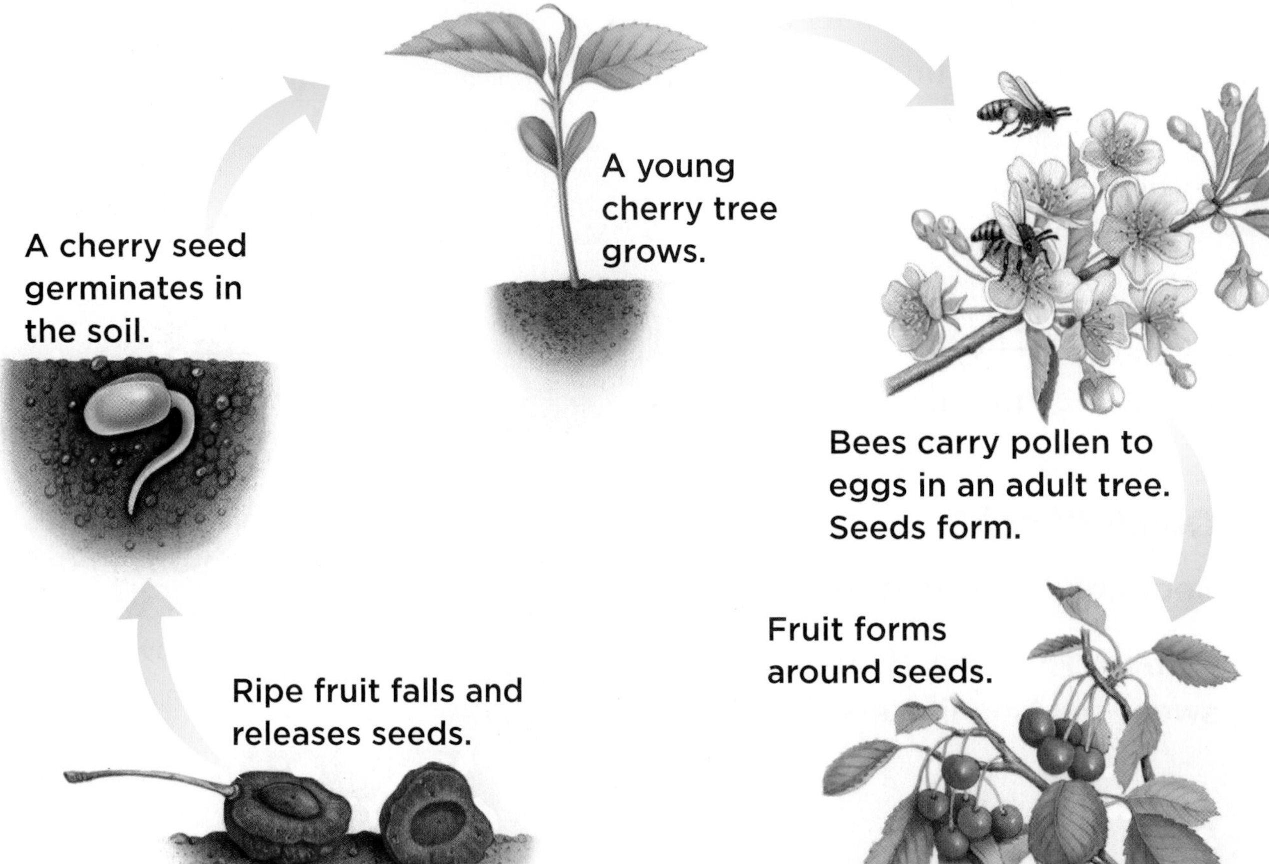

Conifers

Have you ever picked up a pine cone? **Cones** are plant structures that make seeds. Plants that reproduce with cones are called *conifers*. They include pine, spruce, and hemlock trees. Conifers have similar life cycles to flowering plants. Both grow from seeds. Both reproduce and make new seeds through pollination. However, conifers make seeds inside of cones instead of flowers.

Quick Check

Sequence How do conifers form seeds?

Critical Thinking How are flowers and cones alike? How are they different?

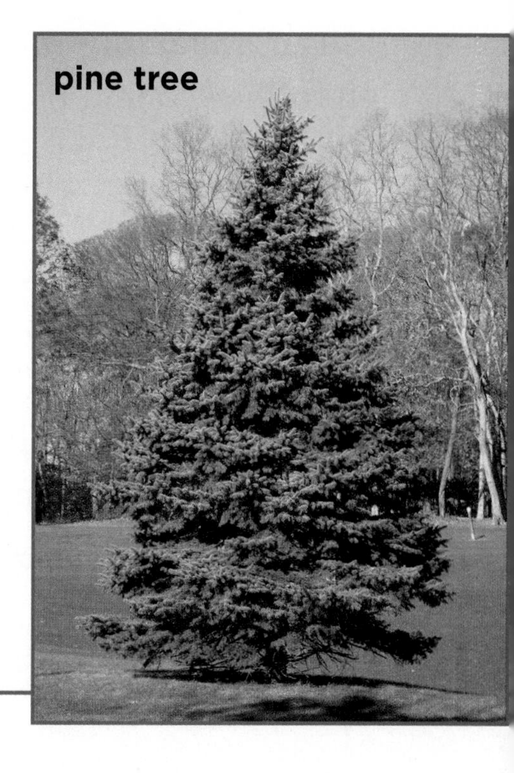

Life Cycle of a Pine Tree

A young pine tree grows.

An adult tree makes male and female cones. Wind blows pollen from the male cones onto the female cones.

Seeds develop inside the female cone.

When the cone is ripe, seeds fall out.

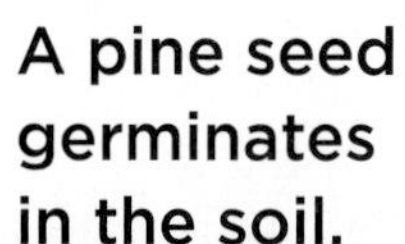

A pine seed germinates in the soil.

Read a Diagram

What are the stages of a conifer's life cycle?

Clue: Arrows help show a sequence.

Science in Motion Watch plant life cycles and learn more at www.macmillanmh.com

How do plants grow without seeds?

Some plants reproduce without making seeds. A type of plant called a fern never makes seeds. Instead it makes *spores*. Like a seed, a spore can fall to the ground. It can grow into a new fern plant. Unlike a seed, a spore does not have stored food.

New plants can also grow from parts of plants. Potato plants can grow from the white spots, or "eyes," on a potato. Other plants grow from an underground stem called a *bulb*. An onion is one type of bulb. Sometimes, a new plant can also grow from a stem or a leaf that is placed in water.

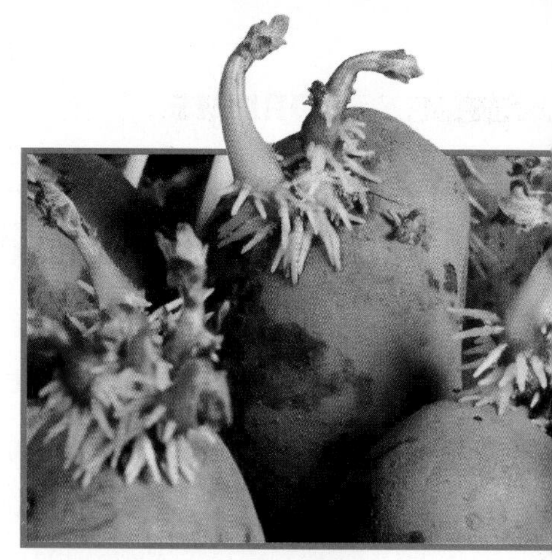

▲ New stems and leaves can grow from the "eyes" of a potato.

Life Cycle of a Fern

Adult ferns grow and release spores.

A spore grows into a small organism with male and female parts.

A young fern grows when cells from the male and female parts join.

fern

Quick Check

Sequence What is the life cycle of a fern?

Critical Thinking Would a fern survive if it landed in soil with few nutrients? Why?

Lesson Review

Visual Summary

	Plants go through a series of **changes as they grow** into adults.
	**Flowering plants and conifers** grow from seeds and have similar life cycles.
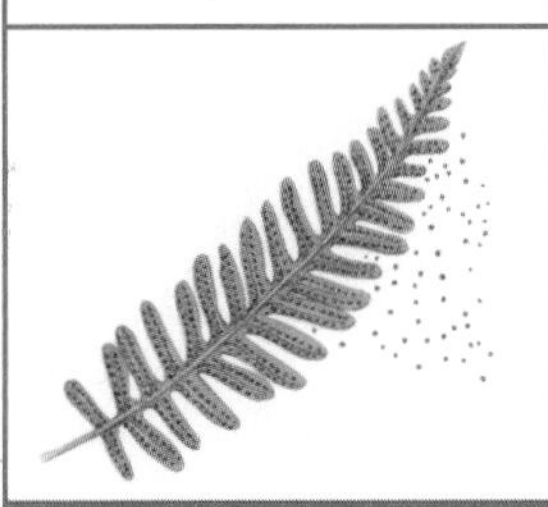	Some plants do not grow from seeds. **Ferns** make spores. **Other plants** grow in different ways.

Make a FOLDABLES® Study Guide

Make a layered-look book. Use it to summarize what you learned about plant life cycles.

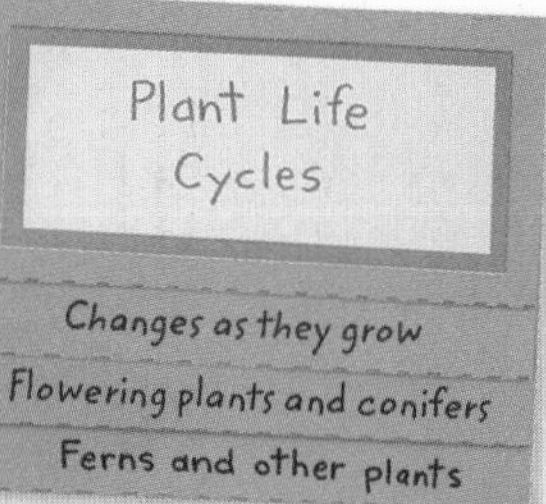

Think, Talk, and Write

1. **Vocabulary** What is fruit?
2. **Sequence** What is the life cycle of a flowering plant?

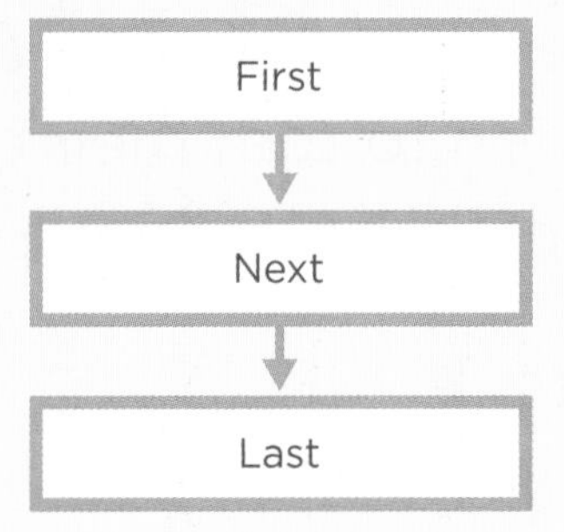

3. **Critical Thinking** How do animals help flowering plants?
4. **Test Prep** **How does a conifer reproduce?**
   - **A** with bulbs
   - **B** with flowers
   - **C** with cones
   - **D** with spores
5. **Essential Question** How do plants grow and reproduce?

Writing Link

Writing That Compares
Which plants grow in your community? Choose four plants around your home. Do they reproduce using flowers, cones, or spores? Write paragraphs comparing and contrasting the plants you found.

Art Link

Paint a Picture
Use library resources to find images of flowers painted by Georgia O'Keeffe. Look closely at two of her paintings of plants. How do they make you feel? Describe them. Then make your own paintings of plants.

e-Review Summaries and quizzes online at www.macmillanmh.com

Inquiry Skill: Form a Hypothesis

You just learned how seeds grow into plants. Can seeds grow when the weather is cold? To answer questions like this, scientists start with what they know about plants. Then they use this information to turn their question into a testable statement. That is, they **form a hypothesis**.

▶ Learn It

When you **form a hypothesis**, you make a statement that you can test by collecting data. Suppose you want to find out if plants need sunlight. Based on what you know, you could **form a hypothesis** like this: If plants do not get sunlight, then they will not grow.

A good hypothesis needs to be testable. You could test the hypothesis above by placing one plant in the dark and one in sunlight. Then you could observe and record what happens. A hypothesis also needs to identify the variables. In the example above, sunlight and plant growth are variables.

▶ Try It

Form a hypothesis about what seeds need to grow. Then test that hypothesis with an experiment.

Materials **water, 2 paper towels, 6 pea seeds, 2 sealable plastic bags, 2 foam cups, ice**

1. Think about what you know about seeds. Now form a hypothesis about this question: *Will pea seeds germinate more quickly in a cold spot or in a warm spot?* Begin with "If I plant a pea seed in the cold, then..."

2. Fold two wet paper towels in half and place three seeds onto each. Place each paper towel into a plastic bag and seal the bags.
3. Fill one foam cup with ice. Place one bag into this foam cup. Place the other bag into the empty foam cup.
4. Make a chart like the one below. Use it to record your observations each day. Do your results support your hypothesis?

Step 4

	Cold	Warm
Day 1		
Day 2		
Day 3		
Day 4		

▶ Apply It

Now that you have learned to think like a scientist, you can answer other questions. Do seeds germinate more quickly in the light or dark? **Form a hypothesis** about this question. Then plan an experiment to test your hypothesis.

Lesson 2

Animal Life Cycles

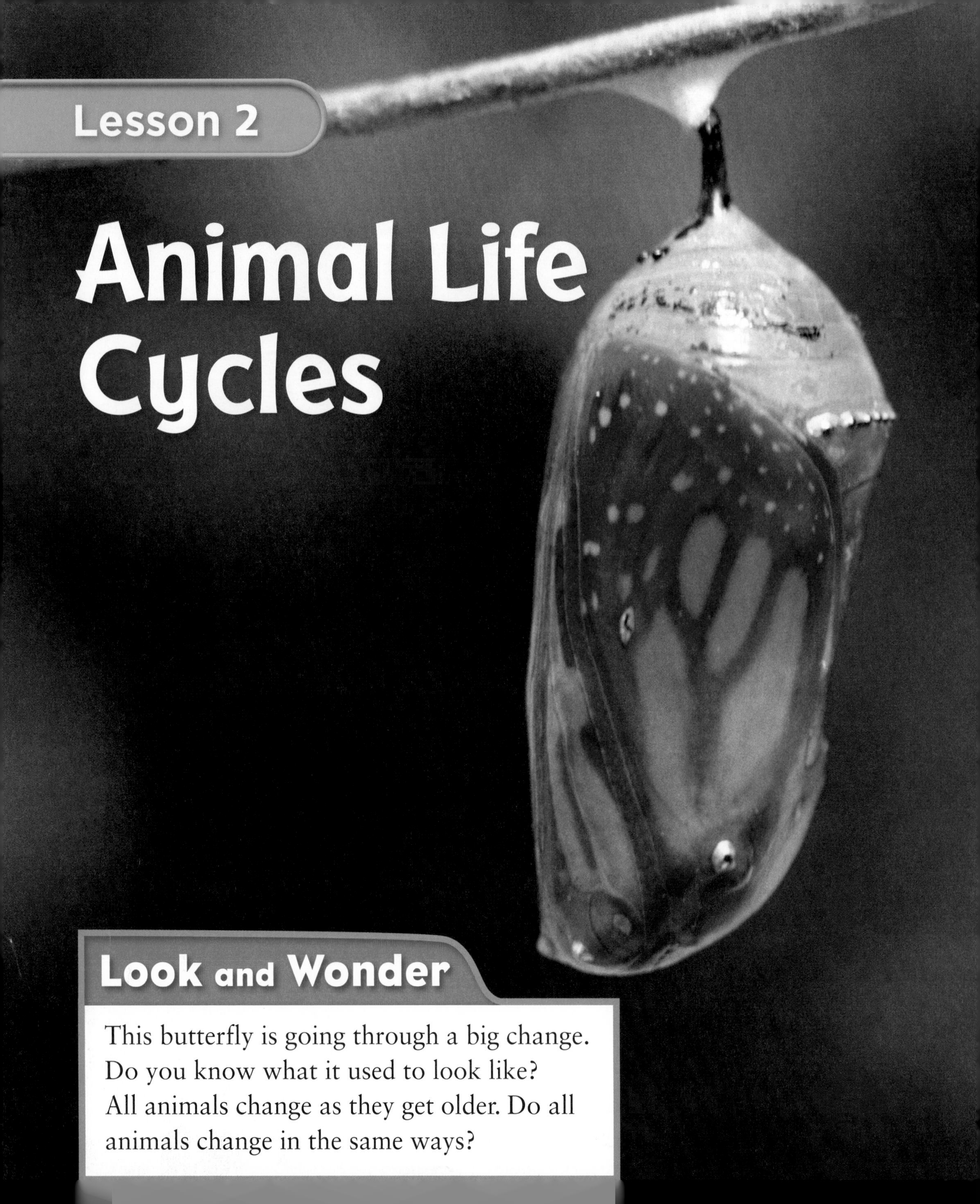

Look and Wonder

This butterfly is going through a big change. Do you know what it used to look like? All animals change as they get older. Do all animals change in the same ways?

Explore

Inquiry Activity

How does a caterpillar grow and change?

Make a Prediction

How does a caterpillar change as it grows? Make a prediction.

Test Your Prediction

1. **Observe** Look at the caterpillar. Draw a picture of it and label all the parts you can see. △ **Be Careful.** Handle animals with care.
2. **Measure** Find the length of your caterpillar. Record the caterpillar's length on your drawing.
3. Put your caterpillar into the kit.
4. **Observe** Once a day, observe your caterpillar and draw a picture of it. Label any changes you observe. If you can measure the caterpillar's length without disturbing it, record the length each day.

Draw Conclusions

5. **Interpret Data** What small changes did the caterpillar go through? What big changes did you observe?
6. **Infer** What are the stages in a butterfly's life cycle?

Explore More

Experiment How do tadpoles change as they grow? Make a plan to test your ideas.

Materials

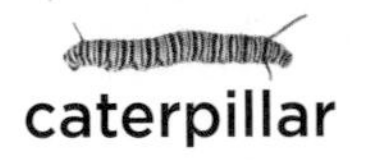
caterpillar

hand lens

ruler

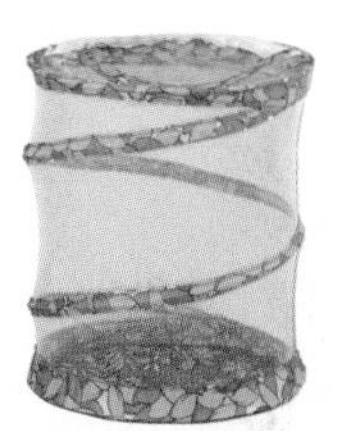
caterpillar kit

Step 1

Step 2

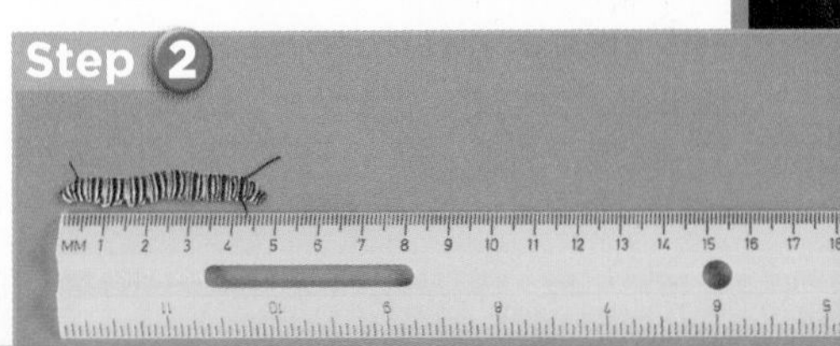

Read and Learn

Essential Question
How do animals grow and reproduce?

Vocabulary
metamorphosis, p. 83
egg, p. 83
larva, p. 83
pupa, p. 83

Reading Skill
Sequence

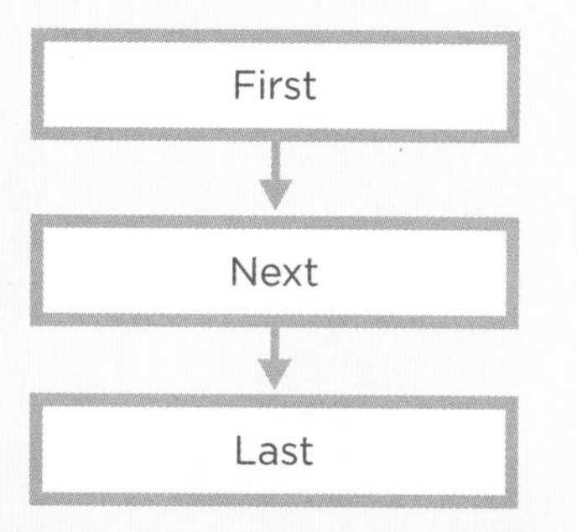

Technology 

e-Glossary and e-Review online at www.macmillanmh.com

What are some animal life cycles?

Did you know that a caterpillar is actually a young butterfly? A tadpole is a young frog. These animals go through big changes as they grow. Do all animals change in the same ways?

Different types of animals change in different ways. Some animals are born looking like their parents. Others are not. These animals might change shape or color as they grow. They might even grow new structures. The way an animal changes with age is part of its life cycle.

An animal is born. It grows. It reproduces as an adult. In time it dies. Its body breaks down and becomes part of the soil. It adds nutrients to the soil that other organisms need to grow.

Life Cycle of a Frog

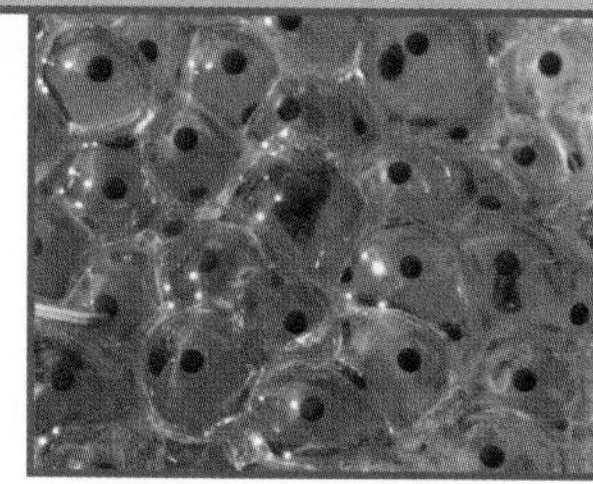

Egg Frogs lay eggs in water.

Tadpole Young frogs, or tadpoles, hatch. Like fish, they swim and breathe with gills.

Becoming an Adult A tadpole starts to grow legs and lungs.

Adult Now the frog looks like its parents. It moves onto land and can reproduce.

Metamorphosis

Some animals change shape through a process called **metamorphosis** (me•tuh•MOR•fuh•sis). Amphibians and most insects go through metamorphosis. Their life cycle begins with an **egg**. Eggs contain food that young animals need. Most have a shell that protects the animal.

When the young animal has grown enough, it *hatches*, or breaks out of the egg. It looks different from adults of its kind. With time, it grows into an adult that can have its own young. Most amphibians and insects do not look after their young. The young can get food on their own.

Quick Check

Sequence Name the stages in a ladybug's life cycle.

Critical Thinking Compare a frog's life cycle to a ladybug's life cycle.

Life Cycle of a Ladybug

Egg A ladybug starts life as an egg.

Larva When an insect hatches, it is called a **larva**. A ladybug larva eats bugs and grows.

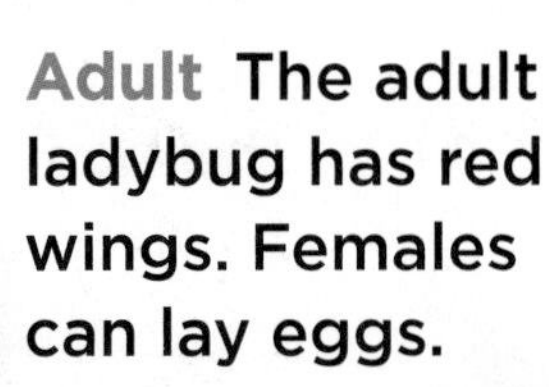

Adult The adult ladybug has red wings. Females can lay eggs.

Pupa The larva changes into a **pupa**. It forms a hard shell. Inside, it grows wings.

How do reptiles, fish, and birds change as they grow?

Reptiles, fish, and birds have similar life cycles. Most of these animals lay eggs. Reptiles lay their eggs on dry land. Fish lay their eggs in water. Birds often build nests to protect their eggs. Most birds sit on their eggs until the eggs are ready to hatch.

An animal grows inside the egg. For a time it gets everything it needs to survive from the egg. When the young animal has grown enough, it hatches. Young reptiles, fish, and birds do not go through metamorphosis. They look similar to adults of their kind when they hatch.

Life Cycle of a Sea Turtle

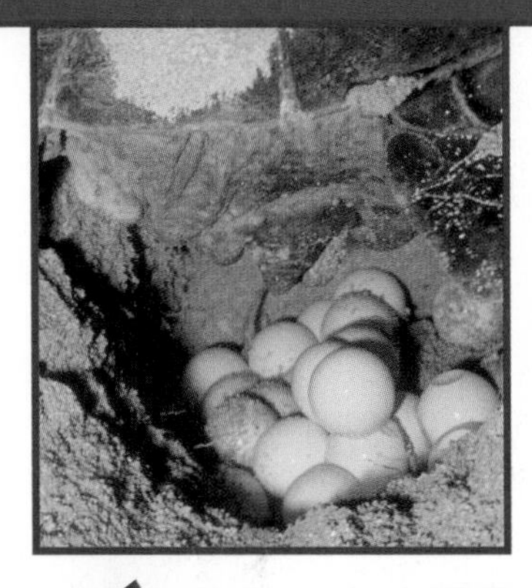

Egg Females crawl to the beach to lay eggs in the sand.

Young Sea turtles hatch on the beach and quickly crawl to the ocean.

Adult Turtles grow to 140 kilograms (300 pounds). Females stay in the sea until they are ready to lay eggs.

In time, young reptiles, fish, and birds grow into adults. They can reproduce and have young of their own. Most reptiles and fish do not look after their young. The young can find food on their own. Birds often raise their young until the young can fly and find food for themselves.

Quick Check

Sequence What happens after a fish lays eggs?

Critical Thinking How is a reptile's life cycle similar to a frog's? How does it differ?

Quick Lab

A Bird's Life Cycle

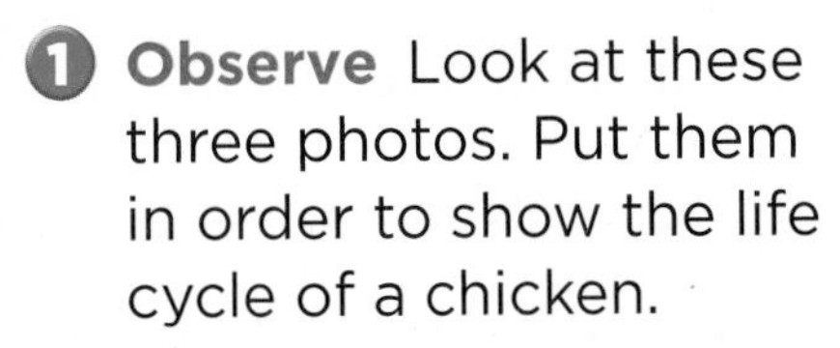

1. **Observe** Look at these three photos. Put them in order to show the life cycle of a chicken.

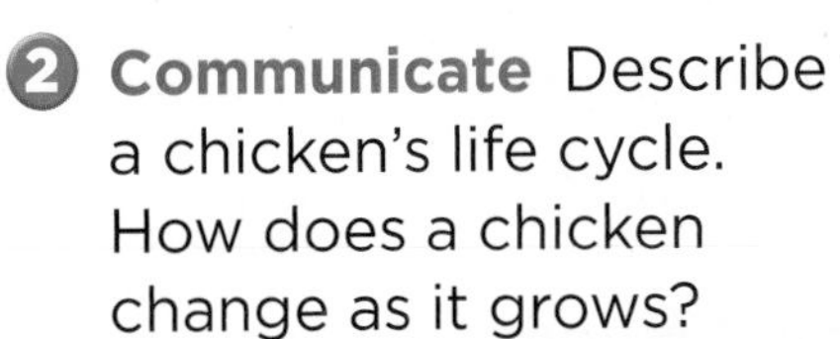

2. **Communicate** Describe a chicken's life cycle. How does a chicken change as it grows?

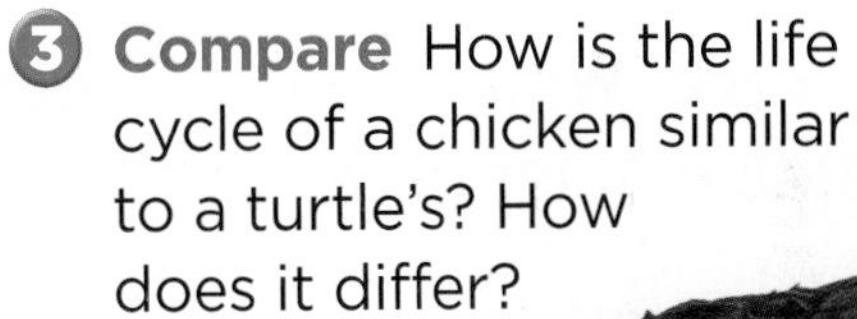

3. **Compare** How is the life cycle of a chicken similar to a turtle's? How does it differ?

Life Cycle of a Trout

Egg Fish eggs may float in water or sink to the bottom.

Young Fish hatch and begin to find food.

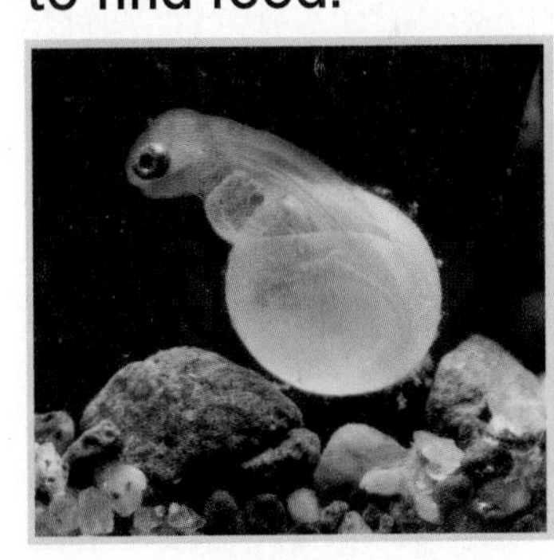

Adult Most fish continue to grow all their lives. Females may lay thousands of eggs each year!

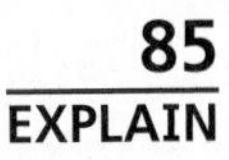

What is the life cycle of a mammal?

Most mammals do not hatch from eggs. Young mammals are born live. They look much like their parents from the start. Adult mammals feed and care for their young.

As they grow, young mammals lose fat and grow stronger. Their faces change to look more like adults. In time, they learn to survive on their own. They grow into adults that can reproduce and have their own young.

Quick Check

Sequence Which does a cheetah do first: reproduce or learn to hunt?

Critical Thinking How might growing bigger help an animal survive?

Life Cycle of a Cheetah

Cub Most female cheetahs have three to five cubs at once. They protect and feed the cubs.

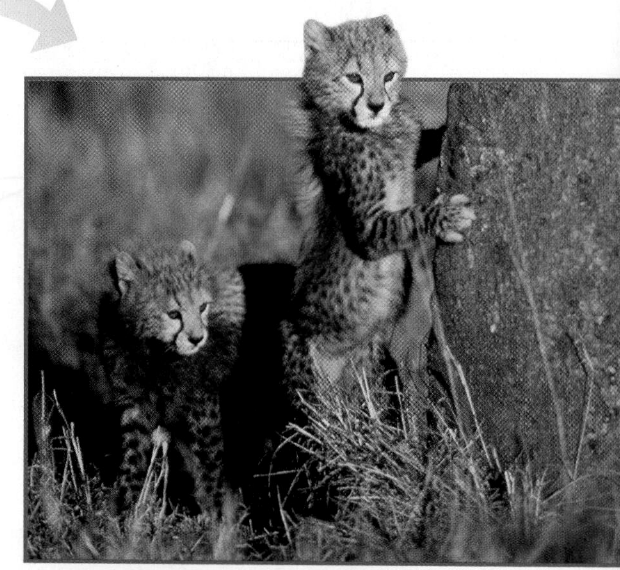

Young Cheetahs learn and practice the skills they will need to hunt.

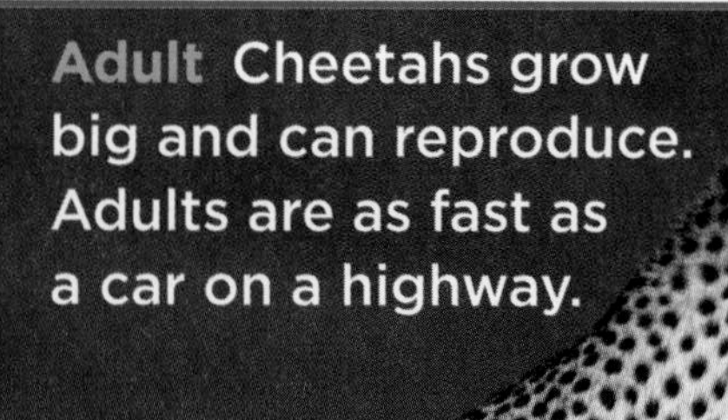

Adult Cheetahs grow big and can reproduce. Adults are as fast as a car on a highway.

Read a Diagram

How does a cheetah change as it grows?

Clue: Arrows help show a sequence.

Lesson Review

Visual Summary

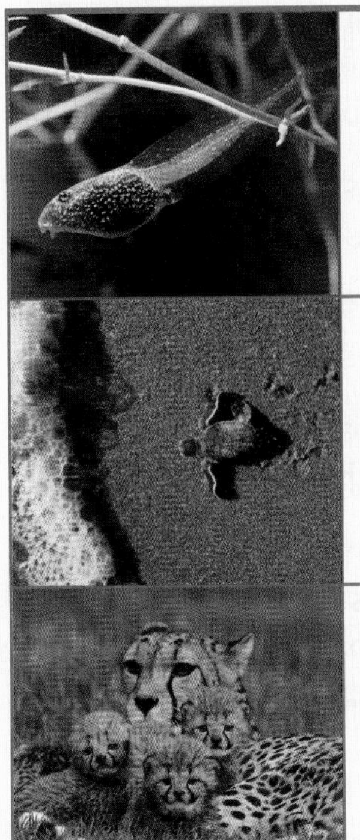

Every type of animal has its own life cycle. **Amphibians** and most **insects** go through metamorphosis.

Most **reptiles, birds, and fish** hatch from eggs. Reptiles and fish do not usually care for their young.

Most **mammals** are born live. They depend on their parents until they can get their own food.

Make a FOLDABLES® Study Guide

Make a layered-look book. Use it to summarize what you learned about animal life cycles.

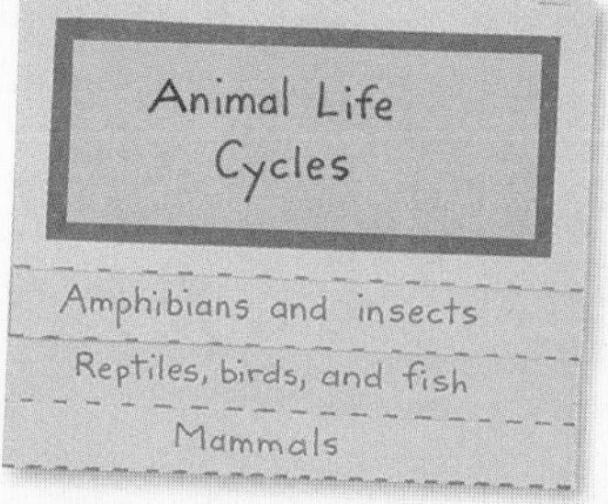

Think, Talk, and Write

1. **Vocabulary** What is metamorphosis?
2. **Sequence** Name three stages in a sea turtle's life cycle. Put them in order.

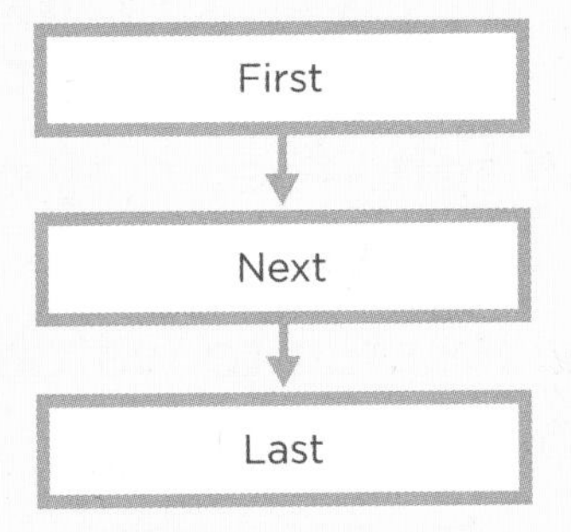

3. **Critical Thinking** Do you go through metamorphosis? How do you know?
4. **Test Prep** **An iguana's life cycle would be most like a**
   - **A** turtle's.
   - **B** cheetah's.
   - **C** fly's.
   - **D** bear's.
5. **Essential Question** How do animals grow and reproduce?

Writing Link

Write a Story
Choose an animal you know about. Pretend you are that animal. Describe how you change and grow as you get older.

Math Link

Solve a Problem
Female cheetahs usually give birth to a minimum of three cubs and a maximum of five cubs. What is the minimum and maximum number of cubs that five female cheetahs could have?

LOG ON e-Review Summaries and quizzes online at www.macmillanmh.com

The Little Lambs

I live on a farm, so I watch animals grow up. We have chickens, cows, and sheep. I like sheep the best. When they are born, they are small. Their wool is curly and soft. They stay close to their parents. When they get bigger, they run and play. It is fun to watch them. Their wool grows longer. Soon, we will cut their wool to make yarn. Next year they will be adult sheep.

Personal Narrative

A good personal narrative

- tells a story from the writer's own experience;
- expresses the writer's feelings;
- tells the events in an order that makes sense;
- uses time-order words, such as *first*, *then*, or *after that.*

These little lambs will grow to look like their parents. ▼

Write About It

Personal Narrative Have you ever watched a plant or animal grow and change? Write about your experience. Describe the changes. Write what you observed and how it made you feel.

-Journal Write about it online at www.macmillanmh.com

Math in Science

Graphing Life Spans

How long do animals live? A fruit fly is likely to live for only about one month. A Galapagos tortoise can live for 150 years! Each type of animal has its own life span. A *life span* is the amount of time an organism usually lives.

You can compare the life spans of different animals. Examine the life-span data in the table below. Use the data to make a bar graph comparing the animals' life spans.

Animal	Average Life Span*
black-tailed deer	10 years
American robin	13 years
rat snake	23 years
fence lizard	4 years
American toad	15 years

* under ideal conditions

American robin

Make a Bar Graph

- ▶ **Title your graph. Label the left side and the bottom.**
- ▶ **Next, list the animals along the bottom of the graph.**
- ▶ **Write numbers up the left side. Start with 0 and count to the longest life span.**
- ▶ **Draw a bar for each animal up to the number that shows its life span.**

Solve It

Make a bar graph using the data from the chart. Then use your bar graph to compare the life spans.

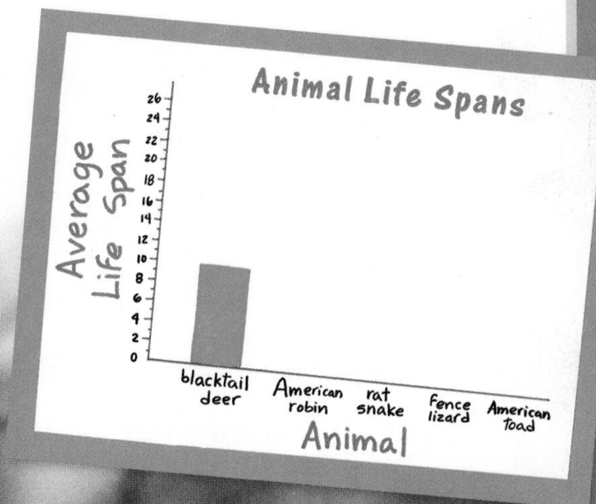

Lesson 3

From Parents to Young

Look and Wonder

How is this young horse like its parents? How is it different? Which traits are passed on from parents to their young?

Explore

Inquiry Activity

Which traits are passed from parents to their young?

Make a Prediction

Which of your traits are inherited, or passed on, from your parents? Is your hair color or hair length inherited? Make a prediction.

Test Your Prediction

1. **Communicate** Make a data table like the one shown. Use your table to describe your traits.
2. **Classify** Some traits have changed since you were little. Others have not changed. Circle the traits that have not changed.
3. **Communicate** Compare tables with a classmate. Which of your classmate's traits have stayed the same over time?

Step 1

Trait Name	Trait Description
Hair Color	
Hair Length	long/short (circle one)
Dimples	yes/no (circle one)
Earlobes	attached/unattached (circle one)
Favorite Food	

Draw Conclusions

4. Which traits did most students classify as traits that stay the same?
5. **Infer** How do you think you got the traits that do not change?
6. **Infer** Some of your traits are inherited from your parents. Underline the traits that you think you inherited. Explain why you chose these traits.

Explore More

Make a trait table that has a column for each member of your family. Which traits do you share with your family members?

Step 3

Read and Learn

Essential Question
How do organisms get their features?

Vocabulary
trait, p. 92
heredity, p. 92
inherited trait, p. 92
learned trait, p. 94

Reading Skill
Main Idea and Details

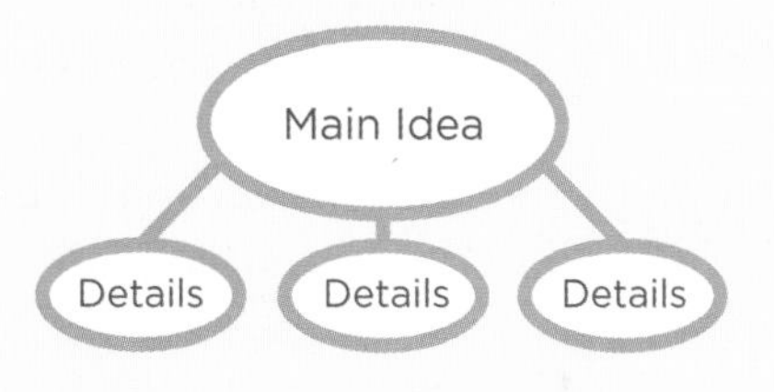

Technology

e-Glossary and e-Review online at www.macmillanmh.com

The offspring of a red tulip and a yellow tulip can be red, yellow, or a mixture of both.

What are inherited traits?

Have you ever wondered why people look the way they do? Why do some people have brown eyes and other people have green eyes, for example? Every organism has traits that make it unique. A **trait** is a feature of a living thing. Eye and hair color are traits. The shape of a plant's flowers, stems, and leaves are traits. Traits help you recognize and describe an organism.

Where do an organism's traits come from? Part of the answer is heredity. **Heredity** is the passing on of traits from parents to young. Traits that come from parents are called **inherited traits.** A flower's shape and color are inherited traits. Your eye color and hair color are inherited traits. The number of arms and legs an animal has is also inherited. Inherited traits make organisms look like their parents.

Most organisms inherit traits from both parents. For this reason, organisms almost never look exactly like either parent. A baby girl might have her mother's hair color but her father's eye color, for example.

Organisms may look more like one parent than the other. This happens when one parent gives its offspring traits that are more visible. *Offspring* are an organism's young. The offspring of a gray dog and a yellow dog may be yellow. The offspring of a red flower and a yellow flower could be red.

Quick Lab

Inherited Traits

1. **Observe** Look at the photo of the family. How are these people all alike? How do they differ?
2. **Communicate** Compare the children to their mother. Discuss which of their traits are similar to their mother's traits.
3. **Communicate** Compare the children to their father. Discuss which of their traits are similar to their father's traits.
4. **Infer** Why do organisms look similar to, but not exactly like, their parents?

Quick Check

Main Idea and Details What are inherited traits?

Critical Thinking Is the length of a dog's tail an inherited trait? Why or why not?

Do these people look like they are related? What traits do they share? ▼

Which traits are not inherited?

Some of your traits come from your parents. Others are learned. People and animals can gain new skills over time. These new skills are called **learned traits**. Riding a bicycle and speaking a language are learned traits.

Some of your traits are affected by your environment. For example, your hair may get lighter from being in sunlight. A plant's green leaves may become yellow if it gets too much water. A rabbit may grow fat when it finds a lot of food. It may grow skinny when food is hard to find.

Learned traits are not passed on from parents to offspring. Your parents may know how to ride a bicycle, but you still had to learn to ride one yourself. Traits affected by the environment are also not passed on. If an animal gets a scar, its offspring will not be born with scars. If a tree loses branches in a storm, its offspring will not grow with missing branches.

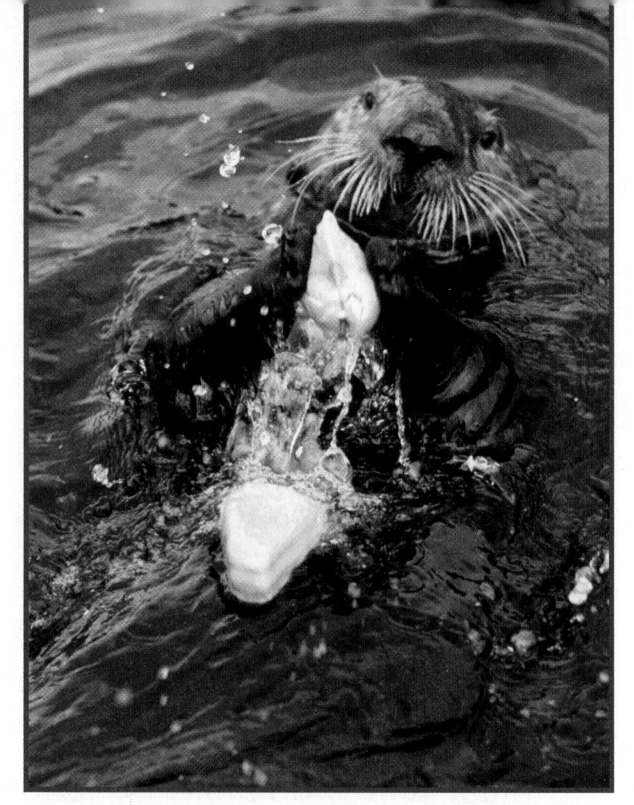

▲ A sea otter smashes a clam shell against a rock. This learned trait helps the otter get food.

Changing Traits

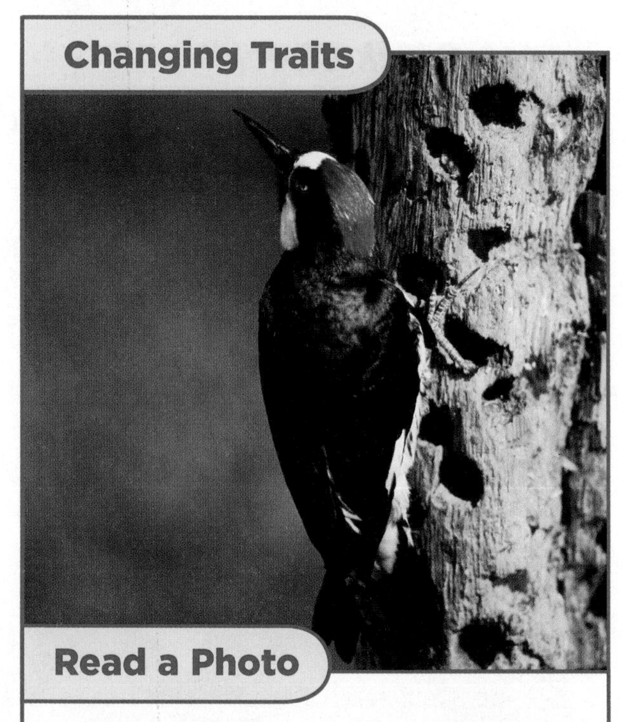

Read a Photo

How has this tree been affected by things in its environment?

Clue: Compare how this tree looks to other trees around you.

Quick Check

Main Idea and Details Name one trait you have that is learned.

Critical Thinking How do learned traits help animals survive?

Lesson Review

Visual Summary

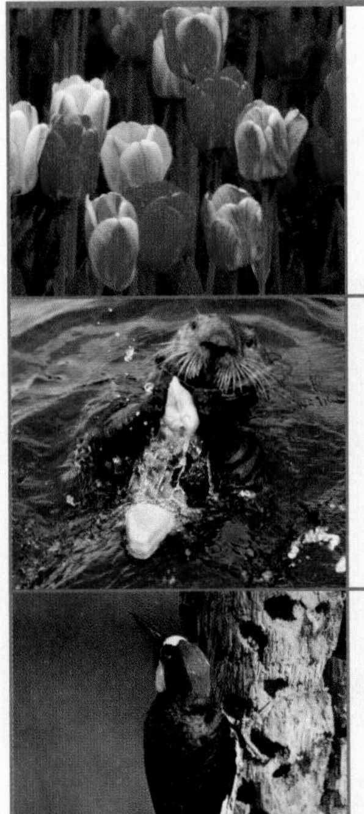

Inherited traits are passed on from parents to offspring.

Learned traits are new skills an organism gains during its life.

Some **traits** are **affected by** an organism's environment.

Make a FOLDABLES® Study Guide

Make a three-tab book. Use it to summarize what you learned about traits in this lesson.

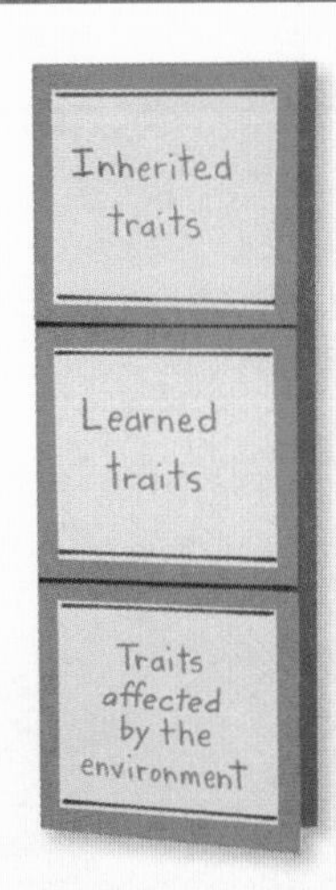

Think, Talk, and Write

1. **Vocabulary** What is heredity?

2. **Main Idea and Details** What is an inherited trait? Give examples.

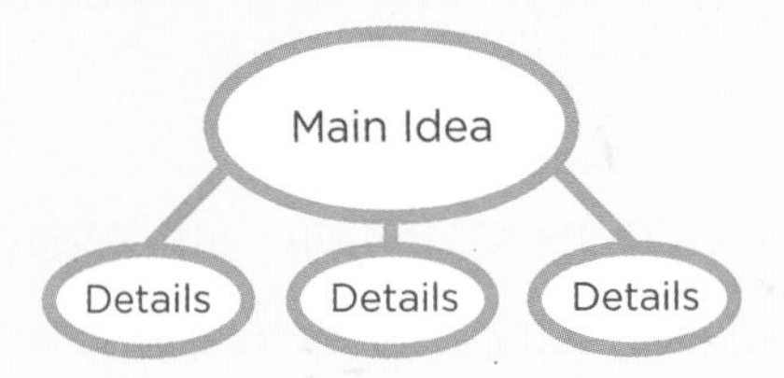

3. **Critical Thinking** Why do you look the way you do?

4. **Test Prep** **A plant loses branches during a storm. This is an example of**
   - **A** an inherited trait.
   - **B** a trait affected by the environment.
   - **C** a learned trait.
   - **D** heredity.

5. **Essential Question** How do organisms get their features?

Math Link

Make a Chart
One surprising inherited trait is the ability to roll your tongue. Count how many people in your class can roll his or her tongue. Make a pie chart to show the results.

Social Studies Link

Make a Family Tree
Collect photos of your family. Arrange them in a family tree. Look for similarities and differences among your relatives. Can you see some inherited traits that you share?

LOG ON e-Review Summaries and quizzes online at www.macmillanmh.com

Meet DARREL FROST

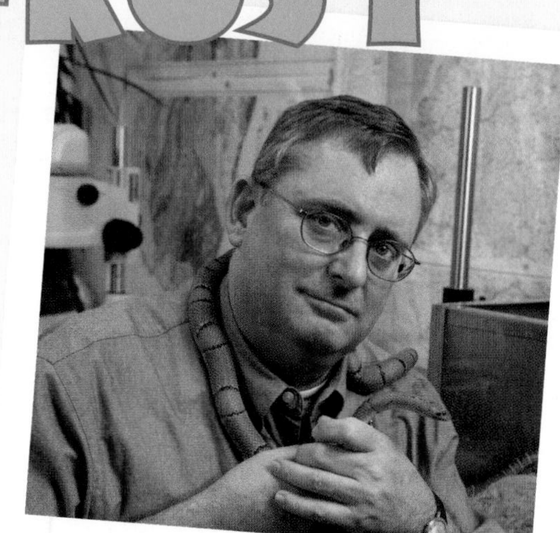
Darrel is a herpetologist, a scientist who studies amphibians and reptiles.

A lizard is soaking up warmth from the Sun when a hawk soars overhead. The hawk spots the lizard, then swoops down to grab a meal. The lizard has no time to scurry away. How can it escape the hawk's claws?

If the hawk catches the lizard's long tail, the tail will break off. The bird will be left holding a wriggling tail while the lizard runs away. In time, the lizard will grow a new tail. Growing new body parts is a trait called regeneration.

This lizard lost the end of its tail. ▼

Connect to AMERICAN MUSEUM OF NATURAL HISTORY at www.macmillanmh.com

Regeneration is one of the many amazing traits that Darrel Frost studies. Darrel is a scientist at the American Museum of Natural History. He travels all over the world to collect different kinds of lizards. Then he observes their traits. Finally, he uses his observations to find out how different kinds of lizards are related.

Sequence

A sequence

- gives events in order;
- tells what happens first, next, and last;
- uses time-order words, such as *first, next, then,* and *finally.*

This lizard's tail broke off near the tip. The new tip looks darker.

Write About It

Sequence Read the article with a partner. Fill in a sequence-of-events chart to show how Darrel learns about lizards. Then use your chart to write a summary about Darrel and his work.

Write about it online at www.macmillanmh.com

CHAPTER 2 Review

Visual Summary

Lesson 1 A life cycle describes how an organism grows and reproduces. Most plants grow from seeds.

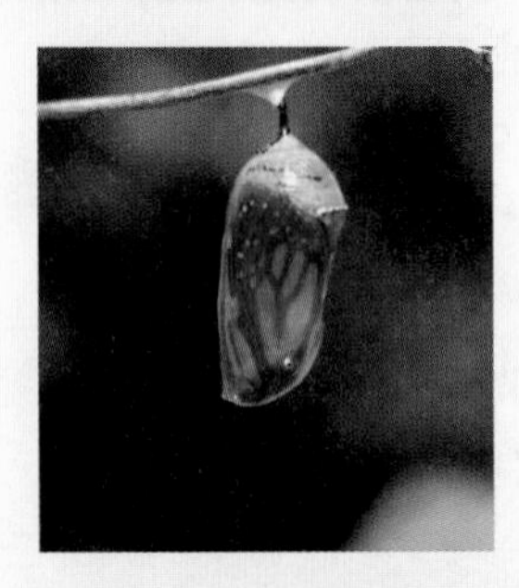

Lesson 2 Animals have different life cycles. Some animals are born looking like their parents. Others change greatly as they grow.

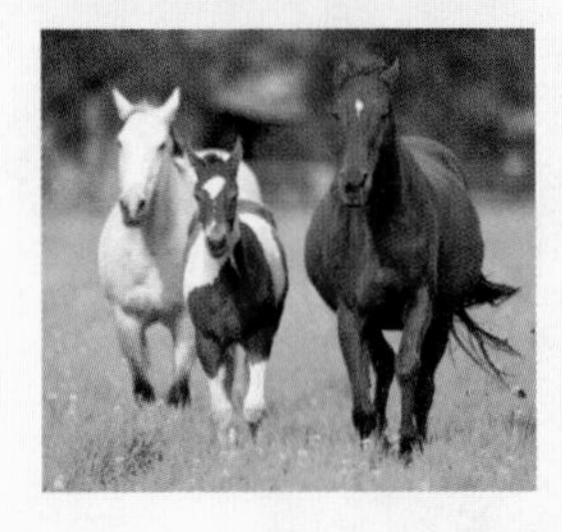

Lesson 3 Organisms have traits that they inherit from their parents. Other traits are learned or affected by the environment.

Make a FOLDABLES® Study Guide

Glue your lesson study guides to a piece of paper as shown. Use your study guide to review what you have learned in this chapter.

Vocabulary

DOK I

Fill each blank with the best term from the list.

cone, p. 75	**life cycle,** p. 74
egg, p. 83	**metamorphosis,** p. 83
heredity, p. 92	**pollination,** p. 73
inherited traits, p. 92	**seed,** p. 70
larva, p. 83	**trait,** p. 92

1. An amphibian begins its life as an ______.
2. A conifer's seeds are made inside a ______.
3. An organism goes through stages that make up its ______.
4. Some organisms, such as caterpillars, go through a ______ in which their bodies change shape.
5. The passing of traits from parents to young is known as ______.
6. A structure that can grow into a new plant is called a ______.
7. Any feature of a living thing is called a ______.
8. Animals and wind help plants reproduce through ______.
9. Traits that an organism gets from its parents are called ______.
10. When a caterpillar hatches from an egg, it is called a ______.

LOG ON e-Glossary Words and definitions online at www.macmillanmh.com

Skills and Concepts

DOK 2–3

Answer each of the following.

11. **Sequence** List the stages of a flowering plant's life cycle in the correct order.

12. **Personal Narrative** Describe how you use learned traits during the course of a typical school day.

13. **Predict** A ripe apple falls to the ground. How can this help an apple tree reproduce?

14. **Critical Thinking** How could the environment affect a bird's life cycle?

15. **Compare and Contrast** A mother sea turtle can have one hundred babies at once. A mother cheetah gives birth to fewer young. How does each mother help its offspring to survive?

16. What is happening in this picture? Which part of a life cycle does this picture show?

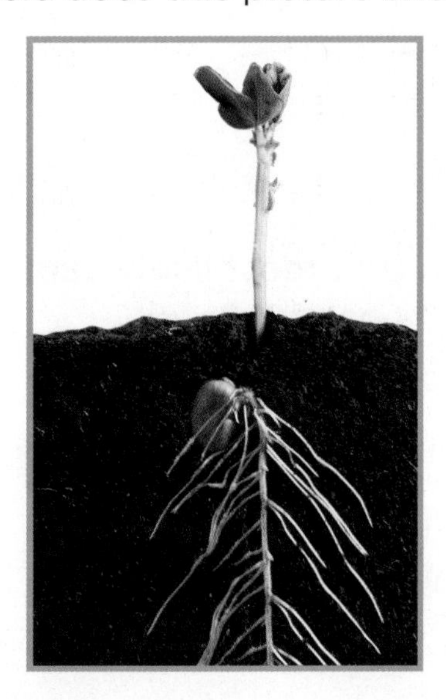

17. **True or False** *A dog "shaking hands" is an inherited trait.* Is this statement true or false? Explain.

18. **True or False** *A young robin goes through metamorphosis.* Is this statement true or false? Explain.

19. Complete the following sentence:

    In flowering plants, the ______ holds the seeds.

    A cone **C** fruit

    B bulb **D** tuber

The Big Idea

20. How do living things change?

Performance Assessment

DOK 2

Make a Life-Cycle Poster

chameleon

robin

- Choose two very different animals, such as a chameleon and a robin. Use this book and do library research to learn about their life cycles.
- Create a poster that shows each animal's life cycle on half the poster. Explain each stage of the life cycle using words and pictures. Compare and contrast the animals' life cycles.

1 **The diagram below shows the inside of a seed.**

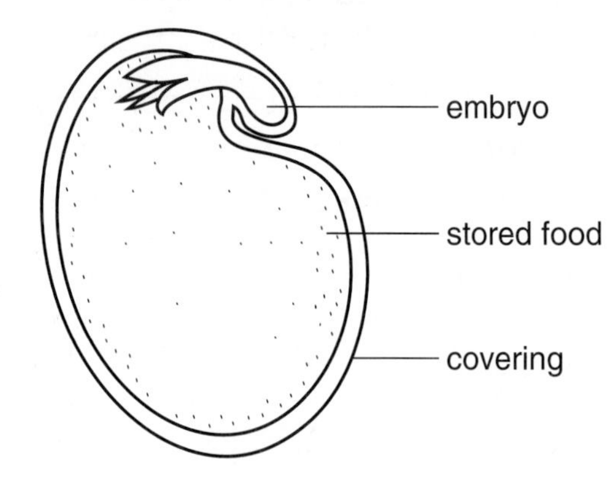

What is the embryo?

A a young plant

B an underground stem

C a structure that makes the seed

D a structure that makes food
DOK 1

2 **What would <u>most likely</u> happen to flowering plants if there were fewer bees?**

A Fewer seeds would be carried to new places.

B Fewer fruits would be made.

C More nectar would be eaten by other insects.

D More fruits would be made.
DOK 2

3 **Look at the diagram of a frog's life cycle.**

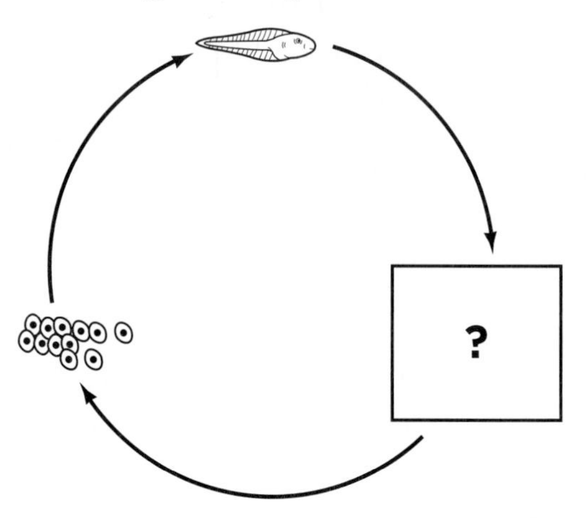

Which stage is missing?

A egg

B tadpole

C young frog

D adult frog
DOK 1

4 **Look at the picture of the dog below.**

Which trait <u>most likely</u> came from its parents?

A spotted fur

B rolling over

C chipped tooth

D wagging tail
DOK 1

5 Look at the chart below.

Inherited Traits		
dimples	hair color	

Which choice best completes the chart?

A bruised elbow

B eye color

C good manners

D hair style

DOK 1

6 A mammal never looks exactly like either of its parents. What is the best explanation for this?

A Inherited traits are affected by the environment.

B Inherited traits come from both parents.

C Learned traits are affected by the environment.

D Learned traits are affected by inherited characteristics.

DOK 2

7 Which uses flowers to reproduce?

A cherry tree

B pine tree

C conifer

D fern

DOK 1

8 Look at the diagrams below. Diagram 1 shows the life cycle of a butterfly with stages A and B missing. Diagram 2 shows the missing stages.

Diagram 1

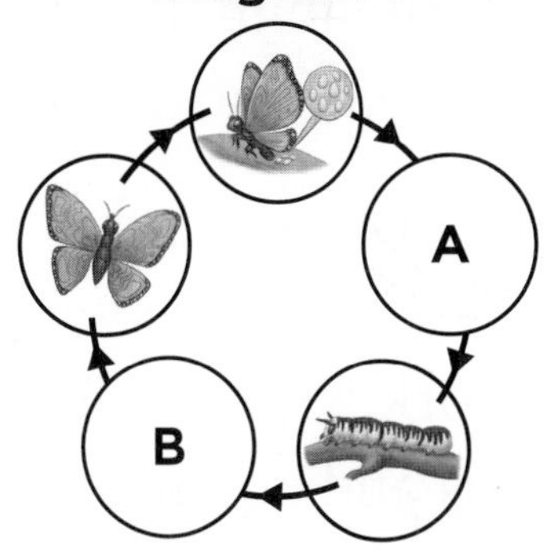

Diagram 2

larva

pupa

Should the larva be placed in stage A or stage B? Why?

Should the pupa be placed in stage A or stage B? Why?

Compare the life cycle of a butterfly with the life cycle of a bird.

DOK 2

Check Your Understanding

Question	Review	Question	Review
1	pp. 70–71	5	pp. 92–94
2	pp. 72–74	6	pp. 86, 92–94
3	p. 82	7	pp. 72–76
4	pp. 92–94	8	pp. 83–85

Animal Trainer

Do you like to be around animals? Do you have fun taking care of pets? Some people train animals as a career. Could you be one of those people?

Animal trainers do many jobs. Some teach dogs to assist people who are blind. Some work with racehorses or animals in zoos and aquariums. Some even work with animals that perform in movies or television shows.

To become an animal trainer, you need to be patient and calm. You need to be a good communicator. You need a strong understanding of animal psychology. Animal psychology is how animals "think" and behave. You also should be in good physical condition.

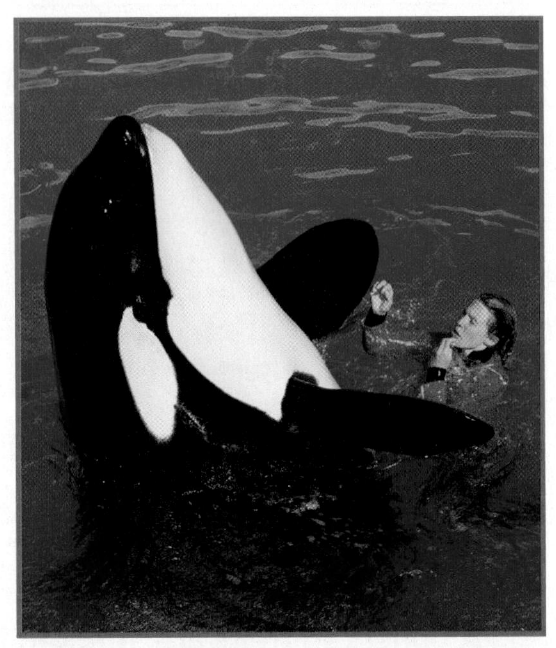

▲ This animal trainer works with killer whales at an aquarium.

Here are some other life science careers:

- veterinarian
- biologist
- farmer
- animal groomer

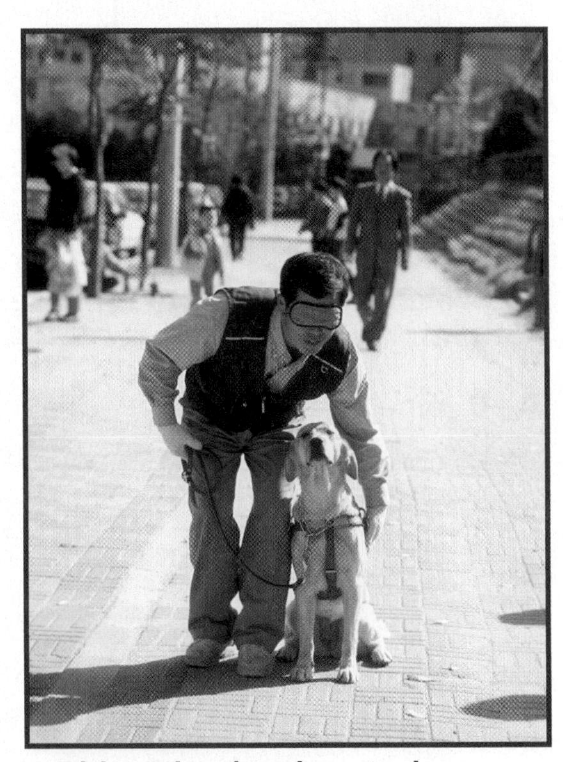

▲ This animal trainer trains seeing eye dogs.

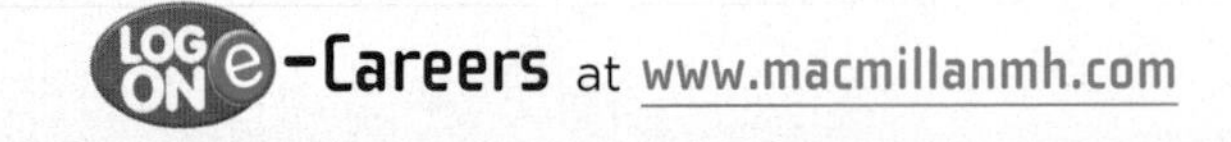

LOG ON e-Careers at www.macmillanmh.com

UNIT B

Ecosystems

Owls can turn their heads around to see behind them.

northern saw-whet owl

Literature

National Wildlife Federation

Ranger Rick

Magazine Article

pileated woodpecker

from *Ranger Rick*

Once Upon a Woodpecker

Tap! Rap-tap-tap! Do you hear that drumming sound? Oh, look, there's a woodpecker tapping on that dead tree. Let's move closer for a better look. If we're quiet we can watch the woodpecker at work.

Woodpeckers have some amazing adaptations that help them survive. They hammer hard into tree trunks with their sharp beaks. Strong muscles in their necks put power behind each blow. These muscles also act as shock absorbers so woodpeckers don't get headaches!

Woodpeckers have extra long tongues for spearing insects they find inside tree trunks. A barb on the tip of their tongues works like a fishhook to reel in meals. Look out bugs! You can't hide from the well-adapted woodpecker!

Write About It

Response to Literature This article tells about special features of woodpeckers that help them survive. What are some special features you have that help you survive? Write about them.

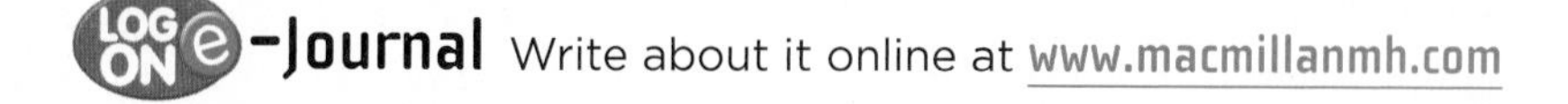

CHAPTER 3

Living Things in Ecosystems

How do living things survive in their environments?

Essential Questions

Lesson 1
How do living things interact?

Lesson 2
How do ecosystems compare?

Lesson 3
How do an organism's traits help it survive?

A vole eats a yew berry.

Big Idea Vocabulary

ecosystem the living and nonliving things that share an environment and interact (p. 108)

producer an organism that makes its own food (p. 110)

consumer an organism that eats other organisms (p. 111)

climate the pattern of weather in a place over a long time (p. 120)

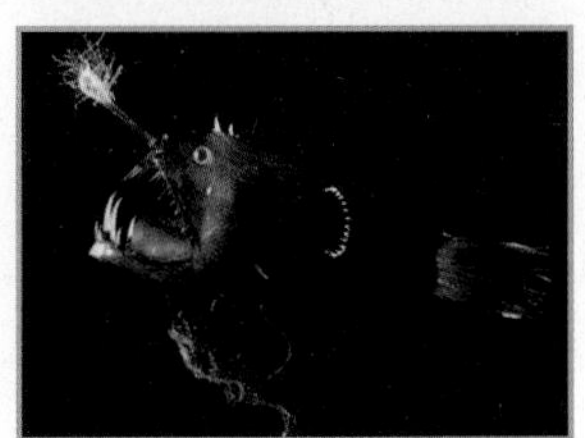

adaptation a structure or behavior that helps a living thing survive in its environment (p. 134)

mimicry when one living thing imitates another in color or shape (p. 139)

Visit www.macmillanmh.com for online resources.

Lesson 1

Food Chains and Food Webs

Look and Wonder

A bald eagle can fly at 200 miles per hour when diving for a fish. Bald eagles depend on fish for food. They also eat turtles, ducks, and other small animals. What do other animals depend on for food?

Explore

Inquiry Activity

What kinds of food do owls need?

Purpose

Find out what an owl eats by studying an owl pellet.

Procedure

1. Work with a partner. Put on plastic gloves. Place your owl pellet onto a paper plate.
2. **Predict** What do you expect to see inside the owl pellet? Write your prediction.
3. Using the tweezers, separate the objects in the owl pellet.
4. **Observe** What is in the owl pellet? Use the hand lens. Record your observations. ⚠ **Be Careful.** Wash your hands when you are done.

Draw Conclusions

5. **Interpret Data** What do the materials inside the owl pellet tell you about what an owl eats?
6. **Infer** What organisms might an owl eat? What might those organisms eat?

Explore More

Interpret Data Keep track of the things you eat in one day. Do most of your foods come from plants or animals?

Materials

plastic gloves

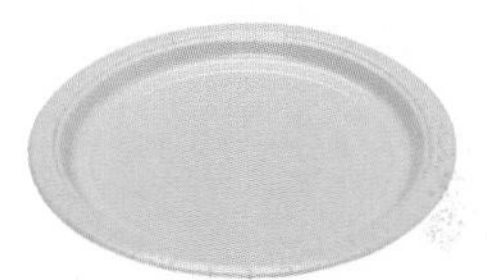

paper plate

owl pellet

tweezers

hand lens

Read and Learn

Essential Question
How do living things interact?

Vocabulary
ecosystem, p.108
habitat, p.109
food chain, p.110
producer, p.110
consumer, p.111
decomposer, p.111
food web, p.112

Reading Skill 
Infer

Clues	What I Know	What I Infer

Technology LOG ON
e-Glossary and e-Review online at www.macmillanmh.com

What is an ecosystem?

Look at the diagram below. Can you see a frog ready to snap up an insect? How about a turtle resting in the Sun? Living things depend on each other. They also depend on nonliving things like sunlight. Living and nonliving things that interact in an environment make up an **ecosystem** (EE•koh•sis•tum). An ecosystem may be a pond, a swamp, or a field. It may be as small as a puddle or as big as an ocean.

A Pond Ecosystem

Crane flies eat plants and algae. They lay eggs in water. ▼

Big or small, ecosystems are made up of living and nonliving things. Frogs, birds, and plants are some living things in a pond. Sunlight, water, and soil are some nonliving things.

Different organisms live in different parts of an ecosystem. Fish live in the water. Water is their **habitat**, or home. A cattail's habitat is along the edge of a pond. Living things get food, water, and shelter from their habitats.

Quick Check

Infer Which pond animals could also survive in a land ecosystem?

Critical Thinking How might an ecosystem change if it suddenly became colder?

Cattails grow well in wet soil. Animals use them as food and shelter. ▶

◀ These turtles climb out of the water to warm up in the sunlight.

◀ Pond snails slide along the bottom looking for plants and algae to eat.

What is a food chain?

All organisms need energy from food to live and grow. Most are a source of energy as well. They pass on energy to organisms that eat them. A **food chain** shows how energy passes from one organism to another in an ecosystem. Look at the diagram below. The arrows show the flow of energy.

The first organism in a food chain is a producer (proh•DEW•sur). A **producer** is an organism that makes its own food. Green plants and algae are two examples. Most producers use energy from the Sun to make their own food. This means that the energy in most food chains starts with the Sun.

The next organisms in a food chain are consumers (kun•SEW•murz). A **consumer** is an organism that eats other organisms. All animals are consumers. A food chain may have many consumers.

Next in the food chain are decomposers (dee•kum•POH•zurz). A **decomposer** is an organism that breaks down dead plant and animal material. Decomposers put nutrients back into the soil. Some worms and bacteria are decomposers.

Quick Check

Infer What might happen to grasshoppers and eagles if turtles were removed from the pond food chain?

Critical Thinking How are these food chains alike?

What is a food web?

One morning a turtle eats a grasshopper. The next day, that same turtle eats a crayfish. Most animals eat several kinds of food. They are part of several food chains. Food chains can connect to form a **food web**.

The diagram below shows a pond food web. Look at the arrows from the largemouth bass to the heron and bald eagle. They show that herons and eagles eat bass. The bass is part of more than one food chain. The heron and the eagle are predators. *Predators* hunt other organisms for food. The organisms they hunt are *prey*.

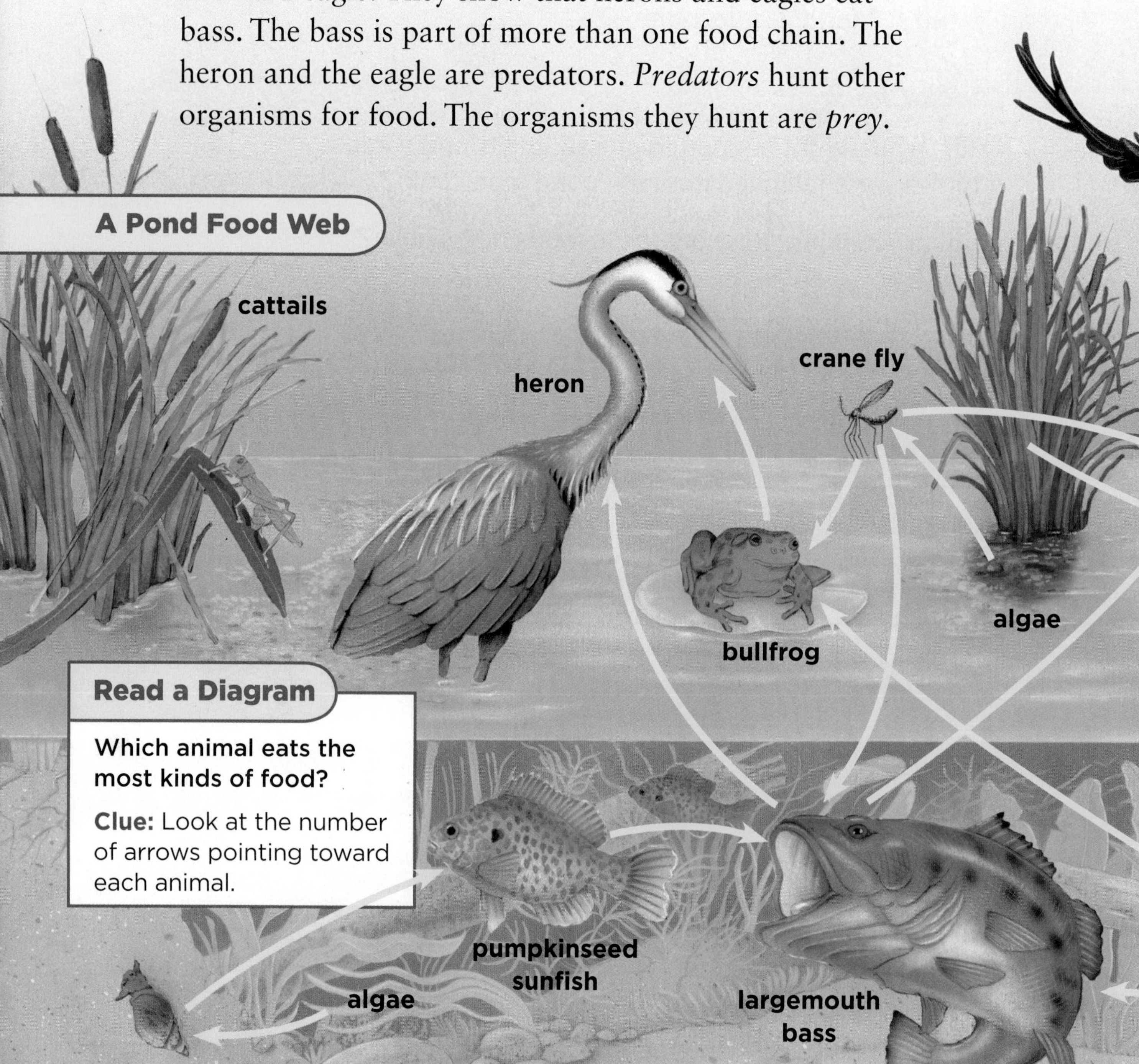

Read a Diagram

Which animal eats the most kinds of food?

Clue: Look at the number of arrows pointing toward each animal.

You can learn about living things by studying food webs. Below, you can see that the snail eats plants. Organisms that eat plants are *herbivores*. Some animals, such as herons, eat other animals. These organisms are *carnivores*. Animals that eat both plants and animals are *omnivores*. Can you find an omnivore below?

Food webs also show how organisms compete for food. Many animals eat crayfish. If snakes eat all the crayfish, the turtles, fish, or bullfrogs might go hungry.

Quick Check

Infer How could a heron survive if there were no more frogs?

Critical Thinking Are you an herbivore, carnivore, or omnivore?

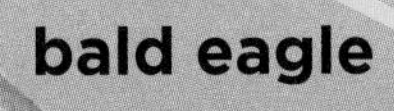

Quick Lab

Observe Decomposers

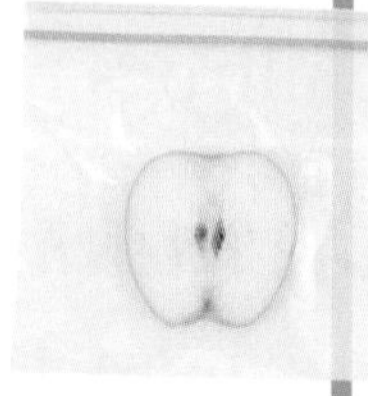

1. Put some apple pieces into a plastic bag. Seal the bag.
   △ **Be Careful.** Do not open the sealed bag.
2. **Observe** Leave the bag in a warm, dark place for a week. Observe the pieces every day. Record the changes you see.
3. **Communicate** What happened to the pieces of apple? How did they change over time?
4. **Infer** What does this activity tell you about decomposers?

Why are decomposers important?

In a pond, dead plant and animal material drifts to the bottom. What keeps the pond from filling with dead organisms? Decomposers!

Decomposers are an important part of ecosystems. Decomposers feed on dead material. As they eat, they release nutrients into the water or soil. These nutrients help plants and other organisms grow. Worms, mold, mushrooms, and some insects and snails are decomposers. Many bacteria are decomposers too.

These leaves will make a good meal for decomposers. ▼

▲ These tiny decomposers were magnified 2,700 times.

Quick Check

Infer How do decomposers help a pond ecosystem?

Critical Thinking What would happen if a forest had no decomposers?

Lesson Review

Visual Summary

The living and nonliving things in an **ecosystem** depend on each other to survive.

Food chains and **food webs** show how energy flows through an ecosystem.

Decomposers play an important role in an ecosystem. They eat dead material and release nutrients.

Make a FOLDABLES® Study Guide

Make a trifold book. Use it to summarize what you learned about food chains and food webs.

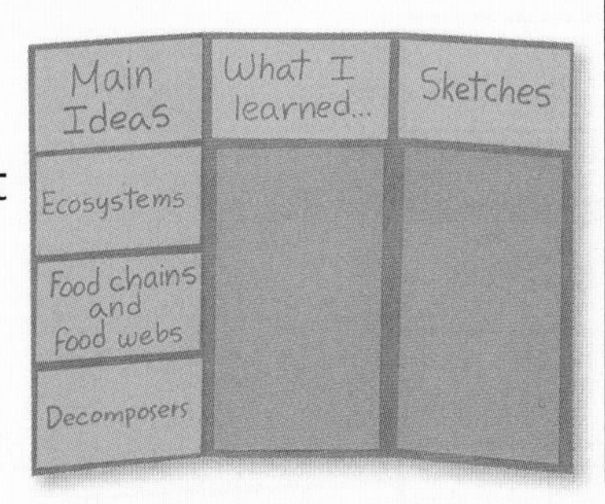

Think, Talk, and Write

1. **Vocabulary** What is a consumer?
2. **Infer** How does it help an animal to be part of more than one food chain?

Clues	What I Know	What I Infer

3. **Critical Thinking** How do both plants and animals depend on decomposers?
4. **Test Prep** **Most producers get their energy from**
   - **A** sunlight.
   - **B** consumers.
   - **C** predators.
   - **D** rocks.
5. **Essential Question** How do living things interact?

Writing Link

Writing That Compares
Choose two animals. Find out where they live and what they eat. Find out what eats them. Then compare the animals in an essay.

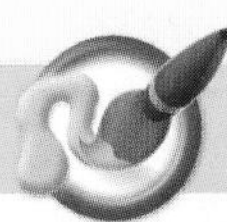

Art Link

Make a Poster
Research an ecosystem near your home. Make a poster to show how organisms in that ecosystem depend on one another.

LOG ON e-Review Summaries and quizzes online at www.macmillanmh.com

Focus on Skills

Inquiry Skill: Communicate

You know that organisms get energy from food. Scientists study ecosystems to learn how different organisms get energy. Then they **communicate**, or share, their observations. Communicating helps people learn about the world.

▶ Learn It

When you **communicate**, you share information with others. Some ways you share information in science are by talking, writing, drawing, or making graphs and charts.

▶ Try It

In this activity you will organize and **communicate** data about a grassland ecosystem. Look at the data table below. It shows how some organisms in a grassland get energy. It also tells how the organisms interact. A table is one way to **communicate** data. You will try some other ways.

Grassland Organisms

Organism	Where Organism Gets Energy
grass	Sun
snake	field mouse
hawk	snake
field mouse	grass

1. One way you can **communicate** the data is by making a food-chain diagram. The photographs on the next page show the start of a food chain. Copy this food chain. Complete it by adding the last three organisms in the correct order.

Skill Builder

2. Next, **communicate** by making a food pyramid. Copy the pyramid shown and fill in the blank spaces.
3. Now, **communicate** by writing a paragraph. In your paragraph, classify each organism as a producer or consumer. Tell where each grassland organism gets its energy.
4. Did all three ways of communicating help you understand the data? Which way did you think worked best? Why?

Sun

grass

se

snake

▶ Apply It

Think of a food chain from another ecosystem. **Communicate** information about this food chain to a partner. Draw a food chain to show where organisms in the ecosystem get energy. Now describe the food chain in words. Discuss what you learned.

Lesson 2

Types of Ecosystems

green sea turtles and longfin batfish

Look and Wonder

What would it feel like to be in this ocean environment? Could humans live in this salty ocean? Why is it a good place for this sea turtle and these fish?

Explore

Inquiry Activity

Can ocean animals live and grow in freshwater?

Make a Prediction

Can brine shrimp grow in freshwater and salt water? Make a prediction.

Test Your Prediction

1. Fill each jar with 480 milliliters of water. Put two tablespoons of sea salt in one jar. Label it *Salt Water*. Label the other jar *Freshwater*.

2. Add one teaspoon of brine shrimp eggs to each jar.
3. **Observe** Watch what develops in each jar over the next few days. Use a hand lens.

Draw Conclusions

4. **Interpret Data** In which jar did the brine shrimp eggs hatch? How could you tell?
5. **Infer** Can all ocean animals live and grow in freshwater? How do you know?

Explore More

Experiment Does temperature affect the hatching of brine shrimp eggs? Design an experiment to find out.

Materials

2 jars

measuring cup and water

sea salt

measuring spoon

brine shrimp eggs

hand lens

This northern ecosystem is called a tundra. Caribou dig for food in the tundra's snow.

Read and Learn

Essential Question

How do ecosystems compare?

Vocabulary

climate, p. 120

soil, p. 120

desert, p. 122

forest, p. 124

ocean, p. 126

wetland, p. 128

Reading Skill

Compare and Contrast

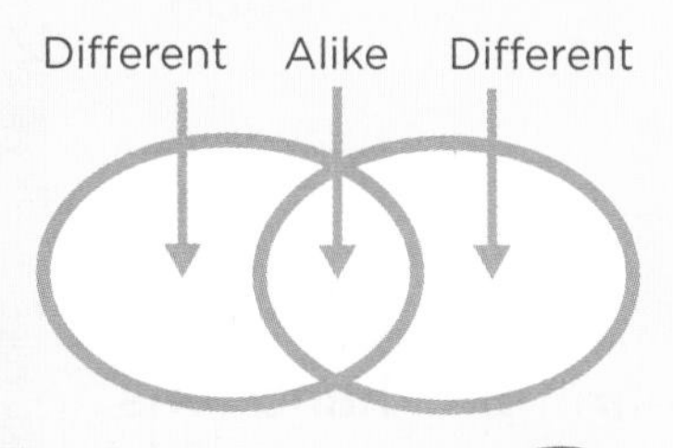

Technology 

e-Glossary and e-Review online at www.macmillanmh.com

How do ecosystems differ?

If you could take a trip around the world, you would learn that Earth has many kinds of ecosystems. Some are dry and sandy with almost no living things. Some are covered with trees or ice. Others are underwater.

Each of Earth's land ecosystems has a certain kind of climate. **Climate** is the pattern of weather in a place over a long time. Some land environments have a warm and wet climate. Others are cold and dry. Some environments are dry at some times and wet at other times.

Different ecosystems have different types of soil. **Soil** is made of bits of rock and humus (HYEW•mus). *Humus* is broken-down plant and animal material. It contains nutrients and soaks up rainwater. Soil rich in humus holds plenty of water and nutrients for plants to use.

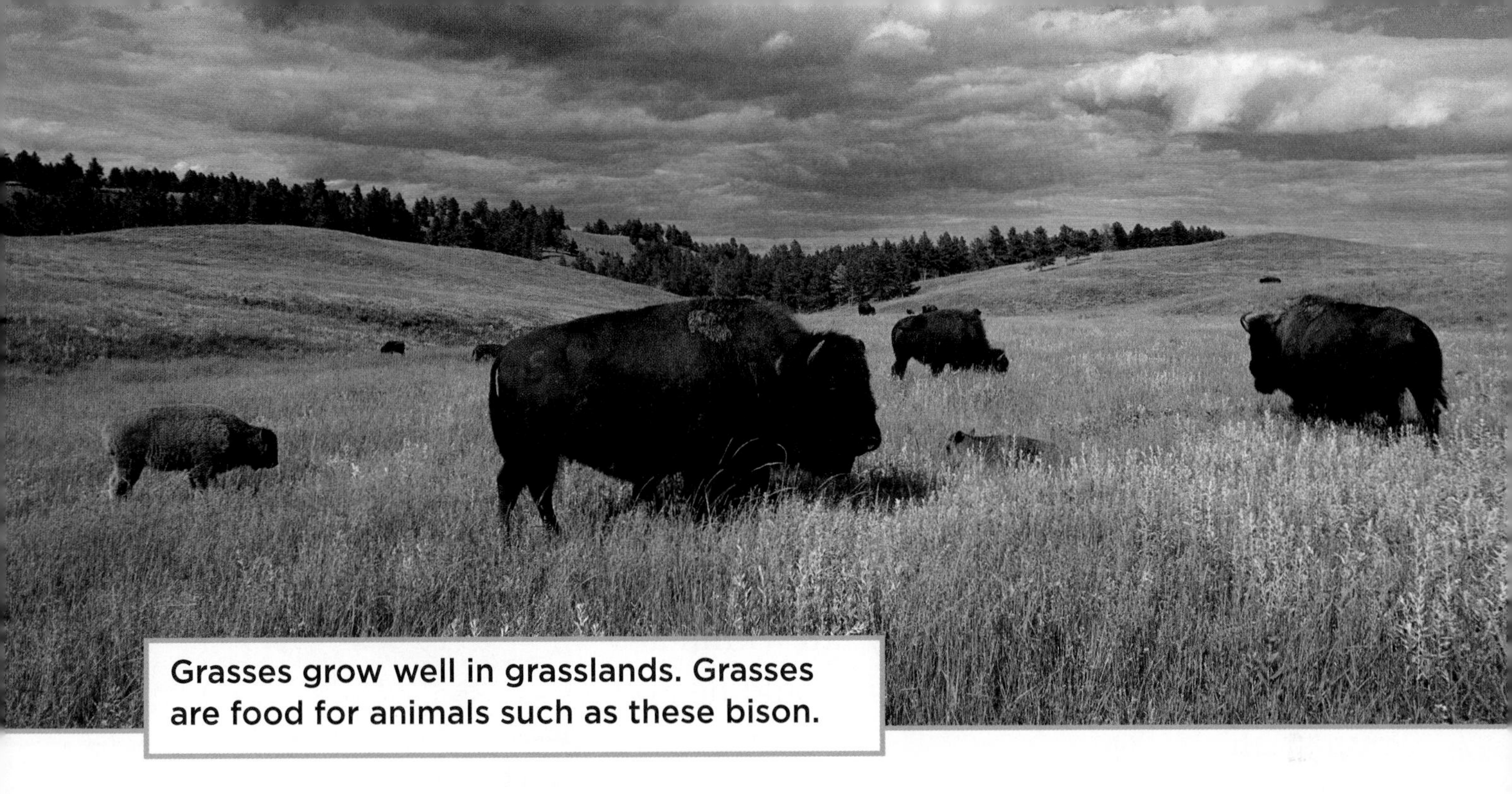

Grasses grow well in grasslands. Grasses are food for animals such as these bison.

Earth's water ecosystems differ in many ways. Some have salt water. Some have freshwater. Water ecosystems may be warm or cool, shallow or deep.

Ecosystems also differ in the types of plants and animals they have. Grasslands are covered in grass while forests are filled with trees. Oceans are filled with fish that can live only in salt water. Ponds are filled with fish that can live only in freshwater.

▲ Ponds are freshwater ecosystems filled with plants, algae, and animals.

Quick Check

Compare and Contrast What are some ways in which ecosystems differ?

Critical Thinking Describe the ecosystem in which you live.

What is a desert?

A wave of heat blasts your body. You take a deep breath and dry air stings your nose. Dust from the sandy ground covers your shoes. You are in the Sonoran Desert. It is one of the largest deserts in North America.

A **desert** is an ecosystem that has a dry climate. Fewer than 25 centimeters (10 inches) of rain fall in a desert each year. Several centimeters of rain may fall within a few days. Then for months there could be no rain at all.

Temperatures in most deserts vary widely between day and night. During the day, heat from the Sun warms the land and air. After the Sun sets, the temperature drops quickly.

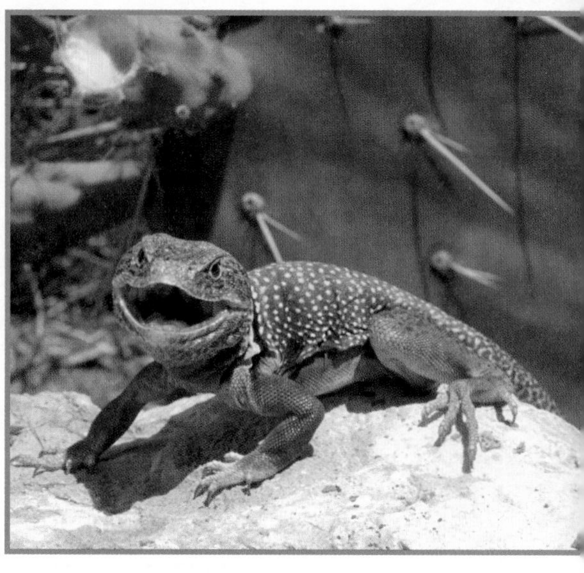

▲ This collared lizard hunts insects and other lizards in the Sonoran Desert.

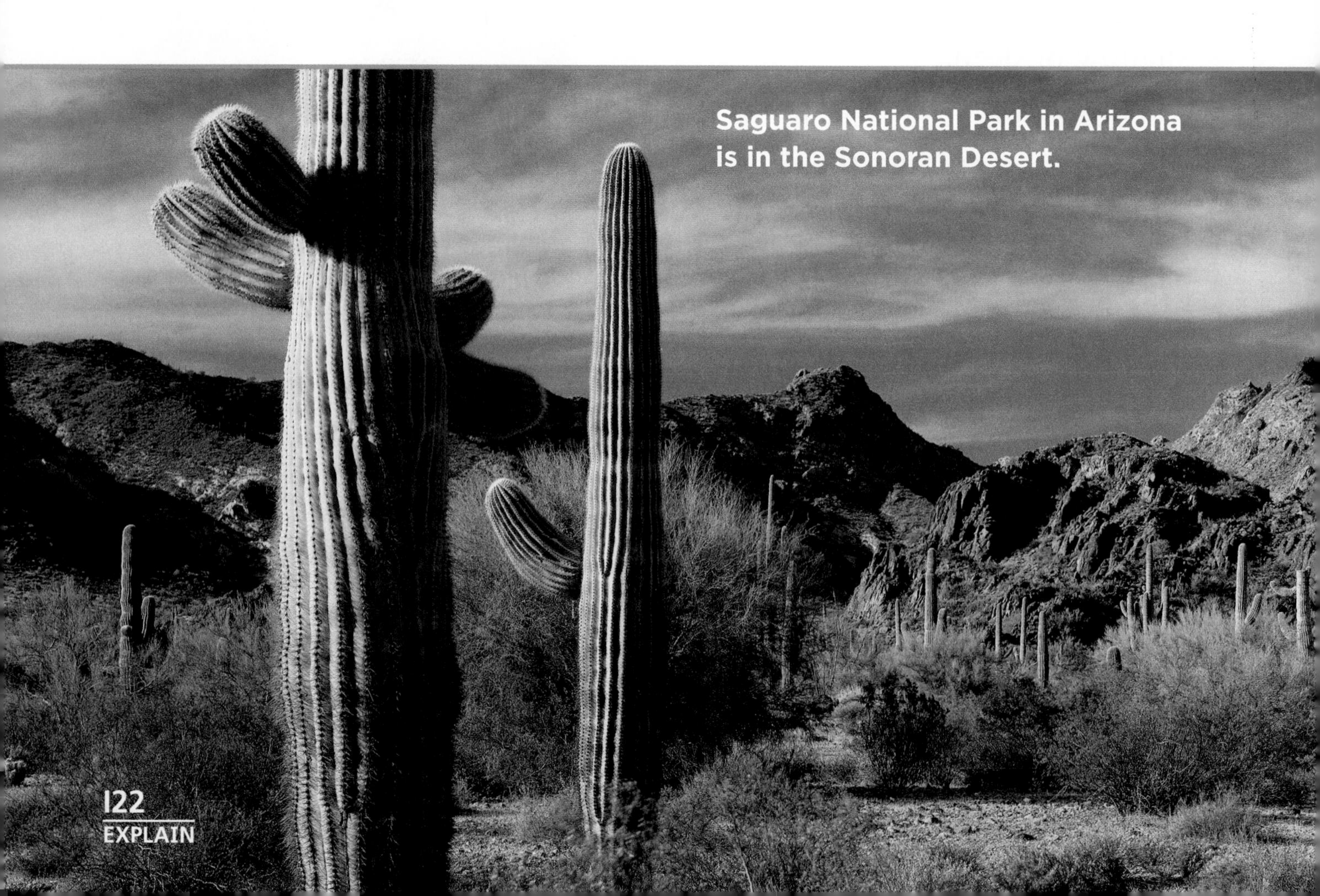

Saguaro National Park in Arizona is in the Sonoran Desert.

The soil in a desert is mostly sand. There is little humus to soak up rainwater. Rainwater trickles down through the sand. It goes deeper than most plants' roots can reach.

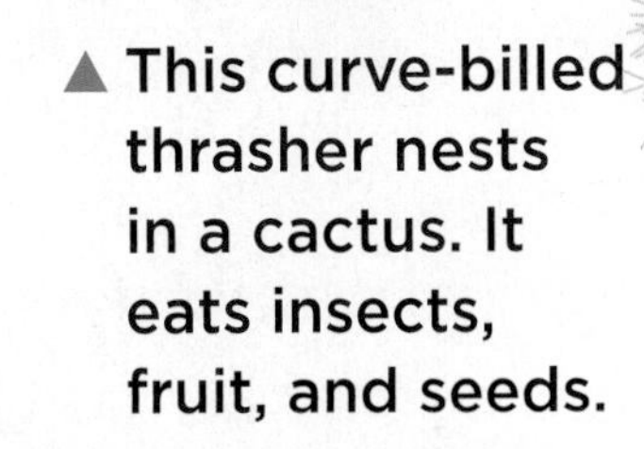

Few plants and animals can survive in deserts. Desert soil has little water and nutrients for many plants. Desert plants that do survive usually grow far apart. Most desert animals find shady spots to rest from the warm Sun. They hunt at night when temperatures are cooler. Jackrabbits, rattlesnakes, and cactus wrens are some common desert animals.

▲ This curve-billed thrasher nests in a cactus. It eats insects, fruit, and seeds.

Quick Check

Compare and Contrast How do a desert's daytime and nighttime temperatures compare?

Critical Thinking Why do deserts generally have few plants?

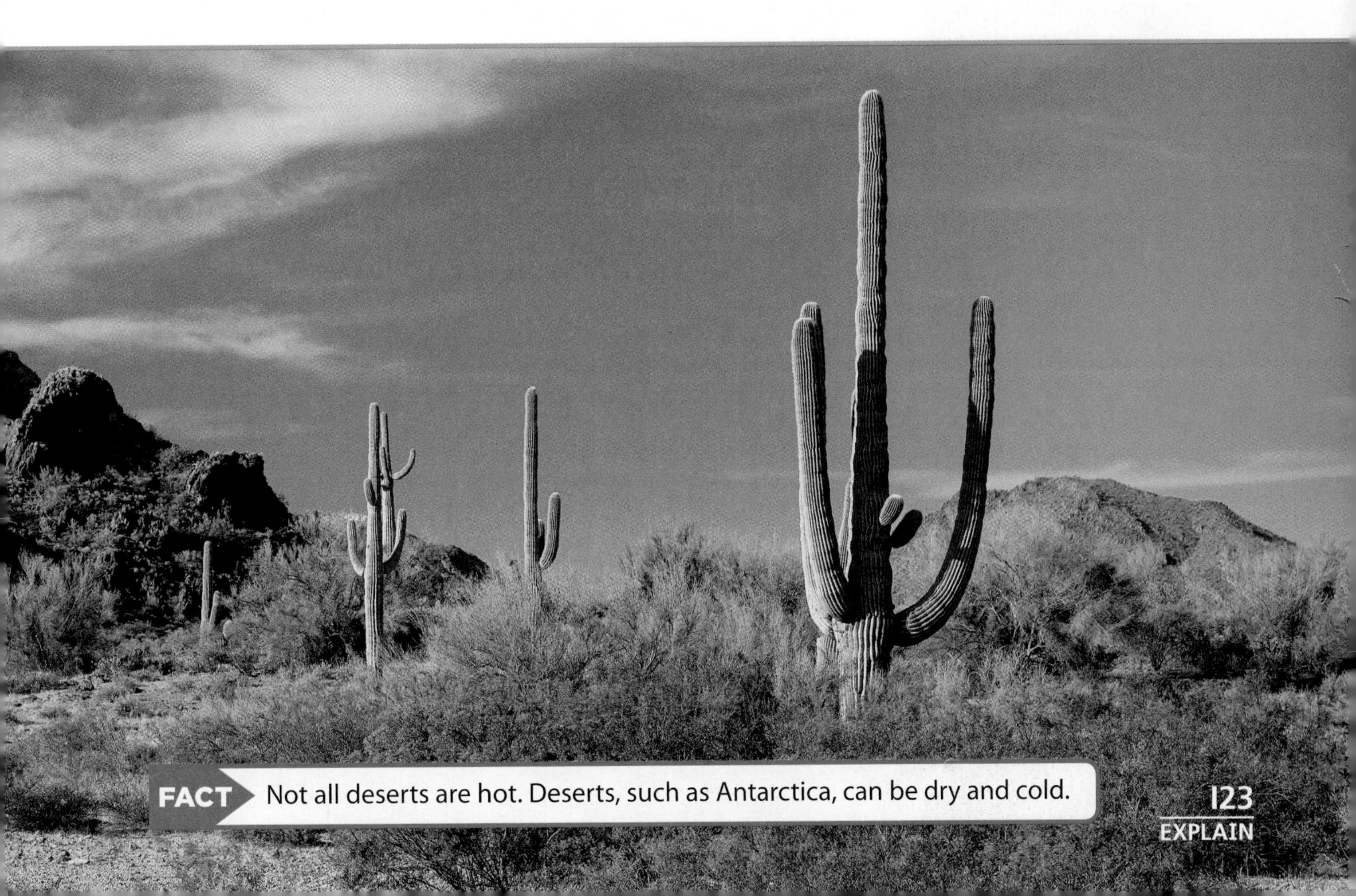

FACT Not all deserts are hot. Deserts, such as Antarctica, can be dry and cold.

What is a forest?

It is dark and damp. Tall trees surround you. Raindrops drip from above. Birds sing. You are in the Amazon Rain Forest.

A **forest** is an ecosystem that has many trees. Different types of forests can be found in different parts of the world.

Tropical Rain Forests

A *tropical rain forest* is a forest that is hot and damp. This climate helps many living things grow. A tropical rain forest has more kinds of living things than any other land environment. Brightly colored birds live in trees along with mammals, insects, and reptiles.

Tropical rain forests are warm all year long. They get twice as much rain as Hawaii and Louisiana, which are the rainiest places in the United States.

The soil in a tropical rain forest is not very rich in nutrients. Rain forest plants quickly absorb any nutrients in the soil.

This tropical forest is warm and damp all year. Toucans eat fruit from rain forest trees and other plants.

toucan

Temperate Forests

Temperate forest ecosystems are found in North America, Europe, and Asia. Bears, deer, foxes, and many other animals find homes in these forests. Unlike tropical rain forests, *temperate forests* have different weather during different seasons. Winters are cold and dry. Summers are warm and wet.

Temperate forests get enough rain for large trees to grow. However, they get less than half as much rain as tropical rain forests. The soil in a temperate forest is rich in humus. It is full of nutrients and soaks up plenty of water.

Quick Check

Compare and Contrast Which forest gets more rain—a tropical rain forest or a temperate forest?

Critical Thinking How do nutrients get added to rain forest soil? What happens to those nutrients?

Trees in a temperate forest can survive a cold winter. Pine martens search the forest floor for squirrels, beetles, and other small animals to eat.

pine marten

What is an ocean?

You are swimming in warm, clear, salty water. Before your eyes is a world of brightly colored animals. There are striped fish, sea stars, and sponges. A ridge made of tiny animals called coral stretches out before you. You are in a coral reef, a beautiful part of Earth's largest ecosystem—the ocean.

An **ocean** is a large body of salt water. Earth has five oceans, which are all connected. They are the Atlantic, Pacific, Indian, Arctic, and Southern oceans. The Pacific Ocean is the largest. It covers about one third of the planet.

Billions of living things are found in Earth's oceans. Almost all ocean organisms live in the shallow waters that are less than 100 meters (about 330 feet) deep. Here the water is lit and warmed by the Sun. Green plants and algae get enough sunlight to grow. They attract animals that depend on these producers for food and shelter. Few creatures can survive in the cold, dark ocean depths, which can be more than 1,500 meters (4,900 feet) deep.

Coral Reef

Coral reefs form in warm, shallow parts of tropical waters. Many fish visit reefs to hunt other sea animals.

Quick Check

Compare and Contrast How is the bottom of the ocean different from the surface?

Critical Thinking Why do more ocean organisms live in shallow water than in deep water?

Quick Lab

Water Temperatures

1. Fill two jars each with 200 mL of salt water. Label one jar *Sunlight* and put it in a sunny place. Label the other jar *No Sunlight* and put it in a very dark place.
2. **Observe** Measure the water temperature in each jar with a thermometer later in the day. Which jar is warmer?

3. **Infer** The two jars model two parts of the ocean. What are those parts? How are they different?

Read a Photo

How do the animals in this coral reef differ?

Clue: The coral growing from the sea floor is a type of animal.

What is a wetland?

You slowly paddle your canoe through dark, muddy water. Tall trees tower above you. Frogs croak and insects buzz. You are in a wetland.

A **wetland** is an ecosystem where water covers the soil for most of the year. Wetlands are often found along the edges of rivers, lakes, ponds, and oceans. They have freshwater or salt water.

Wetlands are important in several ways. Many kinds of animals live there. Soils are usually full of minerals that help plants grow. Wetlands help prevent land from flooding by collecting water. Wetland plants even help clean dirty water.

▲ Alligators in a wetland find fish, snakes, frogs, and turtles to eat.

Quick Check

Compare and Contrast How is a wetland like an ocean? How is it different?

Critical Thinking How could draining the wetlands affect living things?

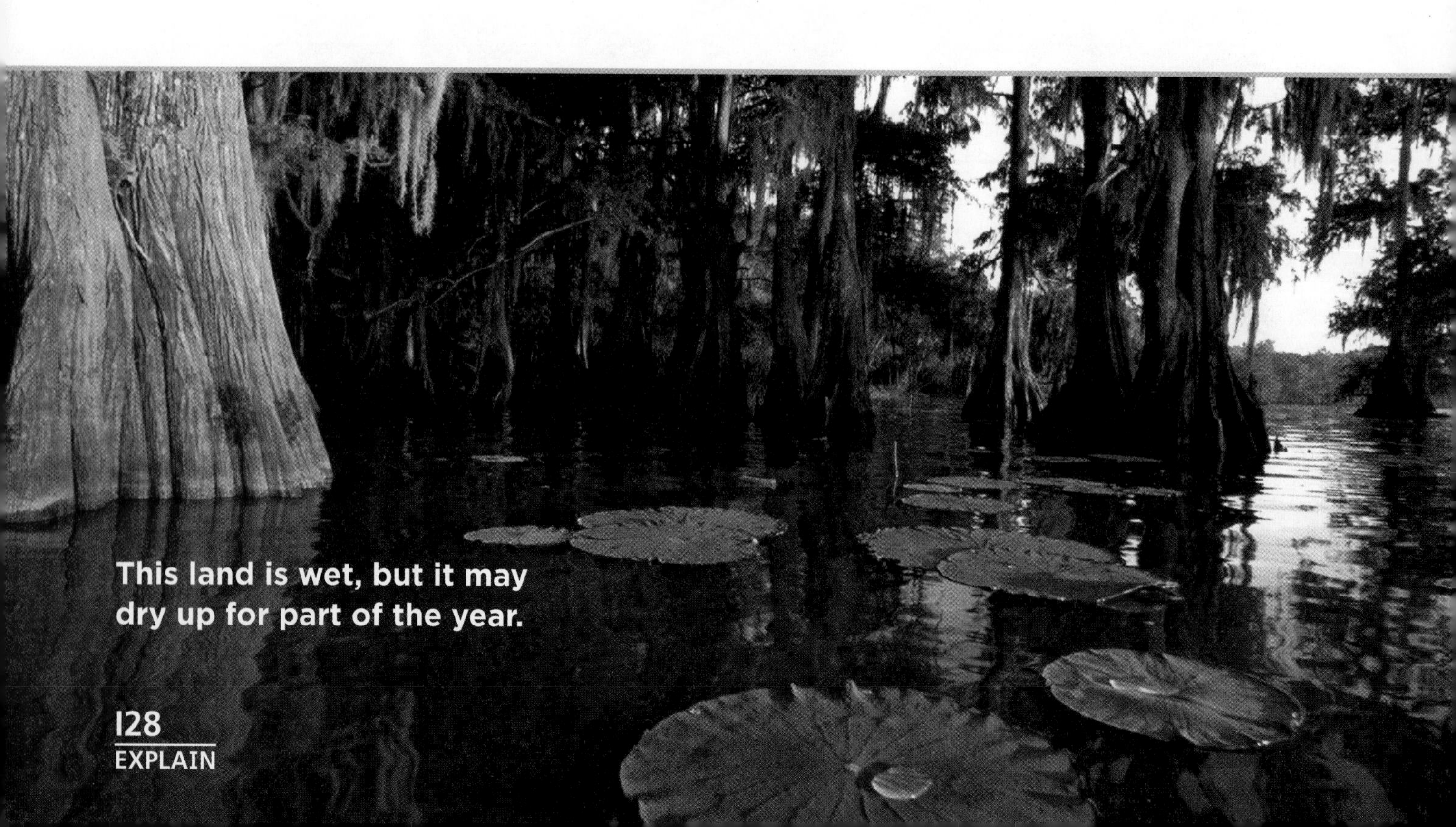

This land is wet, but it may dry up for part of the year.

Lesson Review

Visual Summary

Earth's ecosystems differ by the types of living and nonliving things they have.

Deserts and **forests** are types of land ecosystems.

Oceans and **wetlands** are types of water ecosystems.

Make a FOLDABLES Study Guide

Make a trifold book. Use it to summarize what you learned about ecosystems.

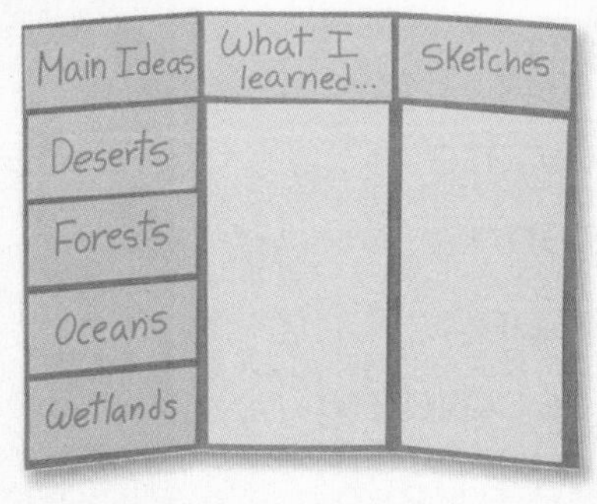

Think, Talk, and Write

1. **Vocabulary** What is a wetland?

2. **Compare and Contrast** How is a forest different from a desert? How are the two ecosystems similar?

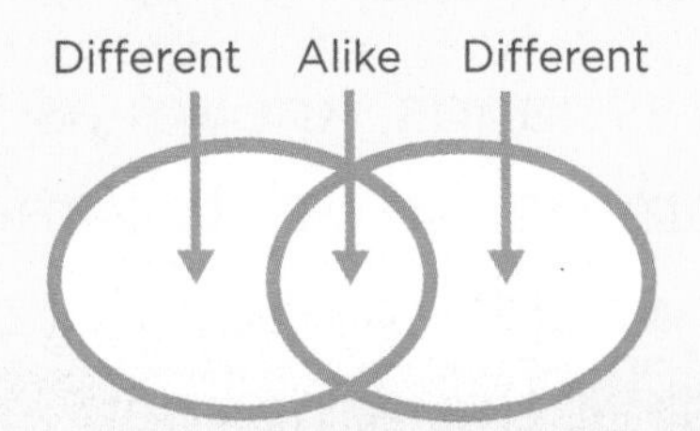

3. **Critical Thinking** You have a plant and need to choose an ecosystem for it. What do you need to know about the plant before you decide where it belongs?

4. **Test Prep** **The soil in a tropical rain forest is poor because**
   - **A** it is too sandy.
   - **B** it has no oxygen.
   - **C** it needs fertilizer.
   - **D** plants use up all the nutrients.

5. **Essential Question** How do ecosystems compare?

Writing Link

Write a Description
Choose one type of ecosystem. Imagine that you are hiking or boating through that environment. Describe what you see, hear, and feel.

Social Studies Link

Do Research
Research the kinds of ecosystems that are found in your state. Draw a map of your state showing the different ecosystems.

LOG ON e-Review Summaries and quizzes online at www.macmillanmh.com

Meet Ana Luz Porzecanski

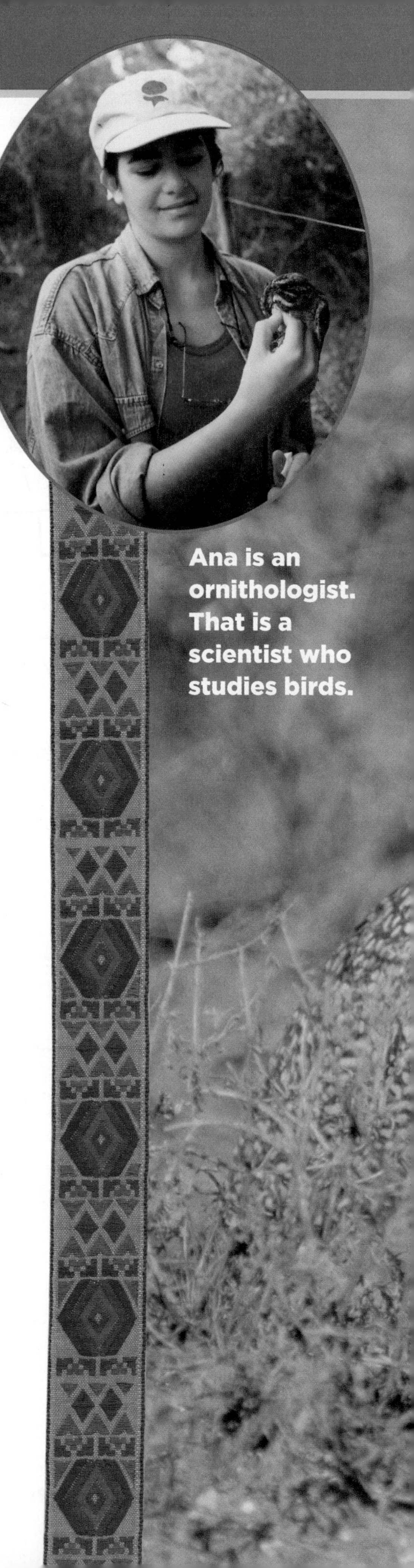

Ana is an ornithologist. That is a scientist who studies birds.

Grasslands, known as pampas, are common in South America. That is where Ana Luz Porzecanski, a scientist at the American Museum of Natural History, grew up.

Ana studies birds of the pampas. Some of the birds she studies are called *tinamous* (TIH•nuh•mewz). Their brown and gray feathers help them blend in with the tall grass, shrubs, and bushes. This helps them hide from predators, such as foxes and hawks, that eat the birds or their eggs.

How does Ana find tinamous if they are so well hidden? She listens for their songs. Each kind of tinamou has a different song. Sometimes she has to sing or play a recording of their song to get the birds to answer back. It takes time, patience, and a little luck.

The tinamous are hard to see, but their shiny green, turquoise, or purple eggs really stand out. Ana wants to know why the eggs are so colorful. What do you think?

Connect to

at www.macmillanmh.com

Compare and Contrast

- First, you explain how things are alike.
- Next, you explain how things are different.
- You use compare words, such as *like* and *both,* and contrast words, such as *unlike* and *but.*

Tinamou eggs are colorful.

Write About It

Compare and Contrast Work with a partner to compare the tinamou with another animal you know about. List ways the animals are alike and different in a Venn diagram. Then use your diagram to write about the animals.

Write about it online at www.macmillanmh.com

Lesson 3

Adaptations

Look and Wonder

These walruses sleep on a bed of ice. Do you think they feel chilly? How do the walruses keep warm in such a cold place?

Explore

Inquiry Activity

How does fat help animals survive in cold environments?

Form a Hypothesis

Can fat help keep your finger warm in cold water? Write a hypothesis. Write your answer in the form "If my finger has a layer of fat, then..."

Test Your Hypothesis

1. Use a paper towel to spread vegetable fat over one index finger. Try to coat it completely. Leave your other index finger uncovered.
2. **Predict** What will happen when you put both index fingers in a bowl of ice water?
3. **Experiment** Put one index finger into the ice water. Ask a partner to time how long you can keep your finger in the water. Repeat with your other index finger. Record the data in a chart.
4. Trade roles with your partner and repeat steps 1 through 3.

Draw Conclusions

5. **Interpret Data** Which finger could you keep in the ice water longer? Why? Did your results support your hypothesis?
6. **Infer** Walruses have a layer of fat under their skin. How does this help them survive?

Explore More

Experiment How could you measure how well fat keeps things warm? Could you use thermometers? Make a plan and test it.

Materials

vegetable fat

paper towel

ice water

stopwatch

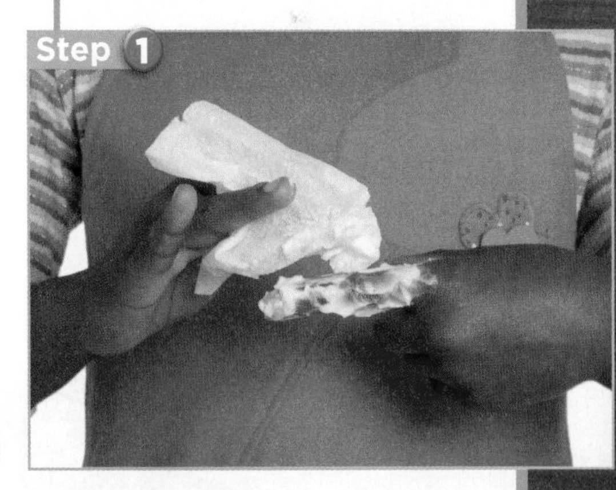

Read and Learn

Essential Question
How do an organism's traits help it survive?

Vocabulary
adaptation, p.134
camouflage, p.134
nocturnal, p.137
mimicry, p.139
hibernate, p.139
migrate, p.141

Reading Skill Predict

What I Predict	What Happens

Technology LOG ON
e-Glossary, e-Review, and animations online at www.macmillanmh.com

How are living things built to survive?

An insect buzzes through a forest. A frog flicks out its sticky tongue. It catches and swallows the insect whole. The frog's sticky tongue is an adaptation. An **adaptation** is a structure or behavior that helps an organism survive in its environment.

Some adaptations help living things get food. Sharp claws help animals such as bears and lions hunt, for example. Flat teeth help animals such as horses chew grass.

Some adaptations help living things stay safe. For example, some animals hide from predators by blending into their environment. Blending in is an adaptation called **camouflage** (KA•muh•flahj). Camouflage can also help animals sneak up on their prey. A polar bear's white fur blends in with the snow and ice. This helps it hunt seals without being seen.

The frog's sticky tongue helps it catch flies for food.

Camouflage helps this snake seem invisible to a hungry hawk flying overhead.

Some adaptations help living things survive in certain climates. Plants near the North Pole might have fuzzy leaves. This keeps frost and snow away from the leaf's surface. Sea lions, walruses, and other animals in cold climates have a layer of *blubber,* or fat, under their skin. This blubber is an adaptation that helps keep them warm. Animals have more blubber in winter than in summer.

This arctic willow plant has a layer of fuzz that shields its leaves from snow. ▼

Quick Check

Predict Could different types of animals have similar adaptations?

Critical Thinking Why don't all animals have the same adaptations?

What adaptations help desert plants and animals survive?

Not all living things can survive in a desert. Organisms that do survive have adaptations that help them live in a dry climate. Desert plants have adaptations for taking in and storing water, for example. Their roots may spread wide to soak up rainwater from a large area. Special stems can help store water. Desert animals eat plants to get water. Spines and thorns protect plants from thirsty animals.

Adaptations of Desert Plants

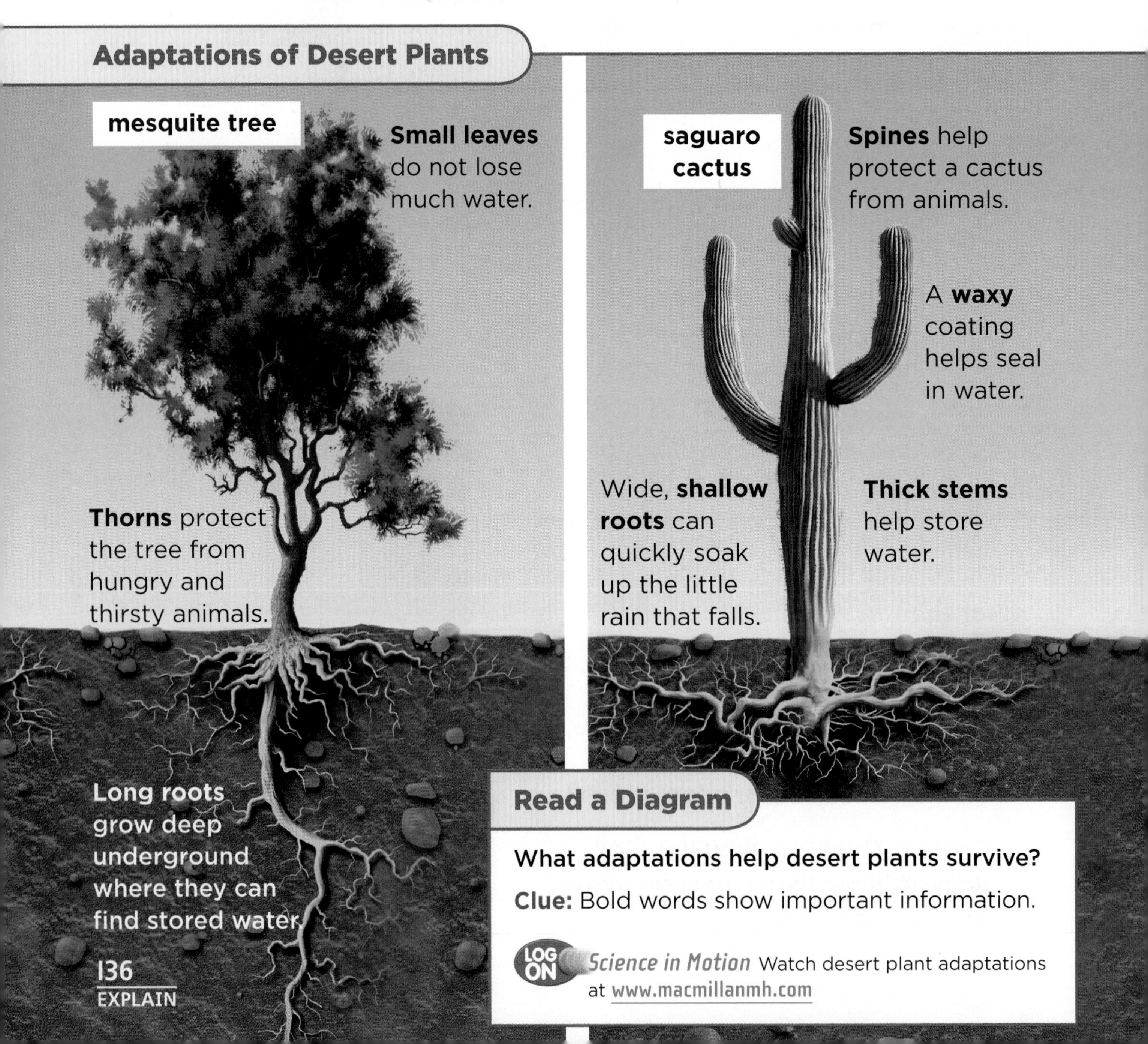

Read a Diagram

What adaptations help desert plants survive?

Clue: Bold words show important information.

LOG ON *Science in Motion* Watch desert plant adaptations at www.macmillanmh.com

▲ This bat is nocturnal. It sleeps in the daytime when the desert is hot. At night, it feasts on fruit.

Many desert animals, such as rattlesnakes and coyotes, are nocturnal (nahk•TUR•nul). **Nocturnal** means they are active at night. They sleep during the day. They come out at night when the desert is cooler.

Large ears and thin bodies help animals, such as the jackrabbit, stay cool. As warm blood flows through an animal's ears, it loses heat. The bigger the ears, the more heat is given off. Pale-colored body coverings keep animals from absorbing too much heat.

Quick Check

Predict Could an animal with a lot of blubber live in a hot desert?

Critical Thinking Would a desert jackrabbit survive better with large or small ears? Why?

Quick Lab

Storing Water

1. **Make a Model** Wet two paper towels. Then wrap one in wax paper. This models a plant that has waxy skin. Use the uncovered towel to model a plant that does not have waxy skin.

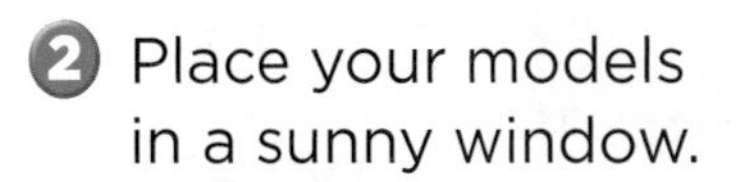

2. Place your models in a sunny window.
3. **Observe** How do the paper towels feel later in the day?
4. **Infer** How does waxy skin help desert plants survive?

▲ Warm blood flows to the jackrabbit's ears and releases some of its heat.

What adaptations help forest plants and animals survive?

In a forest, tall trees grow toward sunlight. Smaller plants grow in shade under the trees. Animals may find food high in trees or on the dark forest floor. Adaptations help forest organisms survive.

Forest Plants

In tropical rain forests, plants on the forest floor get a lot of rain and not much sunlight. Too much water can damage leaves and branches. Some rain forest leaves have grooves and "drip tips" that help rainwater flow off. These leaves are often large. They catch the little sunlight that shines through the trees.

▲ The "drip tip" at the end of each leaf helps rainwater flow off the leaf.

In temperate forests, winters are cold and dry. There is less sunlight for trees to make food. Some trees lose their leaves in fall as the temperature drops. This adaptation helps the trees save water.

These leaves can no longer make food. They fall off and make way for new leaves to grow in the spring.

Forest Animals

Some forest animals blend in by looking like other, very different organisms. This adaptation is called mimicry. **Mimicry** is when one living thing imitates another in color or shape. Like camouflage, mimicry helps an organism stay safe in its environment. It can also help an organism hunt without being seen.

Skunks are temperate forest animals that have unusual ways of staying safe. Skunks spray a stinky chemical if a predator gets too close. This chemical can also sting a predator's eyes.

When winter comes in a temperate forest, food may be difficult to find. Animals like the dormouse survive by hibernating. **Hibernate** means to go into a deep sleep. While hibernating, animals use less energy and do not need to eat. Hibernating is an adaptation that helps some animals survive when seasons change.

Quick Check

Predict What does a predator usually do when a skunk gives its warning?

Critical Thinking How is mimicry different from camouflage?

▲ Can you find the insect in this photo? The thorn-mimic treehopper on the right looks like the thorn on the left.

▲ Dormice curl up when they hibernate. That helps to keep them warm.

When in danger, a skunk gives a warning by raising its tail before spraying. ▶

Kelp is a kind of algae. Kelp makes up this seaweed forest.

What adaptations help ocean plants and animals survive?

Oceans are home to millions of living things. Each has adaptations that help it survive in an ocean's salty water. While salt water would kill most organisms, ocean organisms need salt water. They could not survive in the fresh water of lakes or ponds.

Ocean Algae

The seaweed you see floating in the ocean are plantlike organisms called *algae*. Like plants, algae use sunlight to make their own food. Most algae have leaflike structures that take in sunlight. Some have rootlike structures that attach to the ocean floor. These algae can live only in shallow water where they can get sunlight. Other algae have no roots. They drift near the water's sunlit surface. Balloonlike structures called air bladders help some algae float.

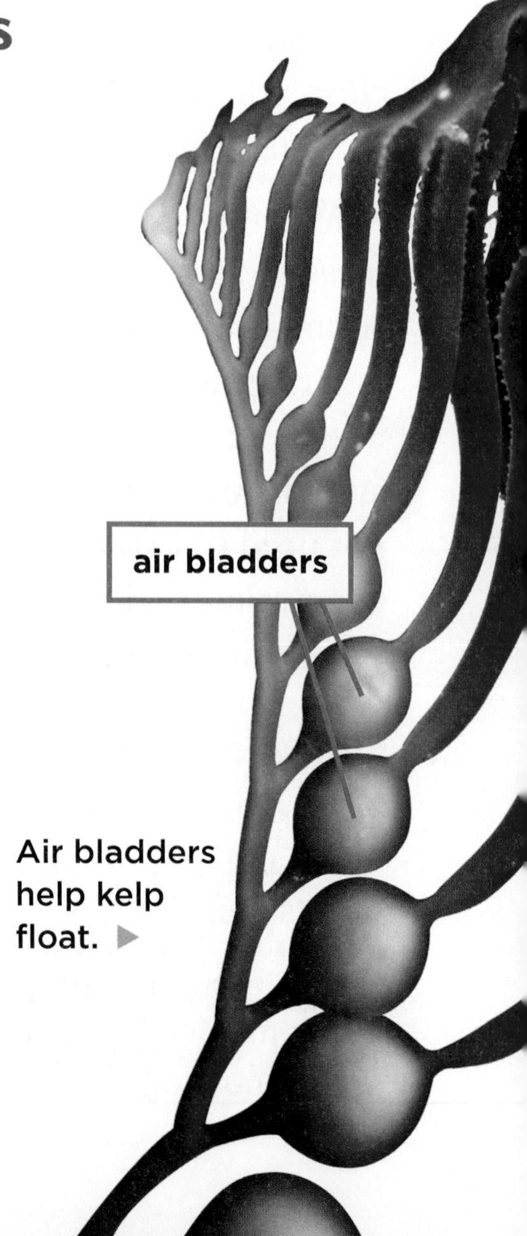

Air bladders help kelp float. ▶

Sperm whales swim thousands of kilometers when they migrate.

Ocean Animals

Ocean animals have adaptations for moving and living in water. A dolphin's fins and tail help it move. A fish's gills help it breathe.

Some ocean animals migrate from one part of the ocean to another in different seasons. **Migrate** means to move from one place to another. Animals might migrate when their environment gets too cold or when food is hard to find.

It is extremely cold and dark deep in the ocean. Few organisms have adaptations for living there. The angler fish is one that does. It has a growth on top of its head that lights up. The light attracts other animals. They swim close, and the angler fish attacks.

▲ The angler fish has a lighted "fishing pole" to attract prey in deep, dark ocean water.

Quick Check

Predict Would the angler fish's adaptation work in sunlit, shallow water? Why or why not?

Critical Thinking How are algae similar to plants?

What are adaptations to a wetland?

Wetland organisms have adaptations to survive changes. One day their environment could be underwater. The next day it could be soggy or dry.

Wetland Plants

Plants in wetlands can survive changing water levels. Mangroves live in wetlands along rivers and oceans. Their roots spread out to get a firm grip on the muddy ground.

▲ The walking catfish can breathe oxygen from the air for short periods.

Wetland Animals

Wetland animals have ways to survive dry seasons. Walking catfish live in wetland ponds that can dry up. The catfish can use its fins to move over land to another body of water.

Quick Check

Predict Could grassland plants survive in a wetland? Why or why not?

Critical Thinking How might the roots of a mangrove tree help it get oxygen?

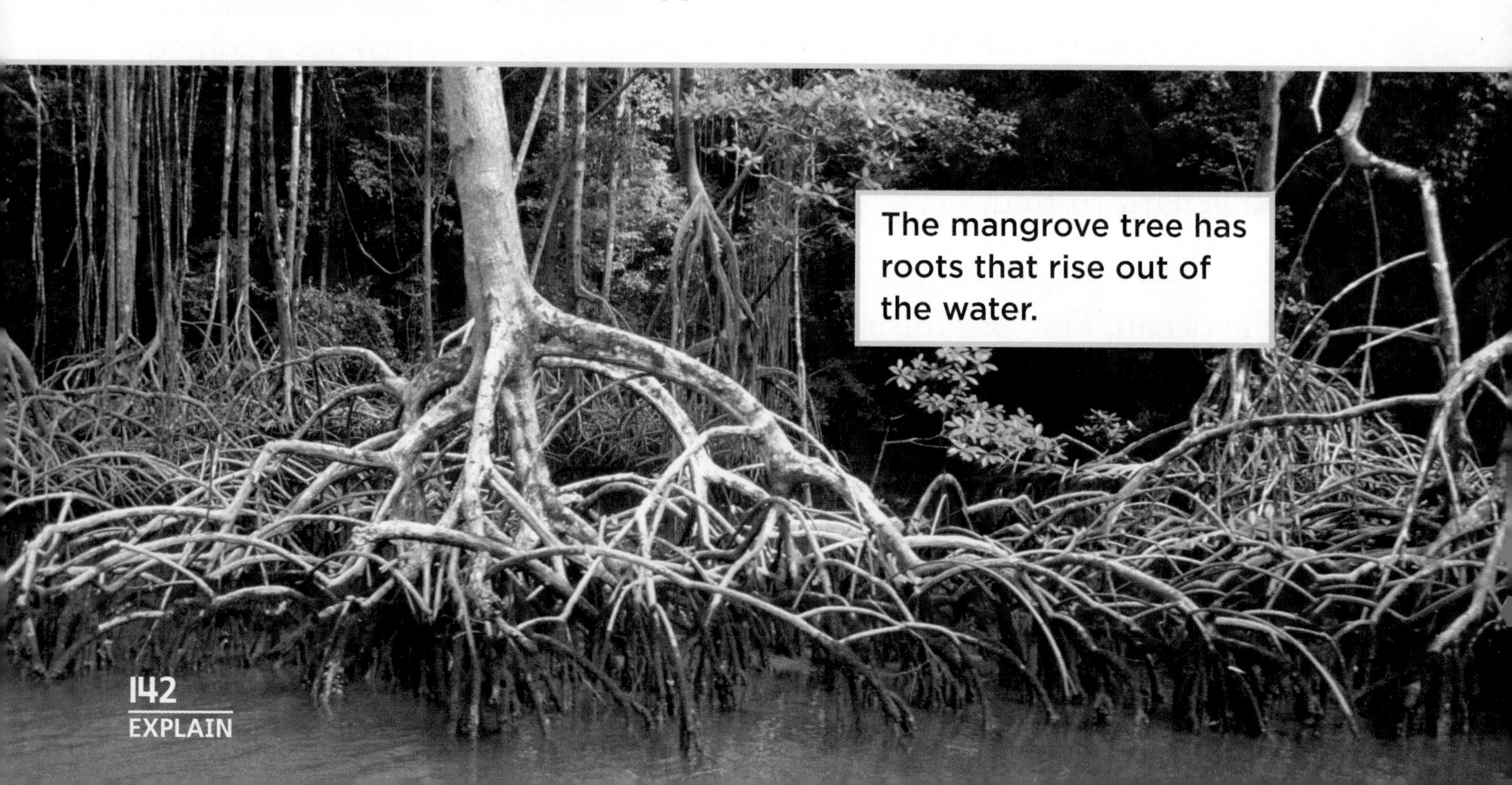

The mangrove tree has roots that rise out of the water.

Lesson Review

Visual Summary

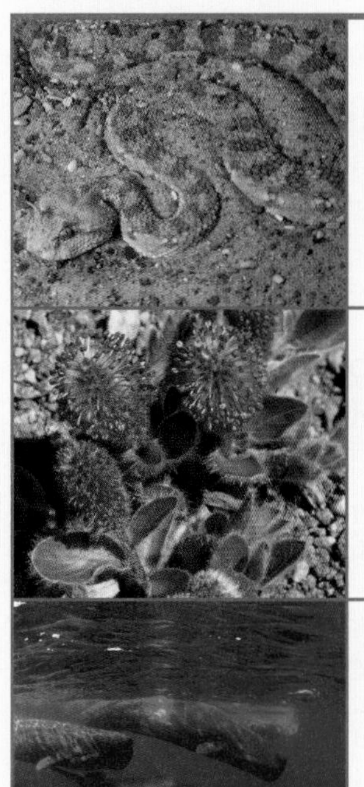

Adaptations are structures or behaviors that help an organism survive in its environment.

Some **plant adaptations** include fuzzy leaves, pointed leaves, and shallow roots.

Some **animal adaptations** include camouflage, mimicry, migrating, and hibernating.

Make a FOLDABLES® Study Guide

Make a three-tab book. Use it to summarize what you learned about adaptations.

Think, Talk, and Write

1. **Vocabulary** What does the word *nocturnal* mean?

2. **Predict** What might happen to an arctic willow plant if you moved it to a tropical rain forest?

What I Predict	What Happens

3. **Critical Thinking** Compare two or more organisms from this lesson. Explain how the organisms are alike and different.

4. **Test Prep** **Why do some animals migrate?**
   A to escape prey
   B to avoid cold weather
   C to find their families
   D to make a change

5. **Essential Question** How do an organism's traits help it survive?

Math Link

Find Distance
The American robin migrates from Iowa to Alaska. It travels about 40 miles in one day. How far can it fly in five days?

Social Studies Link

Migration Map
Research to learn where one type of animal migrates. Draw a map to show the migration path.

Be a Scientist

Materials

yellow paper

brown paper

scissors

stopwatch

Structured Inquiry

How does camouflage help some animals stay safe?

Form a Hypothesis

Which is easier to find, an animal that blends into its environment or an animal that does not blend in? Form a hypothesis. Write your answer in the form "If an animal blends into its environment, then..."

Test Your Hypothesis

1. Cut out 20 yellow circles and 20 brown circles.
2. **Experiment** Spread out the circles on yellow paper to model animals with and without camouflage. Then ask a classmate to pick up as many circles as he or she can in 10 seconds.

3. **Communicate** How many of each color circle did your classmate pick up? Use a chart to record the results.
4. Repeat steps 1 and 2 with two other classmates.

Step 3

Name	number of yellow circles	number of brown circles
David	3	8

Draw Conclusions

5. **Interpret Data** Did your classmates pick up more yellow or brown circles? Which circles were harder to find?

6. **Infer** How does camouflage help animals stay safe?

Guided Inquiry

How do pale colors help some animals survive?

Form a Hypothesis

How do pale body coverings affect a desert animal's temperature? Write a hypothesis.

Test Your Hypothesis

Design a plan to test your hypothesis. Use the materials shown. Write the steps you plan to follow.

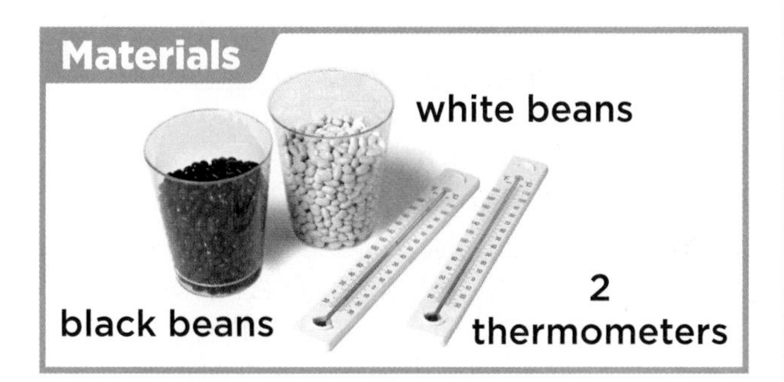

Draw Conclusions

Did your results support your hypothesis? Why or why not? Share your results with your classmates.

Open Inquiry

What other questions do you have about plant and animal adaptations? Discuss with classmates the questions you have. How might you find the answers to your questions?

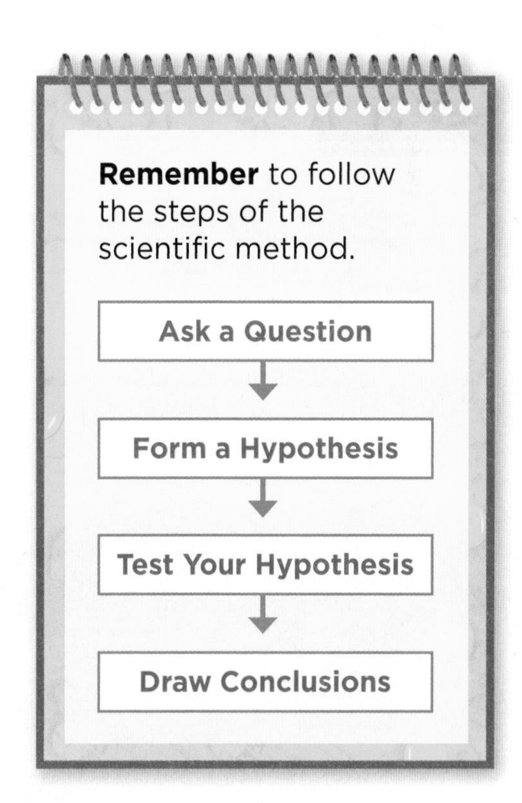

CHAPTER 3 Review

Visual Summary

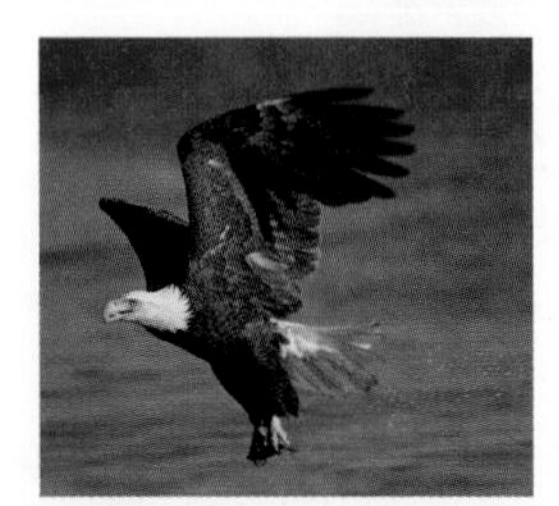

Lesson 1 Food chains and food webs show how organisms in an ecosystem depend on one another.

Lesson 2 Earth has different ecosystems. They are classified by the type of climate, soil, plants, and animals they have.

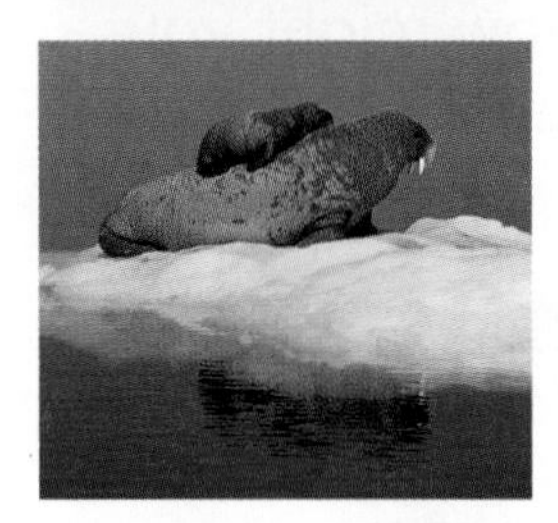

Lesson 3 Plants and animals have adaptations that help them survive in different environments.

Make a FOLDABLES® Study Guide

Glue your lesson study guides to a piece of paper as shown. Use your study guide to review what you have learned in this chapter.

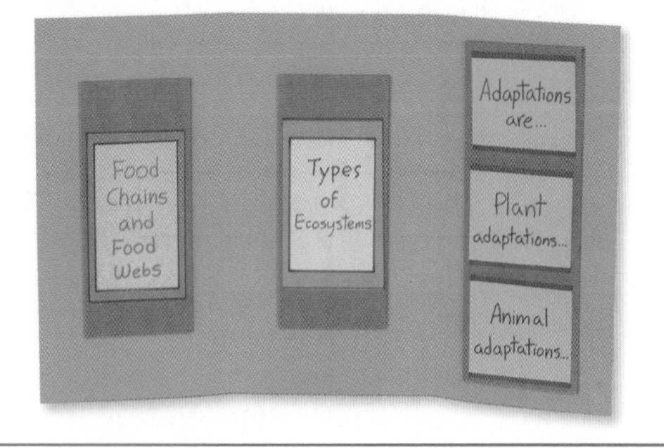

Vocabulary

DOK I

Fill each blank with the best term from the list.

adaptation, p. 134
camouflage, p. 134
climate, p. 120
decomposer, p. 111
desert, p. 122
ecosystem, p. 108
food chain, p. 110
forest, p. 124
nocturnal, p. 137
producer, p. 110

1. An ecosystem that has many trees is called a ________.

2. In an ecosystem a ________ shows how energy passes from one organism to another.

3. An animal that is active at night is ________.

4. A structure or behavior that helps an organism survive in an environment is an ________.

5. An organism that makes its own food is called a ________.

6. Plants often grow far apart in the dry climate of the ________.

7. An adaptation called ________ helps an animal blend in with its environment.

8. Living and nonliving things interacting in their environment make up an ________.

9. An organism that breaks down dead plants and animals is called a ________.

10. The pattern of weather in a place over a long time is its ________.

LOG ON e-Glossary Words and definitions online at www.macmillanmh.com

Skills and Concepts

DOK 2–3

Answer each of the following.

11. **Infer** Is it possible to have more than one producer in a food chain? Could there be more than one consumer?

12. **Writing That Compares** How is a pond ecosystem like a wetland? How is it different? Write about as many similarities and differences as you can.

13. **Communicate** Make a chart with two columns: *Plants* and *Animals.* Record the foods you eat during one day. Place each food in the correct column. If a food contains plant and animal material, list it in both columns.

14. **Critical Thinking** You are taking care of a plant and an animal from a desert ecosystem. What kind of environment would you create for them to live in?

15. **Critical Thinking** How can an organism's color protect it?

16. **Predict** What do you think would happen if a fish that lived in freshwater was placed in an ocean? Explain.

17. What traits would help an animal survive in the cold environment below?

18. **True or False** *All animals are consumers.* Is this statement true or false? Explain.

19. Look at the table shown below.

Death Valley, California	
Climate	• hot and dry • some rainfall each year
Soil	• dry and sandy
Organisms	• cactus • coyote

Death Valley, California, is what kind of ecosystem?

A forest **C** pond

B wetland **D** desert

20. How do living things survive in their environments?

Performance Assessment

DOK 3

A New Organism

- Choose an ecosystem. Make up a new kind of organism that could live in this ecosystem. What adaptations would your organism have? How would it behave? How would it get food?
- Make a poster showing your organism in its habitat. Label and describe its adaptations. Tell how the adaptations help it survive. Draw a food chain that includes your organism.

Test Preparation

1 **Which adaptation most likely helps this cactus survive in a desert ecosystem?**

A large leaves

B small flowers

C a waxy stem

D a waxy flower

DOK 2

2 **The porcupine below has sharp quills.**

How do the porcupine's quills help it survive?

A The quills keep the porcupine warm.

B The quills help the porcupine catch food.

C The quills protect the porcupine from predators.

D The quills allow the porcupine to get oxygen.

DOK 1

3 **Which best describes a green plant's role in an ecosystem?**

A making oxygen and food

B breaking down dead animals

C eating other organisms

D recycling soil

DOK 1

4 **How do animals use camouflage to survive?**

A by standing out from their environments

B by imitating other animals

C by blending in with their environments

D by giving warning calls

DOK 1

5 **The table below shows the number of birds in a wetland ecosystem.**

Season	Number of Birds
summer	700
winter	60

What most likely explains the difference between seasons?

A Many birds drown when the ice melts in the fall.

B Many birds are killed by predators in the spring.

C Many birds die when the winter becomes cold.

D Many birds migrate south for the winter.

DOK 2

6 Carlos made a poster showing an example of mimicry. Which poster below is his?

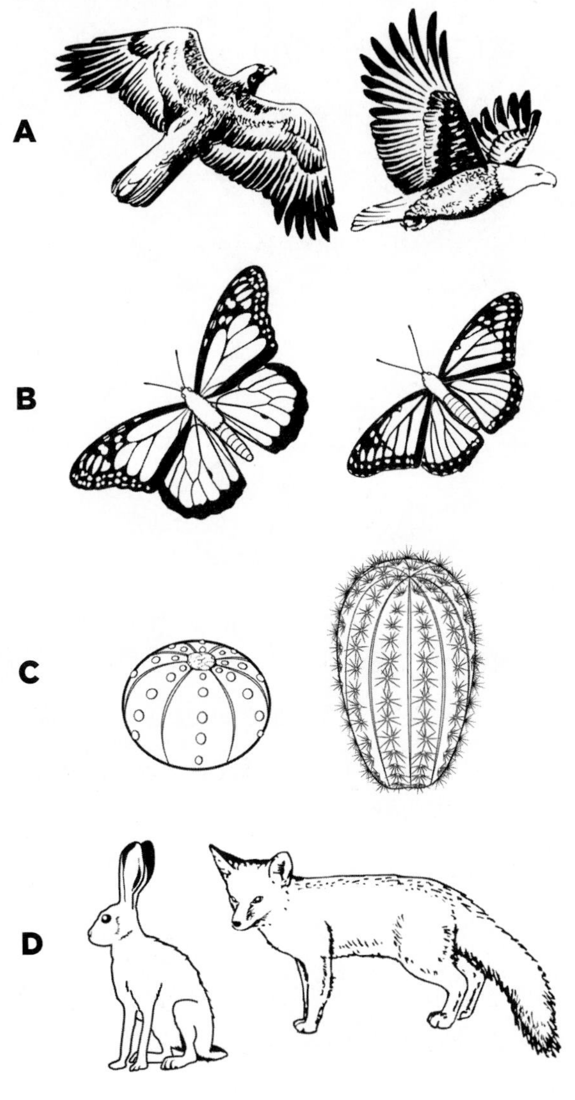

DOK 2

7 Name a producer in an ocean ecosystem.
DOK I

8 Name a consumer in a forest ecosystem.
DOK I

Use the illustration below to answer questions 9–10.

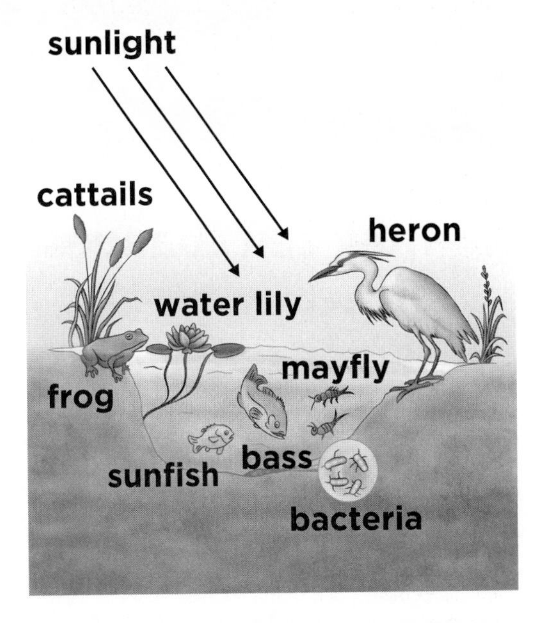

9 Frogs and bass eat mayflies in this pond ecosystem. In spring, temperatures were cold. Few mayflies survived.

How will the pond food web be affected in summer? Explain your answer.
DOK 2

10 How are dead organisms in the pond broken down? Why is this important?
DOK I

Check Your Understanding

Question	Review	Question	Review
1	pp. 122–123, 136–137	6	p. 139
2	pp. 48, 134–139	7	pp. 110, 126–127
3	pp. 110–111	8	pp. 111, 124–125
4	pp. 134–135	9	pp. 112–113
5	pp. 134–142	10	pp. 111, 114

CHAPTER 4

Changes in Ecosystems

How can changes affect living things and their environments?

Essential Questions

Lesson 1

How can people and other living things change their environments?

Lesson 2

How can changes in an environment affect living things?

Lesson 3

What can we learn about living things of the past?

jack pine seedling sprouts after a wildfire

Big Idea Vocabulary

resource something in the environment that helps an organism survive (p. 152)

competition the struggle among living things for water, food, or other resources (p. 153)

pollution what happens when harmful materials get into water, air, or land (p. 154)

endangered when one kind of organism has very few of its kind left (p. 168)

fossil the trace or remains of something that lived long ago (p. 174)

extinct when there are no more of an organism's kind left (p. 174)

Visit www.macmillanmh.com for online resources.

Lesson 1

Living Things Change Their Environments

Look and Wonder

Leaves fall from trees and cover the forest floor. Have you ever wondered what happens to fallen leaves? What makes them disappear?

Explore

Inquiry Activity

How can worms change their environment?

Purpose

All living things change their environments as they get food, water, shelter, and other needs. In this activity, find out how worms change their environment.

Procedure

1. **Make a Model** Put some soil in a plastic container. Then put small stones and leaves on top of the soil. This models a forest floor.
2. Place live worms on the forest floor.
3. **Predict** What will the worms do? Make a short list of the things you might see the worms do.
4. **Observe** Check the worms, soil, leaves, and stones every three to four days. Keep the soil moist. Record your observations.

Draw Conclusions

5. **Infer** What happened to the leaves over time?
6. **Communicate** How do worms change the environment in which they live?

Explore More

Experiment How do other living things change their environments? Make a plan to test your ideas. Then try your plan.

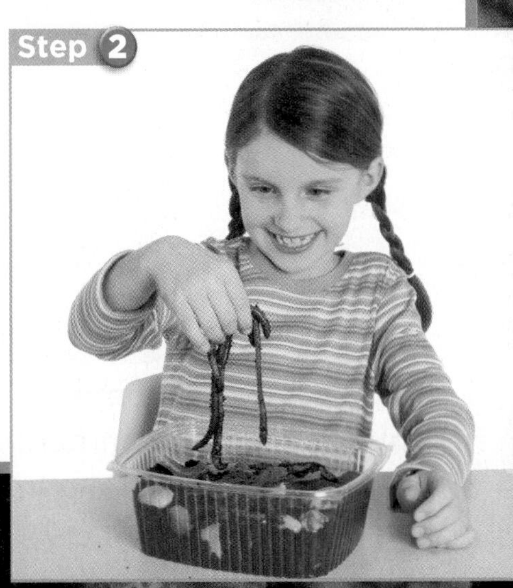

Read and Learn

Essential Question

How can people and other living things change their environments?

Vocabulary

resource, p.152

competition, p.153

pollution, p.154

reduce, p. 156

reuse, p.156

recycle, p.156

Reading Skill

Predict

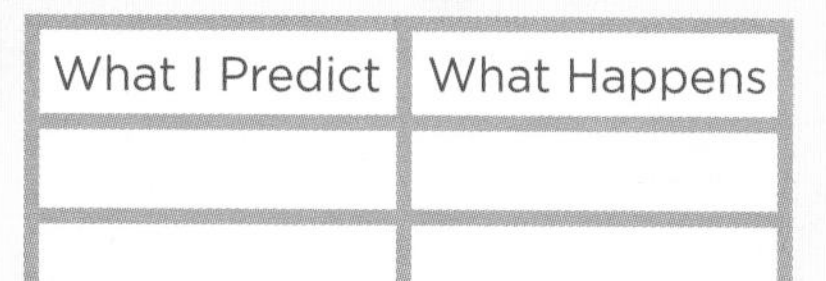

What I Predict	What Happens

Technology 

e-Glossary, e-Review, and animations online at www.macmillanmh.com

How do living things change their environments?

Every living thing changes its environment as it meets its needs. A spider spins a web to catch insects for food. A bird builds a nest for shelter. A plant takes water from the soil. These actions change an environment in small ways.

Other living things make bigger changes to their environments. For example, bacteria, worms, and fungi break down leaves and other dead material. These decomposers return valuable nutrients to the soil. Later, plants can use those nutrients to grow.

All of these living things are trying to get resources. A **resource** is something that helps an organism survive. Food, water, air, space, sunlight, and shelter are some resources.

A Changing Environment

Seeds blow onto bare ground. The environment changes as plants take in water and nutrients.

As more plants grow, animals move to the environment. They use the plants for food and shelter.

Every environment has a limited amount of resources. As a result, living things must compete for them. **Competition** (kahm•puh•TISH•un) is the struggle among living things for resources. Competition can be a cause of environmental change. The diagram below shows how competition can change an environment.

Quick Check

Predict How would a forest change if a big tree fell?

Critical Thinking How do you change your environment?

Read a Diagram

How has this environment changed over time?

Clue: Arrows help show a sequence.

Science in Motion Watch environments change at www.macmillanmh.com

In time the plants grow larger. They compete for water, space, and sunlight. Animals compete for food and water.

Trees block sunlight from reaching smaller plants. These plants may die as trees grow larger.

How do people change their environments?

People change their environments more than any other organism. Changes such as planting trees are helpful. However, other changes can be harmful.

▲ The garbage on this beach is a form of land pollution.

Pollution

People can harm their environments by creating pollution. **Pollution** happens when harmful materials get into the air, land, or water. Cars can pollute the air. Trash pollutes the water and land.

Clearing Land

Sometimes people change natural areas when they build shops and homes. In the past, people drained wetlands and built over them. Wetlands help filter water, so pollution increases without them.

People also cut down trees to make wood products. If forests are removed, living things can be left without a home. Soil can wear away without tree roots to hold it in place.

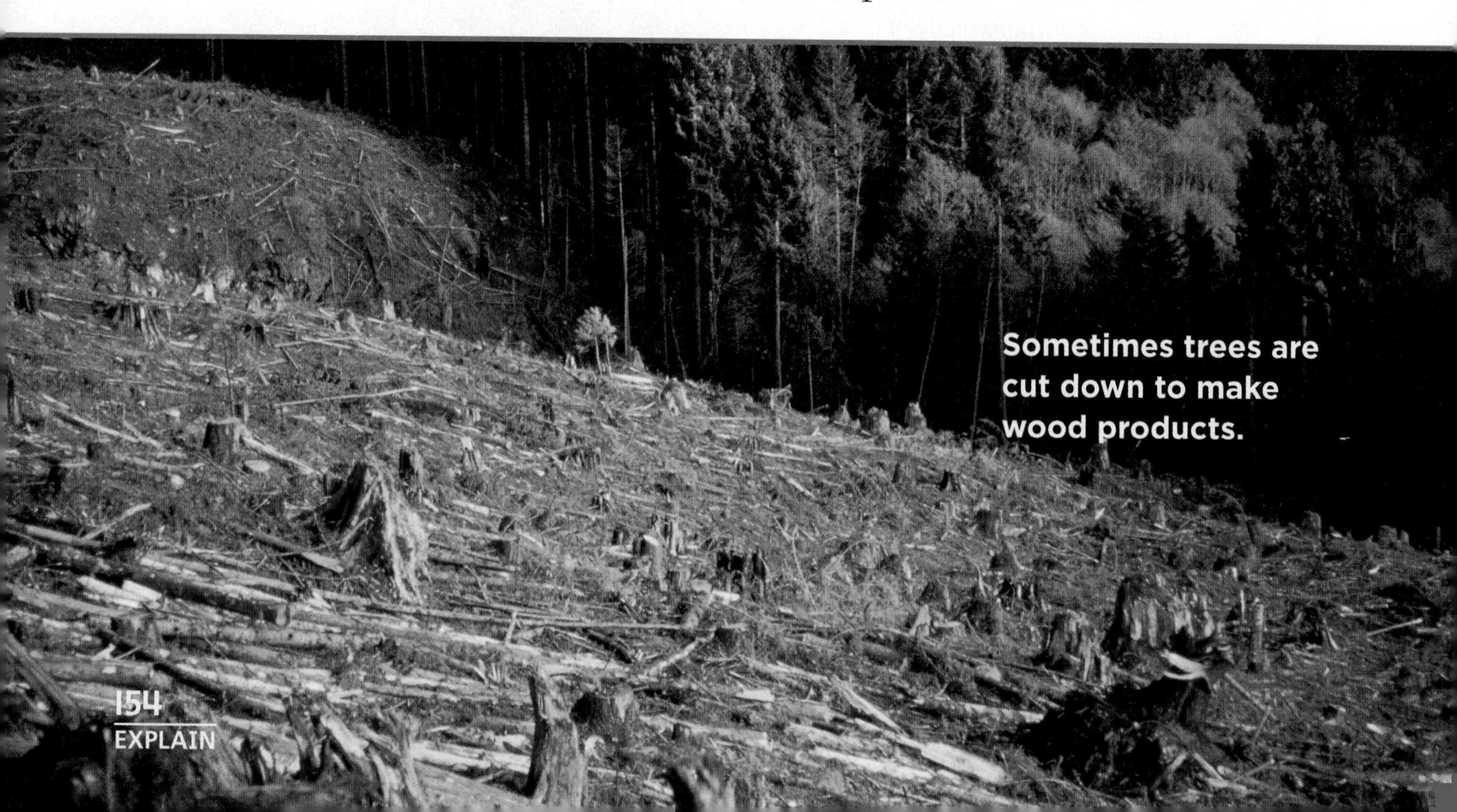

Sometimes trees are cut down to make wood products.

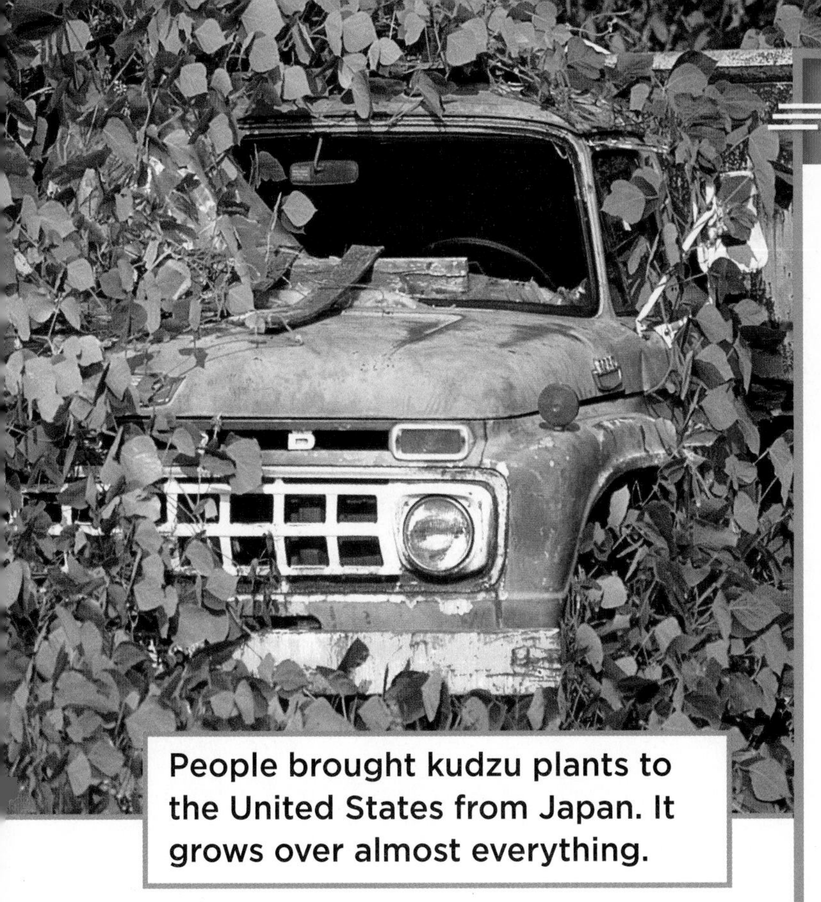

People brought kudzu plants to the United States from Japan. It grows over almost everything.

Creating Competition

Sometimes people bring new organisms into an environment. These new organisms can harm the environment by competing for limited resources. For example, kudzu plants were brought to the United States from Japan. Kudzu grows rapidly. It takes water and nutrients that other plants need.

The new organisms might not be eaten by many animals in their new environment. They might increase in number. If this happens, the new organisms can use up most of the limited resources in an environment.

Quick Lab

Model Pollution

1. **Observe** Look at the shell of a hard-boiled egg. Is it hard or soft? Why do you think the egg has this type of shell?
2. **Make a Model** Fill a large cup with vinegar. This models polluted land or water. Place your egg inside the cup.

3. **Observe** Look at the egg throughout the day. Study the shell of the egg. Do you notice any differences in the egg or its shell?
4. **Infer** After being placed in vinegar, can the shell still protect the egg?
5. **Predict** What might happen to eggs near polluted land or water?

Quick Check

Predict What might happen to plants and animals if their environment is harmed?

Critical Thinking How does pollution affect people?

This house was built from reused bottles.

How can people protect their environments?

People can help protect their environments. One thing people can do is practice the 3 *Rs*—reduce, reuse, and recycle. To **reduce** means to use less of something. To **reuse** means to use something again. To **recycle** means to turn old things into new things. When you practice the 3 *Rs*, you produce less trash and cut down on pollution.

People can also help their environments by planting trees. Trees help environments in many ways. Trees clean the air and provide homes for animals. Their roots help keep soil from washing or blowing away. By planting a tree, you help keep your environment healthy.

Recycling is one way that people can help protect their environments.

Quick Check

Predict How might recycling paper protect your environment?

Critical Thinking List some things you can reuse.

Lesson Review

Visual Summary

Living things change their environments as they meet their needs.

People change their environments more than any other living thing.

People can help their environments by practicing the 3 *R*s.

Make a FOLDABLES® Study Guide

Make a trifold book. Use it to summarize what you learned about environments and change.

Main Ideas	What I learned...	Examples
Living things change their environments...		
People change their environments...		
People can help their environments...		

Think, Talk, and Write

1. **Vocabulary** What is competition?

2. **Predict** What might happen if people do not practice the 3 *R*s?

What I Predict	What Happens

3. **Critical Thinking** What are some things you could reduce your use of?

4. **Test Prep** **People can do all of the following to help the environment <u>except</u>**
   A recycle.
   B pollute.
   C reuse.
   D plant trees.

5. **Essential Question** How can people and other living things change their environments?

Math Link

Make a Bar Graph
Keep track of the paper, metal, plastic, and food scraps that you throw away in one week.
Make a bar graph that shows how many of each item you threw away during that week.

Art Link

Make a Poster
Make a poster about the things people can do to help the environment. Include things you learned from this lesson as well as things you already knew.

Focus on Skills

Inquiry Skill: Use Numbers

The average American changes his or her environment by producing about 2 kilograms (4 pounds) of trash every day! We can not get rid of trash completely. However, we can cut down on the amount we create by practicing the 3 *Rs*. Do students in your school practice the 3 *Rs*? Find out the same way scientists do—**use numbers** to record data.

▶ Learn It

When you **use numbers**, you present data in a way that people can clearly understand. Basic math skills, such as counting and ordering numbers, help you collect and organize information. Often, scientists gather and record data by asking questions or by having people fill out surveys. Then they **use numbers** to put the data into a chart or graph. You can do it too.

▶ Try It

In this activity, you will gather data and **use numbers** to find out how much trash is thrown out by students in your school. You can not survey the whole school, but you can do a mini-survey.

1. Choose five students to survey in the lunchroom.
2. Ask each student questions about how many pieces of trash from lunch he or she threw away today. Ask about the containers used. Will anything be reused or recycled?

3. Use a table like the one shown below to organize your data.

Student's Name	Pieces Reused	Pieces Recycled	Pieces Thrown Away	Total Pieces of Trash
Total				

Now **use numbers** to answer the following questions.

- Did every student throw out some trash or packaging material?
- How many pieces of trash did students recycle? How many pieces did they reuse?
- How many total pieces of trash did these five students create altogether?

Apply It

Use numbers to combine your data with those of your classmates. Add to find the totals for each column. Then make a bar graph to show the results.

Do you predict these same students will throw out more or less trash tomorrow? Plan another survey. Then **use numbers** to compare the new results to your first results just as scientists do!

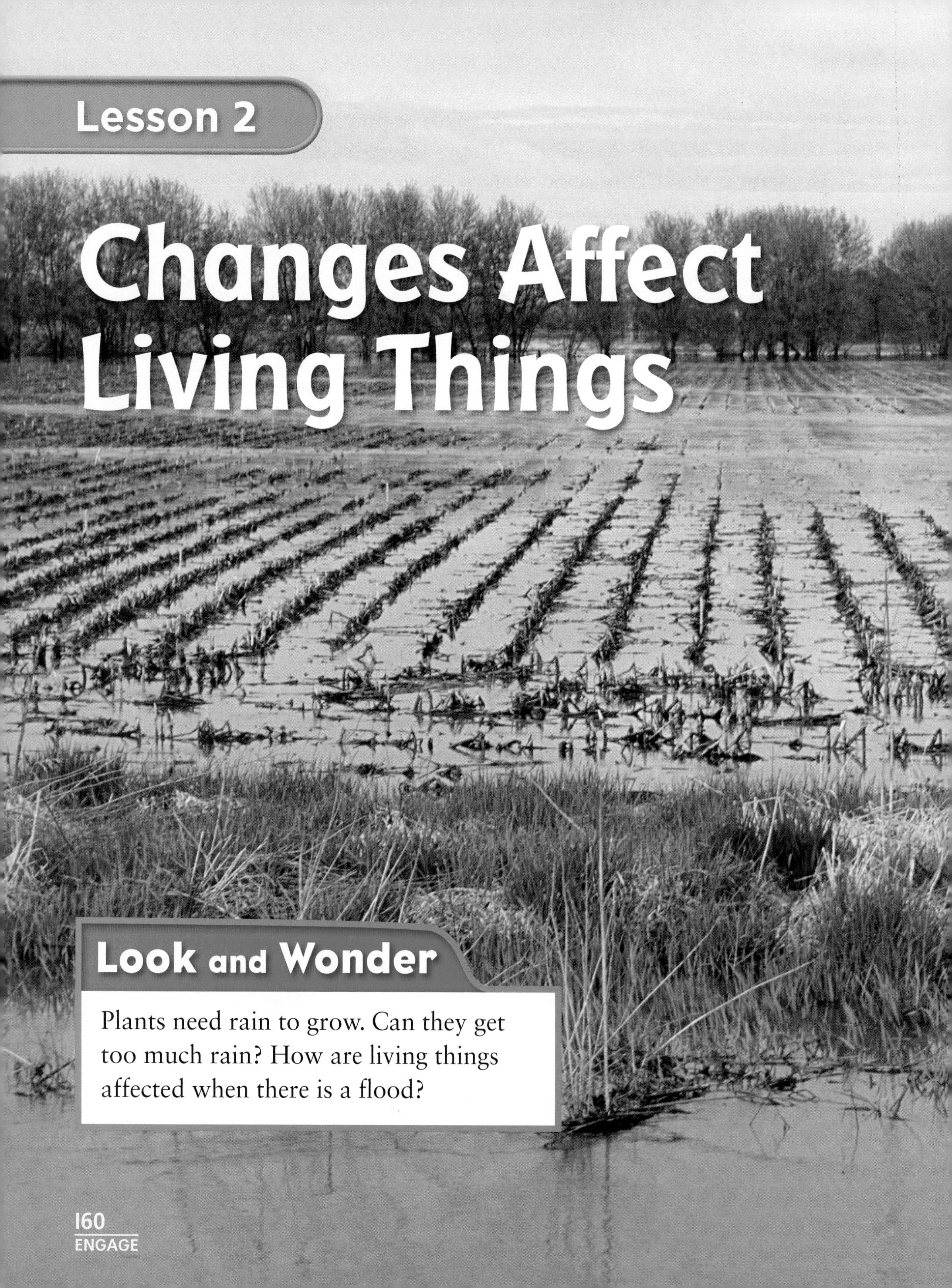

Lesson 2

Changes Affect Living Things

Look and Wonder

Plants need rain to grow. Can they get too much rain? How are living things affected when there is a flood?

Explore

Inquiry Activity

How can a flood affect plants?

Form a Hypothesis

What happens to plants when they get too much water? Write a hypothesis.

Test Your Hypothesis

1. Label three plants *A, B,* and *C.* Water plant *A* once a week with 60 milliliters of water. Water plant *B* every day with 60 mL of water. Water plant *C* every day with 120 mL of water.
2. **Predict** Which plant will grow to be the tallest? Record your prediction.
3. **Observe** Monitor your plants every few days. Measure how tall they grow. Record how they look with words and pictures.

Draw Conclusions

4. **Interpret Data** How did the plants change over time? Which plant grew the tallest? Which do you think is the healthiest?
5. **Infer** What happens to some plants when there is a flood?

Explore More

Experiment Could your plant recover from a flood? Stop watering plant *C* for one week. How does the plant change?

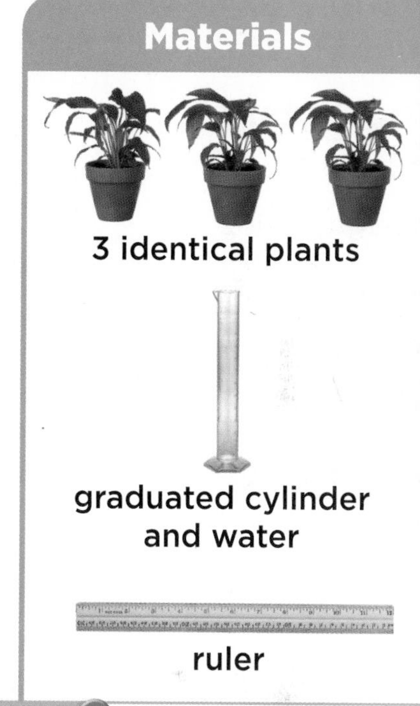

Step 1

Step 3

Read and Learn

Essential Question
How can changes in an environment affect living things?

Vocabulary
flood, p. 162

drought, p. 162

population, p. 166

community, p. 166

endangered, p. 168

Reading Skill
Cause and Effect

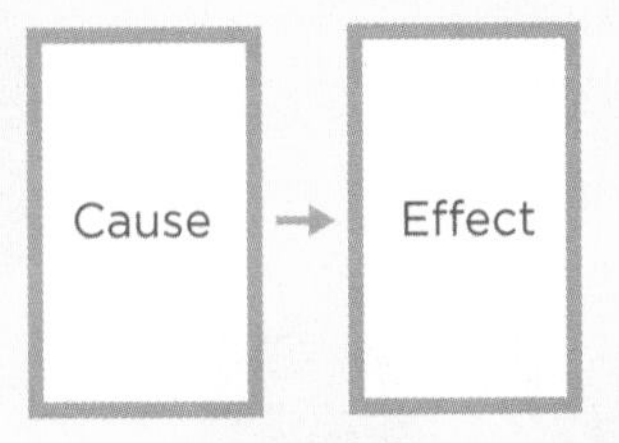

Technology

e-Glossary and e-Review online at www.macmillanmh.com

What are some ways environments change?

You learned some ways that living things change their environments. Environments can be changed in other ways too. Natural disasters and diseases can change an environment.

A flood is one type of natural disaster. A **flood** happens when dry land becomes covered with water. Heavy rains and other storms can cause a flood. Floods change an environment by washing away soil and plants. They can cause animals to lose their habitats.

A drought (DROWT) is the opposite of a flood. A **drought** happens when there is no rain for a long time. Without rain, rivers and lakes can dry up. Soil can also dry out. Living things need water to survive, so droughts can harm living things.

Before a drought set in, this area was a lake. Many living things once made their homes here.

Wildfires can destroy natural environments like this forest.

▲ New plants can grow after a fire.

Droughts can lead to wildfires. If a dry part of a forest or grassland is struck by lightning, a wildfire can start. Wildfires can harm plants and the habitats of many animals. Smoke from the fires pollutes the air.

Environments can also be changed by diseases. Many things, such as bacteria, mold, and mildew, can cause diseases. Some diseases spread easily and harm many living things. An entire forest, for example, can eventually be destroyed if one tree catches a disease.

The black spots on these rose leaves are a sign of a disease. ▼

Quick Check

Cause and Effect What can cause sudden changes in an environment?

Critical Thinking Can living things return to an environment after a natural disaster?

FACT Wildfires can help some plants.

How do organisms respond to changes?

Environmental changes can affect living things. For example, each year some grasslands in Africa go through a dry season. When this happens, plants and animals get less water. Watering holes can dry up. Tall grasses can dry out. Living things respond to these changes in different ways.

When an environment changes, some living things must move. They must find a new habitat where they can live. Elephants, for example, migrate in search of water and grasses.

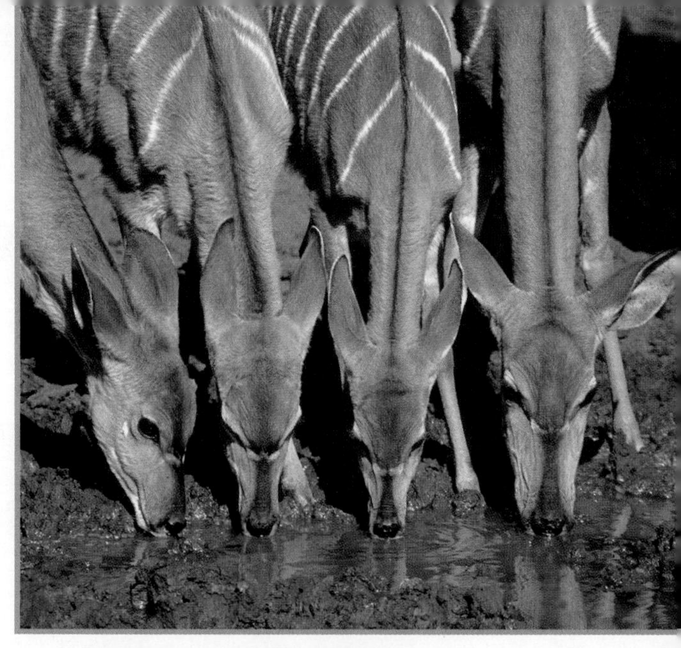

▲ Animals, such as these springbok, depend on watering holes.

Some living things adjust to survive. Predators might hunt different prey or eat whatever is available. Some animals might hunt at night if it makes it easier to find food.

Other living things have adaptations that help them survive changes. Some frogs and fish are adapted to burrow into mud when their environment becomes too dry. They go into a deep sleep and do not eat. They come out when the environment is wet again.

Living things that are not able to travel or change could die. Plants, such as savanna grasses, can not move to new places. Without rain, these grasses begin to dry up. If the drought lasts for too long, they might die.

▲ Many frogs are adapted to burrow underground when their environment becomes dry.

Quick Check

Cause and Effect What environmental changes might cause an animal to move to another place?

Critical Thinking What might happen to an elephant if its habitat suddenly becomes too cold?

◀ In a dry season, these African elephants migrate. They find a new habitat that can provide water, food, and shelter.

How do environmental changes affect an entire community?

Underneath the grasslands of the central United States lies an unseen world! Beneath the grasses is a complex system of tunnels. Prairie dogs build these tunnels and live in them. They come to the surface to eat grasses.

Prairie dogs are one population (pah•pyuh•LAY•shun) in a prairie. A **population** is all the members of one kind of organism in an ecosystem. All the coyotes are another population.

Populations in an ecosystem depend on one another. Burrowing owls and snakes make homes in the tunnels that prairie dogs leave behind. Eagles and coyotes feed on prairie dogs. If anything happens to prairie dogs, the whole community (kuh•MYEW•nuh•tee) is affected. A **community** is all the populations in an ecosystem.

A Prairie Community

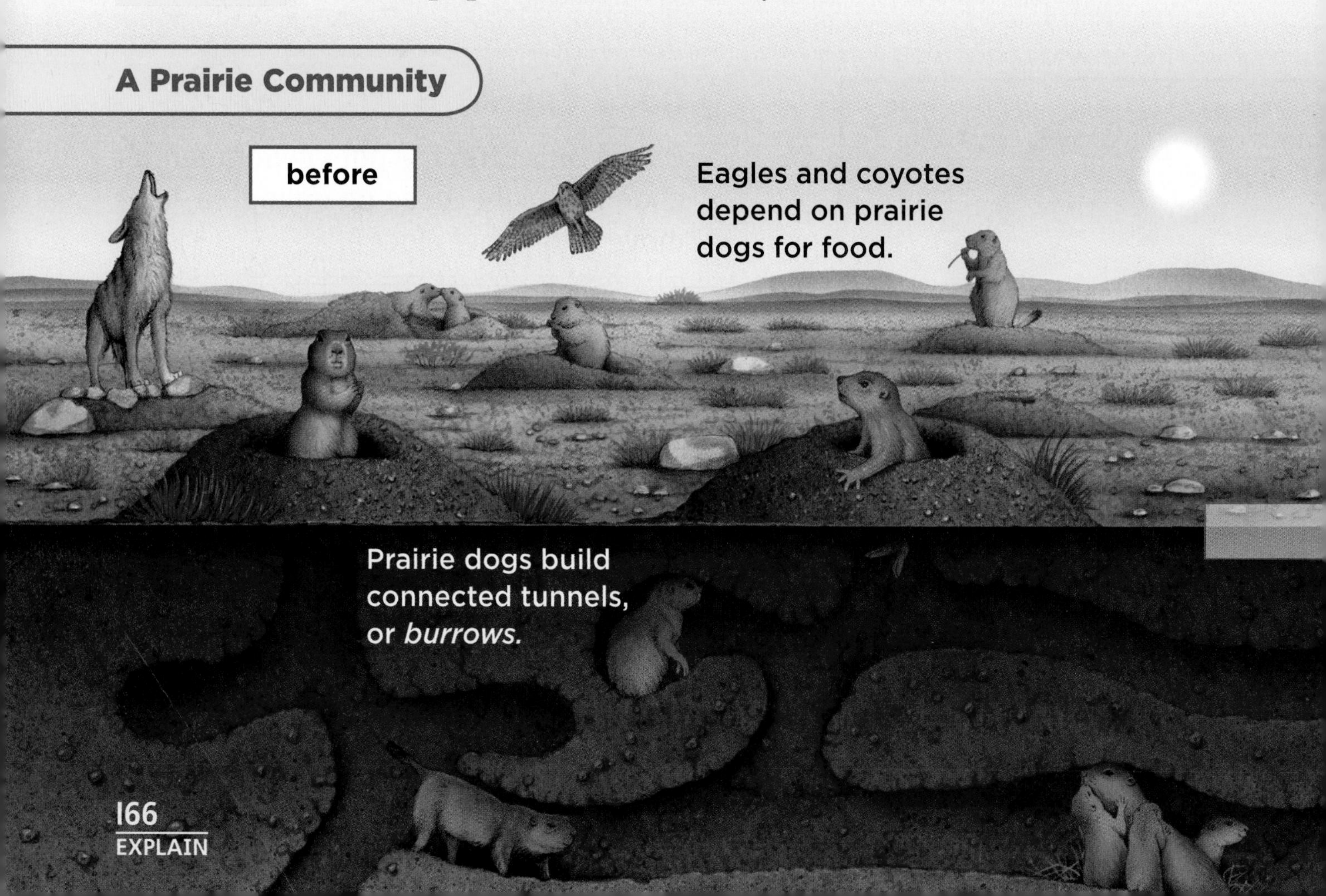

Sometimes, prairie dogs are harmed by disease. If the disease spreads, many prairie dogs could die. Animals that eat prairie dogs would lose a food source. Mice and snakes might not find homes.

The loss of prairie dogs could help some living things. Grasses might grow taller and thicker. Other animals might move into the area to feed on the new grasses.

Quick Check

Cause and Effect How could a disease that kills prairie dogs harm coyotes that live nearby?

Critical Thinking What might happen to prairie dogs if eagles left the ecosystem?

Quick Lab

A Changing Ecosystem

1. Make five character cards. Label the cards: *prairie dog, snake, burrowing owl, eagle,* and *coyote.*

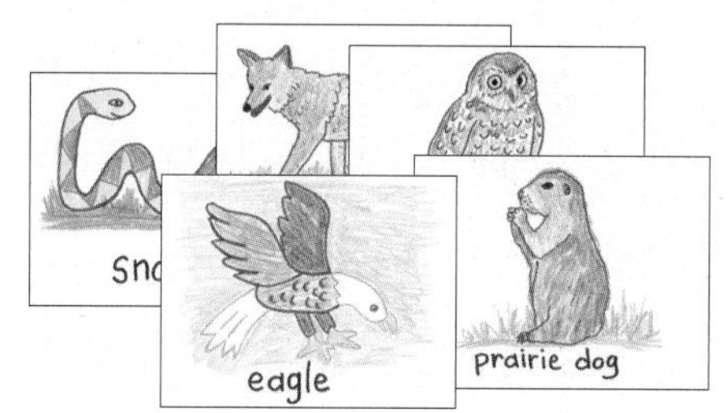

2. Paste the cards onto a large sheet of paper.
3. Draw an arrow from each animal to the organisms it depends upon for food or shelter.
4. **Infer** What would happen if prairie dogs disappeared?
5. **Infer** What would happen if eagles disappeared?

after

Wild horses might join the community to feed on grasses.

If disease kills prairie dogs, eagles and coyotes lose a source of food.

Snakes and mice move into abandoned prairie dog holes.

Read a Diagram

What happens when prairie dogs leave a prairie ecosystem?

Clue: A before-and-after diagram shows change.

How does a living thing become endangered?

When an environment changes, organisms must migrate or adjust to their new surroundings to survive. Those organisms that can not migrate or adjust might become endangered (en•DAYN•jurd). An organism is **endangered** when there are only a few living members of its kind left.

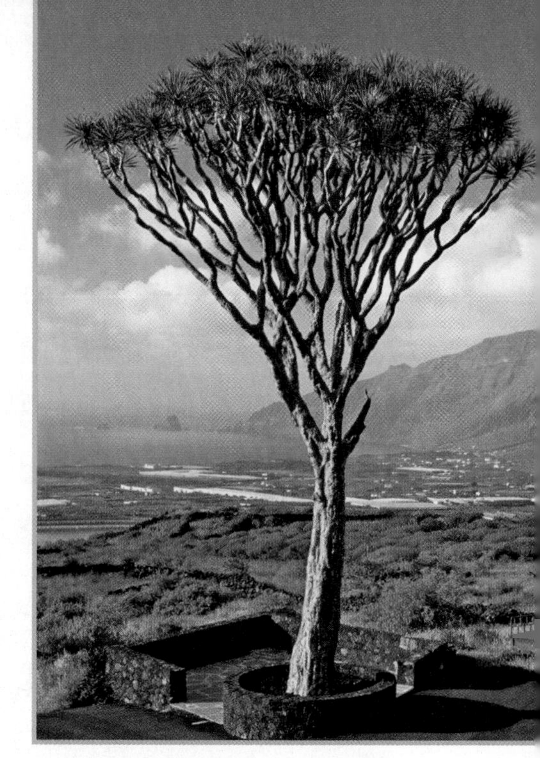

▲ Dragon trees might become endangered. Their environment is drying out.

Saharan cypress trees are endangered. They are found in the mountains of the Saharan Desert. Scientists think that Saharan cypress trees are adapted for a wet climate. Yet, their environment has become dry. Saharan cypress trees are endangered because they can not adjust to the hot, dry weather.

Organisms can also become endangered as a result of people. Bengal tigers are endangered because hunters have killed many of them for their fur. Panda bears are also endangered, partly because people are destroying their forest habitat.

Quick Check

Cause and Effect What can cause an organism to become endangered?

Critical Thinking How can people help protect endangered organisms?

Bengal tigers are hunted for their fur. ▼

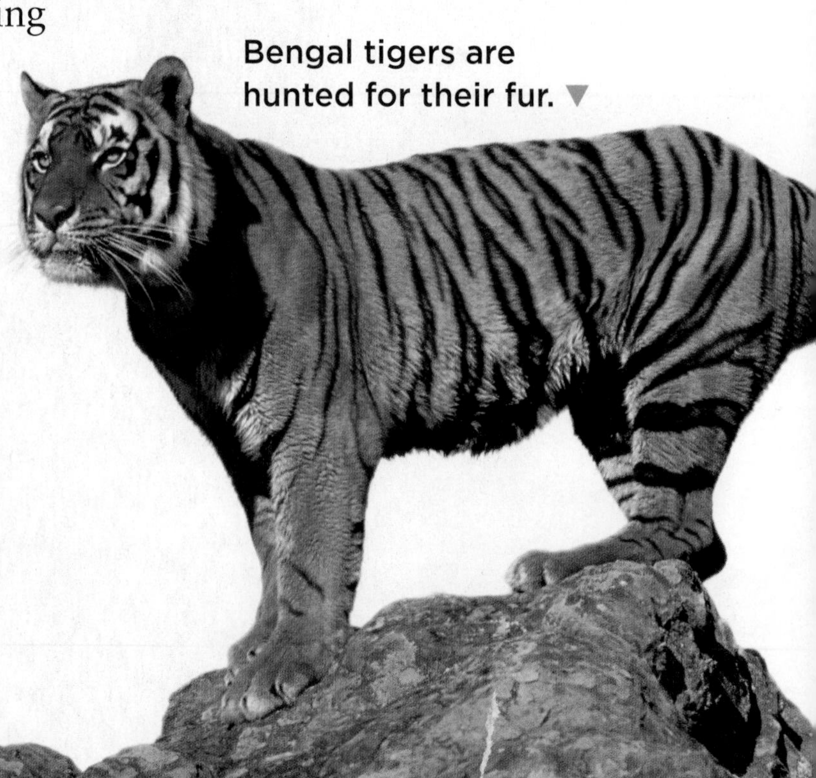

Lesson Review

Visual Summary

Natural disasters and disease are **ways environments can change.**

Living things are affected by change. Some might be harmed. Others might move or adjust.

When one population changes, it can affect other living things.

Make a FOLDABLES® Study Guide

Make a three-tab book. Use it to summarize what you learned about how changes affect living things.

Think, Talk, and Write

1. **Vocabulary** What does the word *endangered* mean?

2. **Cause and Effect** What are some of the effects of a drought?

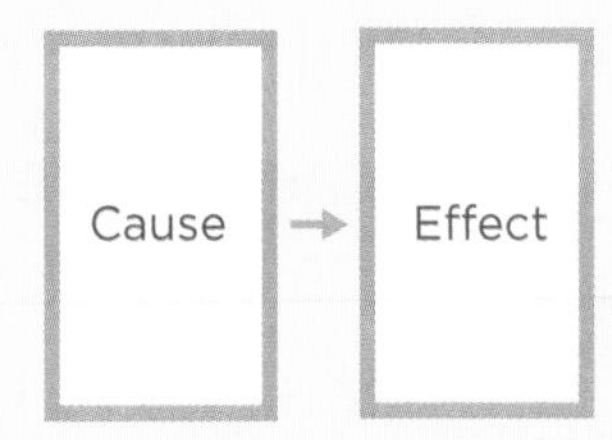

3. **Critical Thinking** Why should people take special care when they change a natural environment by building or farming? Explain.

4. **Test Prep** **All of these are natural disasters except**
   - **A** wildfires.
   - **B** floods.
   - **C** mold.
   - **D** droughts.

5. **Essential Question** How can changes in an environment affect living things?

Writing Link

Write an Essay
Learn about an environment that has recently changed. Then make a cause-and-effect chart. List what caused the environment to change and what happened as a result. Use your chart to write an essay.

Social Studies Link

Research Mount St. Helens
Use research materials to learn about Mount St. Helens. How did the eruption change the environment around Mount St. Helens? Make a poster or write a report.

LOG ON e-Review Summaries and quizzes online at www.macmillanmh.com

Save the Koala Bears

I believe it is very important to save the koala bears. We can not let them die. Eucalyptus forests are where koala bears live and find food. People cut down these forests. They build new houses. Today, the forests are getting smaller. It is harder for koalas to find food. The death of one kind of animal can hurt other plants and animals around it. That is why it is important to save the koala bear.

Persuasive Writing

Good persuasive writing

- clearly states an opinion;
- uses reasons to convince the reader to agree;
- organizes the reasons in a logical order;
- includes opinion words such as *I believe.*

Koala bears live in Australia. They eat leaves from eucalyptus trees. ▶

Write About It

Persuasive Writing Choose an endangered animal you care about. Research to find out why this animal is in trouble. Write a paragraph to convince readers that this animal should be saved. Be sure to end with a strong argument.

Write about it online at www.macmillanmh.com

SUBTRACTING LARGE NUMBERS

Whooping cranes are endangered. There are very few of them left in the wild. Like many endangered animals, whooping cranes are protected. That means that people can not hunt them or harm their habitat.

Look at the information in the chart. It shows how some living things grow in number when they are protected.

Name of Animal	Year of Original Count	Original Count	2005 Count
whooping crane	1941	16	341
snow leopard	1960	1,000	6,105
California condor	1986	17	200
giant panda	1965	1,000	1,817
humpback whale	1966	20,000	35,105

Subtract Multi-Digit Numbers

- First, subtract the ones. Regroup if necessary.

$$\begin{array}{r} {\scriptstyle 3\ 11} \\ 3\not{4}\not{1} \\ -\ 16 \\ \hline 5 \end{array}$$

- Then, subtract the tens. Regroup if necessary.

$$\begin{array}{r} {\scriptstyle 3\ \ } \\ 3\not{4}\not{1} \\ -\ 16 \\ \hline 25 \end{array}$$

- Continue until you have subtracted the numbers in all the places.

$$\begin{array}{r} 3\not{4}\not{1} \\ -\ 16 \\ \hline 325 \end{array}$$

whooping crane

Solve It

Use the chart above. Subtract the original count from the 2005 count for each kind of animal. This tells you how much each animal population grew.

Lesson 3

Living Things of the Past

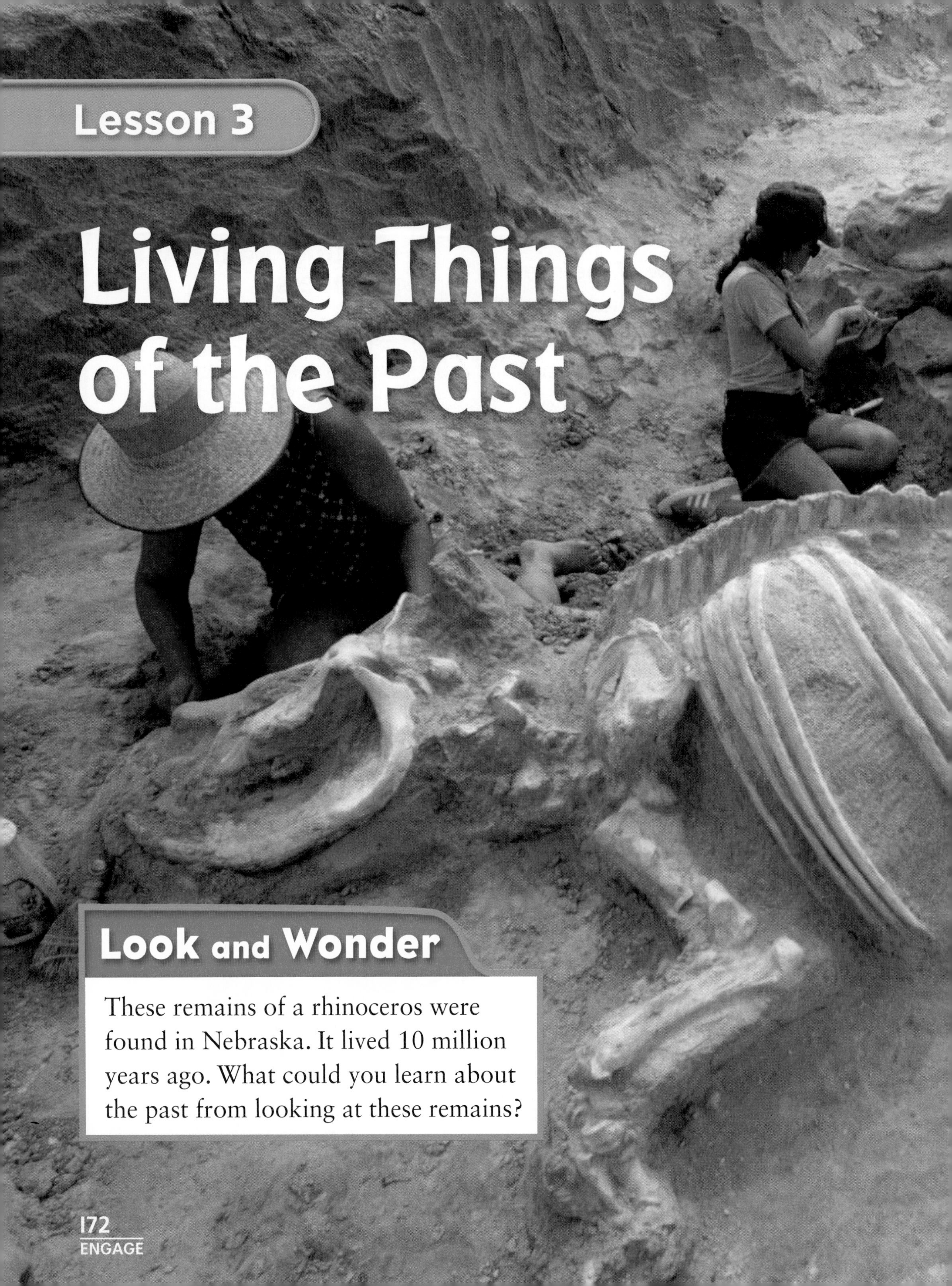

Look and Wonder

These remains of a rhinoceros were found in Nebraska. It lived 10 million years ago. What could you learn about the past from looking at these remains?

Explore

Inquiry Activity

How do fossils tell us about the past?

Purpose

Find out how fossils can teach us about the past.

Procedure

1. Mix a little glue and water in a measuring cup.
2. **Make a Model** Pour a thin layer of colored sand into a paper cup. Add a "fossil" object. Cover the object with sand of the same color. Add a little water and glue to "set" this layer. This models a fossil in rock.
3. Repeat step 2 with different objects and different colors of sand. Make three layers in all. Allow the layers to dry.
4. **Observe** Trade cups with another group. Carefully peel the paper cup away. Use the brush to find the fossils. Start at the top layer. Work your way down.
5. **Communicate** Record in a table the order in which each fossil object was found.

Draw Conclusions

6. **Interpret Data** Which fossil was buried first? Last? Which fossil is oldest?
7. **Infer** What can layers of rock tell us about Earth's past?

Explore More

What are some other ways you could model a fossil? Make a plan and try it.

Materials

measuring cup and water

glue

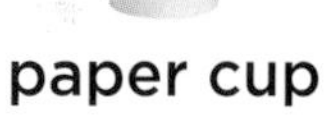

colored sand

paper cup

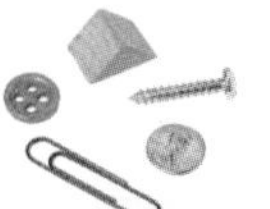

"fossil" objects

brush

Step 4

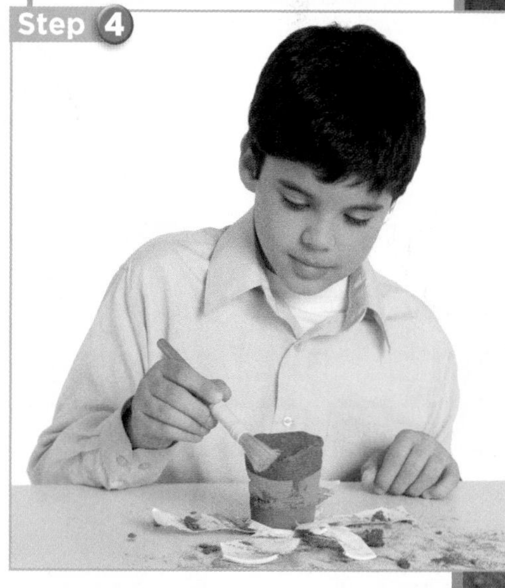

Step 5

Layer	Fossil
top	
middle	
bottom	

Read and Learn

Essential Question
What can we learn about living things of the past?

Vocabulary
fossil, p.174
extinct, p.174

Reading Skill 
Draw Conclusions

Text Clues	Conclusions

Technology
e-Glossary and e-Review online at www.macmillanmh.com

What can happen if the environment suddenly changes?

Did you know that dinosaurs once lived in North America? Millions of years ago, dinosaurs might have been roaming through the land that is now your town! All that is left of dinosaurs today are their fossils (FAH•sulz). **Fossils** are the remains of organisms that lived long ago.

Many scientists think that dinosaurs became extinct (ihk•STINGT) after a meteor hit Earth a long time ago. A living thing is **extinct** when no more of its kind are alive.

Dinosaurs, such as this triceratops, once roamed Earth. ▼

FACT Most dinosaurs were herbivores.

After the dinosaurs, large animals such as saber-toothed cats made their homes in North America. These animals lived more than ten thousand years ago during the Ice Age. Huge sheets of ice covered much of the land at that time. Then the climate changed. Temperatures began to rise, and the ice started to melt. The animals that the cats fed on could not survive. The cats lost their main food source. In time, the cats became extinct.

▲ Saber-toothed cats had large front teeth. They used the teeth to pierce the thick skin of animals that they ate.

Some plants and animals are becoming extinct even today. Some scientists think that up to 100 kinds of organisms become extinct each day! In 1996 a type of mammal called the red gazelle became extinct. It was hunted too often by humans. The St. Helena Olive tree used to grow on an island off the coast of Africa. It became extinct in 2004 because of disease and dry weather.

Quick Check

Draw Conclusions What are some reasons that living things become extinct?

Critical Thinking If high temperatures cause polar ice to melt, what might happen to polar bears?

The St. Helena Olive tree, shown here, became extinct in 2004.

How can we learn about things that lived long ago?

You can learn about plants and animals that lived long ago by studying fossils. Some fossils give clues about a living thing's size and shape. The large skeleton of a *Tyrannosaurus rex* tells us that the animal was about 5 meters (16.4 feet) tall.

Fossils can also tell us what an animal ate. Animals with pointed teeth were probably meat eaters. Animals with flat teeth were probably plant eaters.

Other fossils can show how an animal moved. A fossil with fins shows that the animal could move through water. A fossil with wings shows that the animal could fly.

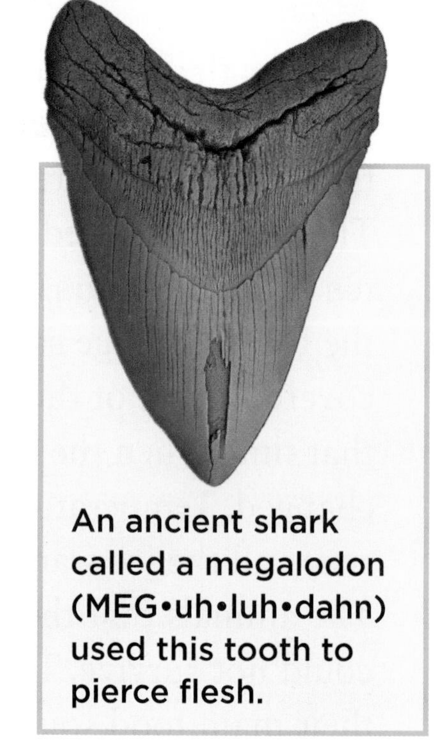

An ancient shark called a megalodon (MEG•uh•luh•dahn) used this tooth to pierce flesh.

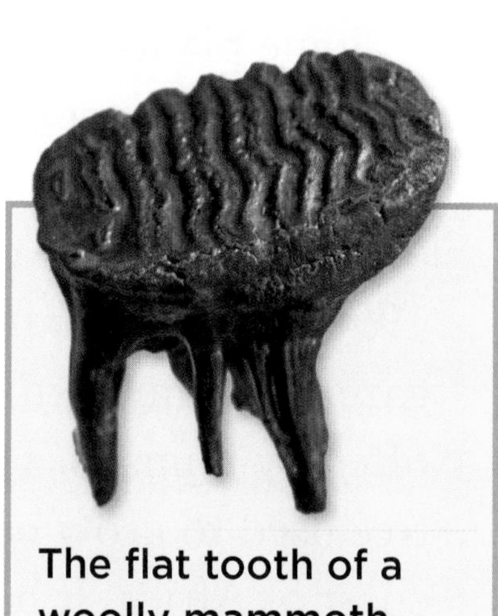

The flat tooth of a woolly mammoth was used to grind leaves and grasses.

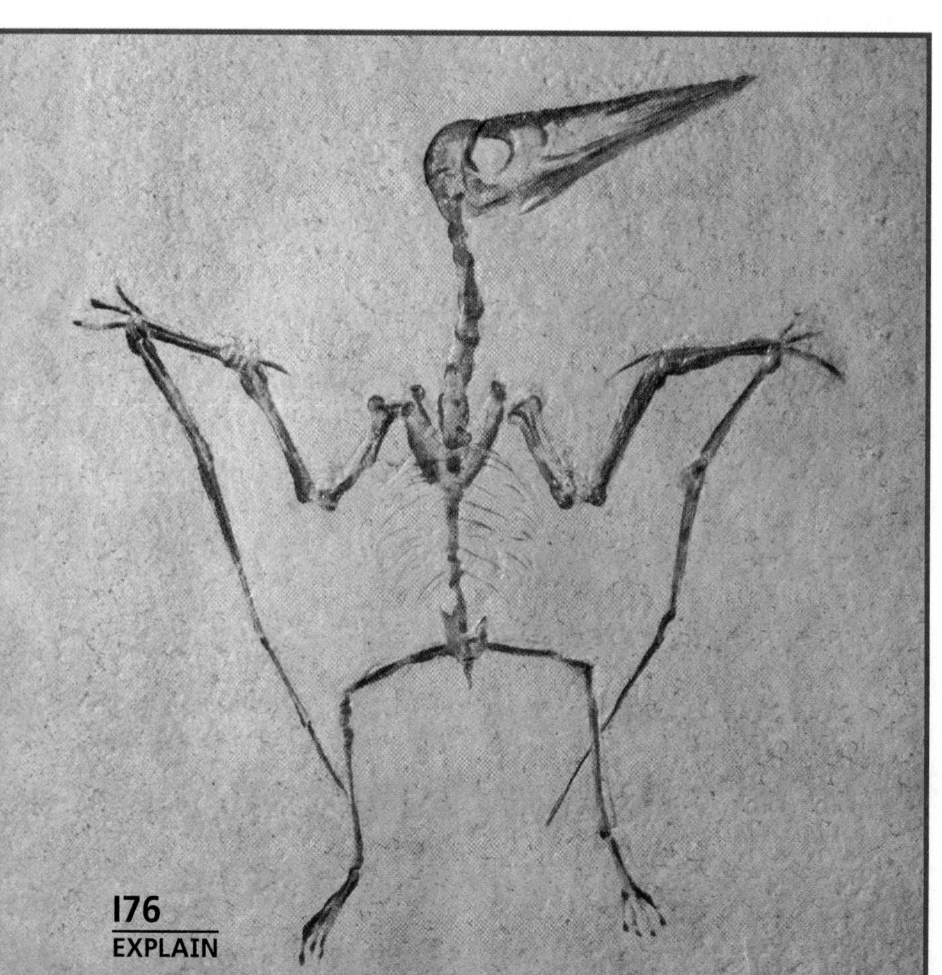

◀ This pterodactyl (ter•uh•DAK•tul) fossil shows that ancient reptiles might be related to modern birds.

Scientists also use fossils to find out how Earth and living things have changed over time. Many fish fossils are found on land. This means that millions of years ago, that land was covered with water. Over time, the land rose above the water. Fossils remained in the rock and soil that had been underwater.

How deep a fossil is buried gives clues about when an organism lived. Fossils found closest to the surface are usually youngest. Fossils found in deeper layers are older.

Quick Check

Draw Conclusions What does a fossil with fins tell you about that animal?

Critical Thinking Why do you think scientists study fossils?

Plant fossils tell us where plants grew in the past.

Quick Lab

Fossil Mystery

1. **Make a Model** Choose a favorite animal. Then use the key below to make fossil marks for your animal on some modeling clay.

If your animal is a . . .	then shape the clay into a . . .
mammal	circle
bird	square
amphibian	rectangle
reptile	triangle
fish	ball

2. Use the key below to make more fossil marks.

If your animal . . .	then mark your clay with . . .
lives in water	fins
lives on land	feet
lives both in water and on land	fins and feet
is a carnivore	pointed teeth
is an herbivore	flat teeth
is an omnivore	pointed and flat teeth

3. Trade your model fossil with the person sitting to your right.
4. **Infer** What can you learn about the animal that your classmate chose? How do scientists use fossils to learn about extinct animals?

How are living things of today similar to those that lived long ago?

The woolly mammoth became extinct thousands of years ago. Fossils tell us that it had a large trunk and tusks. Yet, fossils cannot tell us how this animal used its body parts. Instead, scientists learn how ancient animals used their body parts by studying similar animals living today.

Elephants are similar to woolly mammoths. They use their trunks to grasp and smell. As a result, scientists think that the woolly mammoth used its trunk in similar ways.

Many organisms living today look similar to those that lived long ago. Some modern birds resemble ancient reptiles. Eagles look very similar to flying reptiles called pterodactyls. Pterodactyls had long wingspans and large beaks. Scientists think pterodactyls used their beaks and claws to catch fish just like eagles do!

Connecting with the Past

Read a Diagram

What do the woolly mammoth and elephant have in common? What is different about them?

Clue: Compare the features of the animals in the pictures above.

Quick Check

Draw Conclusions Why do scientists study elephants when they want to learn more about woolly mammoths?

Critical Thinking How are organisms of the past similar to organisms living today?

Lesson Review

Visual Summary

Living things can become extinct when their environment suddenly changes.

Fossils tell us about living things and environments of the past.

You can learn about extinct organisms by studying organisms of today.

Make a FOLDABLES Study Guide

Make a layered-look book. Use it to summarize what you learned about living things of the past.

Think, Talk, and Write

1. **Vocabulary** What is a fossil?

2. **Draw Conclusions** What are some reasons an animal might become extinct?

Text Clues	Conclusions

3. **Critical Thinking** Why do people study fossils?

4. **Test Prep** **All fossils**
   - **A** are living things.
   - **B** are found only in cold places.
   - **C** are the remains of living things.
   - **D** were created millions of years ago.

5. **Essential Question** What can we learn about living things of the past?

Writing Link

Write a Report

Use research materials to learn about woolly mammoths. When did they live? What was their environment like? Write a report to share what you learn.

Math Link

Estimate

The bottom of a fossil bed has a fern plant that is 400 million years old. The top has a 300-million-year-old fern fossil. Between the two layers is a scorpion fossil. About how old do you think the scorpion fossil is?

LOG ON e-Review Summaries and quizzes online at www.macmillanmh.com

Looking at DINOSAURS

Dinosaurs were once common on Earth. Many dinosaurs became extinct millions of years ago. New evidence is helping scientists find out how dinosaurs lived and why they might have disappeared. Take a look at how ideas about dinosaurs have changed based on new evidence.

1842

Dinosaurs Are Named
In 1842 British scientist Richard Owen named the group of large, extinct reptiles "dinosauria." The name came from Greek words meaning "fearfully great lizard." People once thought dinosaurs' strange bones came from dragons or giants.

Dinosaur Nests Are Found
In 1923 American scientists Roy Chapman and Walter Granger found dinosaur nests in the Gobi Desert in China. The nests prove that dinosaurs laid eggs.

Dinosaurs Don't Drag Their Tails
In 1995 the American Museum of Natural History changed its *T. rex* skeleton. Instead of standing upright, the new skeleton is displayed with its head low and its tail off the ground. This change was based on studies of fossils, dinosaur tracks, and how different animals move.

Connect to

at www.macmillanmh.com

Fact and Opinion

- A fact is a true statement that you can prove.
- An opinion is how you feel about something.

Dinosaurs Have Feathers
In 2000 a team of Chinese and American scientists found a 130-million-year-old fossil dinosaur covered with feathers. Now most scientists agree that birds are living dinosaurs.

2000

Today
Scientists continue to find new fossils. They use new tools to discover more about dinosaurs.

Write About It

Fact and Opinion What animal do scientists think are living dinosaurs? Why do scientists think this? What animal do you think dinosaurs are like?

-Journal Write about it online at www.macmillanmh.com

CHAPTER 4 Review

Visual Summary

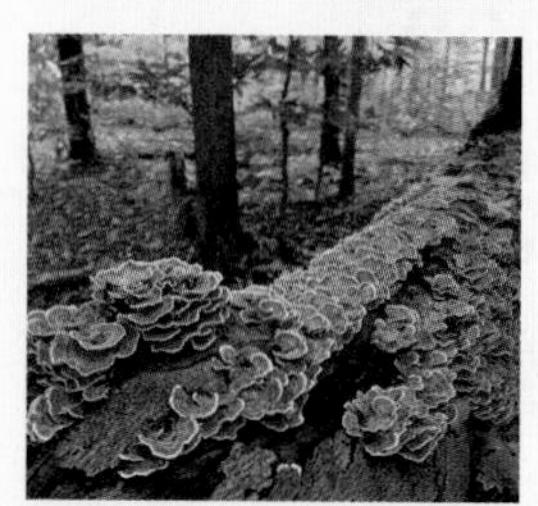

Lesson 1 Living things change their environments as they meet their needs. These changes can be helpful or harmful.

Lesson 2 Diseases and natural disasters can change environments. Living things respond to these changes in different ways.

Lesson 3 We can study fossils to learn about ancient plants and animals and their environments.

Make a FOLDABLES® Study Guide

Glue your lesson study guides to a piece of paper as shown. Use your study guide to review what you have learned in this chapter.

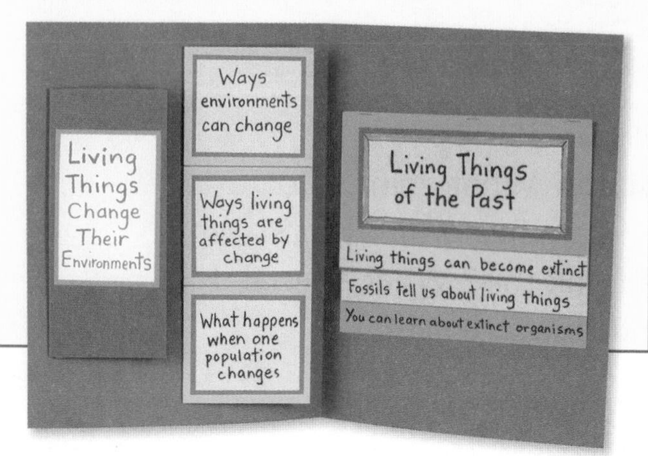

Vocabulary

DOK I

Fill each blank with the best term from the list.

drought, p. 162	**pollution,** p. 154
endangered, p. 168	**recycle,** p. 156
extinct, p. 174	**reduce,** p. 156
flood, p. 162	**resource,** p. 152
fossil, p. 174	**reuse,** p. 156

1. If you ________ your old aluminum cans, companies can make them into new cans.

2. When there is little or no rain for a long time, a ________ occurs.

3. Food is an example of a ________ that living things need to survive.

4. You could ________ a water bottle by filling it up with water again.

5. Heavy rains can cause a ________ if water covers normally dry land.

6. A kind of organism is ________ when there are no more left alive.

7. When harmful materials are put into an environment, it is called ________.

8. By taking shorter showers, you ________ how much water you use.

9. A kind of organism is ________ when there are only a few of those organisms left.

10. Scientists can use a ________ to learn more about a living thing from the past.

LOG ON e-Glossary Words and definitions online at www.macmillanmh.com

Skills and Concepts

DOK 2–3

Answer each of the following.

11. Predict How might cutting down trees in the rain forest affect people living there?

12. Persuasive Writing Write an advertisement that convinces people to visit a museum's new fossil exhibit. Include information about why fossils are important.

13. Use Numbers In 1963 there were only 417 pairs of bald eagles in the United States. In 2008 there were 9,789 pairs. How many more pairs were there in 2008?

14. Critical Thinking Cardinals live in forests. What might happen to a cardinal if a wildfire burns its forest habitat?

15. Critical Thinking How do plants in your neighborhood compete for resources?

16. Infer An ocean plant fossil is found in a local park. What does this most likely tell you?

17. What modern animal does this fossil resemble? What traits do both animals have in common?

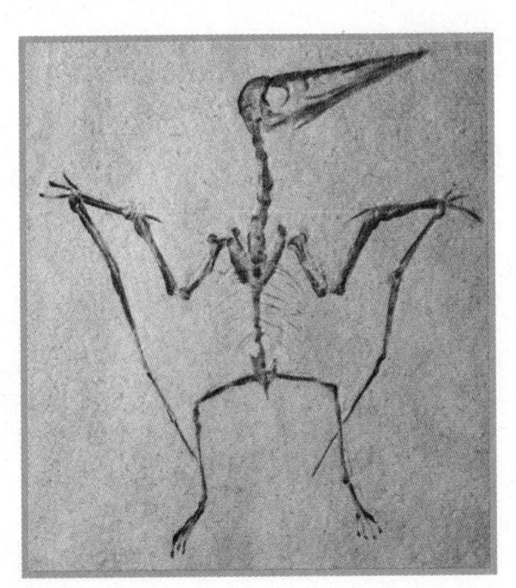

18. Predators are animals that

A hunt other animals for food.

B are hunted by other animals for food.

C are plant eaters.

D exist only as fossils.

19. True or False *New organisms introduced into an environment can be harmful to other organisms.* Is this statement true or false? Explain.

20. How can changes affect living things and their environments?

Performance Assessment

DOK 2

Conservation Cards

- Make three cards that show how people can protect the environment. Make one card for each of the 3 *Rs*.
- On the top of each card, write either *Reduce, Reuse,* or *Recycle.* Under each word, write a plan that helps people conserve resources. Then add a drawing to show how your plan works.
- On the back of the card, explain how your plan helps the environment.

Test Preparation

1 **Which could you compare to learn more about living things of the past?**

A wolf and tiger

B hummingbird and eagle

C elephant and woolly mammoth

D frog and lizard

DOK 1

2 **What would most likely happen if prairie dogs were removed from a grassland ecosystem?**

A Mice would have more habitats.

B Coyotes would have more food.

C Eagles would have more food.

D Grasses would become extinct.

DOK 2

3 **The fossil below was found at a local park.**

What can you most likely infer from this finding?

A The fossil was lost by someone in the past.

B The park was once under water in the past.

C Fossils are difficult to find.

D Fossils of fish are different from fish that live today.

DOK 2

4 **A type of rhinoceros lived in the United States about 5 million years ago.**

Which object can help scientists learn about this animal from the past?

A drawings

B fossils

C minerals

D rocks

DOK 1

5 **What causes droughts?**

A diseases

B wildfires

C no rain for a long time

D heavy rains and other storms

DOK 1

6 **How can a flood change an environment?**

A It can wash away plants and soil.

B It can cause plants to grow quickly.

C It can dry out rivers and lakes.

D It can cause a wildfire.

DOK 1

7 Look at the deer shown below.

Which living thing most likely competes with the deer for food?

A bats

B owls

C shrubs

D other deer

DOK 2

8 Which word describes fossils found in deeper layers of rock?

A larger

B older

C smaller

D younger

DOK 1

9 What would most likely cause an organism to become extinct?

A loss of habitat

B loss of predators

C increase in habitat size

D increase in resources

DOK 2

10 In the rain forest below, some trees have just been cut down.

How will the changed ecosystem compare with the original rain forest?

DOK 2

11 Sometimes people bring new plants or animals into an environment. Describe how a new organism in an environment might affect other organisms.

DOK 2

12 Describe how a drought affects plants and animals. Which organisms have a better chance of survival during a drought?

DOK 2

Check Your Understanding

Question	Review	Question	Review
1	p. 178	7	pp. 152–153
2	pp. 166–167	8	p. 177
3	pp. 176–177	9	pp. 152–155, 162–168, 174–175
4	pp. 176–177	10	pp. 152–154
5	pp. 162–163	11	p. 155
6	pp. 162–163	12	pp. 162–165

Wildlife Manager

Do you like to learn about plants and animals? Do you want to help keep environments clean and healthy? Then one day you might become a wildlife manager.

Wildlife managers help take care of animals and their environments. They keep track of the plants and animals in places such as wildlife parks. They look for ways to help living things. They also teach people about wildlife and why it is important to take care of environments.

To become a wildlife manager, you must care about environments and living things. Plan to study science in high school and college. You will also need a degree in a field such as biology or environmental science.

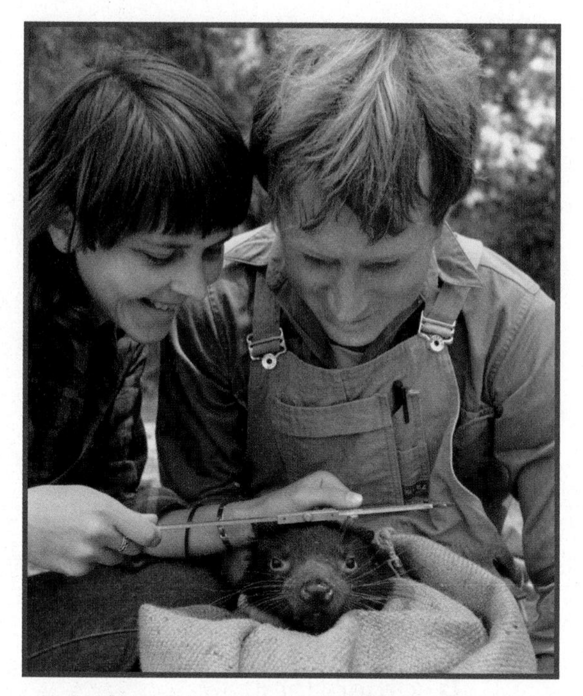

▲ Animal rescue workers measure a Tasmanian devil.

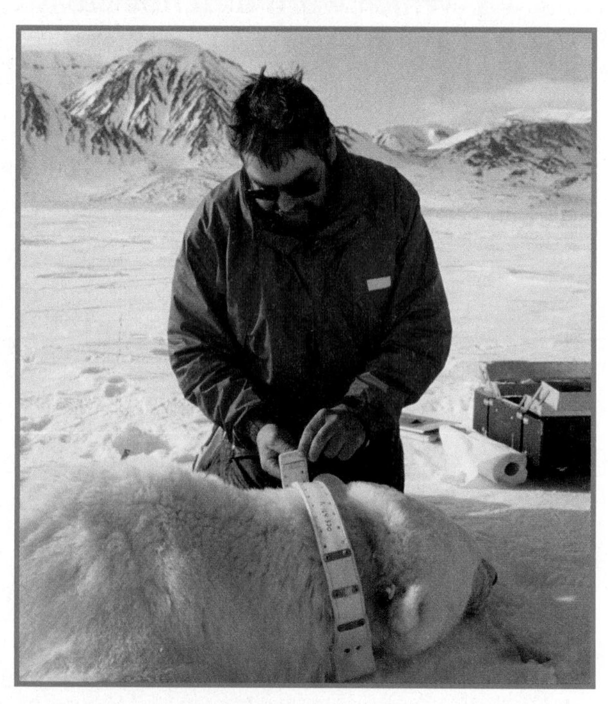

▲ A park ranger places a satellite collar on a polar bear to track its movements.

Here are some other life science careers:

- emergency medical technician
- animal rescue worker
- gardener
- park ranger

Earth and Its Resources

Rainwater wore away limestone underground and formed this *cenote*, or water-filled cavern.

underground cavern in Dzitnup, Mexico

Magazine Article

Liv and Ann each pulled a sled that weighed 267 pounds on their journey across Antarctica.

from *Time for Kids*

ONE COOL ADVENTURE

February 23, 2001

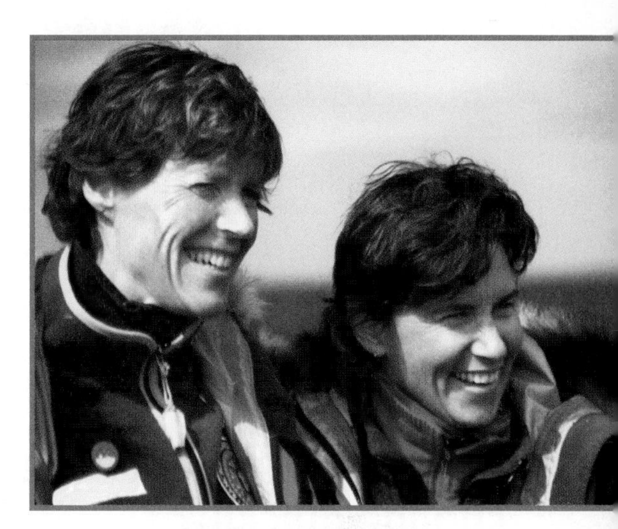

▲ Liv Arnesen and Ann Bancroft

Whew! They made it! On February 11, former teachers Ann Bancroft of the U.S. and Liv Arnesen of Norway reached the Ross Ice Shelf in Antarctica. They became the first women to cross Antarctica's land mass on skis! It took the explorers 90 days to ski and parasail across 1,688 miles of ice. They kept going despite bitter cold, injuries, ripped sails, and broken sleds.

The pair had hoped to cross the Ross Ice Shelf, but lack of wind forced them to shorten their trek. They flew back to McMurdo Station to catch a ship before icy waters made the trip home dangerous.

Write About It

Response to Literature This article tells about the first women to cross Antarctica on skis. What do you know about Antarctica or other places on Earth? Suppose you took a trip around the world. What kinds of things might you see? Write about it.

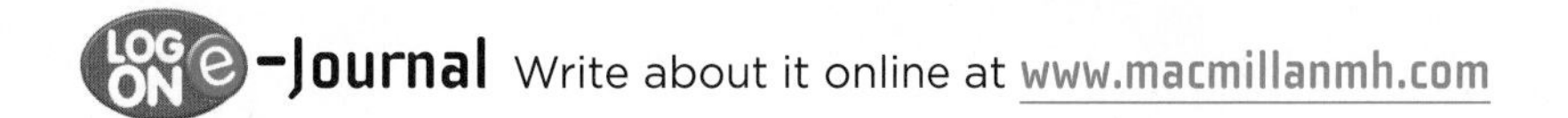

CHAPTER 5

Earth Changes

What can cause Earth's features to change?

Essential Questions

Lesson 1
What shapes can the land take?

Lesson 2
How can Earth's surface change quickly?

Lesson 3
How can Earth's surface change slowly?

Grand Canyon National Park, Arizona

Big Idea Vocabulary

ocean a large body of salt water (p. 192)

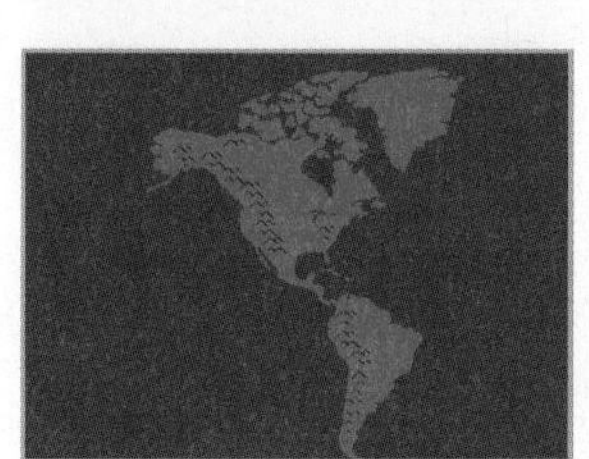

continent a great area of land on Earth (p. 193)

earthquake a sudden movement of the rocks that make up Earth's crust (p. 204)

volcano a mountain that builds up around an opening in Earth's crust (p. 206)

weathering the breaking down of rocks into smaller pieces (p. 214)

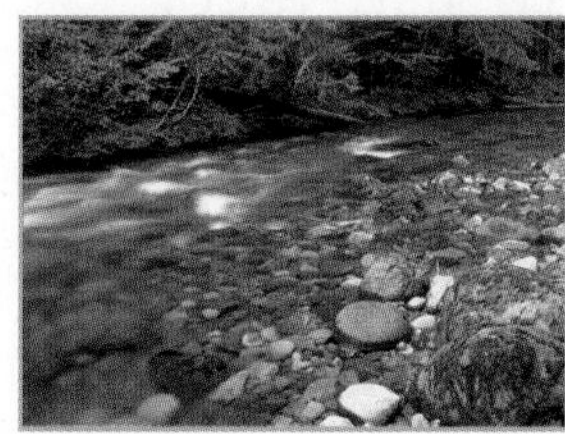

erosion the wearing away and movement of weathered rock (p. 216)

Visit www.macmillanmh.com for online resources.

Lesson 1

Earth's Features

Boardman State Park, along the Oregon coast

Look and Wonder

Both land and water cover Earth's surface. Which one covers more of Earth?

Explore

Inquiry Activity

Does land or water cover more of Earth's surface?

Make a Prediction

Do you think that there is more land or more water on Earth's surface? Write your prediction.

Test Your Prediction

1. Make a table like the one shown for ten spins.
2. **Experiment** Slowly spin a globe. Do not look at it. Touch your finger to the globe to stop it.
3. **Observe** Did your finger stop on land or water? Record the information on the chart.
4. Repeat steps 2 and 3 nine more times.
5. **Use Numbers** How many times did you touch water? How many times did you touch land?

Draw Conclusions

6. **Infer** Is there more land or more water on Earth? How do your results compare with the results of others?

Explore More

Experiment Which covers more of Earth—rivers or oceans? Make a plan to find out.

Materials

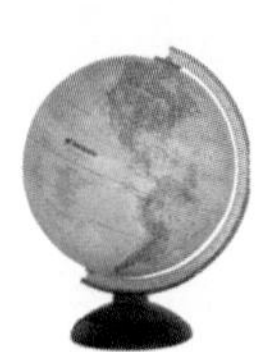

globe

Step 1

Spin	Land	Water
1		
2		
3		
4		

Step 2

Read and Learn

Essential Question
What shapes can the land take?

Vocabulary
ocean, p.192
continent, p.193
landform, p.194
crust, p.198
mantle, p.198
core, p. 198

Reading Skill
Main Idea and Details

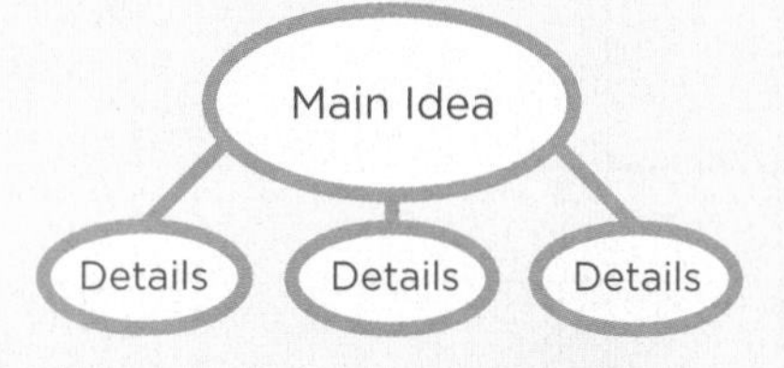

Technology
e-Glossary and e-Review online at www.macmillanmh.com

What covers Earth's surface?

If you could see Earth from space, it would look mostly blue. That is because almost three fourths of Earth is covered by water. Most of this water is in oceans (OH•shunz). **Oceans** are large bodies of salt water.

Rivers, streams, glaciers, and ponds are some other water features on Earth. These water features are made up of freshwater. *Freshwater* is water that is not salty. Lakes are another water feature. Most lakes have freshwater. Some have salt water.

Oceans and Continents

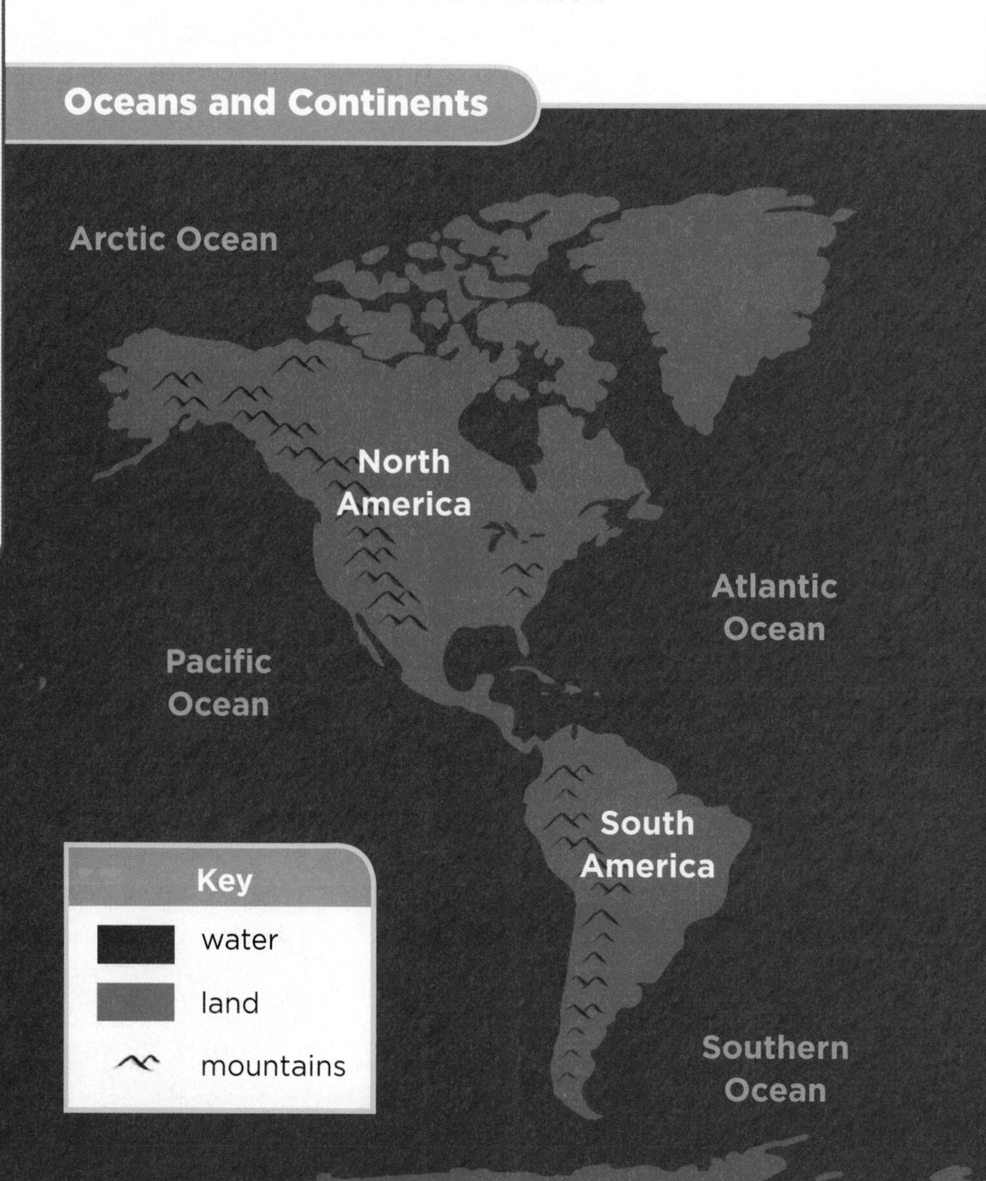

Earth also has seven great areas of land called **continents** (KAHN•tuh•nunts). North America is the continent you live on.

A map can show Earth's land and water features. To read a map, look at its key. A *key* shows what a map's colors and shapes mean. Can you find North America on the map below?

Quick Check

Main Idea and Details What covers Earth's surface?

Critical Thinking About how much of Earth is covered by land?

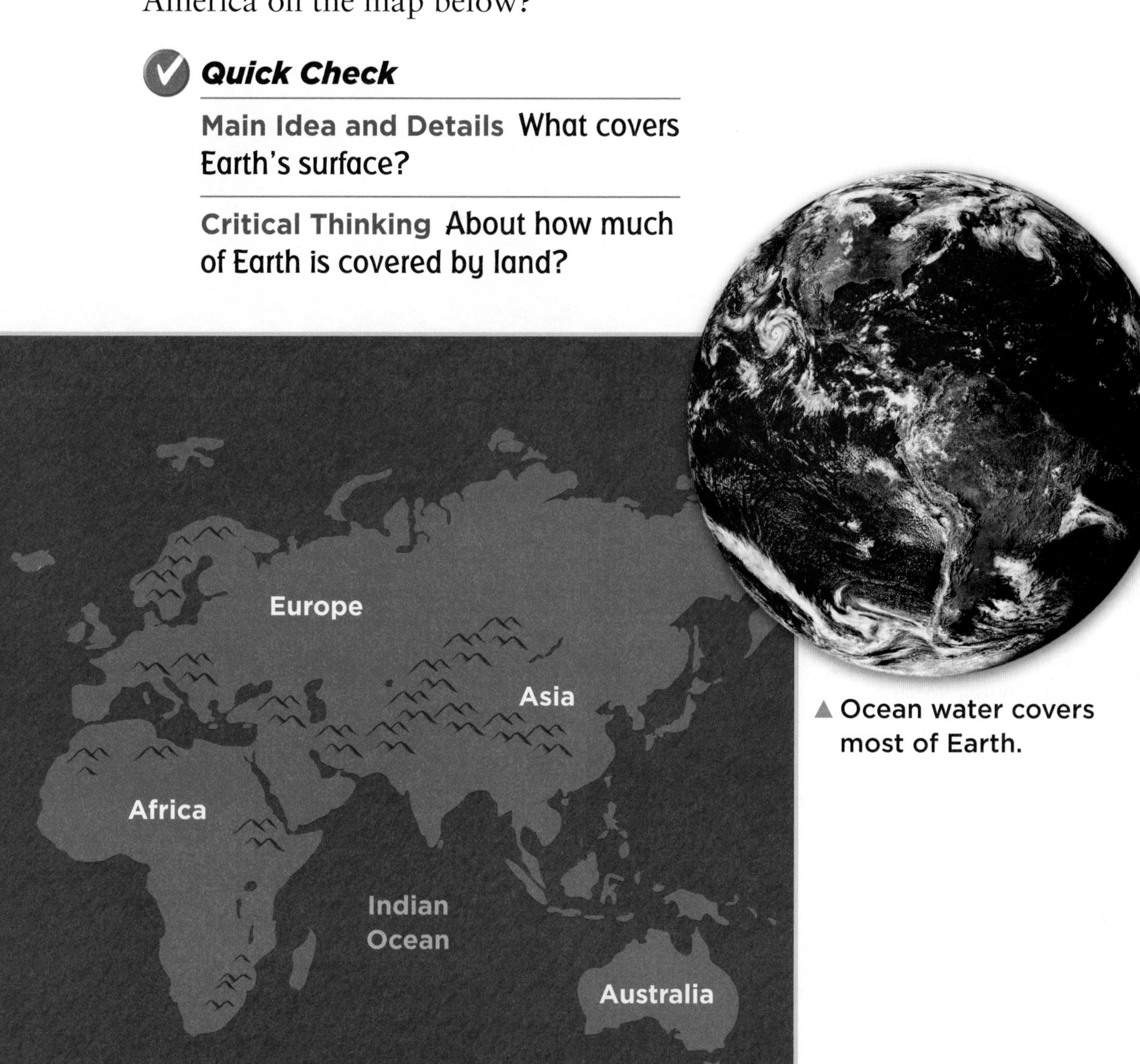

▲ Ocean water covers most of Earth.

FACT The oceans are really one big body of water.

What are some of Earth's land and water features?

There are many land and water features on Earth. Land features are called **landforms** (LAND•formz). This diagram shows a few of Earth's features.

Features of Earth

1. A *mountain* is the tallest landform. It often has steep sides and a pointed top.
2. A *valley* is the low land between hills or mountains.
3. A *canyon* is a deep valley with steep sides. Rivers often flow through them.
4. A *plain* is land that is wide and flat.
5. A *lake* is water that is surrounded by land.
6. A *river* is a large body of moving water.
7. A *plateau* (pla•TOH) is land with steep sides and a flat top. It is higher than the land around it.
8. A *coast* is land that borders the ocean.
9. A *peninsula* is land surrounded by water on three sides.
10. An *island* is land with water all around it.

Quick Check

Main Idea and Details What are landforms?

Critical Thinking How could you tell a mountain from a plain?

Quick Lab

Your State's Features

1. **Make a Model** Draw a map of your state. Decide how to show your state's land and water features. Then make a key and complete the map.
2. **Observe** Where is your town or city located? Draw a large dot there. Which landforms and water features are found in your town or city? How do these features compare with those found in other parts of your state?

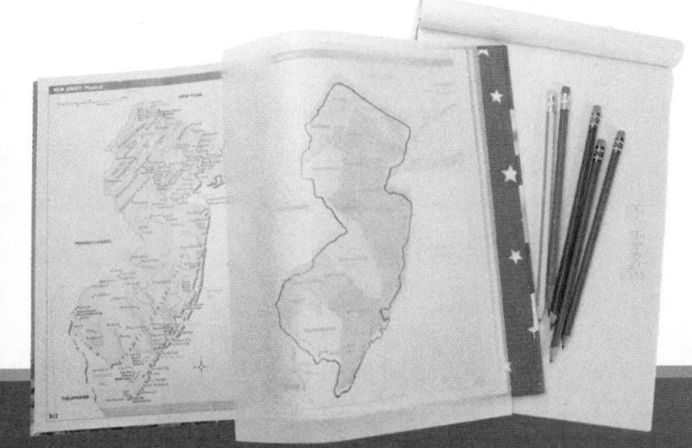

What land features are in the oceans?

Did you know that there is land below the ocean? The land below the ocean is called the *ocean floor*. The ocean floor has many features similar to land features. If you could travel there, you would find mountains, valleys, and canyons. You would even see plains.

The ocean floor begins at a coast where dry land borders the water. Here you find a continental shelf. A *continental shelf* is like a huge plateau. It lies under the ocean at the edge of a continent. About 80 kilometers (50 miles) away from the coast, the continental shelf slopes down steeply.

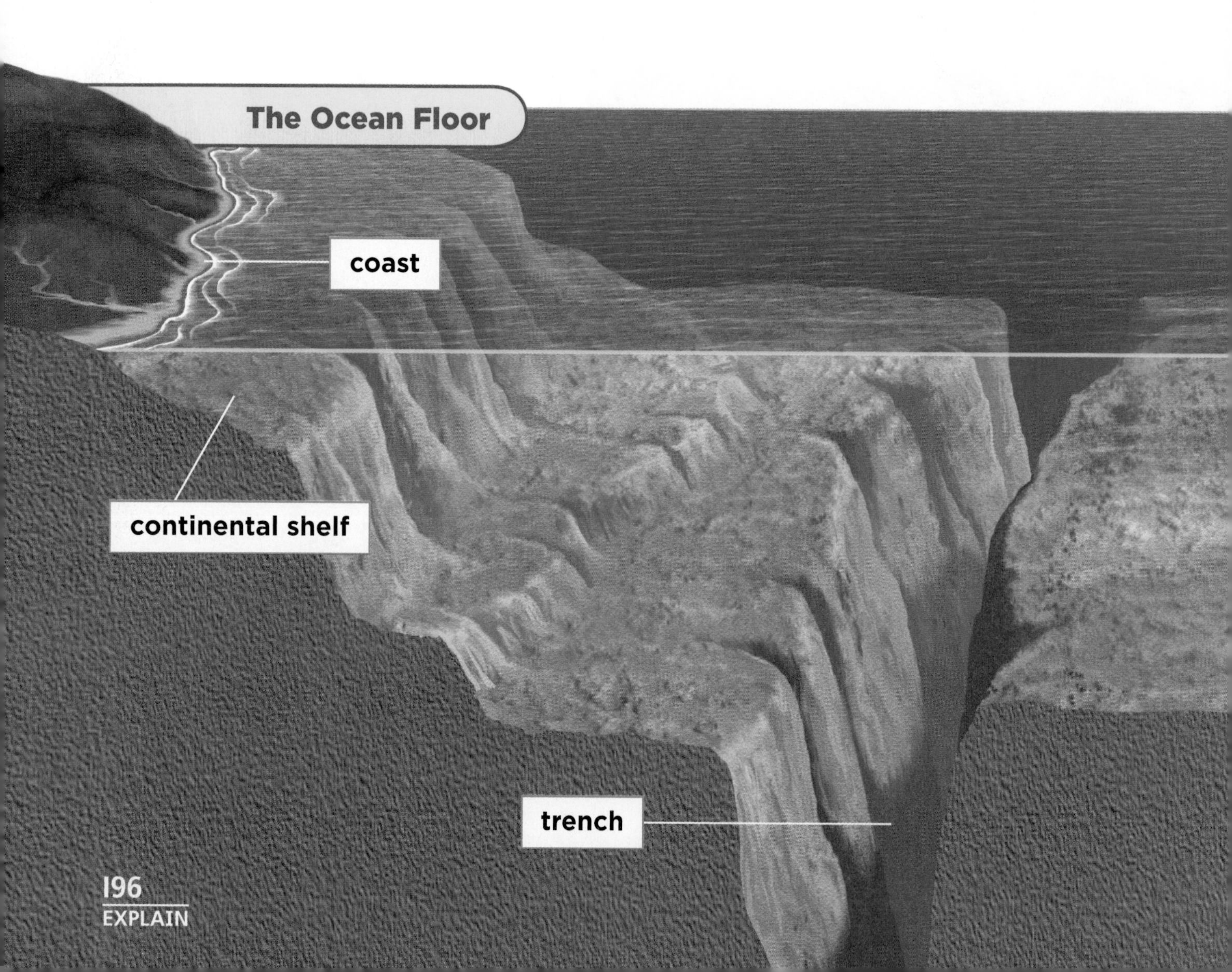

An abyssal plain (uh•BIH•sul PLAYN) begins a little farther out. An *abyssal plain* is wide and flat. It stretches thousands of kilometers across the ocean.

A trench is another feature you might recognize. A *trench* is a canyon on the ocean floor. Trenches are the deepest parts of the ocean floor. The deepest trench is the Mariana Trench in the Pacific Ocean. It is almost 11 kilometers (7 miles) deep.

Quick Check

Main Idea and Details How is the ocean floor like the land of the continents?

Critical Thinking What do you think you would find on the abyssal plain? Hint: Think about what covers a river's bottom.

Read a Diagram

Which feature of the ocean floor is a type of mountain?

Clue: Think about a mountain's shape.

What are the layers of Earth?

Have you ever eaten a hard-boiled egg? If so, you know that an egg has several layers. It has a thin shell, a white part, and a yolk.

Like an egg, Earth has several layers. The continents and ocean floor make up Earth's outermost layer, called the **crust**. The crust is Earth's thinnest and coolest layer.

The layer below the crust is the **mantle**. Part of the mantle is solid rock. Part is nearly melted rock that is soft and flows. It is a lot like putty.

At the center of Earth is the core. The **core** is the deepest and hottest layer of Earth. The *outer core* is melted rock. The *inner core* is solid rock.

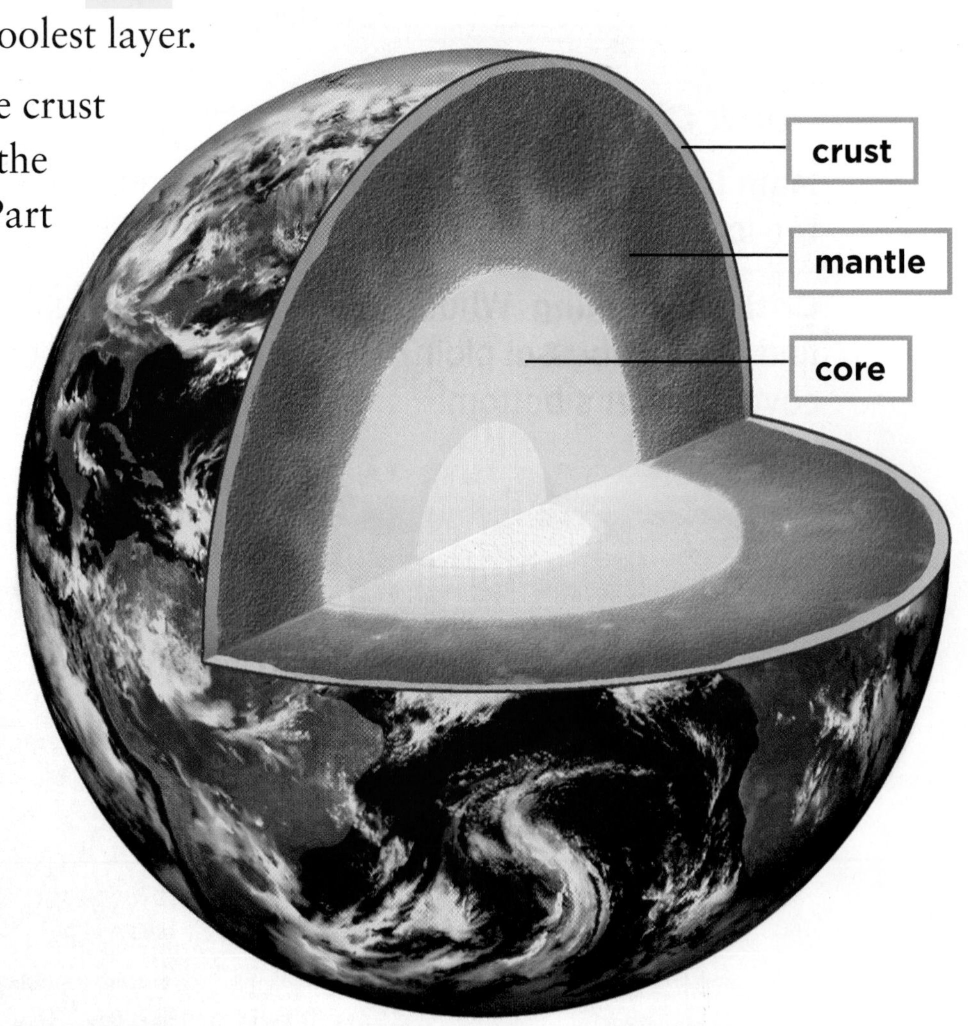

Quick Check

Main Idea and Details What is Earth's deepest layer called?

Critical Thinking Which of Earth's layers is like the shell of an egg? Why?

Lesson Review

Visual Summary

Earth has many **land features** and **water features**. Most of Earth is covered by water.

The **ocean floor** has features similar to Earth's land features.

Earth has three main **layers**—the crust, the mantle, and the core.

Make a FOLDABLES Study Guide

Make a layered-look book. Use it to summarize what you learned about Earth's features.

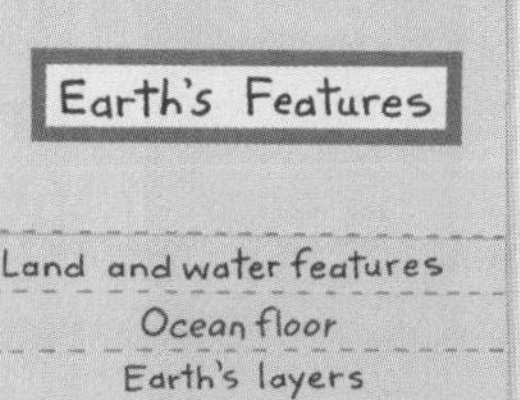

Think, Talk, and Write

1. **Vocabulary** Which landform is a deep, narrow valley with steep sides and a river flowing through it?

2. **Main Idea and Details** What are the layers of Earth?

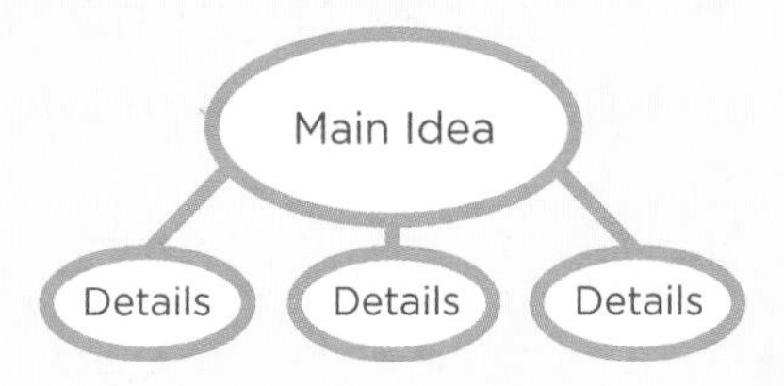

3. **Critical Thinking** Where would you be if you were at the deepest place on Earth's crust?

4. **Test Prep** **All of the following are landforms except**
   - **A** an island.
   - **B** a canyon.
   - **C** a plain.
   - **D** a river.

5. **Essential Question** What shapes can the land take?

Writing Link

Compare Numbers

Here are the lengths of some coastlines in miles. Write the states in order from shortest coastline to longest coastline.

Oregon: 296
Georgia: 100
New Jersey: 130
South Carolina: 187
New York: 127
Maryland: 31

Math Link

Do Research

Some people use stories, called *myths*, to explain how mountains formed. Research a myth that tells how mountains formed. Write a report about the myth.

Focus on Skills

Inquiry Skill: Make a Model

You just learned about many landforms. Some of them are found on land. Some lie under the ocean. In some places a limestone cave forms below the ground. It forms when water seeps into the ground and changes rock. This can take millions of years. You can **make a model** to show a cave.

▶ Learn It

When you **make a model**, you build something to represent, or stand for, a real object or event. A model can be bigger or smaller than the real thing. Models help you learn about objects or events that are hard to observe directly. Maps and globes are two examples of models.

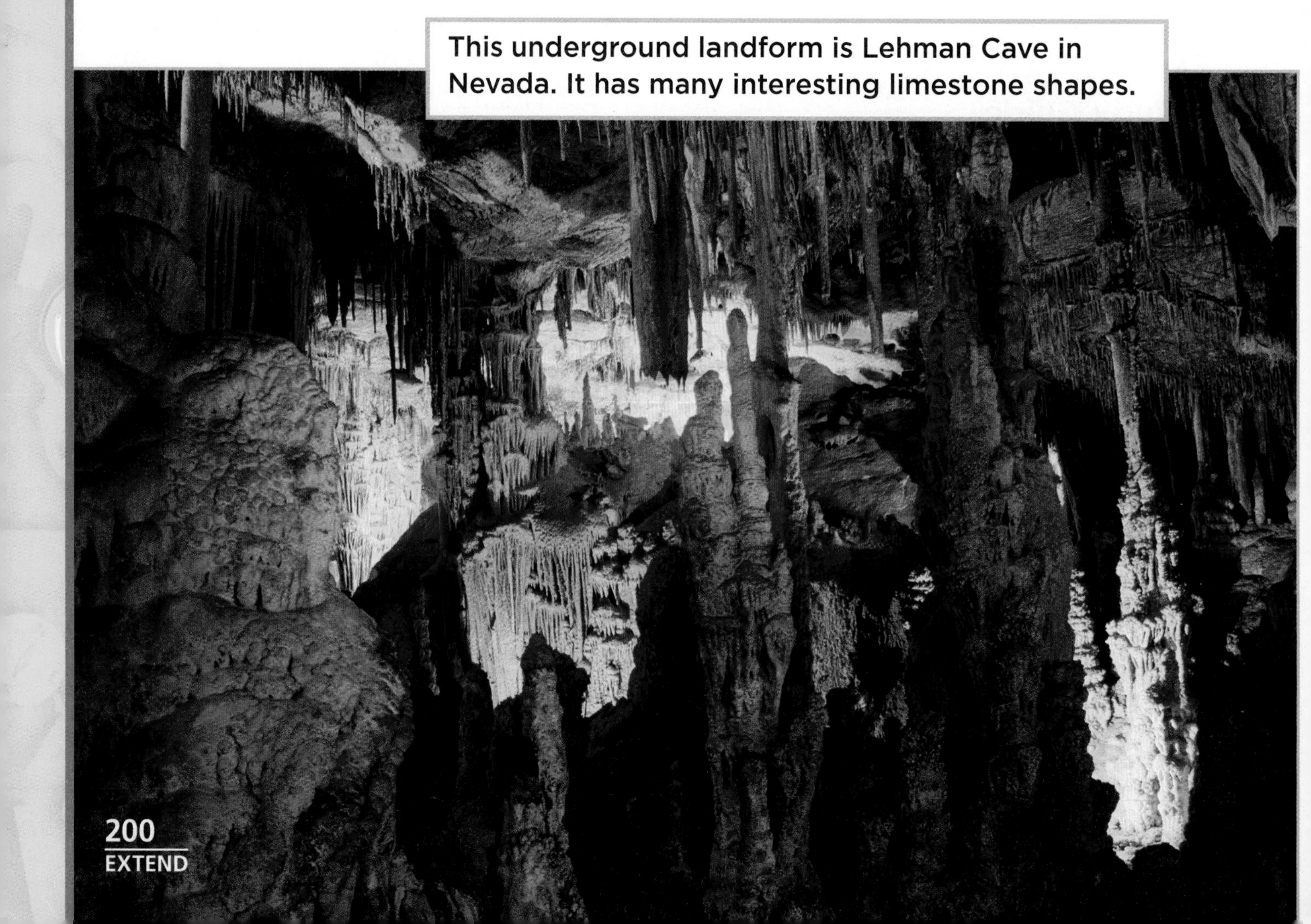

This underground landform is Lehman Cave in Nevada. It has many interesting limestone shapes.

▶ Try It

In this activity, you will **make a model** of a cave.

Materials **ruler, scissors, tan or white construction paper, crayon, shoe box or other small box, clear tape**

Step 4

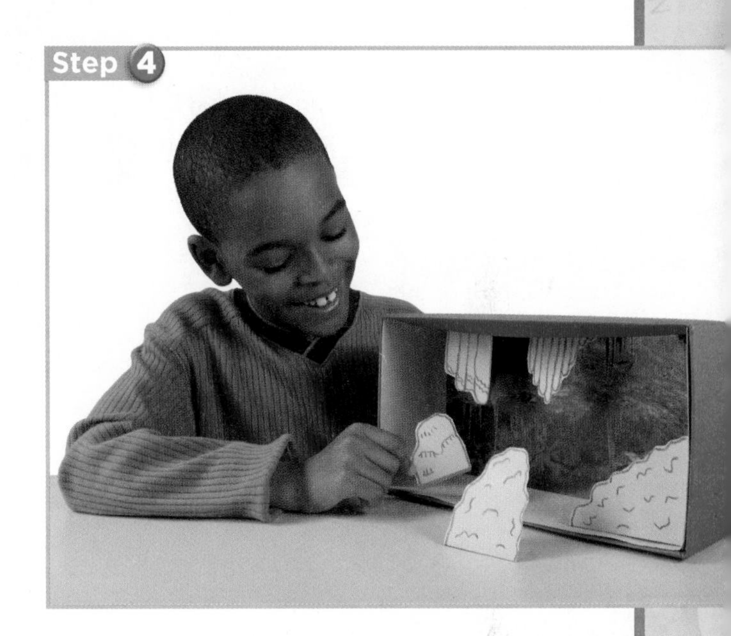

1. △ **Be Careful.** Cut a piece of construction paper that is a little smaller than the size of the back wall of the box.
2. On the paper draw limestone rocks like the ones shown. Tape the paper to the box's back wall.
3. Draw more limestone rocks on another piece of construction paper. Draw a flap for each rock.
4. △ **Be Careful.** Cut out each rock and its flap. Bend the flap for each rock. Tape each rock inside the box. Use the photo of the model to help you.

Now use your model to answer these questions:

- ▶ How would you describe the shapes of rocks in a limestone cave?
- ▶ Where do the rocks form?

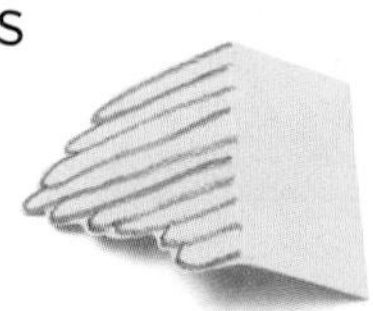

▶ Apply It

Make a model of a landform that you learned about. It can be a landform on the ocean floor or one on land. What details do you want to show? Which materials will you use to help you model these details?

Lesson 2

Sudden Changes to Earth

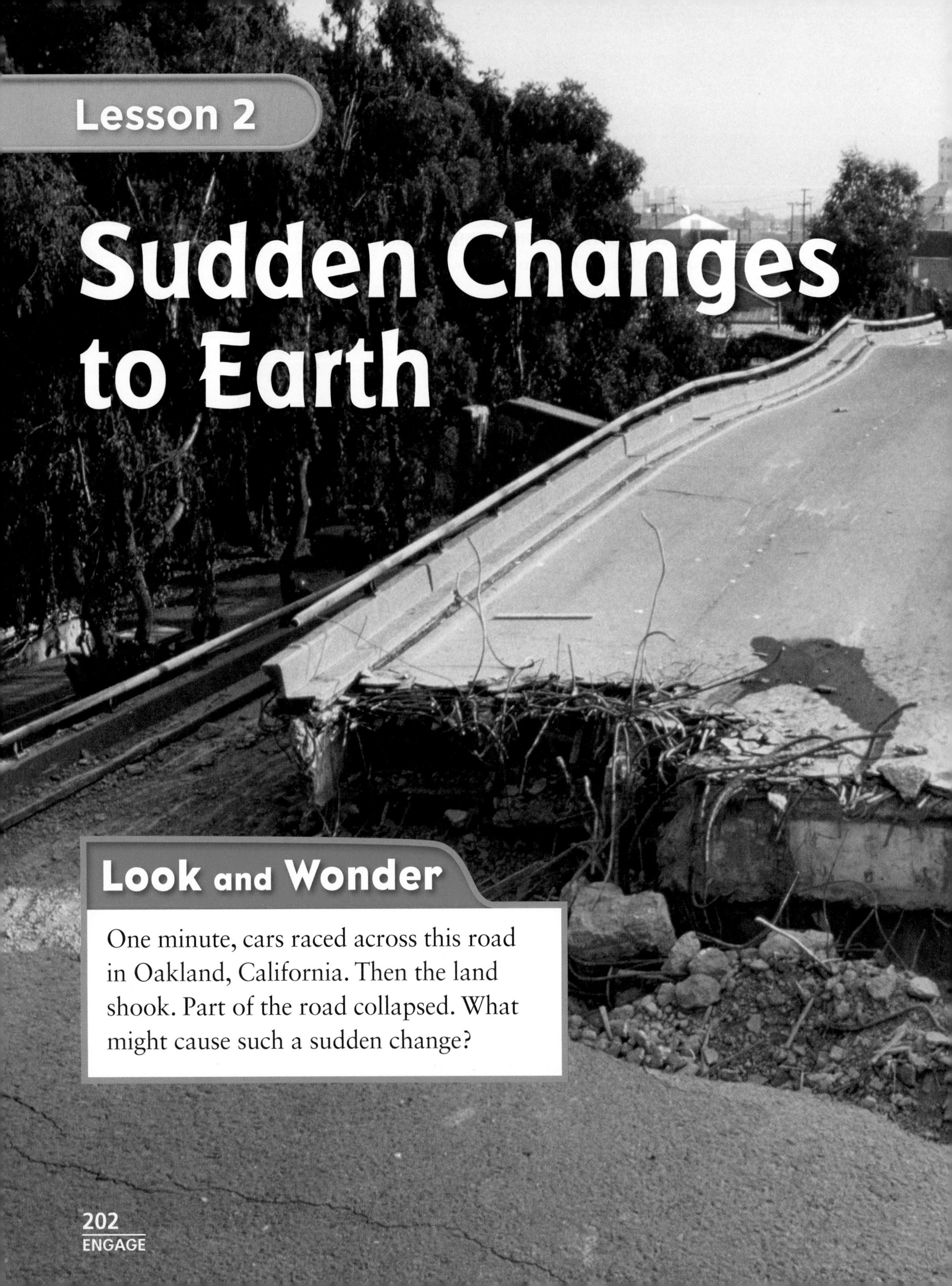

Look and Wonder

One minute, cars raced across this road in Oakland, California. Then the land shook. Part of the road collapsed. What might cause such a sudden change?

Explore

Inquiry Activity

How does sudden movement change the land?

Purpose

Model what happens when the land suddenly moves.

Procedure

1. **Make a Model** Fill a pan halfway with sand. Make a mountain with the sand.
2. Place blocks in the sand to model buildings. Add twigs to model trees.
3. **Communicate** Draw the land's surface.
4. **Experiment** What will happen if you tap the pan gently? Try it.
5. **Experiment** What will happen if you tap the pan harder? Try it.

Draw Conclusions

6. **Infer** How can the sudden movement of land change the land?

Explore More

Experiment Different rocks and soils make up land. Does sudden movement change all land the same way? Make a plan to find out. Try it.

Materials

aluminum pan

sand

assorted blocks

twigs

Step 2

An earthquake broke apart Earth's crust.

Read and Learn

Essential Question
How can Earth's surface change quickly?

Vocabulary
earthquake, p. 204
volcano, p. 206
magma, p. 206
lava, p. 206
landslide, p. 208
flood, p. 208

Reading Skill
Cause and Effect

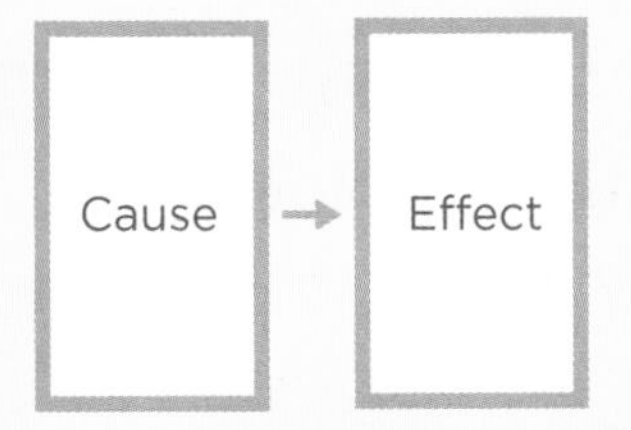

Technology
e-Glossary, e-Review, and animations online at www.macmillanmh.com

What are earthquakes?

Some events can change Earth's surface in less than a minute. One example is an earthquake (URTH•kwayk). What causes an earthquake? Why does it change the land? The answers are found under the ground.

Earth's Moving Crust

Earth's outside layer, the crust, is made up of huge slabs of rock. You might think that slabs of rock can not move. They do move, however. Rocks deep below the ground can slowly slide past each other. They can press against each other. They can pull apart too. These movements can cause rock to bend and snap back like a bent stick. This causes an earthquake. An **earthquake** is a sudden movement of the rocks that make up Earth's crust.

When an earthquake happens, the ground shakes, or vibrates. The vibrations travel out from the earthquake's center through the land. Some earthquakes are very weak. They are not even noticed. Some feel like a truck rumbling by. Others are very strong. Earthquakes can crack roads. They can cause buildings and bridges to fall. They can even cause parts of mountains to collapse.

Quick Check

Cause and Effect What can happen when huge slabs of rock in Earth's crust move?

Critical Thinking You drop a pebble in water. What happens to the water? How is this similar to what happens to the crust during an earthquake?

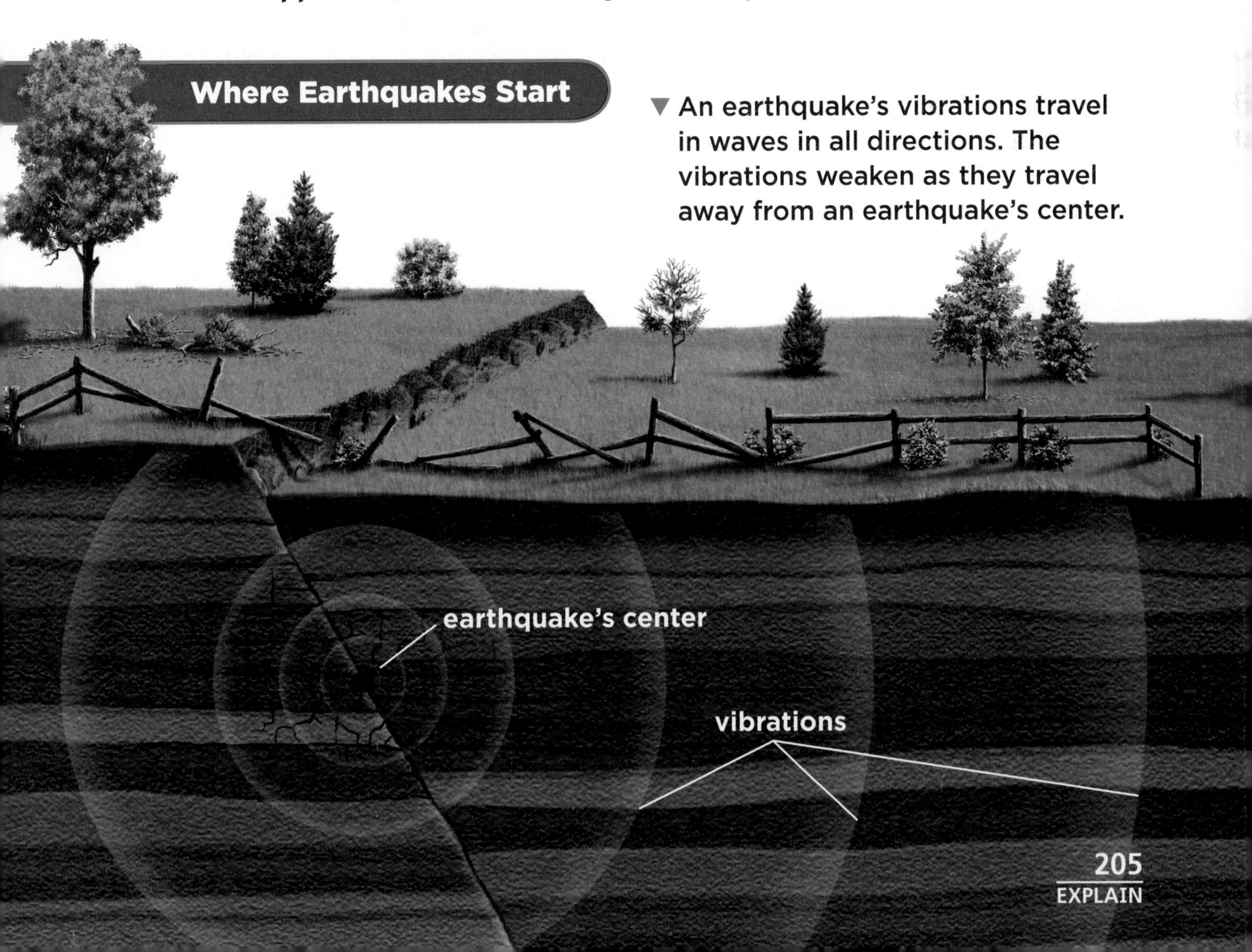

▼ An earthquake's vibrations travel in waves in all directions. The vibrations weaken as they travel away from an earthquake's center.

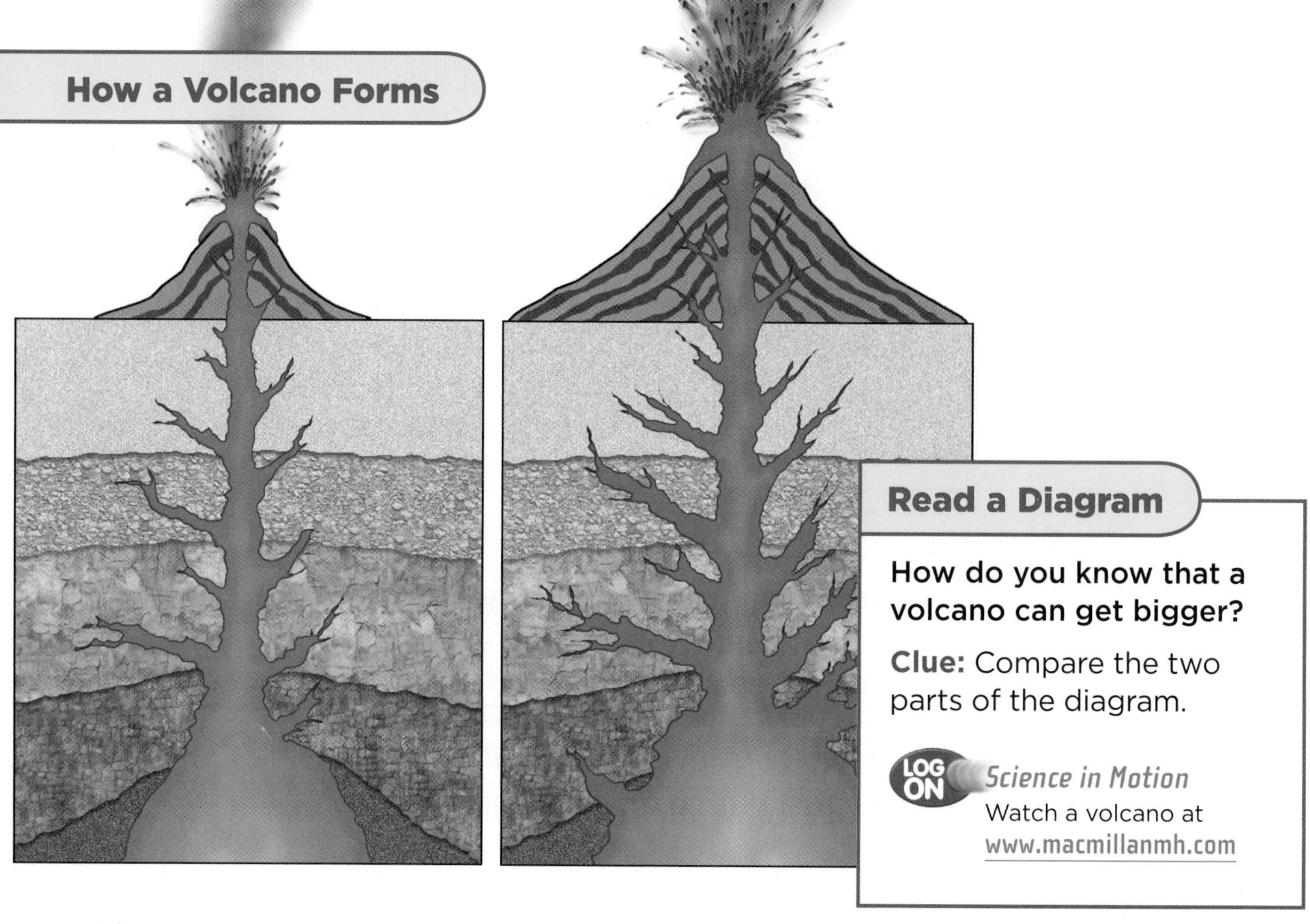

Read a Diagram

How do you know that a volcano can get bigger?

Clue: Compare the two parts of the diagram.

LOG ON *Science in Motion* Watch a volcano at www.macmillanmh.com

What are volcanoes?

A **volcano** (vahl•KAY•noh) is a mountain that builds up around an opening in Earth's crust. Sometimes a volcano explodes. Like an earthquake, this event can change the land quickly.

Volcano Formation

You learned about Earth's layers, the crust, mantle, and core. Parts of the mantle and crust have melted rock called **magma**. Sometimes magma moves up through a large crack in the crust and flows onto land. Melted rock that flows onto land is called **lava**. Lava, rocks, and ash are forced out onto Earth's surface. They pile up in layers and form a mountain. Sometimes a volcanic mountain forms in just a few years.

FACT Volcanoes are not always active.

Effects of Volcanoes

Sometimes, lava oozes from a volcano slowly. The lava hardens and the mountain gets bigger. At other times, lava is forced out of a volcano in an explosion. When this happens, a large part of the mountain can be blown away. Materials from volcanoes can cause a lot of damage to buildings. They can harm living things too.

Quick Check

Cause and Effect What happens when lava flows out of an opening in Earth's crust?

Critical Thinking Why are some volcanoes a danger to people?

Quick Lab

A Model Volcano

1. **Make a Model** Cover a desk with newspaper. Place a small tube of toothpaste on the desk to model Earth's surface.
2. Carefully make a hole in the tube on the side opposite the cap. This represents an opening in Earth's surface.
3. **Observe** Press on the tube near the cap. What happens by the hole? What do you think the toothpaste is a model of?
4. **Communicate** Did the same thing happen to everyone's tube? What was different? Why were there differences?

◀ Lava shoots out of this volcano in Hawaii.

What are landslides and floods?

Have you ever seen a pile of rocks at the bottom of a mountain? How did the rocks get there? Part of the answer is gravity. *Gravity* is a pulling force that acts on all objects. Gravity can cause a landslide. A **landslide** is the rapid movement of rocks and soil down a hill. A landslide can cause a hill or mountain to change quickly.

Heavy rains and melting snow can quickly fill a river. When water flows over a river's banks, or sides, there is a flood. A **flood** is water that flows over land that is usually dry. Flood waters are very strong. They can change land quickly by washing it away.

▲ This mountain was quickly changed by a landslide.

◀ Hurricane Wilma caused flooding in Florida in 2005.

Quick Check

Cause and Effect What effect do landslides have on land?

Critical Thinking Explain how an earthquake can cause a landslide to happen.

Lesson Review

Visual Summary

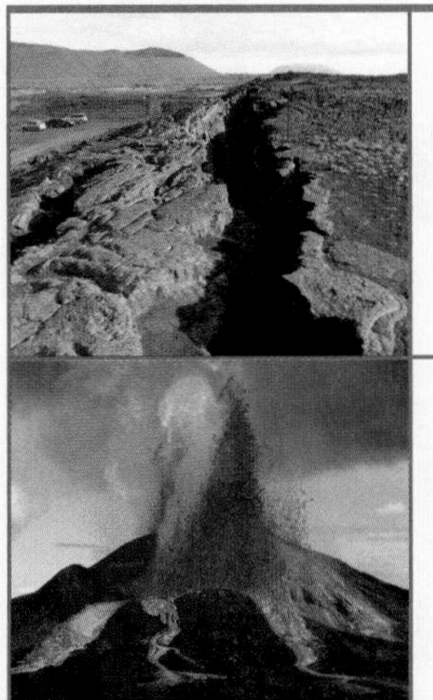

Earthquakes happen when rocks in the crust move. They can change the land quickly.

When lava, ash, and rock are forced from a **volcano,** the land can change quickly.

Landslides can quickly change the shape of a hill or mountain. **Floods** can wash land away.

Make a FOLDABLES Study Guide

Make a three-tab book. Use it to summarize what you learned about how Earth changes quickly.

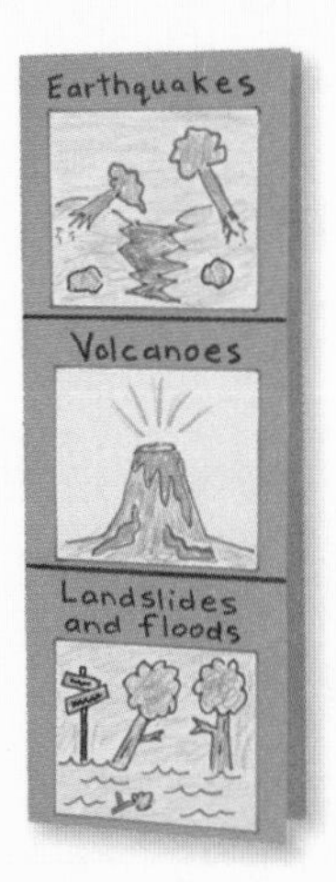

Think, Talk, and Write

1. **Vocabulary** What is a volcano?

2. **Cause and Effect** What causes earthquakes to happen?

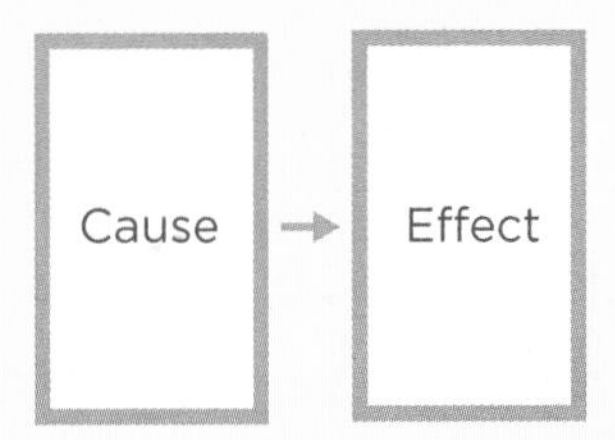

3. **Critical Thinking** What do earthquakes, volcanoes, landslides, and floods have in common?

4. **Test Prep Which event can be caused by heavy rains?**
   - **A** flood
   - **B** earthquake
   - **C** volcanic eruption
   - **D** drought

5. **Essential Question** How can Earth's surface change quickly?

Writing Link

Write a Story
Think about what it must be like to experience an earthquake. Write a story about it. Be sure to mention what the earthquake does to land.

Math Link

Make a Bar Graph
The Richter scale rates earthquakes by how strong they are. Research the strengths of five major earthquakes in recent years. Make a bar graph to compare their strengths.

LOG ON e-Review Summaries and quizzes online at www.macmillanmh.com

SLIDE on the Shore

The western coast of the United States is a beautiful place to live. The views from its cliffs are awesome. Heavy rains, melting snow, and construction can weaken these cliffs, however. Then landslides may happen.

There are some things people can do to help prevent landslides. People can carve steps of land called *terraces* into the cliffs. Rocks and water stay on the terraces and do not flow to the bottom of the cliff.

People can also use drains and covers to keep the land dry. They can plant shrubs and other plants to help keep the soil in place.

People can build things to help keep the soil from moving down a hill. For example, walls of rock and concrete can support a cliff from below. Ditches can direct water around buildings. All of these things help keep people living on or around cliffs safe.

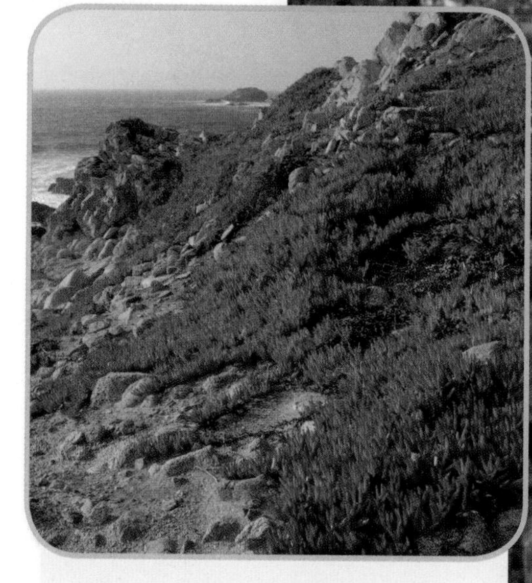

These ice plants help to control erosion on California's coast.

This rock wall will keep soil from moving down this hillside.

Connect to

at www.macmillanmh.com

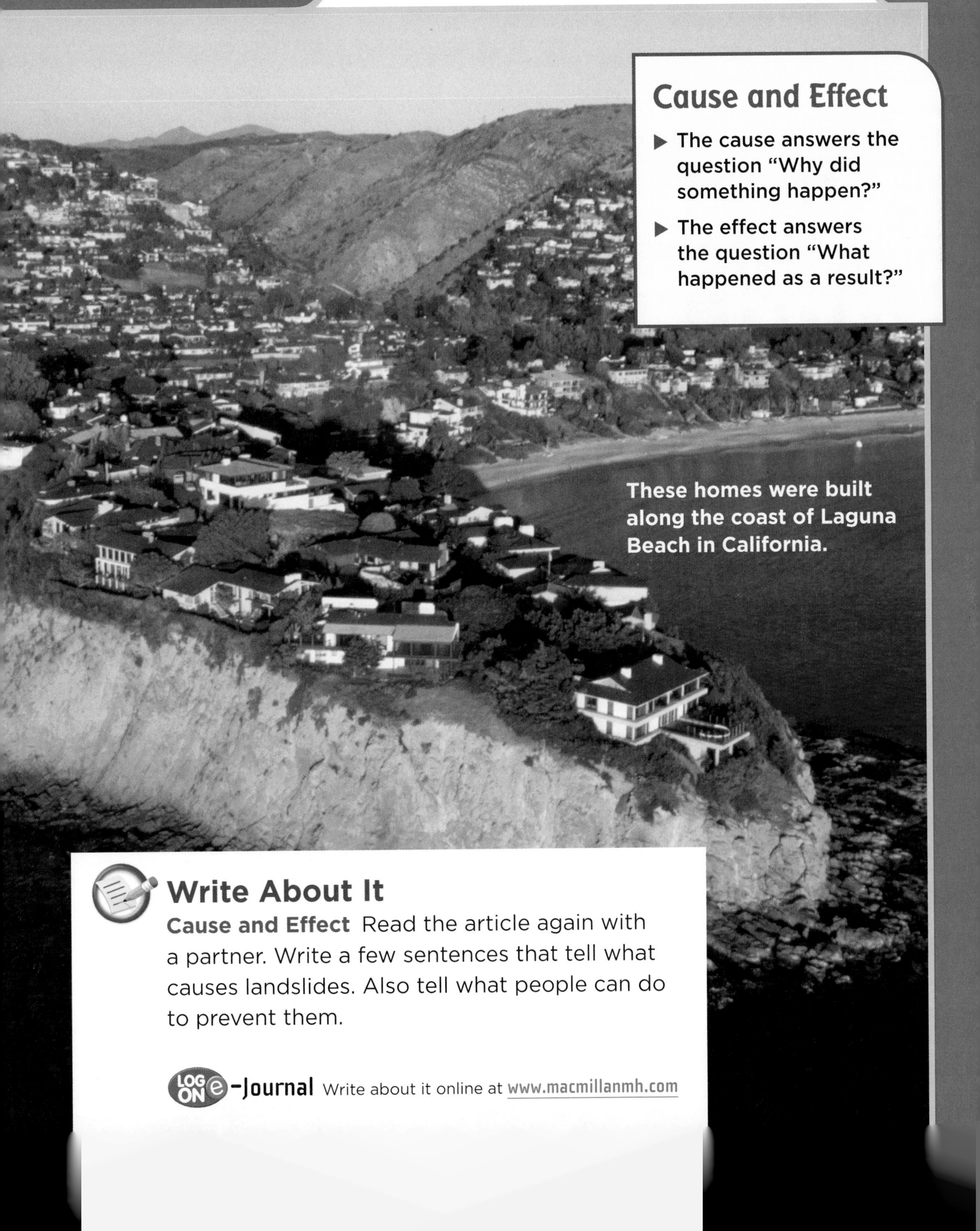

Cause and Effect

- The cause answers the question "Why did something happen?"
- The effect answers the question "What happened as a result?"

These homes were built along the coast of Laguna Beach in California.

Write About It

Cause and Effect Read the article again with a partner. Write a few sentences that tell what causes landslides. Also tell what people can do to prevent them.

LOG ON e-Journal Write about it online at www.macmillanmh.com

Lesson 3

Weathering and Erosion

Grand Canyon National Park, Arizona

Look and Wonder

This canyon was once flat land. Today, parts of the Grand Canyon are nearly one mile deep. How do canyons form?

Explore

Inquiry Activity

How can rocks change in moving water?

Form a Hypothesis

What happens to rocks when they move around in water? Write your answer in the form "If I shake rocks in water, then..."

Test Your Hypothesis

1. **Measure** Label three jars *A, B,* and *C.* Put the same number of similar-sized rocks in each jar. Using the measuring cup, fill each jar with the same amount of water. Put a lid on each jar.
2. Let jar *A* sit. Do not shake it.
3. **Use Variables** Shake jar *B* hard for 2 minutes. Then let the jar sit.
4. **Use Variables** Shake jar *C* hard for 5 minutes. Then let the jar sit.
5. **Observe** Use a hand lens to observe the rocks in each jar. What happened? Did the results support your hypothesis?

Draw Conclusions

6. **Infer** How can rocks change in moving water?

Explore More

Experiment Would the results be the same if different rocks were used? Make a plan and try it.

Materials

sandstone rocks

measuring cup

3 plastic jars with lids

stopwatch

hand lens

Step 1

Step 3

Read and Learn

Essential Question
How can Earth's surface change slowly?

Vocabulary
weathering, p. 214
erosion, p. 216
glacier, p. 216
deposition, p. 216

Reading Skill
Draw Conclusions

Text Clues	Conclusions

Technology
e-Glossary and e-Review online at www.macmillanmh.com

What is weathering?

You might think that hard rocks can not change or break, but they do. Large rocks break into smaller rocks. Small rocks break down into sand and soil. The breaking down of rocks into smaller pieces is called **weathering** (WETH•ur•ing). Weathering usually happens so slowly that you do not notice it. The weathering of rocks can take millions of years.

What causes weathering? Running water, wind, rain, and temperature changes are some things that break down rocks.

Running water and wind pick up small rocks. These rocks scrape against other rocks. This scraping slowly wears away rocks.

This rock, called a ventifact, has been weathered by wind.

▲ These hoodoos have been worn mostly by water that freezes and then thaws inside cracks in the rocks.

This tree continues to break this rock apart. ▼

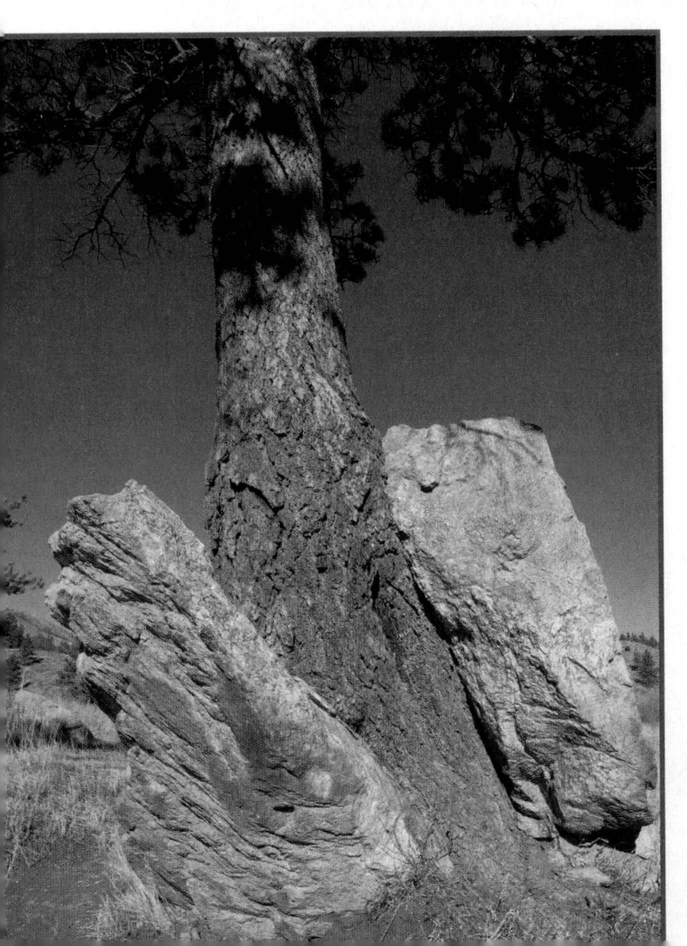

Rain and melting snow can enter the small cracks in rocks. When the water freezes, it *expands,* or takes up more space. This widens the cracks. Then the ice thaws and becomes liquid water again. Over time, repeated freezing and thawing breaks rocks apart.

Living things can cause weathering. Plants can grow in the cracks of rocks. Their roots eventually split rocks apart. When animals dig in the ground, they can uncover buried rocks. The uncovered rocks can then begin to weather.

Quick Check

Draw Conclusions A sidewalk crack became wider during a cold winter. Why?

Critical Thinking Explain how people can cause weathering.

▲ The rocks in this stream were carried here by moving water.

What is erosion?

Broken rocks are sometimes moved to other places. **Erosion** (ih•ROH•zhun) is the wearing away and movement of weathered rock. Moving water, wind, and glaciers (GLAY•shurz) all cause erosion. A **glacier** is a mass of ice that moves slowly across the land. Gravity also causes erosion. Gravity pulls weathered materials downhill.

Erosion can happen slowly or quickly. Floods can cause erosion quickly when rivers overflow their banks. Glaciers cause erosion slowly as they move across the land.

Moving Water and Wind

Moving water in rivers, streams, and ocean waves picks up rocks and sand. The rocks and sand can be carried far away. Then they are dropped in new places. **Deposition** (de•puh•ZIH•shun) is the dropping off of weathered rock.

Wind also picks up small bits of weathered rocks. When the wind slows down, they are deposited.

The rocks here were pulled down by gravity. ▼

Glaciers

As it moves, a glacier picks up and carries away rocks of all sizes. The ice at the bottom of a glacier freezes onto rocks. As the glacier moves, it tears rocks out of the ground. A glacier can move rocks the size of a house. As a glacier melts, it leaves the rocks in a new place.

Quick Check

Draw Conclusions What causes erosion?

Critical Thinking Compare erosion and weathering.

Quick Lab

Materials Settle

1. **Make a Model** Pour one cup each of sand, soil, and pebbles into a jar. Fill the jar almost to the top with water. Seal the jar tightly.
2. Shake the jar ten times. Then let it sit. Draw what you see.
3. **Interpret Data** In which order do the materials settle?
4. **Infer** What happens to eroded materials in a river as the river gradually slows down?

The long sheet of ice shown here is Turner Glacier in Alaska.

Building a Canal

▲ In 1913 the Culebra Mountain in Panama was carved out to build the Panama Canal.

Read a Photo

How did people change the land here?

Clue: Compare the "after" photo with the "before" photo.

How can people change the land?

People change the land too. Some changes are very small, like digging a hole in your backyard. Other changes are much larger.

In some places trees are cut to build roads, stores, and homes. If trees are not replanted, soil can wash away. In other places ponds and swamps are drained. The dry soil left behind can blow away. In still other places, land is dug up to reach valuable rocks.

 ***Quick Check***

Draw Conclusions What effect might planting trees have on the land?

Critical Thinking How are people changing the land where you live?

Lesson Review

Visual Summary

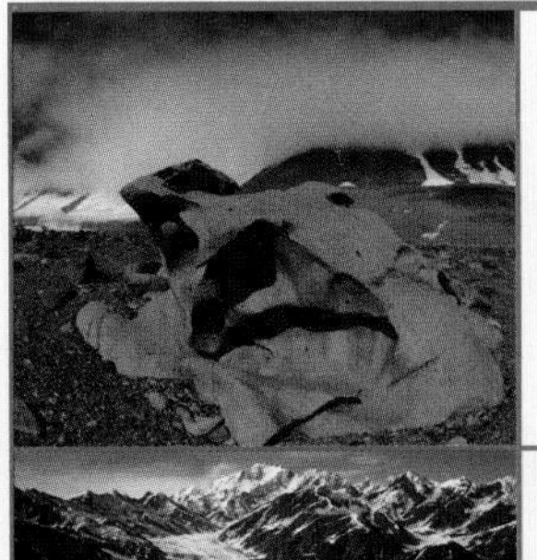

Weathering breaks down larger rocks into smaller rocks.

Erosion is the weathering and movement of weathered rock from one place to another.

People change the land in many ways.

Make a FOLDABLES® Study Guide

Make a trifold book. Use it to summarize what you learned about weathering and erosion.

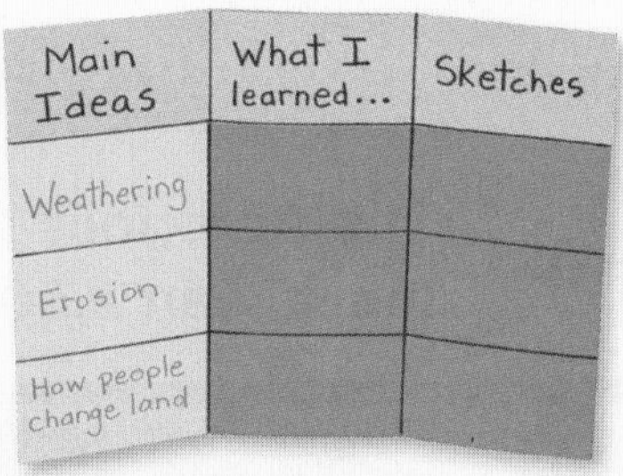

Think, Talk, and Write

1. **Vocabulary** What is deposition?

2. **Draw Conclusions** How do rocks and soil erode?

Text Clues	Conclusions

3. **Critical Thinking** How do weathering and erosion together change land?

4. **Test Prep** **All of the following can cause weathering to rocks except**
   - **A** ice.
   - **B** light.
   - **C** wind.
   - **D** plants.

5. **Essential Question** How can Earth's surface change slowly?

Writing Link

Write a Story
Suppose you are a small rock in a stream. Write a story about what happens to you due to weathering and erosion.

Social Studies Link

River Deltas
Research river deltas. Find out what they are and how they form. What are some famous deltas? Write your findings in a report.

Missing Noses

Rocks are constantly changed by weathering and erosion. However, not all weathering happens the same way.

What happened to the noses on these statues? Did someone break them off? No, something else happened.

It all started when certain gases were released into the air. Many of these gases came from cars, trucks, and factories. The gases combined with rainwater. A weak acid formed. The acid rain chemically changed the minerals in the rock. This is called chemical weathering. The rock broke down. Then rain washed the changed minerals away.

One day, this ancient place could weather and erode completely. All it takes is rain, gases in air, and lots of time.

Expository Writing

Good expository writing

- has a topic sentence that tells the main idea;
- supports the main idea with facts and details;
- draws a conclusion based on the facts.

Write About It

Expository Writing Write a paragraph to describe other causes of weathering. Remember to start with a topic sentence and end with a conclusion.

The Acropolis in Greece is more than 2,500 years old. ▼

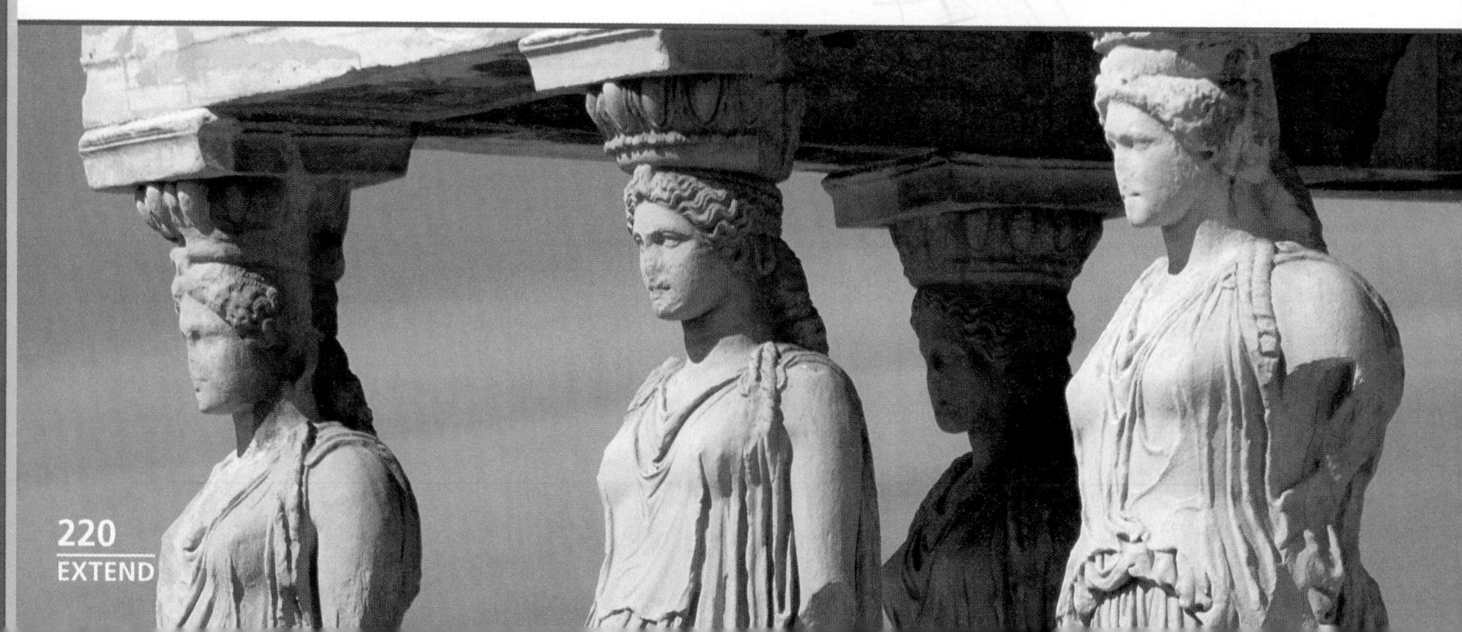

Math in Science

Estimate a Glacier's Change

Sometimes a glacier reaches the ocean and floats on top of it. This long, thin mass of floating ice is called an ice tongue.

The Mertz Glacier is in Antarctica. It has a tongue. Since 1963, melting has caused the glacier's tongue to get longer. It "grows" about 0.9 kilometers each year. If this rate stays the same, about how much should the tongue grow over the next 5 years?

Make Estimations

- An estimate is a number that tells about how much or how many. To estimate the tongue's growth, round 0.9 to the nearest whole number: 0.9 km rounds to 1.0 km.
- To estimate the change over 5 years, multiply the amount of change per year by the number of years.

  1 km per year × 5 years = 5 km

  The glacier's tongue will grow about 5 km in 5 years.

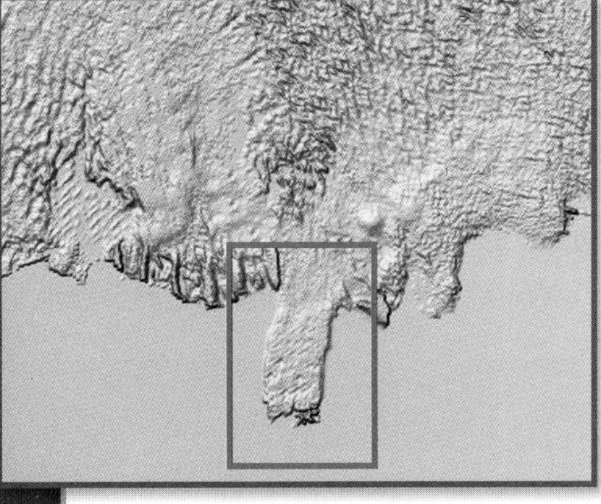

▲ The Mertz Glacier's tongue is about 72 km long.

Solve It

About how much should the glacier's tongue grow in 20 years? If the tongue grows longer than you estimated, what might this tell you about the rate at which the tongue is growing?

CHAPTER 5 Review

Visual Summary

Lesson 1 Earth's surface has many land and water features.

Lesson 2 Earthquakes, volcanoes, landslides, and floods cause Earth's surface to change quickly.

Lesson 3 Weathering and erosion usually cause slow changes to Earth's surface.

Make a FOLDABLES Study Guide

Glue your lesson study guides on a sheet of paper as shown. Use your study guides to review what you have learned in this chapter.

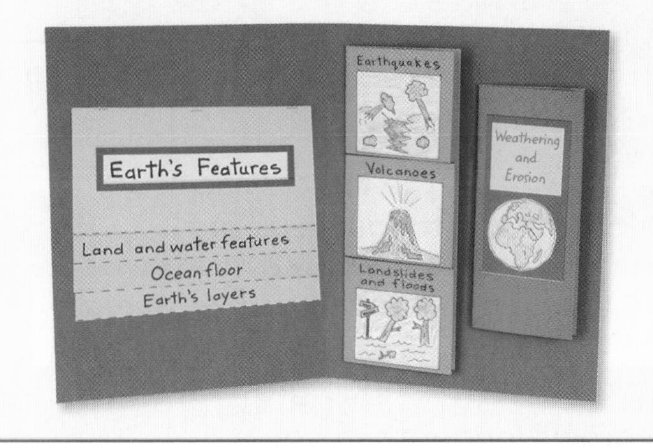

Vocabulary

DOK I

Fill each blank with the best term from the list.

continent, p. 193	**landform**, p. 194
core, p. 198	**landslide**, p. 208
crust, p. 198	**magma**, p. 206
earthquake, p. 204	**volcano**, p. 206
erosion, p. 216	**weathering**, p. 214

1. Each of the seven great land areas on Earth is called a ________.
2. The breaking down of rocks into smaller pieces is called ________.
3. A mountain is an example of a ________.
4. A mountain that builds up around an opening in Earth's crust is a ________.
5. The sudden movement of rocks in Earth's crust might cause an ________.
6. Melted rock below Earth's crust is called ________.
7. The movement of weathered rock by such things as wind, moving water, and glaciers is known as ________.
8. Earth's deepest, hottest layer is the ________.
9. Earth's cool, thin top layer is called the ________.
10. The rapid movement of rocks and soil downhill is a ________.

LOG ON e-Glossary Words and definitions online at www.macmillanmh.com

Skills and Concepts

DOK 2–3

Answer each of the following.

11. **Cause and Effect** What causes landslides?

12. **Descriptive Writing** Describe what the ocean floor looks like.

13. **Make a Model** Suppose you want to show the difference between a plateau and a mountain. Explain how you could build a model to show the difference.

14. **Critical Thinking** What might cause a volcanic mountain to form quickly?

15. How can erosion be caused by a stream or a river?

16. **Experiment** Would clay or sand be washed away more easily by rain? Make a prediction. How could you test your prediction?

17. **True or False** *All mountains are volcanoes.* Is this statement true or false? Explain.

18. **True or False** *Earth's core has melted and solid rock.* Is this statement true or false? Explain.

19. Which of the following most likely causes weathering?

A animals
B rocks
C wind
D soil

20. What can cause Earth's features to change?

Performance Assessment

DOK 3

The Changing Earth

NEWSPAPER

Rain Leads To Flooding

Twenty straight days of rain in Savannah, Georgia have led to severe flooding. Parts of the city are under as much as 6 feet of water.

- Research a recent natural event that happened somewhere in the world. It could be an earthquake, a flood, or a volcanic eruption.
- Find out when and where the event occurred. What caused the event to take place? Did it change the land? How did it affect the people, other living things, or buildings in the area?
- Write a short news report presenting the information you found.

LOG ON e-Review Summaries and quizzes online at www.macmillanmh.com

Test Preparation

1 Look at the diagram below.

The arrow points to which landform?

A mountain

B plateau

C peninsula

D valley

DOK 1

2 Which of these is most likely a slow process?

A a beach flooding

B a volcano erupting

C a rock weathering

D an earthquake

DOK 1

3 Which bodies of water cover most of Earth's surface?

A oceans

B lakes

C rivers

D ponds

DOK 1

4 The melted rock under Earth's crust is called

A lava.

B rocks.

C ash.

D magma.

DOK 1

5 Look at the diagram showing Earth's layers.

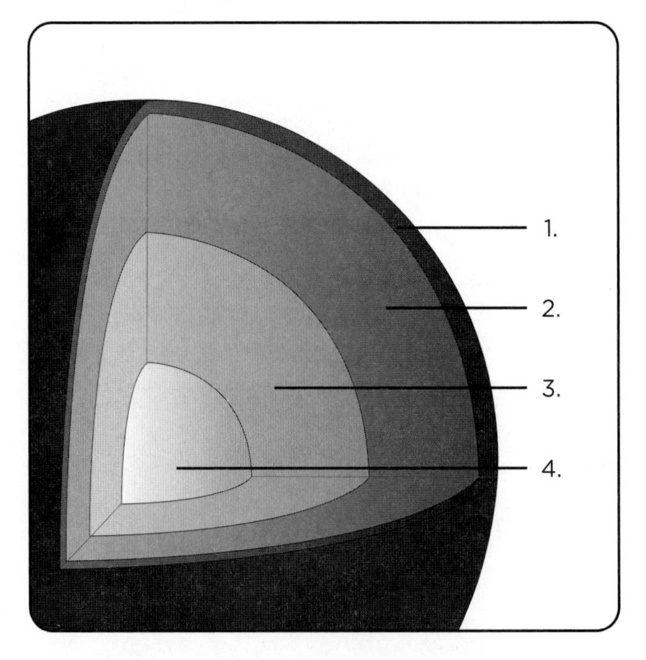

In which layer does an earthquake happen?

A 1

B 2

C 3

D 4

DOK 2

6 Which landform has steep sides with a flat top?

A a peninsula

B an island

C a plateau

D a valley

DOK 1

7 A scientist recorded the number of earthquakes in the United States for four years. She recorded this information in a chart.

Earthquakes in the United States	
Strength	**Number of Earthquakes**
great	0
major	1
strong	2
moderate	32
light	245
minor	800

Which inference can she make from this information?

A A minor earthquake is likely to happen somewhere in the United States every year.

B A great earthquake is likely to happen somewhere in the United States every year.

C A moderate earthquake is not likely to happen anywhere in the United States.

D A light earthquake can never happen in the United States.

DOK 2

8 Which feature of the ocean floor is like a canyon?

A seamount

B abyssal plain

C continental slope

D trench

DOK 1

Use the picture below to answer questions 9–10.

9 The land along this beach is eroding. Name two possible causes of the erosion shown here.

DOK 1

10 Describe one way that this erosion could be slowed down.

DOK 3

Check Your Understanding

Question	Review	Question	Review
1	pp. 194–195	6	pp. 194–195
2	pp. 214–215	7	pp. 13, 204–205
3	pp. 192–193	8	pp. 196–197
4	p. 206	9	pp. 208, 216–217
5	pp. 198, 204–205	10	pp. 216, 218

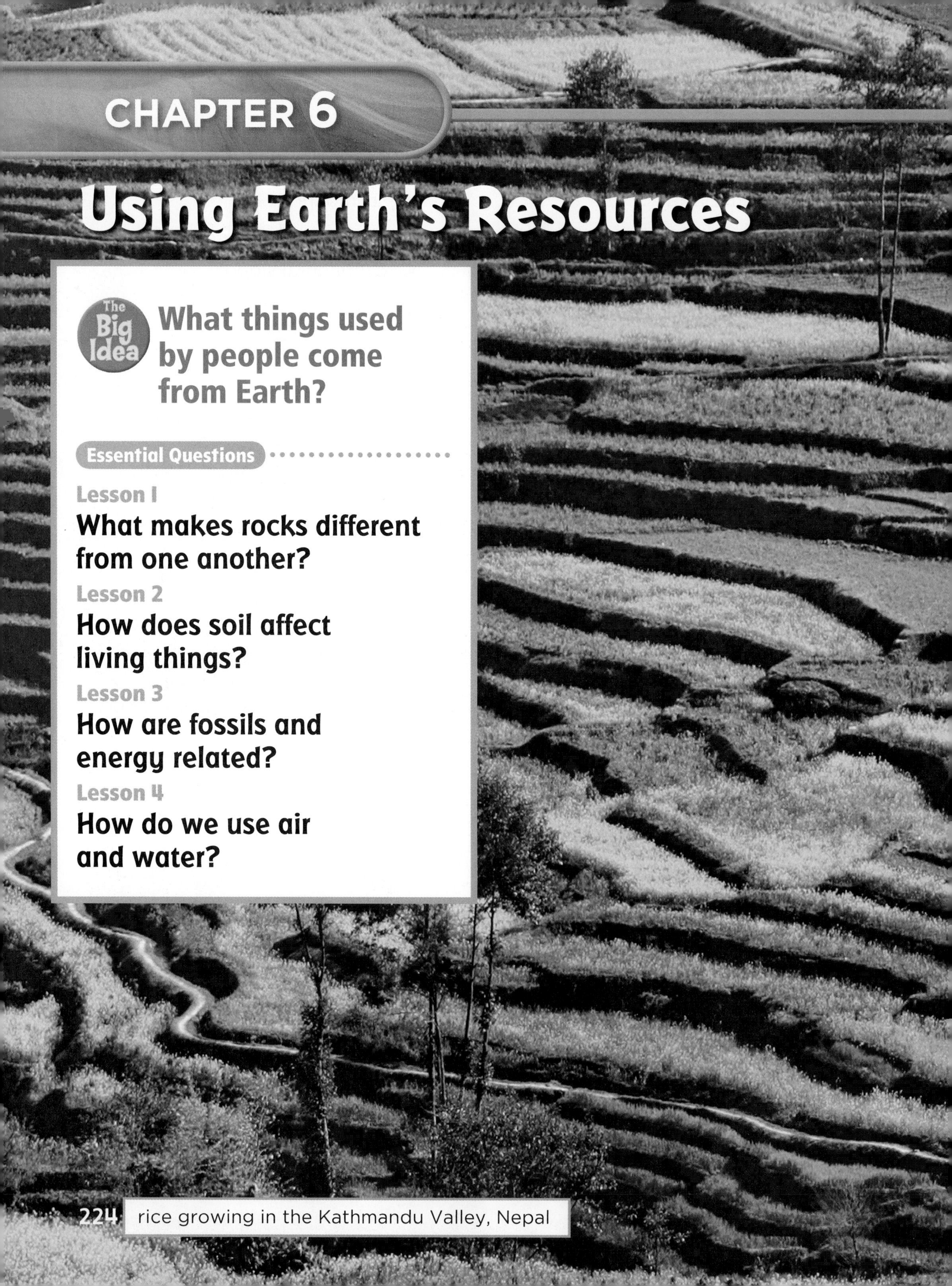

CHAPTER 6

Using Earth's Resources

The Big Idea

What things used by people come from Earth?

Essential Questions

Lesson 1

What makes rocks different from one another?

Lesson 2

How does soil affect living things?

Lesson 3

How are fossils and energy related?

Lesson 4

How do we use air and water?

rice growing in the Kathmandu Valley, Nepal

Big Idea Vocabulary

mineral a solid, nonliving substance found in nature (p. 228)

rock a nonliving material made of one or more minerals (p. 230)

soil a mixture of minerals, weathered rocks, and other things (p. 240)

natural resource a material on Earth that is necessary or useful to people (p. 244)

fuel a material that is burned for its energy (p. 252)

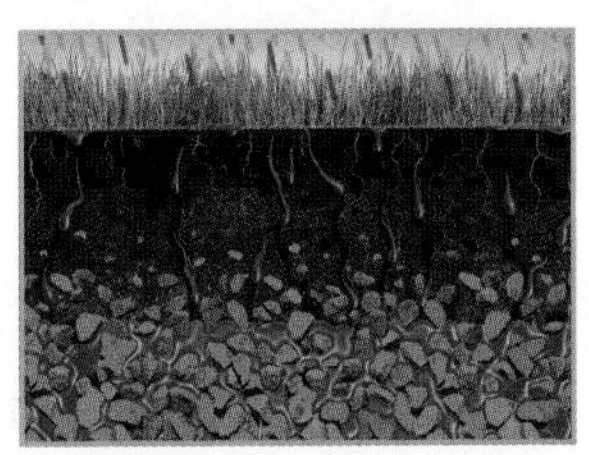

groundwater water that is held in rocks below the ground (p. 261)

Visit www.macmillanmh.com for online resources.

Lesson 1

Minerals and Rocks

Look and Wonder

This mineral looks like gold, but don't be fooled! It's really pyrite, or "fool's gold." How can you tell fool's gold from the real thing?

Explore

Inquiry Activity

How do a mineral's color and mark compare?

Make a Prediction

Some minerals leave a mark behind when you rub them on a white tile. Is the mark left behind always the same color as the mineral?

Test Your Prediction

Materials

minerals

white tile

1. Make a table like the one shown.

Mineral Color	Color Left Behind

2. **Observe** Look at one mineral. Record its color in the table.
3. **Experiment** Rub the mineral across the tile. What color is left behind? Record the color in the table.
4. Repeat steps 2 and 3 for each mineral.

Draw Conclusions

5. **Interpret Data** How did the colors and marks of the minerals compare?
6. **Infer** When might you use mineral marks to help you tell minerals apart?

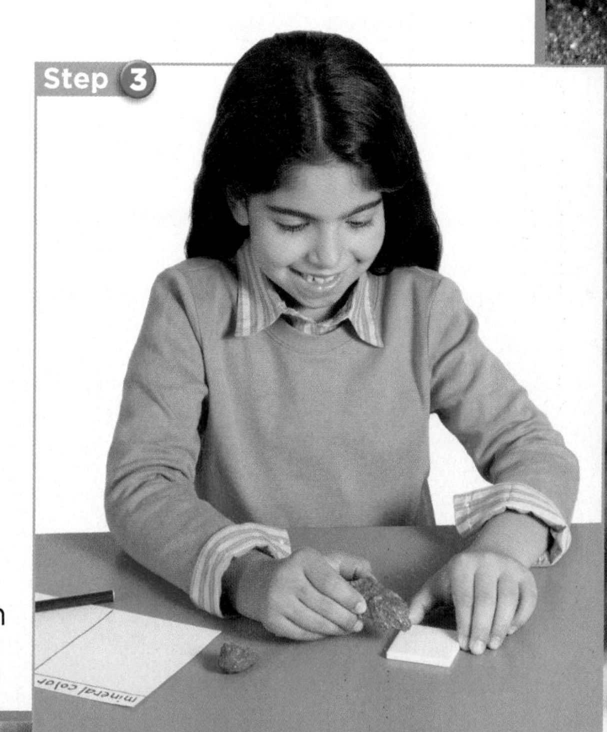

Explore More

Experiment Are some minerals harder than others? Make a plan to find out. Then try it.

Read and Learn

Essential Question
What makes rocks different from one another?

Vocabulary
mineral, p. 228
rock, p. 230
igneous rock, p. 231
sediment, p. 232
sedimentary rock, p. 232
metamorphic rock, p. 233

Reading Skill 
Classify

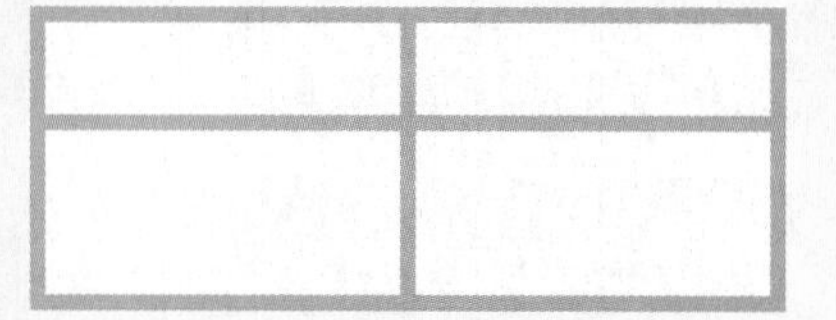

Technology 
e-Glossary, e-Review, and animations online at www.macmillanmh.com

What are minerals?

Many common substances found on Earth are made of minerals (MIH•nuh•rulz). A **mineral** is a solid, nonliving substance found in nature. Table salt, gold, and iron are minerals. The graphite in your pencil is a mineral too. Minerals are the building blocks of rocks. They are found underground and in soil. They are even in the ocean and on the ocean floor.

There are more than 3,000 different kinds of minerals. Each mineral has its own properties. You can use the properties of minerals to tell them apart.

Color

It is easy to observe a mineral's color. Most minerals are only one color. However, some, like quartz, come in many colors. Some, like gold and pyrite, are the same color. You can not use color alone to identify a mineral.

▼ Minerals come in many colors.

turquoise

feldspar

quartz

Streak

Streak is another property used to identify minerals. *Streak* is the color of the powder left when a mineral is rubbed across a white tile. A mineral's streak may or may not be the same as the mineral's color.

Luster

Luster describes how light bounces off a mineral. Some minerals are shiny like metal. Others are not. Luster is another property used to identify a mineral.

Hardness

The *hardness* of a mineral describes how easily it can be scratched. Some minerals, like talc and gypsum, are soft. They can be scratched with a fingernail. Other minerals, like quartz, are much harder. Not even a steel file can scratch quartz.

Quick Check

Classify What are some properties that help you identify a mineral?

Critical Thinking Is wood a mineral? Explain your answer.

▲ Pyrite might look like gold, but its streak is different. Pyrite's streak is greenish-black. Gold's streak is yellow.

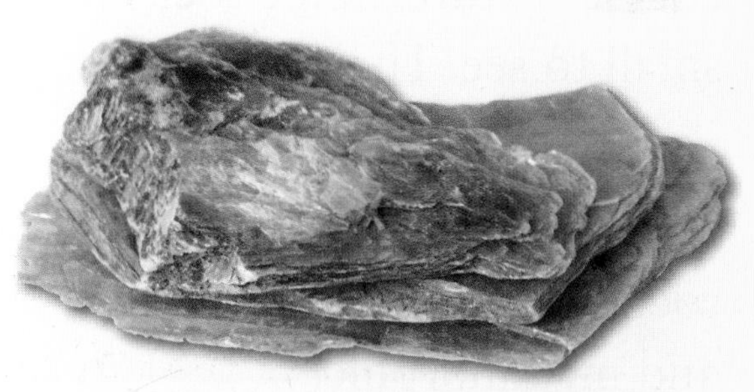

▲ Mica can have a pearly luster.

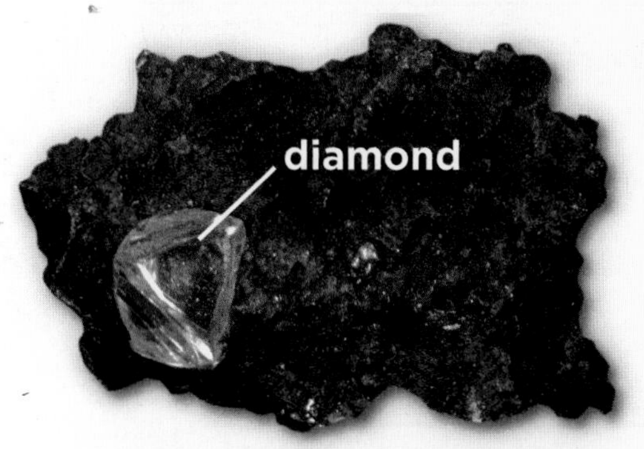

▲ Diamond is the hardest mineral. No other mineral or object can scratch it.

What are rocks?

A **rock** is a nonliving material made of one or more minerals. There are hundreds of different types of rocks. Some rocks, like granite (GRA•nit), are made of several minerals. Some rocks, like limestone, are made mostly of one mineral. A rock's color gives clues about the minerals that make it up.

Rocks are made of mineral pieces called grains. To a person who studies rocks, a rock's *texture* (TEKS•chur) is how its grains look. Some rocks have large grains you can easily see. These rocks have a coarse texture. Some rocks have grains that are too small to see. These rocks have a fine texture.

Rocks are classified by how they form. There are three kinds of rocks—igneous, sedimentary, and metamorphic.

Igneous Rock Formation

Read a Diagram

Where does granite form?

Clue: A cutaway diagram can show what happens below the ground.

Science in Motion Watch igneous rocks form at www.macmillanmh.com

Igneous Rocks

An **igneous** (IHG•nee•us) **rock** forms when melted rock cools and hardens. Inside Earth, melted rock called *magma* cools and hardens very slowly. A rock with large mineral grains forms. Granite is an example.

Melted rock that flows onto Earth's surface is called *lava*. Lava cools and hardens quickly. A rock with small mineral grains forms. Basalt is an example.

Quick Check

Classify What kind of rock is basalt?

Critical Thinking What is the difference between a mineral and a rock?

Quick Lab

Classify Rocks

1. **Observe** Use a hand lens to observe a few igneous rocks. What color are they? Are their grains large or small? Do they have a coarse texture or a fine texture?
2. **Classify** Put the rocks into groups that are alike.
3. **Infer** Which of the rocks do you think formed below Earth's surface? Which formed above Earth's surface? Explain why.

Basalt has a fine texture. It forms when lava cools quickly above Earth's surface.

Granite has a coarse texture. It forms when magma cools slowly beneath Earth's surface.

What are sedimentary and metamorphic rocks?

Some rocks are formed from sediment (SE•duh•munt). **Sediment** is tiny bits of weathered rock or once-living animals or plants. A **sedimentary** (se•duh•MEN•tuh•ree) **rock** is a kind of rock that forms from layers of sediment. Sandstone, shale, and limestone are some kinds of sedimentary rocks.

▲ Shale is a sedimentary rock made up of bits of weathered materials.

Sedimentary rocks form where weathered and eroded materials are dropped. This often happens at the bottom of rivers, lakes, and oceans. Over time, sediment piles up. The top layers press on layers below. They squeeze the water and air from the lower layers and press the sediment together. In time the sediment becomes cemented together and forms rock.

▲ Fossils are often found in the sedimentary rock limestone. Limestone can form from the remains of once-living things.

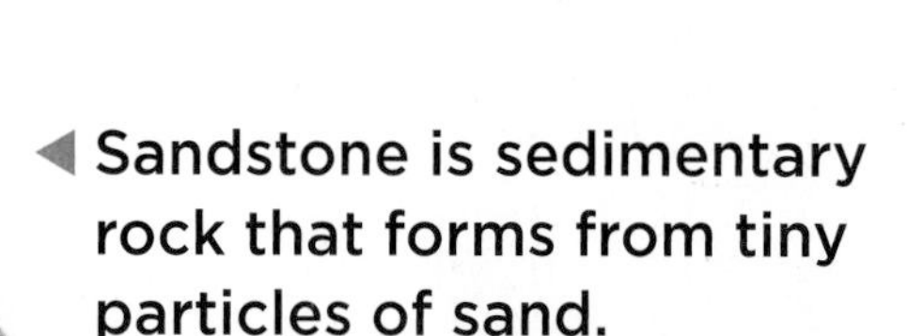

◀ Sandstone is sedimentary rock that forms from tiny particles of sand.

A third kind of rock is metamorphic (me•tuh•MOR•fik) rock. A **metamorphic rock** is a rock that has been changed by heating and squeezing.

Deep inside Earth, rocks heat up and "bake." They also get squeezed by the weight of the rocks above them. All this heating and squeezing can cause a rock's minerals to change into new minerals. A new rock forms with properties that are different from the original rock.

Quick Check

Classify What kinds of rocks are limestone and gneiss?

Critical Thinking What kinds of rocks can change into metamorphic rocks?

Metamorphic Rocks

▲ Gneiss is a metamorphic rock. It forms from granite.

▲ Slate is a metamorphic rock. It forms from shale.

▲ Phyllite is a metamorphic rock. It forms from the metamorphic rock slate.

How do we use minerals and rocks?

Did you write with a pencil today? If so, you used the mineral graphite. Did you eat any food with salt? If so, you ate the mineral halite. Many of the things we use every day come from minerals.

Telephone wires are made with the mineral copper. Some baseball bats are made with aluminum. Aluminum comes from minerals. In fact, most of the metals we use come from minerals. Minerals are even used to make glass, chalk, and toothpaste.

Gold, silver, and iron are minerals we use for jewelry. Other minerals, such as diamonds, topazes, and rubies, are *gems*. People value gems for their beauty.

Rocks are used for building roads, houses, and statues. Limestone is used to make cement. Coal is burned for heat.

▲ Rubies and diamonds are gems.

▲ Marble is a hard rock. It weathers very slowly.

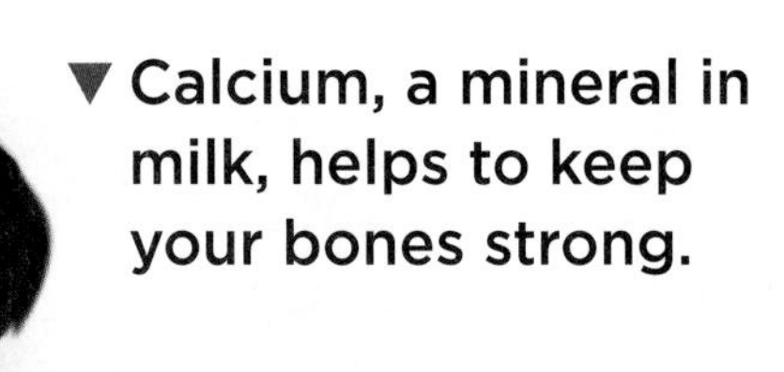

▼ Calcium, a mineral in milk, helps to keep your bones strong.

Quick Check

Classify Is salt a gem?

Critical Thinking What minerals have you used today?

Lesson Review

Visual Summary

Properties such as color, hardness, luster, and streak are used to tell **minerals** apart.

Igneous, sedimentary, and metamorphic **rocks** form in different ways.

People **use rocks and minerals** for many things.

Make a FOLDABLES Study Guide

Make a three-tab book. Use it to summarize what you learned about minerals and rocks.

Think, Talk, and Write

1. **Vocabulary** What is a mineral?
2. **Classify** How could you classify granite, sandstone, basalt, and limestone?

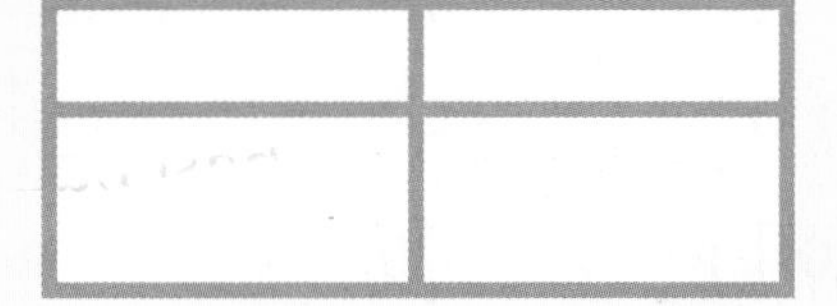

3. **Critical Thinking** Explain the relationship between rocks and minerals.
4. **Test Prep Which of the following is a sedimentary rock?**
   - **A** sandstone
   - **B** basalt
   - **C** diamond
   - **D** granite
5. **Essential Question** What makes rocks different from one another?

Writing Link

Write a Poem
Write a poem about an igneous, sedimentary, or metamorphic rock. Tell how the rock formed or describe how it looks. Be sure to include a title.

Art Link

Make Rock Art
Find something you can decorate with rocks, such as a box. Collect small rocks of different colors and sizes. Glue the rocks to your box.

Marble Memorials

The Lincoln Memorial and the Jefferson Memorial honor past U.S. presidents. These two marble buildings are in Washington, D.C. Each has a statue of a president inside it.

The Lincoln Memorial has the shape of a rectangle. It has white columns. The Jefferson Memorial has columns too, but it is round.

Descriptive Writing

A good description

- uses details to create a picture for the reader;
- includes words that compare, such as *both*, *like*, and *too;*
- includes words that contrast, such as *but* and *unlike.*

Lincoln Memorial

Jefferson Memorial

Write About It

Descriptive Writing Choose two objects made from rock. Write a paragraph that describes and compares them.

e-Journal Write about it online at www.macmillanmh.com

Finding Fractions

This table shows the different rocks in a rock collection.

My Rock Collection		
Igneous	**Sedimentary**	**Metamorphic**
2 basalt	1 coquina	1 schist
3 granite	2 sandstone	1 slate

Fractions

- To find a fraction, use the total number of rocks as the denominator. Use the number of rocks for a particular kind of rock as the numerator.
- Example: What fraction of igneous rocks in the collection are granite?

  $\frac{3}{5}$ 3 ← granite rocks; 5 ← number of igneous rocks
- Example: What fraction of rocks in the collection are metamorphic?

  $\frac{2}{10}$ 2 ← metamorphic rocks; 10 ← total number of rocks

granite

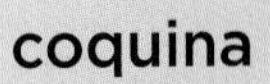
coquina

schist

sandstone

Solve It

What fraction of rocks in the collection are igneous rocks? What fraction of sedimentary rocks in the collection are sandstone?

Lesson 2

Soil

prairie dogs in their burrow

Look and Wonder

Plants, animals, and people could not live without soil. What is in soil? Why is it important to many living things?

Explore

Inquiry Activity

What makes up soil?

Purpose

Find out what soil is made of.

Procedure

1. Use a spoon to spread the soil on the plate.

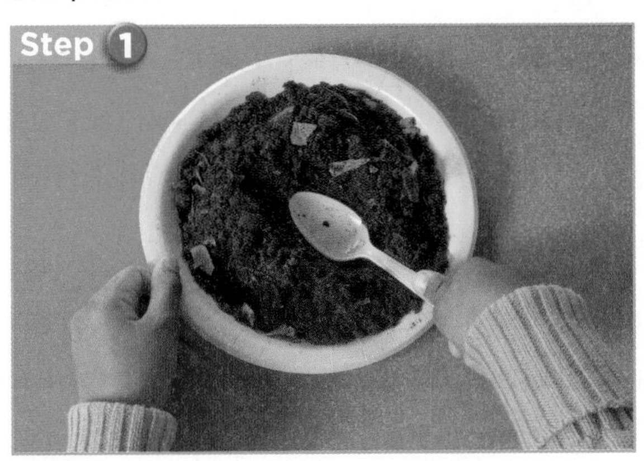
Step 1

2. **Observe** Use the hand lens to observe the soil. Is soil made of small bits of material? What are the shape and color of these small particles? Wash your hands. Record what you see.
3. **Communicate** Talk with others about what the tiny bits in soil may be.

Draw Conclusions

4. **Infer** What kinds of things make up this soil?

Explore More

Experiment Is all soil the same? Make a plan to find out. Then try your plan.

Materials

plastic spoon

soil

paper plate

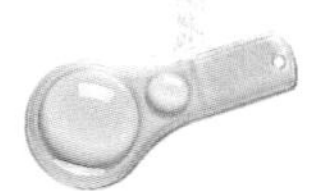
hand lens

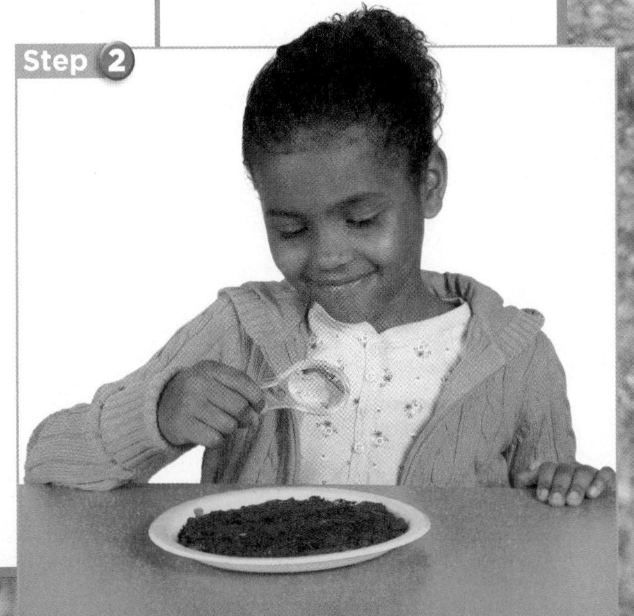
Step 2

Read and Learn

Essential Question
How does soil affect living things?

Vocabulary
soil, p. 240
humus, p. 240
natural resource, p. 244

Reading Skill
Problem and Solution

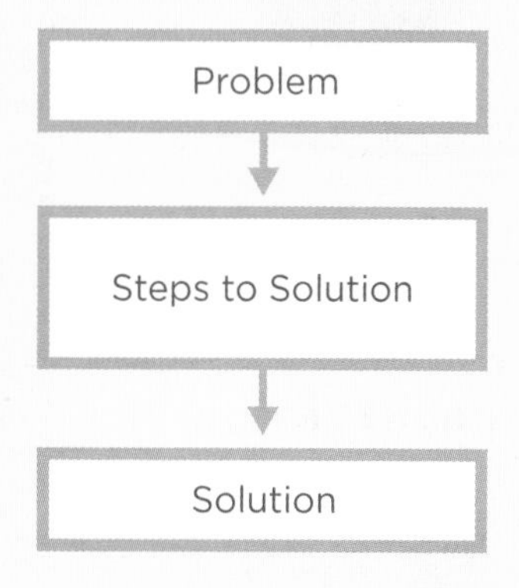

Technology
e-Glossary and e-Review online at www.macmillanmh.com

What is soil?

Soil is a mixture of minerals, weathered rocks, and other things. It has bits of decayed plants and animals called **humus** (HYEW•mus). Humus looks dark. It adds nutrients to soil. Plants then use these nutrients. Humus works like a sponge to soak up rainwater and keep the soil moist. Water, air, and living things are also found in soil.

Living Things in Soil

If you dig away a chunk of soil, you might find roots. A plant's roots take in water and minerals from the soil. They also hold the soil in place and help slow erosion.

You might also find animals living in soil. Animals such as ants, earthworms, and moles break up soil. Their burrows help air and water get into the soil.

Ants and earthworms are just a few of the organisms that live in soil. ▼

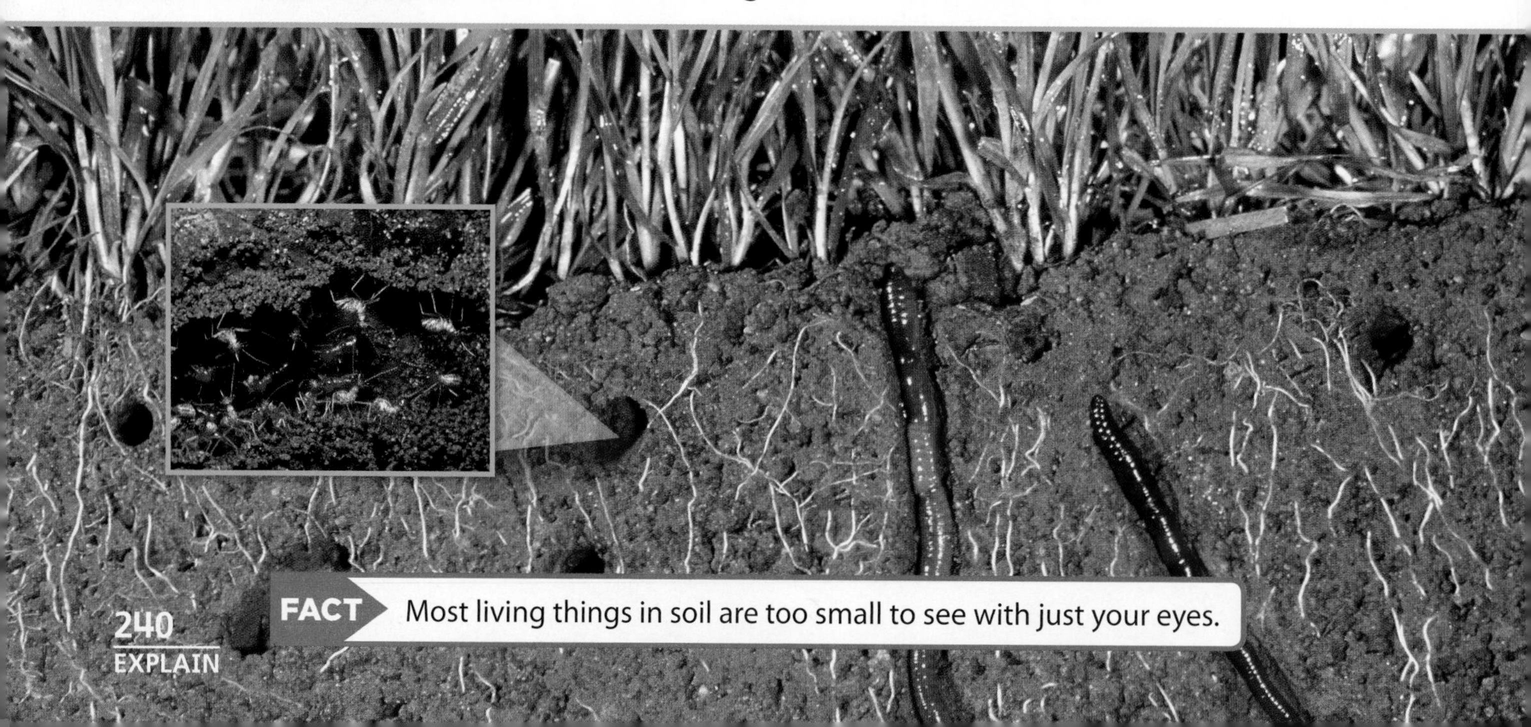

FACT Most living things in soil are too small to see with just your eyes.

How Soil Forms

The making of soil starts with weathering. Weathering causes rocks to break down into smaller and smaller pieces. The tiny bits of weathered rock build into layers. Living things die and decay in the weathered material and become humus. Over time, layers of soil form. The top layer is called *topsoil*. Topsoil is dark and has the most humus and minerals. Below the topsoil is *subsoil*. This layer is lighter in color and has less humus. Below the subsoil is *bedrock*, or solid rock.

Soil takes a long time to form—up to 1,000 years for just 1 centimeter! That is why people try to prevent soil erosion. They add minerals and humus to soil to keep it healthy.

The Layers of Soil

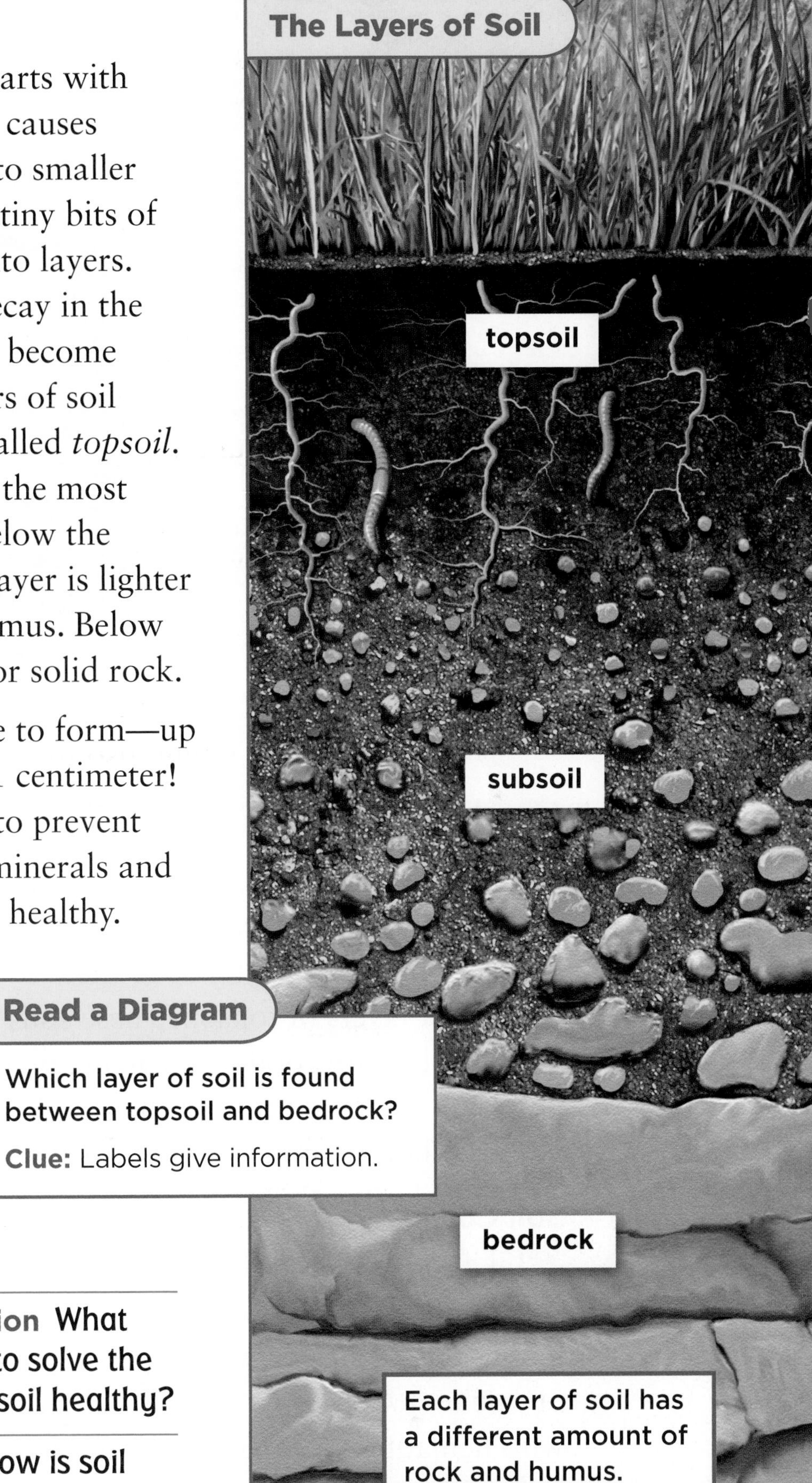

Read a Diagram

Which layer of soil is found between topsoil and bedrock?

Clue: Labels give information.

Each layer of soil has a different amount of rock and humus.

Quick Check

Problem and Solution What do people do to try to solve the problem of keeping soil healthy?

Critical Thinking How is soil a habitat?

▲ This red soil is rich in iron.

▲ This dark soil is rich in humus.

How are soils different?

Different soils are found in different places. They are made of different rocks and minerals. They have different amounts of humus in them too. Some soils have thick layers of topsoil. These soils are rich in humus. They are good for growing plants. Some soils have thin layers of topsoil. These soils have little humus. They are not as good for growing plants.

Soil Color

Like rocks, soils differ in color and texture. A soil's color depends on what is in it. Soil rich in humus looks dark brown or black. Soil with a lot of calcite (KAL•site) in it looks lighter in color. Soil with hematite (HEE•muh•tite) in it looks red. That is because hematite contains iron.

Soil Texture

Soil texture describes how big the pieces, or grains, of soil are. *Sandy soil* has a lot of small grains called sand. *Silty soil* has grains smaller than sand called silt. *Clay soil* has the smallest grains called clay. *Loam* is soil made from a mixture of sand, silt, and clay.

Soil texture affects how much water soil can hold. Clay soil holds a lot of water. Sandy soil holds very little water. Many plants grow best in loam. It is neither too wet nor too dry.

Quick Lab

Classify Soils

1. **Observe** Look at the sandy soil and clay soil in plastic bags. How are they alike? How are they different?
2. **Observe** Use a hand lens to look closely at each soil. Which soil has larger grains?
3. **Classify** Which soil is sandy soil? Which is clay soil? How do you know?

clay soil

sandy soil

loam

Quick Check

Problem and Solution What if plants could not grow well in your neighborhood? What might be the problem? How might you solve it?

Critical Thinking A cactus plant grows best in dry soil. Which soil would be best for a cactus?

Why is soil important?

Soil is a natural resource (NA•chu•rul REE•sors). A **natural resource** is a material on Earth that is necessary or useful to people. Without soil, most plants could not grow. People and animals would not have food to eat. There would be no cotton to make clothes. There would be no wood to build houses or burn for heat. There would be fewer medicines.

It is important to keep soil healthy. It is also important to prevent soil erosion. We can farm in ways to help keep soil from eroding. We can help soil stay healthy by keeping it clean. We can put nutrients into the soil for plants to use.

▲ The bark and leaves of the willow tree were once used to make aspirin.

Quick Check

Problem and Solution How can people keep soil healthy and prevent soil erosion?

Critical Thinking Are rocks and minerals natural resources? Why or why not?

Contour farming helps prevent soil erosion.

Lesson Review

Visual Summary

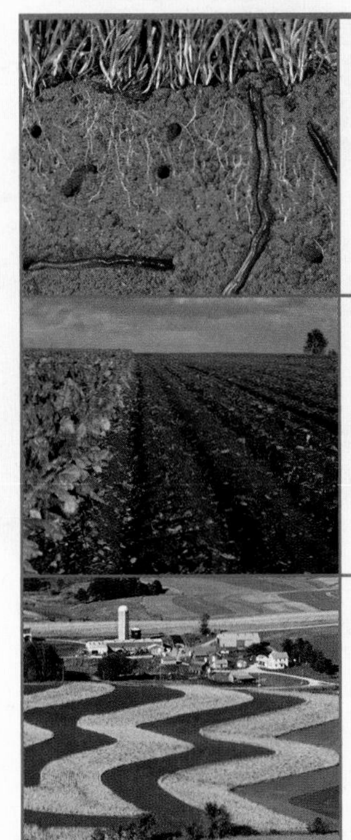

Soil is made up of weathered rocks, minerals, and once-living things.

Soils are different in color and texture. They also hold different amounts of water.

Soil is a natural resource. **Soil is important** to many living things.

Make a FOLDABLES® Study Guide

Make a trifold book. Use it to summarize what you learned about soil.

Think, Talk, and Write

1. **Vocabulary** What is humus?
2. **Problem and Solution** What problems might occur if we do not protect soil?

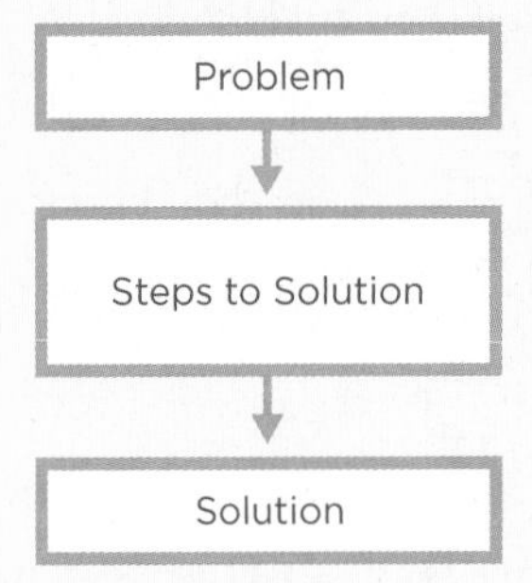

3. **Critical Thinking** Can soil form below Earth's surface? Explain your answer.
4. **Test Prep Which helps soil hold water?**
   - **A** humus
   - **B** air
   - **C** bedrock
   - **D** animals
5. **Essential Question** How does soil affect living things?

Math Link

Solve a Problem

Suppose it takes 1,000 years for 1 cm of soil to form. How long would it take 5 cm of soil to form?

Health Link

Medicines from Plants

Research a medicine that people get from plants. If possible, find out what kind of soil the plant grows best in. Share your findings with the class.

LOG ON e-Review Summaries and quizzes online at www.macmillanmh.com

Focus on Skills

Inquiry Skill: Use Variables

Soils differ from place to place. They contain different amounts of humus and are made up of different kinds of rocks. Do all soils hold the same amount of water? To answer this question, you can **use variables** to test how water moves through different soils.

▶ Learn It

When you **use variables**, you identify things in an experiment that can be changed. Soil type is a variable, for example. The amount of soil you use in an experiment is also a variable. It is important that you change only one variable at a time when you experiment. You should keep all other variables the same. That way you can tell what caused the results.

▶ Try It

You will **use variables** to answer this question: Does sandy soil or potting soil hold more water?

Materials **pencil, 4 disposable cups, potting soil, measuring cup, water, sandy soil, watch or clock**

1. Use a pencil point to poke three tiny holes in the bottom of a cup.
2. Put 250 milliliters of potting soil into the cup. Pack the soil firmly.
3. Fill a measuring cup with 100 mL of water.

4. Hold the cup of potting soil over an empty cup without holes. Slowly pour the water over the soil. Wait for two minutes. Write your observations in a table like the one shown.
5. Pour the water that drained out into the measuring cup. Record the volume in your table.
6. Repeat steps 1–5 using sandy soil in place of potting soil. Record the results.
7. Which soil held more water? How did changing the variable change the results?

Variable	My Observations	Volume That Drained

▶ Apply It

Now **use variables** to experiment more. Choose one of the following variables to test. List the variable in a table and record the results of your experiment. Did changing the variable change the results? If so, how?

- Do not pack the potting soil firmly.
- Mix clay into the sandy soil.
- Mix larger rocks into the potting soil.
- Poke larger holes in the cups.

Lesson 3

Fossils and Fuels

Look and Wonder

This winged ant was trapped in amber millions of years ago. Now it is a fossil. It looks exactly as it did when it was alive. How do you think this fossil formed?

Explore

Inquiry Activity

How do some fossils form?

Purpose

Find out how some living things of the past became fossils.

Procedure

1. **Make a Model** Hold a spoon over a paper towel. Squeeze a small amount of glue onto the spoon. Let the glue set for ten minutes. This models sticky tree resin.
2. **Make a Model** Place a thin apple slice on top of the glue. This models an organism trapped in tree resin. Slowly add more glue until the apple slice is completely covered.
3. **Use Variables** Put the spoon on a paper towel. Place another apple slice next to the spoon.
4. **Observe** Look at the apple slices throughout the day. Record any changes you observe.

Draw Conclusions

5. **Interpret Data** Compare the two apple slices. What differences do you notice?
6. **Infer** What caused any differences you observed?
7. **Infer** How do some fossils form?

Explore More

Experiment Could an organism become a fossil in ice? Make a plan to find out.

Materials

plastic spoon

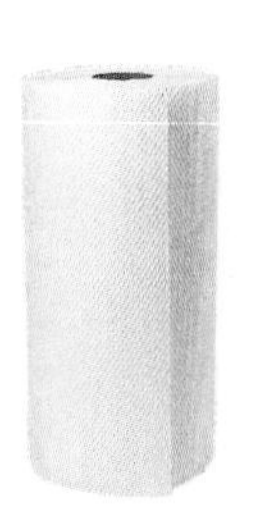

paper towel

glue

2 apple slices

Step 2

These stony fossils of bones were once real dinosaur bones.

Read and Learn

Essential Question

How are fossils and energy related?

Vocabulary

fossil, p. 250

fuel, p. 252

renewable resource, p. 253

nonrenewable resource, p. 253

solar energy, p. 254

Reading Skill

Draw Conclusions

Text Clues	Conclusions

Technology 

e-Glossary and e-Review online at www.macmillanmh.com

How are fossils formed?

A **fossil** (FAH•sul) is the trace, or remains, of something that lived long ago. Shells, bones, skin, leaves, and footprints can become fossils.

Imprints

Sometimes living things leave marks, or *imprints,* in materials like mud. In time the materials can harden into rock. The imprints are saved in the rock.

Actual Organisms and Stony Fossils

Some fossils are actual organisms trapped in amber, tar, or ice. Others look like actual plant or animal remains but are not. They are only stony fossils in sedimentary rock.

◀ This imprint of a dinosaur's foot was left in mud. The mud turned to solid rock.

FACT People did not live when dinosaurs did.

Sometimes, sediment buries an organism that has died. The organism becomes a fossil as the sediment turns into rock. Slowly, water with minerals seeps into the hard parts of the organism. Minerals replace the hard parts of the organism. What is left is a stony fossil.

Molds and Casts

Shells leave fossils called molds. A *mold* is an empty space in rock where something once was. A mold forms after a shell is buried in sand or mud. As water seeps in, it breaks down the shell. A shell-shaped space is left. Minerals can seep into this space and harden. They can form a copy of the mold's shape called a *cast*.

Quick Check

Draw Conclusions What can we learn from fossils?

Critical Thinking What would be more likely to form a fossil—a worm or a shell? Why?

Quick Lab

Model Imprints

1. Break a small chunk of clay into two pieces. Roll each piece to form a ball.
2. **Make a Model** Take one clay ball. Press the front of your thumb into it. Take the other clay ball. Press the back of your thumb into it.

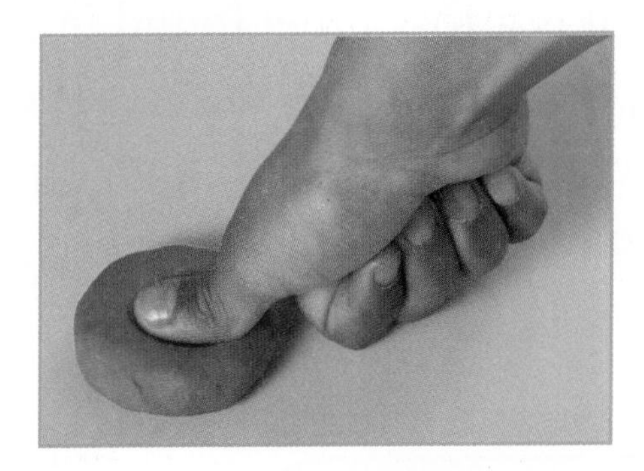

3. **Communicate** Switch clay balls with someone. How are the imprints similar to yours? How are they different?
4. **Infer** What can we learn by comparing fossil imprints?

The fossil on the right is the mold. The fossil on the left is the cast. ▶

What are fossil fuels?

The energy to heat homes and run cars and airplanes comes from fuels. A **fuel** is a material that is burned for its energy. Coal, oil, and natural gas are fossil fuels. A *fossil fuel* is a fuel that forms from the remains of ancient plants and animals. Fossil fuels can be used to make electricity.

Oil is a fossil fuel found in rocks deep below Earth's surface. People use huge drills to dig deep underground for oil. Pumps are used to bring oil to the surface.

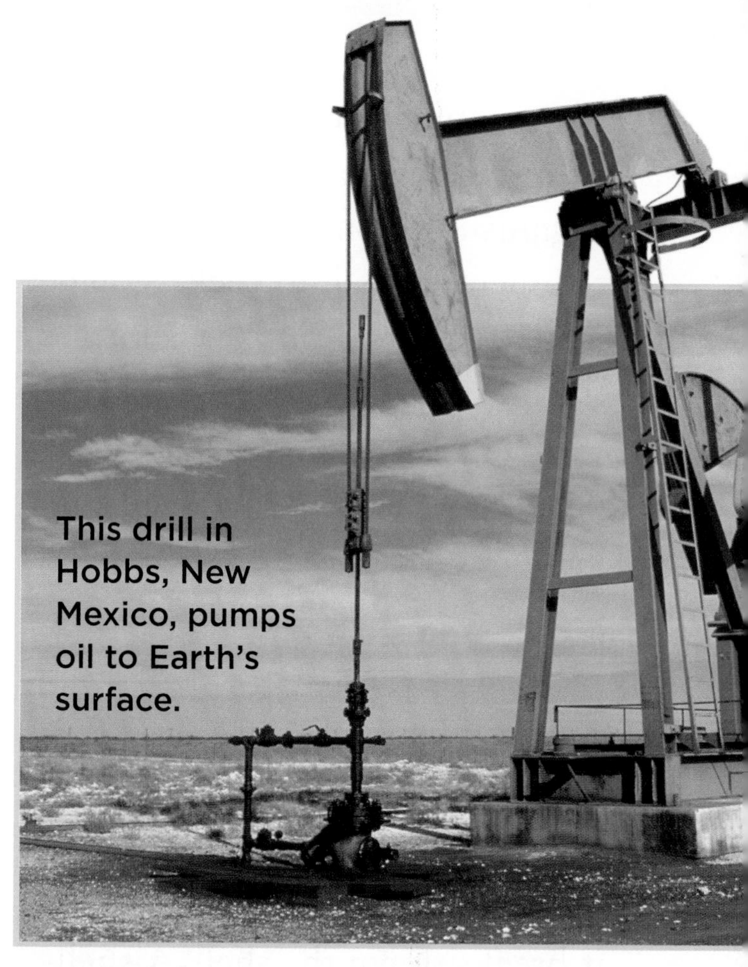
This drill in Hobbs, New Mexico, pumps oil to Earth's surface.

How Coal Forms

① Millions of years ago, swamps covered large parts of Earth's land. Over time, the swamp plants died.

② Layers of decayed plants formed a fuel called ***peat.*** Then the peat was buried under sediment.

Fossil fuels are natural resources. Plants, animals, water, and air are too. Plants, animals, water, and air can be replaced. New plants are grown. New animals are born or hatched. Rain and snow bring more water. Plants put oxygen back into the air. Plants, animals, water, and air are renewable resources (ree•NEW•uh•bul REE•sors•es). A **renewable resource** is a resource that can be replaced or used again and again.

Fossil fuels, however, are nonrenewable (nahn•ree•NEW•uh•bul). A **nonrenewable resource** is a resource that can not be replaced or reused easily. Fossil fuels take millions of years to form. After they are used up, they are gone forever.

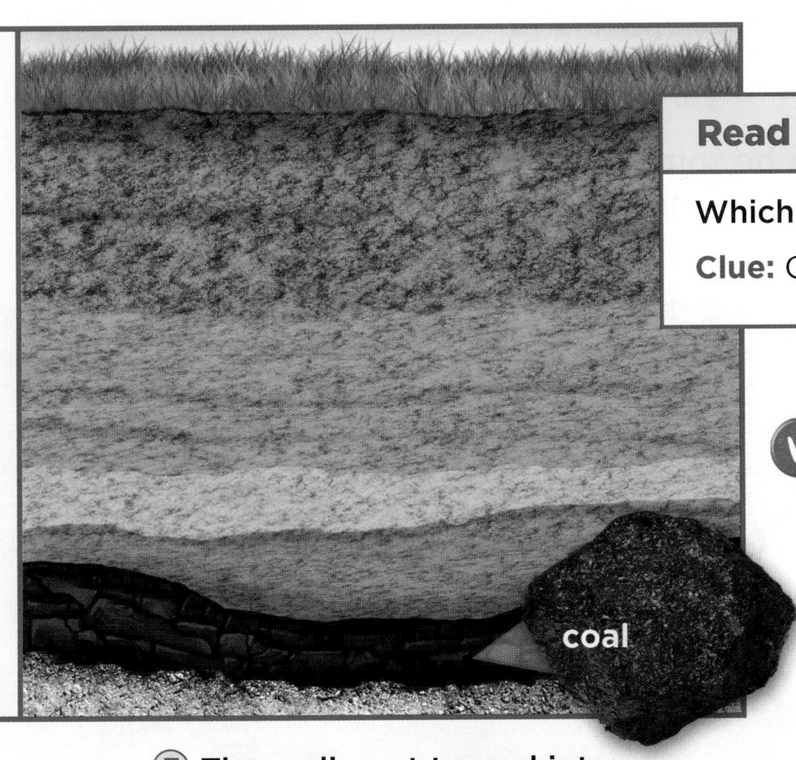

③ The sediment turned into sedimentary rock. Slowly the peat changed into the sedimentary rock *coal.*

Read a Diagram

Which fuel forms before coal forms?

Clue: Captions give information.

Quick Check

Draw Conclusions Why should we be careful not to use up fossil fuels?

Critical Thinking What are other nonrenewable resources?

What are some other sources of energy?

Fossil fuels are just one source of energy. A *source* is where something comes from. Fossil fuels are nonrenewable. They can be used up. For this reason, we need to use renewable sources of energy.

Solar energy is a renewable source of energy. **Solar energy** is energy from the Sun. Moving water and wind are also renewable sources of energy. Underground heat is too. Solar energy, moving water, wind, and underground heat can all be used to make electricity.

▲ Someday you might drive a car powered by solar energy.

Quick Check

Draw Conclusions Why are the Sun, wind, and moving water good sources of energy to use?

Critical Thinking Where might be some good places to use wind for making electricity?

Underground heat is used to make electricity in Iceland.

Lesson Review

Visual Summary

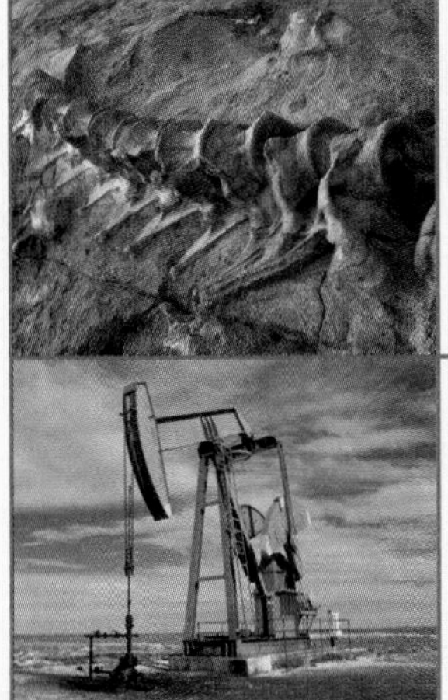

There are different **kinds of fossils.** Each forms in a different way.

Fossil fuels are nonrenewable sources of energy.

The Sun, wind, and moving water are some **renewable sources of energy.**

Make a FOLDABLES® Study Guide

Make a trifold book. Use it to summarize what you learned about fossils and fuels.

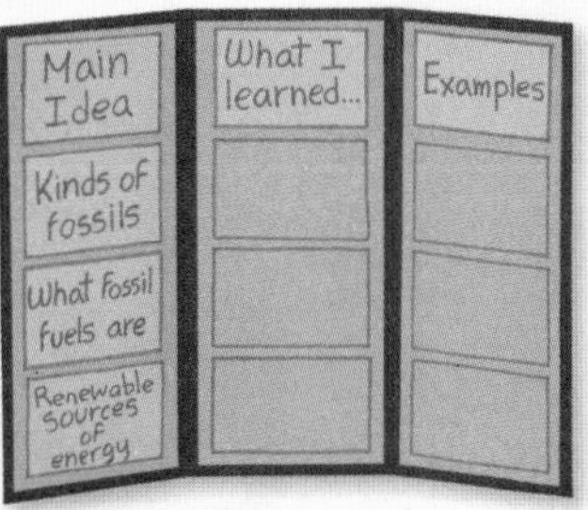

Think, Talk, and Write

1. **Vocabulary** What is a fossil? Give two examples.

2. **Draw Conclusions** Is it possible to use up fossil fuels? Explain.

Text Clues	Conclusions

3. **Critical Thinking** How do you use fossil fuels?

4. **Test Prep** **Which one of the following is a nonrenewable resource for energy?**
   - **A** water
   - **B** air
   - **C** plants
   - **D** coal

5. **Essential Question** How are fossils and energy related?

Math Link

Write a Number Sentence
Ultrasaurus was about 30 meters long. Triceratops was about 8 m long. How much longer was Ultrasaurus than Triceratops? Write the number sentence that shows how you solved the problem.

Social Studies Link

Your State Fossil
Do research about your state fossil. Tell how the fossil formed. What was the organism like? Write this information in a report.

LOG ON e-Review Summaries and quizzes online at www.macmillanmh.com

Reading in Science

Turning the Power On

People use a lot of energy. We need it to power our cars, to heat our homes, and to run many of the machines we use each day. The energy sources we use most—coal and oil—are nonrenewable resources. They will be used up one day and will be gone forever. Other energy sources are renewable. The time line shows how people have developed renewable sources of energy.

Hydropower Energy The first plant in the U.S. opened in Wisconsin. River current was used to turn a turbine to produce electricity.

Wind Energy Wind turbines were invented in Denmark. They used the energy of the wind to produce electricity.

1904

Geothermal Energy Heat energy from geysers was used in Italy. Steam from this hot water that shoots up from the ground was used to turn turbines to produce electricity.

Connect to

at www.macmillanmh.com

Renewable energy sources can be replaced in a short time. The renewable energy sources used most often are hydropower (water), wind, geothermal, solar, and biomass. No matter what energy source you use, it is important to conserve energy.

Draw Conclusions

When you draw conclusions,

- you explain the answer to a question;
- you use what you already know;
- you look for clues in the article.

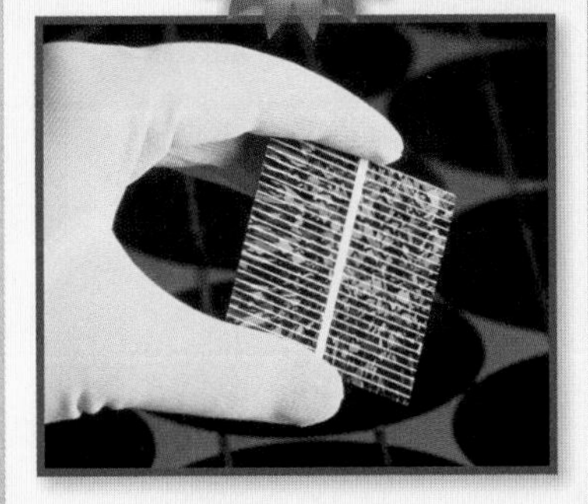

Solar Energy Russell Ohl invented a solar cell. It used light from the Sun to produce electricity.

1985

Biomass Energy This energy source was first used in California. Materials such as dead trees, leftover crops, and animal waste were burned to produce heat, steam, and electricity.

Write About It

Draw Conclusions Why is it important for people to use renewable energy sources? Use what you already know and what you read in the article to draw a conclusion.

Write about it online at www.macmillanmh.com

Lesson 4

Air and Water Resources

Niagara Falls, New York

Look and Wonder

Water is an important natural resource. Moving water from the Niagara River is used to make electricity. How can people use water wisely and keep it clean?

Explore

Inquiry Activity

How is Earth's water made clean?

Purpose

Find out how Earth's water can be made clean.

Procedure

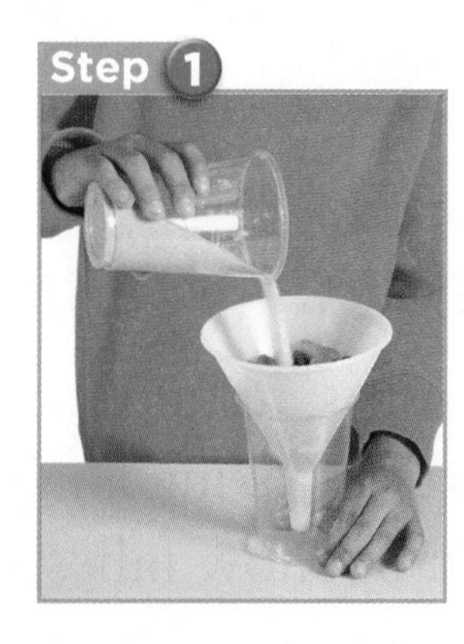

Step 1

1. **Make a Model** Place a funnel inside a large, clear plastic cup. Use a spoon to fill the funnel with gravel half way. Then fill the rest of the funnel with coarse sand. What do you think these layers model?
2. **Observe** Mix a little soil and some crushed leaves into a small, clear cup of water. Draw what you see.
3. **Observe** Slowly pour the water, soil, and leaves into the funnel. Draw what you see in the cup under the funnel.

Draw Conclusions

4. How did the water in step 3 compare with the water in step 2?
5. How can Earth's water be made clean?

Explore More

Experiment Can food coloring be removed from water this way? Repeat to find out.

Materials

funnel

large, clear plastic cup

spoon and gravel

coarse sand

small, clear plastic cup of water

soil and crushed leaves

Step 3

Read and Learn

Essential Question
How do we use air and water?

Vocabulary
groundwater, p. 261
pollution, p. 264
conserve, p. 266

Reading Skill
Fact and Opinion

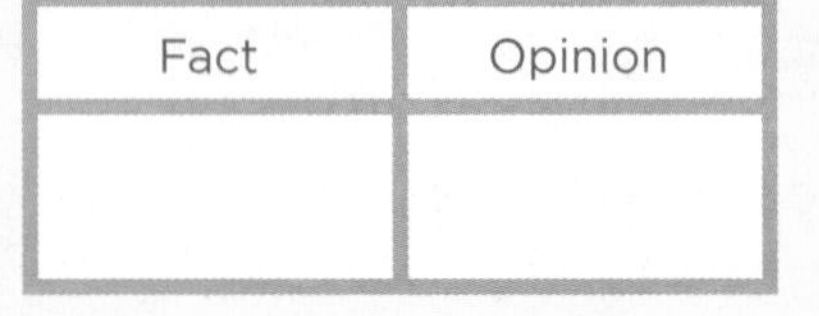

Technology

e-Glossary and e-Review online at www.macmillanmh.com

Rafting on a river can be fun.

How do we use air and water?

Air is an important resource. Air surrounds Earth. Moving air brings clouds from which rain or snow falls. Moving air is used to help make electricity. People need the oxygen in air to breathe.

Water is another important resource. People use water for farming, cooking, drinking, and traveling. Factories use water to make things. People swim, fish, and play in water too.

Air and water are renewable resources. The oxygen in air is replaced by plants when they make food. The water we use is replaced when rain or snow falls from clouds.

◀ Sea ice is freshwater. Ocean water is salty. ▼

Freshwater Sources

Most of Earth's water is in the salty oceans. People can not use it to drink. They must drink freshwater, or water that is not salty. Freshwater is found in rivers, ponds, streams, and most lakes. It is also found underground. Water that is held in rocks below the ground is called **groundwater**.

Glaciers, icebergs, and sea ice are other examples of freshwater. However, this water is frozen. It would not be easy to change this frozen water to liquid water. Only a small part of freshwater is usable. Freshwater is a limited resource.

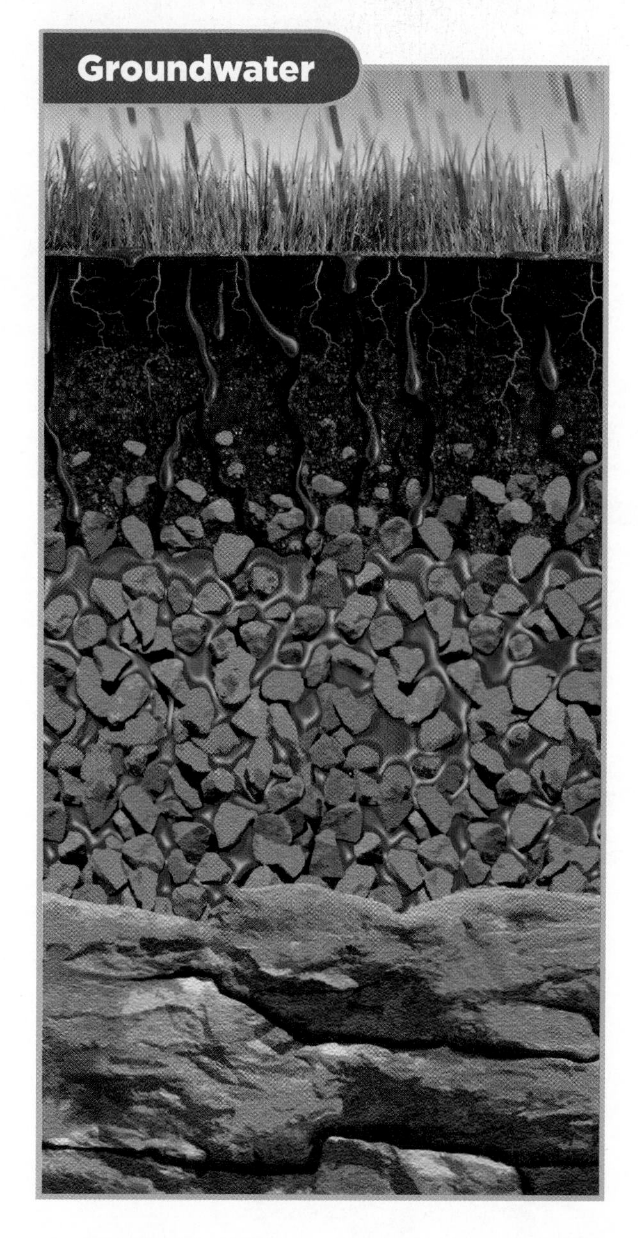

Water seeps into the ground. It collects in the spaces of underground rocks. ▶

Quick Check

Fact and Opinion *Air is an important resource.* How is this a fact?

Critical Thinking What if it never rained or snowed again? What would happen to Earth's freshwater?

▲ Glen Canyon Dam in Arizona forms a reservoir.

▲ Wells bring up water that soaks deep into the ground.

How do people get water?

Not everyone lives near freshwater. How can people move water to where they need it?

Aqueducts and Reservoirs

In some places pipes or ditches, called *aqueducts* (AH•kwuh•dukts), carry water to where it is needed. In other places, people build a wall, called a *dam,* across a river. The water behind the dam gathers in a kind of lake, called a *reservoir* (REH•zuh•vwor). The reservoir stores water for people to use.

Wells

To reach groundwater, people dig *wells,* or deep holes. Water fills the well. It is pumped up from the well when it is needed.

Treating Water

In many cities and towns, water is made safe for people to drink. This happens at a *water treatment plant*. First, water sits still in a tank. Heavy things in water fall to the bottom of the tank. Water is then moved through layers of gravel and sand. These layers trap mud and other tiny things. Chemicals can be added to kill any harmful living things in the water. The clean water is then piped to people.

Quick Check

Fact and Opinion Do you think adding chemicals to water is a good idea? Why or why not?

Critical Thinking What things in nature act as settling tanks and filtering tanks?

Water Treatment Plant

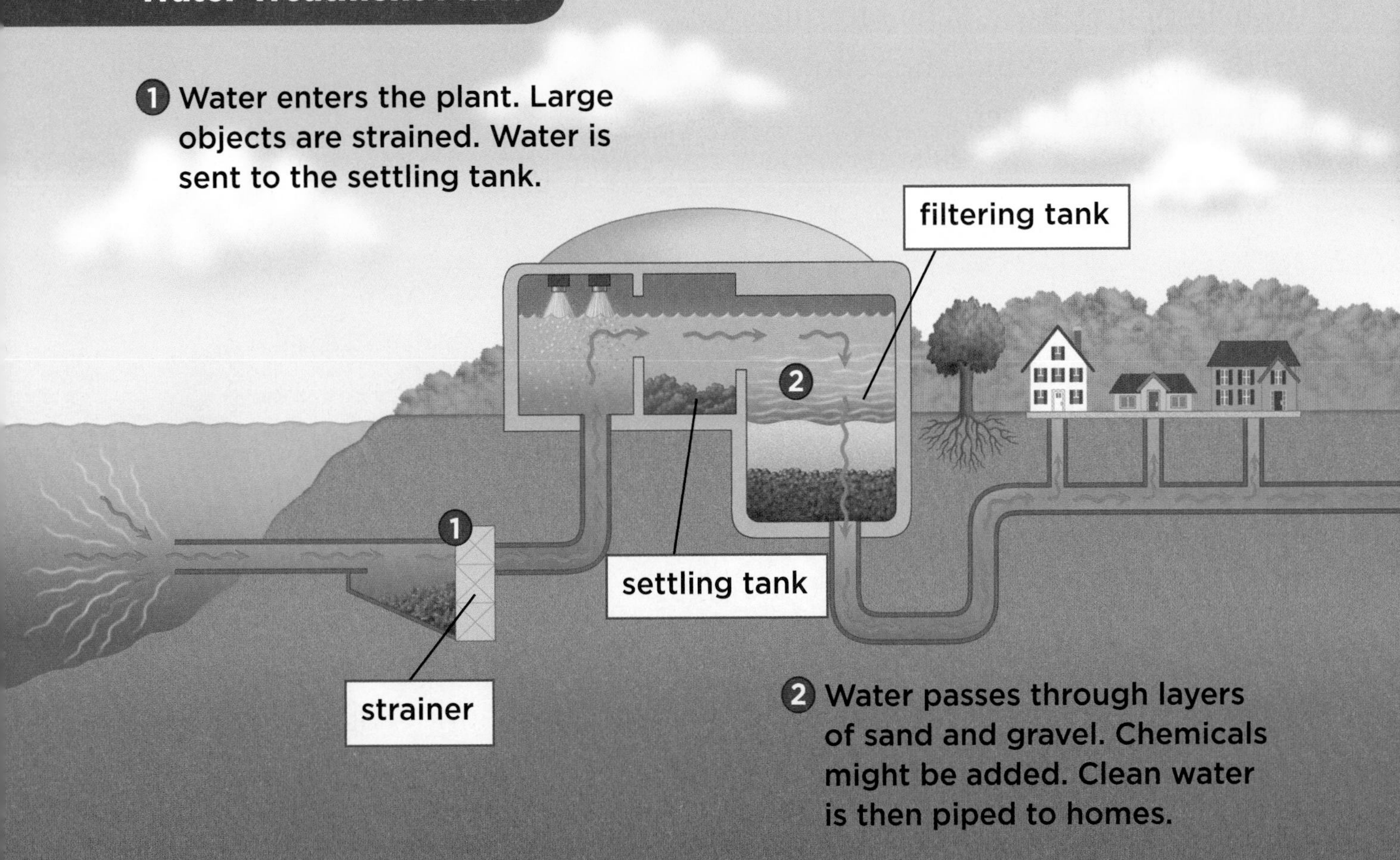

What can happen to air and water resources?

Air and water can be polluted. Water can be used up. Then these resources can not be used.

Pollution

Pollution (puh•LEW•shun) is what happens when harmful things get into water, air, and land. Pollution can make living things sick.

Pollution happens naturally. Volcanoes and wildfires pollute air with dust, gases, and ash. Eroded soil can wash into water and pollute it.

People cause pollution too. When people burn fossil fuels, they pollute the air. Trash, oil, and fertilizers (FUR•tuh•li•zurz) can pollute the land and water. People use fertilizers to help plants grow. When fertilizers soak into the ground, they can pollute groundwater.

▲ Oil can harm living things. This duck is covered with oil that spilled into the ocean.

The smoke from a wildfire pollutes the air. ▶

Using Up Resources

Suppose that you turned on the faucet and no water came out. It could happen. People might use up water if they waste it.

People waste water when they allow leaky faucets to drip. People waste water when they take long showers. Running a dishwasher that is not full also wastes water. Can you think of other ways that people waste water?

Quick Check

Fact and Opinion Why do you think it is important to take care of resources?

Critical Thinking What can you do to help take care of water?

◀ A leaky faucet wastes about 56 liters (15 gallons) of water a day.

Quick Lab

Record Water Use

1. **Communicate** Make a table like the one shown. Record each activity in which you use water.

Activities	Water Use

2. **Use Numbers** Use the table below to see how much water each activity uses. Record the amounts in your table.

Ways People Use Water

Activity	Normal Daily Use
showering	95 L
taking a bath	150 L
brushing teeth	18 L
washing hands	8 L
running a dishwasher	60 L
washing clothes	220 L

3. **Use Numbers** How much water do you use each day? Do you use a lot of water? How could you use less water?

How You Can Help

What You Can Do	How It Helps
• ride a bike instead of using a car • turn down the heat and wear a sweater • turn off lights when leaving a room	• conserves fossil fuels
• take a quick shower instead of a bath • turn off water as you brush your teeth	• conserves water
• plant trees and other plants	• conserves soil • helps put oxygen back into the air
• dispose of chemicals and trash properly	• protects land and water

Read a Table

What can you do to conserve water?

Clue: Find "conserves water" in the column "How It Helps." Then look to the left.

How can you conserve resources?

Everyone can make resources last longer if they conserve (kun•SURV). To **conserve** means to use resources wisely.

You do not have to be a scientist to conserve resources. You do not even have to be an adult. There are many things you can do to use fewer resources. There are things you can do to help protect air, land, and water.

Quick Check

Fact and Opinion Name two facts to explain why we should plant trees.

Critical Thinking How could you use less paper and conserve trees?

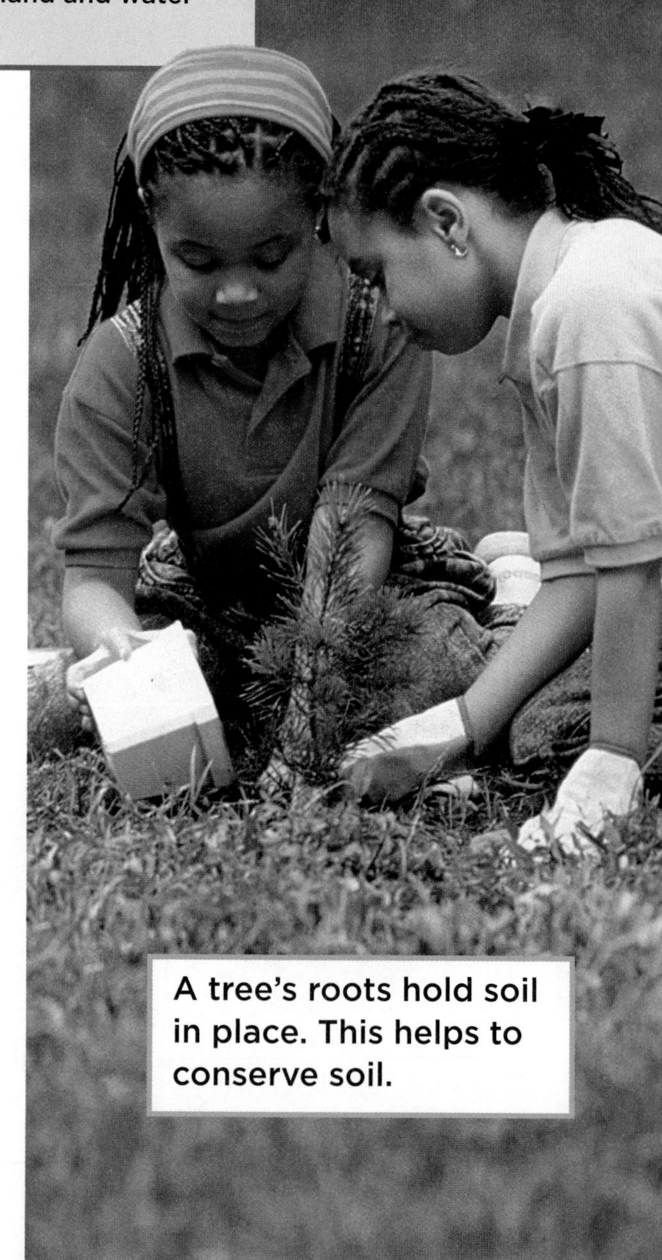

A tree's roots hold soil in place. This helps to conserve soil.

Lesson Review

Visual Summary

People **use air and water** for many things. Air and water are important renewable resources.

Freshwater is moved to where people need it.

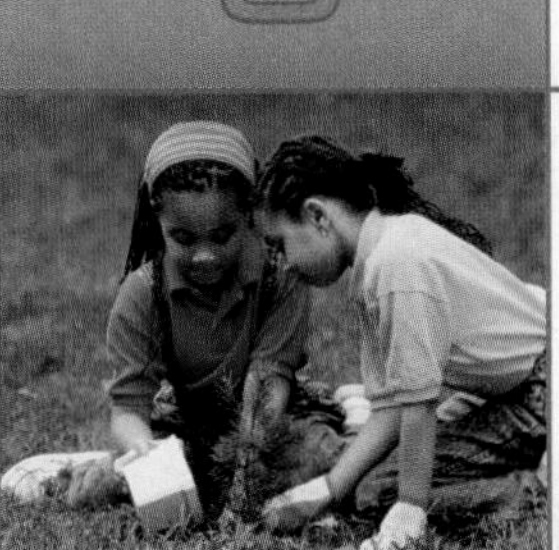

People sometimes pollute air and water. People can **conserve resources.**

Make a FOLDABLES® Study Guide

Make a three-tab book. Use it to summarize what you learned about air and water resources.

Think, Talk, and Write

1. **Vocabulary** What is pollution?
2. **Fact and Opinion** Why do you think people should conserve water?

Fact	Opinion

3. **Critical Thinking** Is groundwater a renewable resource? Explain your answer.
4. **Test Prep** **All of the following are sources of freshwater except**
   A rivers.
   B oceans.
   C streams.
   D groundwater.
5. **Essential Question** How do we use air and water?

Math Link

Solve a Problem

How much garbage does your family throw away in one week? Each day, weigh your family's garbage. Use the weekly total to determine the totals for one month and for one year.

Social Studies

Make a Poster

Do research to learn about Earth's freshwater. On a poster list the places where freshwater is found. For each source, include a drawing or picture from a magazine.

Be a Scientist

Materials

3 sheets of white paper

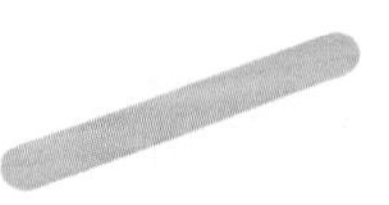

craft stick

petroleum jelly

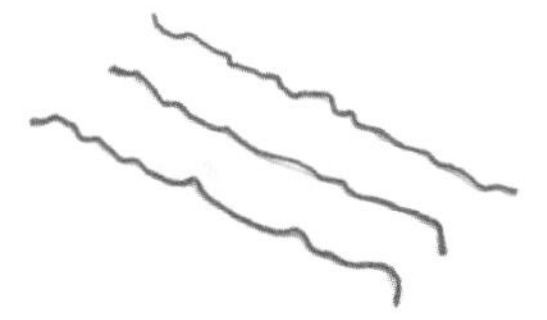

3 pieces of string

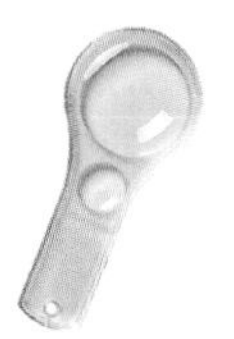

hand lens

Structured Inquiry

What things pollute the air?

Form a Hypothesis

What things are carried in air? Are the same things found in air everywhere? Record your hypothesis. Write your answer in the form "If I hang white papers covered with petroleum jelly in several locations, then..."

Test Your Hypothesis

1. **Use Variables** Choose different locations to hang three sheets of white paper. Some examples are near a heating or cooling vent, near grass or trees, and near a sidewalk. Write your name and a location on the bottom of each paper.
2. Dip a craft stick into petroleum jelly. Use the stick to smear a thin coat of jelly on a large area of each white paper.
3. Get an adult to help you attach the strings to the papers and hang the papers in different locations. Leave the papers for a few hours.
4. **Communicate** Make a table titled "What is in the air?" On it, record each location.
5. **Observe** Use a hand lens to observe each paper. Record your observations in the table.

Step 2

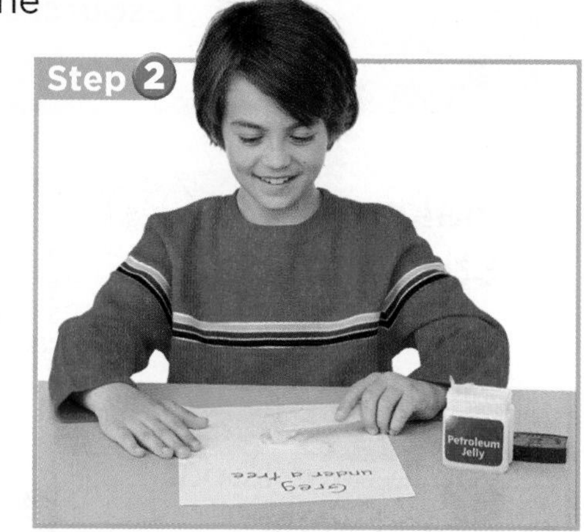

Step 4

What is in the air?	
near a tree	
near a heating vent	
near a sidewalk	

Draw Conclusions

6. **Interpret Data** What kinds of things pollute the air? What differences, if any, were there among the things found on the papers?

7. **Infer** Might any of the things carried in air make some people sick? Which ones?

Guided Inquiry

What is in water?

Form a Hypothesis

What kinds of things are found in water? Are the same things found in different sources of water? Write a hypothesis.

Test Your Hypothesis

Design a plan to test your hypothesis. Use the materials shown. Write the steps you plan to follow.

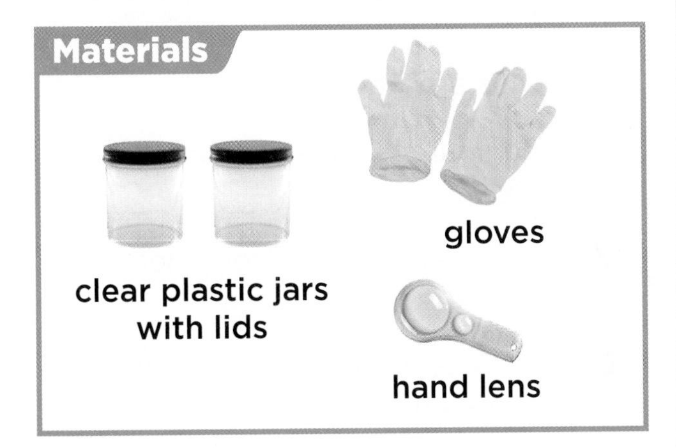

Draw Conclusions

Did your results support your hypothesis? Why or why not? Share your results with your classmates.

Open Inquiry

What other questions do you have about air pollution or water pollution? Share your questions with other classmates. How might you find the answers to your questions?

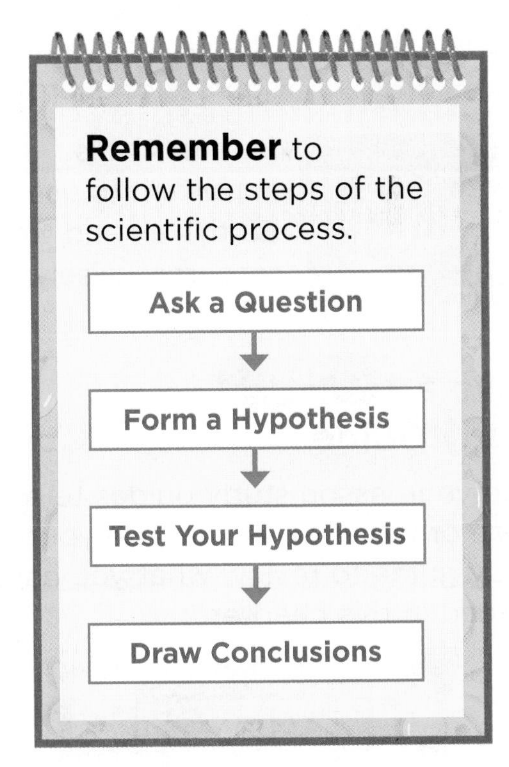

CHAPTER 6 Review

Visual Summary

Lesson 1 Rocks are made of minerals. Rocks are classified as igneous, sedimentary, or metamorphic.

Lesson 2 Soil is made up of weathered rocks, minerals, and once-living things. Many living things need soil to survive.

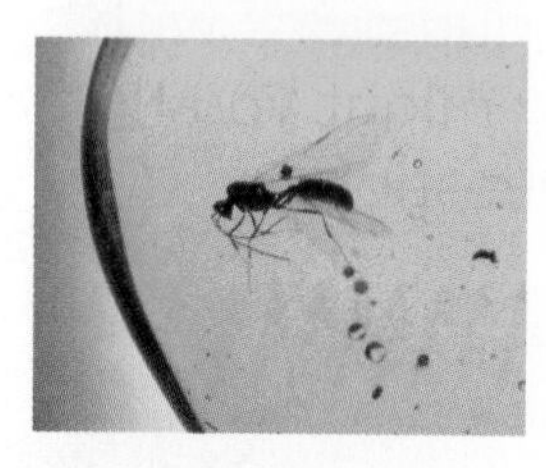

Lesson 3 Fossil fuels come from living things of long ago. They are nonrenewable sources of energy.

Lesson 4 It is important to use Earth's air and water resources wisely.

Make a FOLDABLES® Study Guide

Glue your lesson study guides to a piece of paper as shown. Use your study guide to review what you have learned in this chapter.

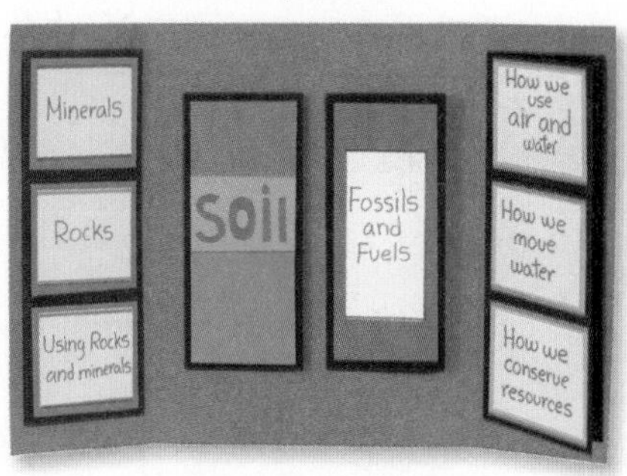

Vocabulary

DOK I

Fill each blank with the best term from the list.

conserve, p. 266	**metamorphic rock,** p. 233
fossils, p. 250	**minerals,** p. 228
groundwater, p. 261	**natural resource,** p. 244
humus, p. 240	**soil,** p. 240
igneous rock, p. 231	**solar energy,** p. 254

1. A rock that is changed by heating and squeezing is a ________.
2. Imprints and stony models are types of ________.
3. Rocks are made up of one or more ________.
4. Energy from the Sun is called ________.
5. Soil is an example of a ________.
6. One source of freshwater is ________.
7. To use resources wisely means to ________.
8. The decayed plants and animals in soil are known as ________.
9. Rock that forms when melted rock cools and hardens is ________.
10. A mixture of minerals, weathered rock, and other things is ________.

LOG ON e-Glossary Words and definitions online at www.macmillanmh.com

Skills and Concepts

DOK 2–3

Answer each of the following.

11. Draw Conclusions Scientists are developing fuels from plants such as corn. What kind of fuels would they be—renewable or nonrenewable? Explain.

12. Descriptive Writing Think of a sedimentary rock you learned about. Describe it.

13. Use Variables You want to find the best materials for filtering dirty water. You make one filter from paper and another from rocks and sand. You observe as you pour dirty water through them. Which variables changed? Which were the same?

14. Critical Thinking Which are more important to conserve—renewable or nonrenewable resources? Explain your answer.

15. Critical Thinking How does soil help plants grow?

16. Infer Where would you find freshwater?

17. Which layer of soil has the most humus in it? Why?

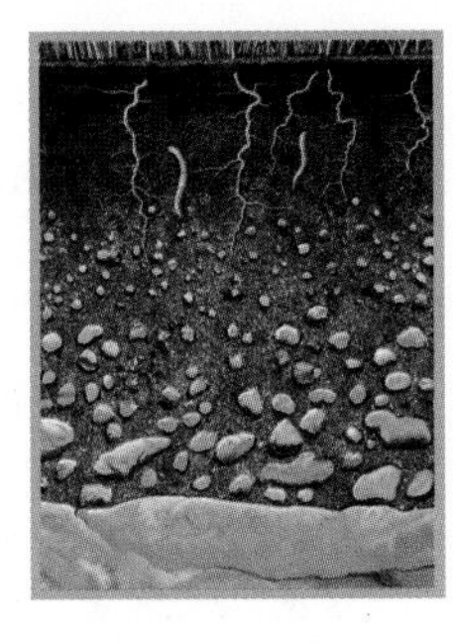

18. True or False *Most of Earth's water can not be used as drinking water.* Is this statement true or false? Explain.

19. Which mineral property is being tested in this photograph?

A color

B streak

C luster

D hardness

The Big Idea

20. What things used by people come from Earth?

Performance Assessment

DOK 3

Make a Poster

- Make a poster encouraging conservation of at least three different kinds of natural resources.
- Explain how you use each natural resource in your daily life. Why do you need each resource?
- Suggest ways you can conserve each of these resources.

Test Preparation

1 Look at the picture below.

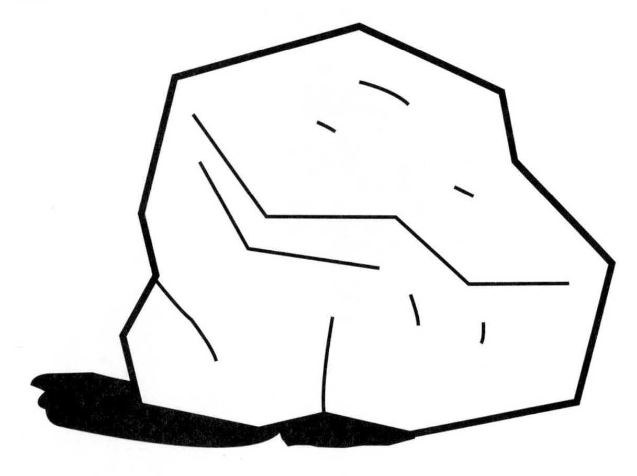

Which property would scientists most likely use to classify this object?

A size

B hardness

C temperature

D smoothness

DOK 2

2 Which type of rock forms when magma cools and hardens?

A metamorphic rock

B sedimentary rock

C igneous rock

D weathered rock

DOK 1

3 The layer of soil that contains the most humus is called

A topsoil.

B subsoil.

C bedrock.

D fertilizer.

DOK 1

4 Ronaldo poured water on three soil samples. He recorded the amount of time it took for water to flow through each sample.

Soils	
Type of Soil	**Time**
sand	20 seconds
silt	40 seconds
clay	60 seconds

How long would it take for water to flow through loam?

A about 20 seconds

B about 60 seconds

C more than 60 seconds

D between 20 and 60 seconds

DOK 2

5 All rocks are made of

A calcite.

B minerals.

C pebbles.

D sediment.

DOK 1

6 Which action would help conserve natural resources?

A Reuse fossil fuels.

B Recycle aluminum cans.

C Burn trash.

D Bury trash.

DOK 1

7 Kyla wants to know the texture of the soil in her garden. What should she do?

A Observe the color of the soil.

B Identify the minerals in the soil.

C Observe the size of the soil pieces.

D Measure how deep the soil is.

DOK 2

8 Which of these is a fossil fuel?

A oil

B wind

C electricity

D groundwater

DOK 1

9 Which most likely tells you that a soil will be good for growing crops?

A black color

B red color

C contains hematite

D contains animals

DOK 2

10 Which type of rock most likely forms in layers?

A metamorphic rock

B sedimentary rock

C igneous rock

D weathered rock

DOK 1

11 Where can you find groundwater?

A in a glacier

B on the ground

C under Earth's crust

D in a well

DOK 2

12 The picture below shows animal fossils in rock.

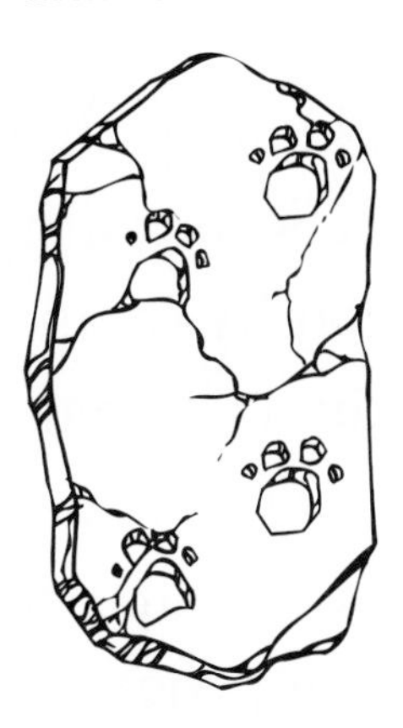

Describe or draw a different fossil.

DOK 1

Describe how your fossil formed.

DOK 2

Check Your Understanding

Question	Review	Question	Review
1	pp. 228–229	7	pp. 242–243
2	pp. 230–231	8	pp. 252–253
3	pp. 240–241	9	pp. 240–243
4	p. 243	10	p. 232
5	p. 228	11	p. 261
6	pp. 253, 265–266	12	pp. 250–251

Mapmaker

Do you like working on puzzles with small pieces? Are you good at giving directions or describing places? You might think about becoming a mapmaker.

Scientists who make maps have many different skills. Some gather data about the geography of an area. Others make three-dimensional models of landforms. Still others use data and models to draw the maps with computerized mapping programs.

There are things you can do right now to prepare for this job. Learn about Earth's land and water. Play games that require you to solve a problem. In high school take math, science, and computer classes. Then, get a college degree.

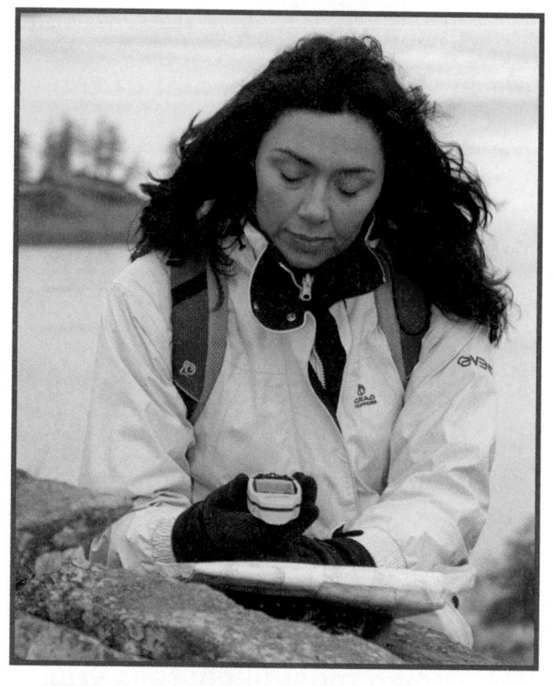

▲ This scientist is gathering data about landforms.

Here are some other earth science careers:

- oceanographer
- miner
- jewelry designer
- geologist

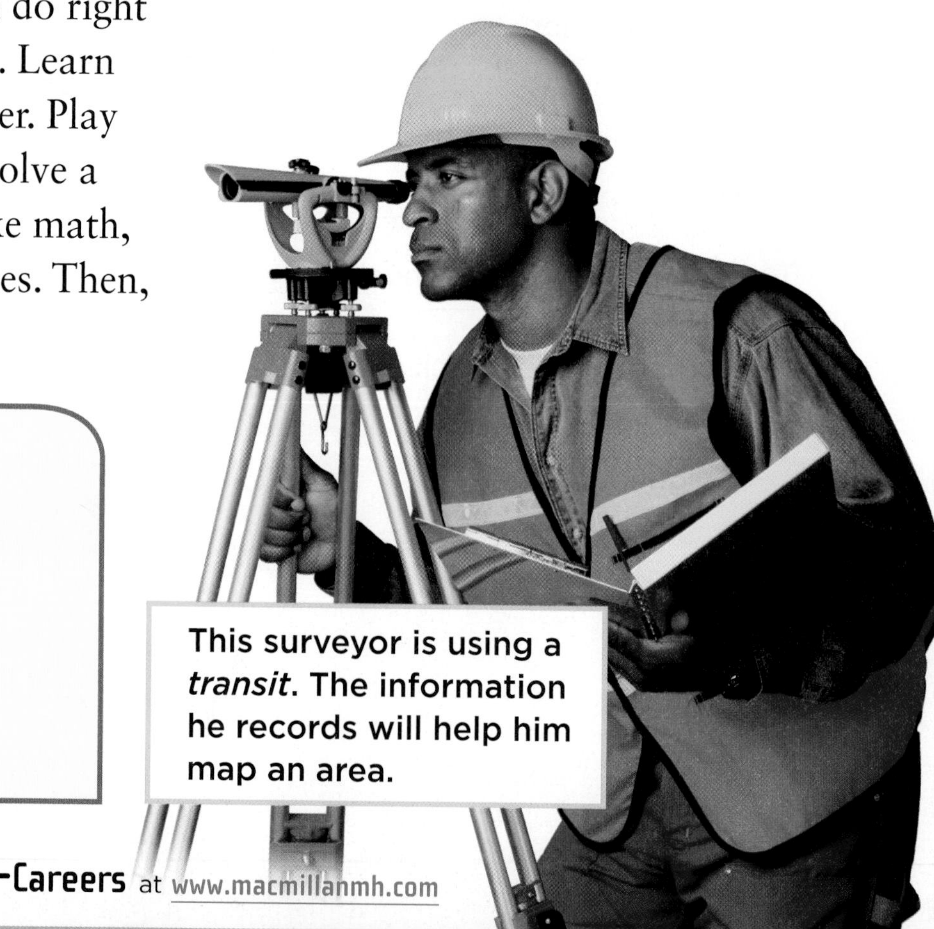

This surveyor is using a *transit*. The information he records will help him map an area.

LOG ON e-Careers at www.macmillanmh.com

UNIT D

Weather and Space

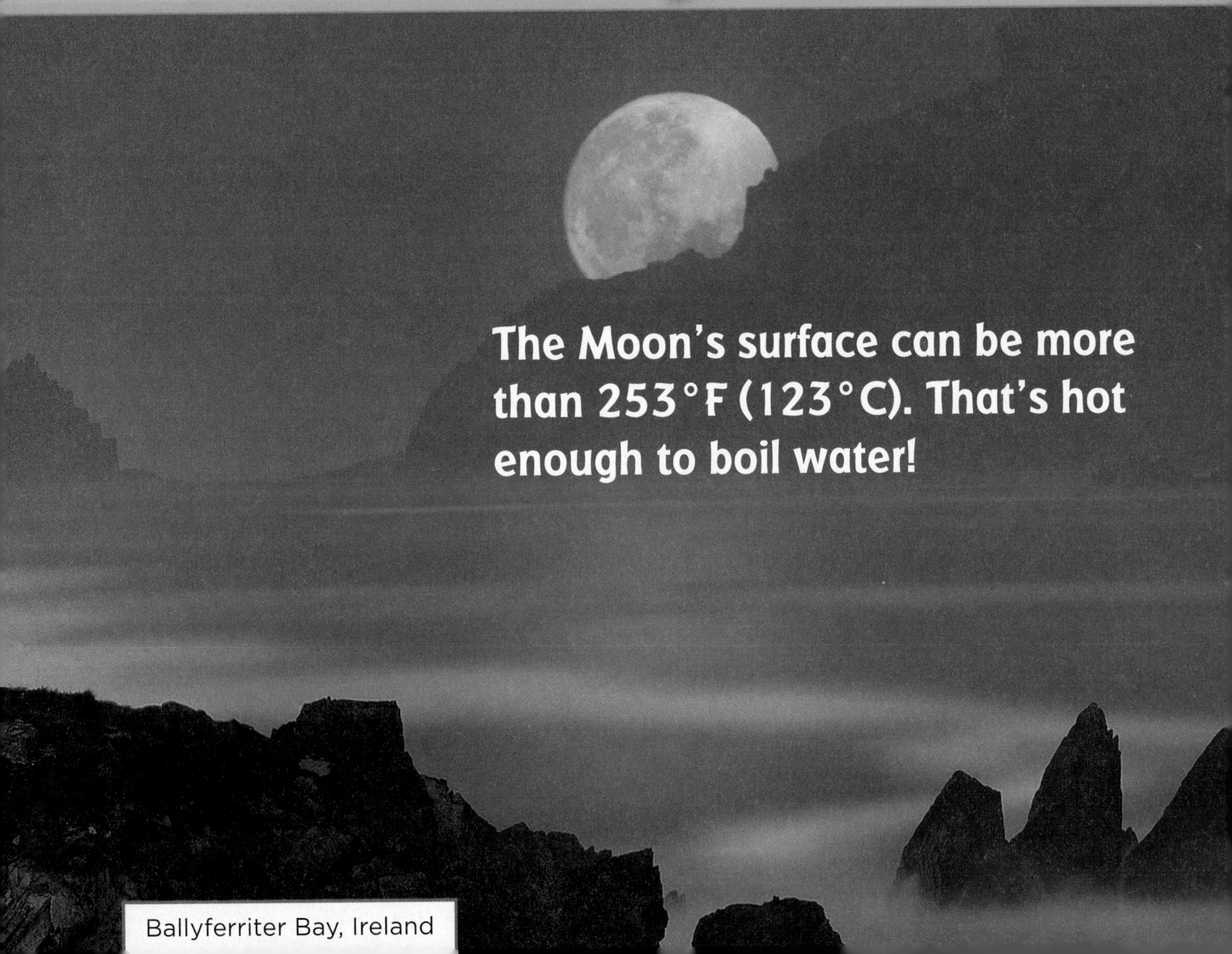

The Moon's surface can be more than 253°F (123°C). That's hot enough to boil water!

Ballyferriter Bay, Ireland

These geese are migrating in V formation over New York's Adirondack Mountains.

from *Ranger Rick*

What a Difference Day LENGTH Makes

How do you know spring is coming? Maybe you hear birds singing and see leaves growing on trees. How do birds and trees know it's time to do these things? They don't have calendars or wear wristwatches, of course. Their bodies keep track of the changing length of the days.

In the fall, shorter days tell plants and animals to start getting ready for winter. Trees lose their leaves. Some birds fly south.

The longer days of spring tell some birds to migrate north, sing songs to attract mates, and begin building nests. Trees and other plants begin growing leaves and flowers.

Do you do different things at different times of the year? For many living things, how long a day is makes a big difference!

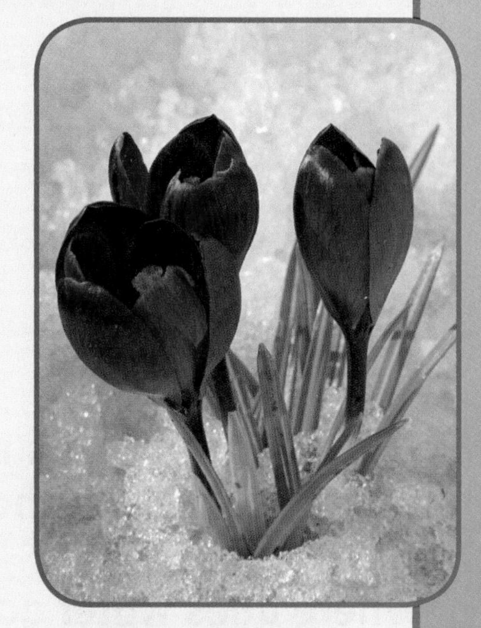

▲ Crocuses grow in early spring.

Chipmunks gather nuts in fall. ▶

Write About It

Response to Literature Animals respond to changing seasons in many ways. What are some ways you have seen nature change from season to season? Write about it.

CHAPTER 7

Changes in Weather

How does the weather where you live change throughout the year?

Essential Questions

Lesson 1
What information is used to predict the weather?

Lesson 2
Where does water go?

Lesson 3
How do weather patterns change?

lightning in Chimayo, New Mexico

Big Idea Vocabulary

weather what the air is like at a certain time and place (p. 280)

precipitation water that falls to the ground from the atmosphere (p. 282)

cloud a collection of tiny water drops or ice crystals in the air (p. 290)

water cycle the movement of water between Earth's surface and the atmosphere (p. 294)

climate the pattern of weather at a certain place over a long time (p. 304)

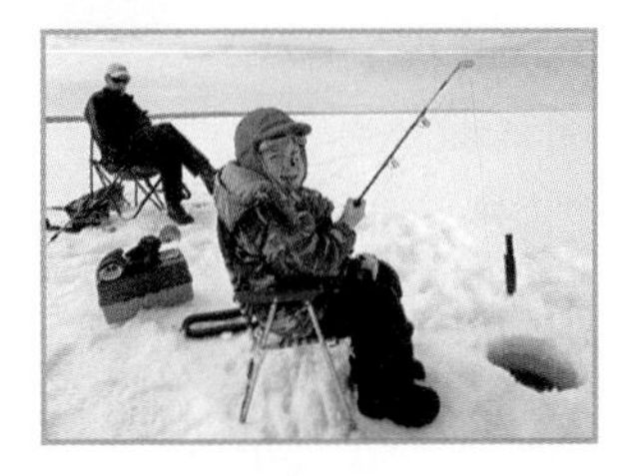

season a time of the year with different weather patterns (p. 308)

Visit www.macmillanmh.com for online resources.

Lesson 1

Weather

Look and Wonder

You can not see, taste, or smell it. What is it? It's air. How do you know air is around you?

Explore

Inquiry Activity

How can you tell air is around you?

Make a Prediction

Can air keep a paper towel inside a cup from becoming wet? Make a prediction.

Test Your Prediction

1. Fill a container about two thirds with water. Stuff a dry paper towel in the bottom of a cup.
2. **Experiment** Hold the cup upside down over the water. Push the cup straight down into the bottom of the container. Do not tilt the cup.
3. **Observe** Lift the cup out of the water. Do not tilt it. How does the paper towel feel?
4. **Observe** Repeat step 2. Slowly tilt the cup. Remove it from the water. What do you observe?

Draw Conclusions

5. **Infer** What escaped from the cup in step 4? How did this affect the paper towel?
6. **Infer** How do you know that air is around you?

Materials

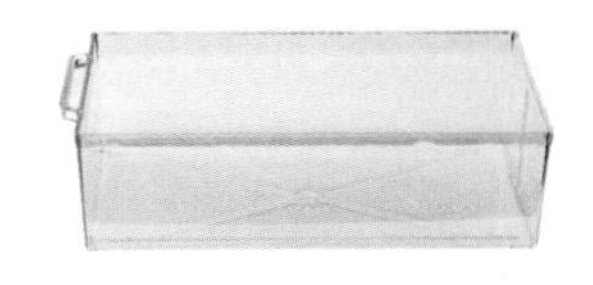

plastic container

water

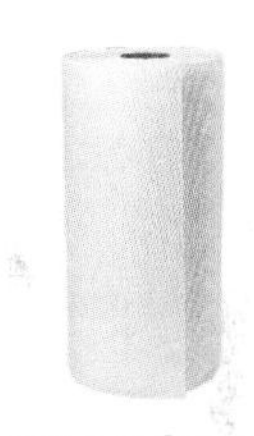

paper towel

plastic cup

Explore More

Experiment How else could you show that air is around you? Make a plan to find out.

Step 2

Read and Learn

Essential Question
What information is used to predict the weather?

Vocabulary
atmosphere, p. 280
weather, p. 280
temperature, p. 280
precipitation, p. 282
wind, p. 283
air pressure, p. 283

Reading Skill
Predict

What I Predict	What Happens

Technology
e-Glossary and e-Review online at www.macmillanmh.com

What is weather?

Have you ever filled a balloon with air? Air is made up of gases such as oxygen and nitrogen. You can not see, smell, or taste air. You know it is around you because it takes up space and can move things.

The air that surrounds Earth is part of the atmosphere (AT•muh•sfeer). The **atmosphere** is a blanket of gases and tiny bits of dust that surrounds Earth. The atmosphere has several layers. The layer closest to Earth is where weather (WE•thur) takes place. **Weather** is what the air is like at a certain time and place.

Air Temperature

Suppose a friend asks, "What is the weather like today?" You might describe the air temperature (TEM•puh•ruh•chur). **Temperature** is a measurement of how hot or how cold something is. A *thermometer* is a tool that measures temperature.

The weather can be warm or cool. It can be cloudy or sunny. It can be windy or still. What is the weather like here?

The Sun's energy heats Earth's land and water. Then the land and water heat the air. The Sun heats the land and water more at midday than at sunrise or sunset. This causes the air temperature to change throughout the day.

Quick Check

Predict How will the air temperature change today?

Critical Thinking How could you find out how much the air temperature changes during the school day?

Measuring Air Temperature

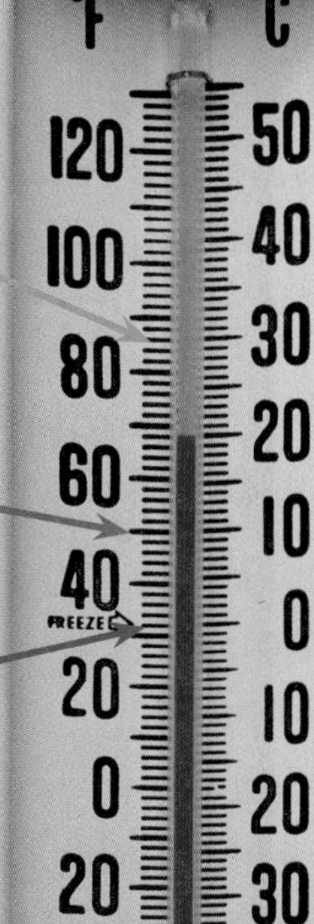

Read a Photo

What temperature is shown on the thermometer? Give the temperature for °F and for °C.

Clue: Look to the left and right of the top of the colored liquid.

How can you describe the weather?

Air temperature is just one thing you can describe about the weather. Precipitation (prih•sih•puh•TAY•shun), wind, and air pressure also describe the weather. When one of these changes, so does the weather.

Precipitation

Precipitation is water that falls to the ground from the atmosphere. You have probably seen rain and snow. There are other kinds of precipitation. *Sleet* is rain that freezes as it falls. *Hail* is lumps of ice that fall during a thunderstorm.

▲ Hail can be as large as a golf ball or even larger.

Weather Tools

A **rain gauge** (GAYJ) measures the amount of precipitation. ▶

A **weather vane** points out wind direction. ▶

Wind

Have you ever been pushed by moving air? **Wind** is moving air. On a windy day, air moves fast. On a calm day, air moves slowly. Weather tools are used to find the direction and speed of wind.

Air Pressure

Air not only takes up space, it has weight too. Its weight presses down on Earth. **Air pressure** is the weight of air pressing down on Earth. It affects our daily weather.

Quick Check

Predict What kind of precipitation might fall on a very cold day?

Critical Thinking Suppose it is sleeting. The air temperature rises above freezing (32°F or 0°C). What will happen?

Quick Lab

Make a Windsock

1. Bend wire to make a circle about 10 centimeters across.
2. Cut a sleeve from an old long-sleeved shirt. Staple the sleeve's large opening around the wire. Cut a small opening so you can tie some string to the wire.
3. Tape a small rock across from the string.
4. **Observe** Tie the string to a tree branch. Observe the windsock during the day. Keep a record of what you see.
5. **Infer** What can you tell about the wind from what you observed?

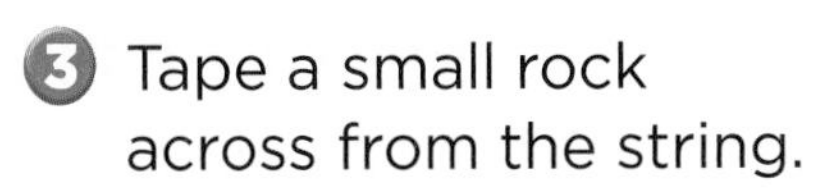

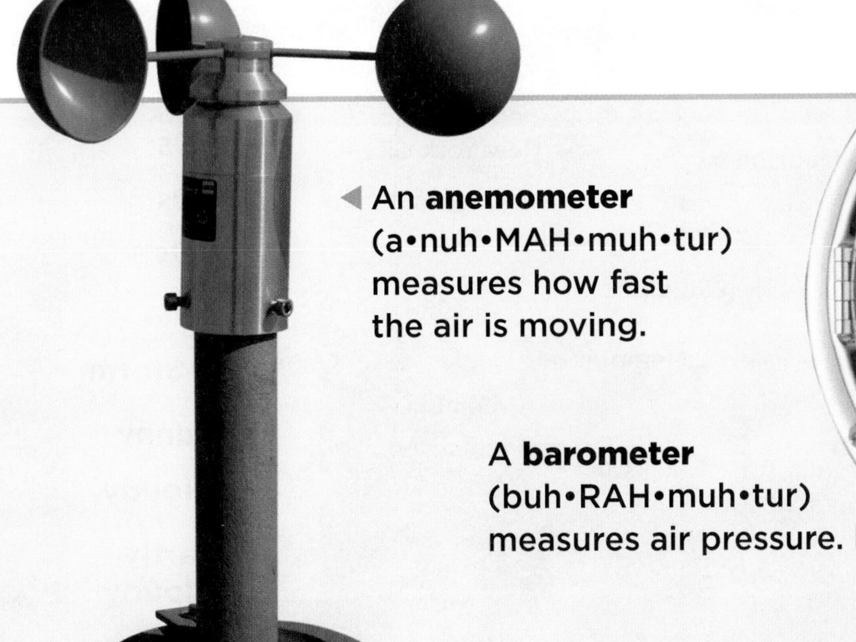

An **anemometer** (a•nuh•MAH•muh•tur) measures how fast the air is moving.

A **barometer** (buh•RAH•muh•tur) measures air pressure.

How do we predict weather?

Scientists use special tools to gather weather data. Weather balloons gather data about the atmosphere. Satellites observe weather from above Earth's surface. The data they collect is used to predict future weather. Weather conditions are shown on maps like the one below.

Who needs to know about weather? We all do. You need to know what to wear. Farmers need to know when to plant and pick crops. Pilots need to know about weather to fly airplanes safely.

▲ Weather balloons are used to gather data about weather.

Quick Check

Predict Low air pressure often means rain. What do you think high air pressure means?

Critical Thinking Who needs to know about weather? Why?

Read a Map

What weather is shown for Chicago?

Clue: Find Chicago. Then use the key and weather symbols.

A Weather Map

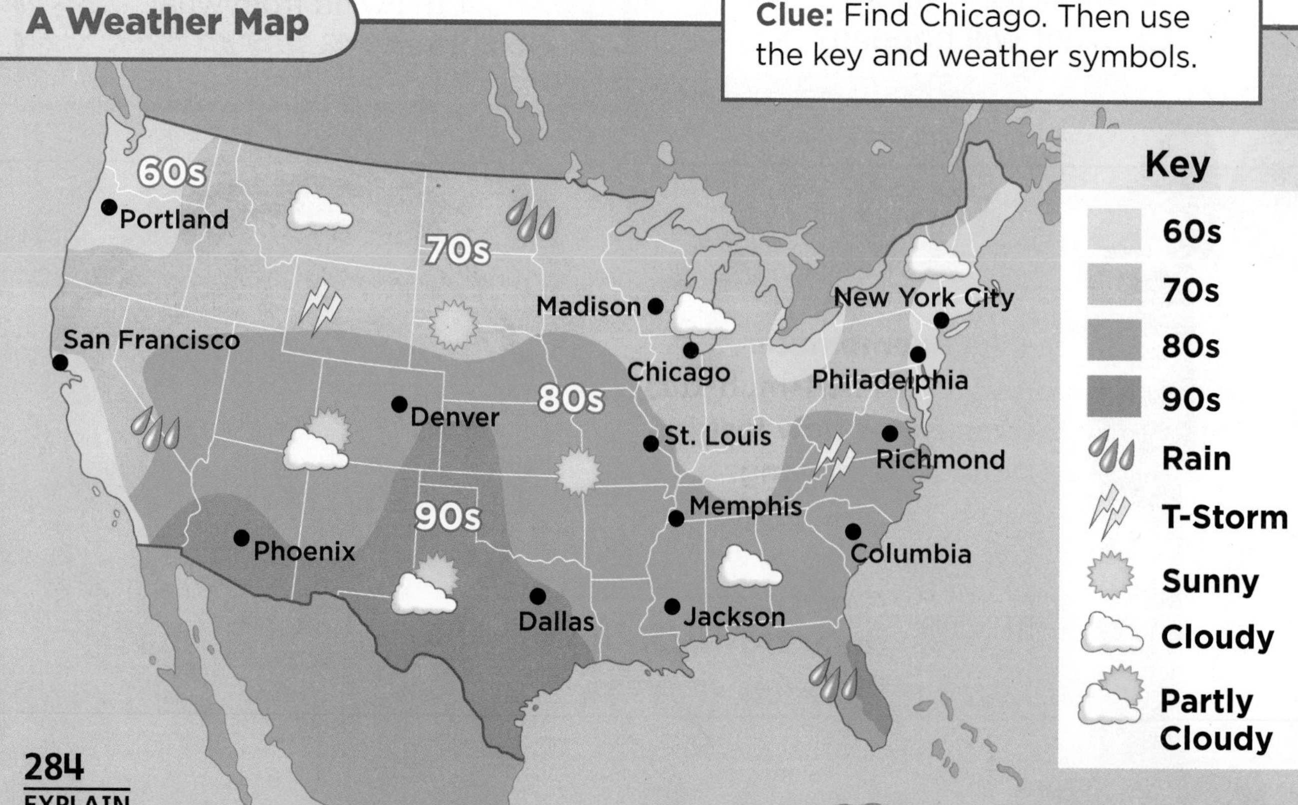

Lesson Review

Visual Summary

Weather is what the air is like at a certain time and place.

You can **describe weather** using air temperature, precipitation, wind, and air pressure.

Data gathered about weather can be used to **predict weather.**

Make a FOLDABLES® Study Guide

Make a layered-look book. Use it to summarize what you learned about weather.

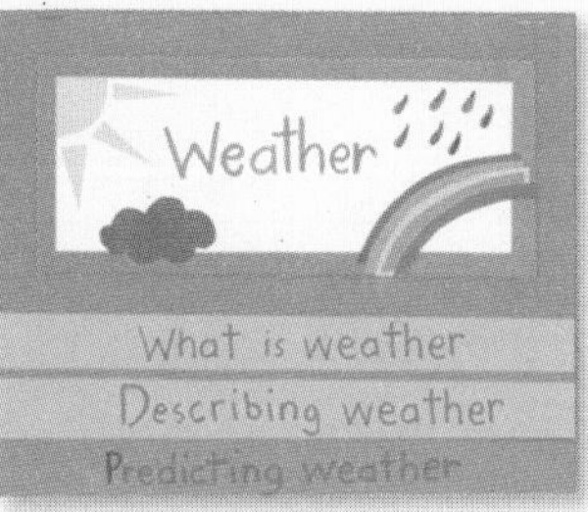

Think, Talk, and Write

1. **Vocabulary** What is precipitation? Give some examples.

2. **Predict** Suppose the air temperature is 70°F (21°C). It is also a dark, cloudy day. What precipitation might fall?

What I Predict	What Happens

3. **Critical Thinking** Why are weather predictions sometimes wrong?

4. **Test Prep** **Which tool measures temperature?**
   - **A** barometer
   - **B** anemometer
   - **C** thermometer
   - **D** wind vane

5. **Essential Question** What information is used to predict the weather?

Math Link

Compare Numbers

Watch a weather report. Find out what the high and low temperatures were for your area today. Do this for three days. Record your data. Which day was warmest? Which day was coldest?

Art Link

Draw the Weather

Observe the weather in the morning. Draw what you observe. Label your drawing with the date and time. Observe and draw the weather in the afternoon. Label your drawing. Did the weather change? How?

LOG ON e-Review Summaries and quizzes online at www.macmillanmh.com

Inquiry Skill: Interpret Data

Have you ever noticed that some months are warmer or wetter than others? This is generally true from year to year. How did scientists figure this out? One way is to **interpret data** from past years.

▶ Learn It

When you **interpret data**, you use information that has been gathered to answer questions or to solve problems. It is easier to **interpret data** when it is in a table or a graph. That way you can quickly see differences in the data.

▶ Try It

Scientists collect information about air temperatures in certain places. They use the data to figure out the average air temperature of a certain place for each month of the year. The data here shows the average monthly air temperatures for the city of Atlanta, Georgia. You can organize and **interpret data** to draw conclusions too.

Average Air Temperature (in °C)

Jan.	Feb.	Mar.	Apr.	May	June	July	Aug.	Sept.	Oct.	Nov.	Dec.
5	7	12	16	21	24	26	26	23	17	18	7

First, organize the data by making a bar graph. Follow these steps to make your bar graph.

1. List the months in order along the bottom of the graph. Label the bottom *Month.*
2. Write the numbers for the temperatures along the left side of the graph. Write the numbers *0, 2, 4, 6, 8,* and so on. End with the number *26.* Label this side and write a title for the graph.
3. Draw a bar to match each of the numbers from the data.
4. Now answer these questions: Which months are warmest? Which month is coolest?

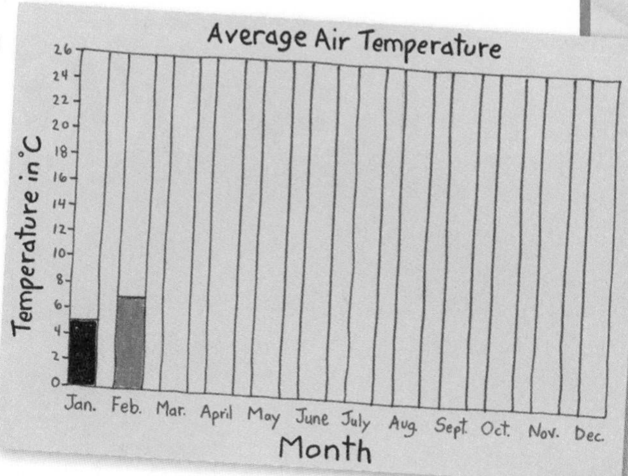

▶ Apply It

Now it is your turn to collect and **interpret data**. Measure the air temperature every hour for one school day. Begin at 9:00 A.M. and end at 2:00 P.M. Record your data in a table. Use the table to make a bar graph.

Use your bar graph to interpret your data. When is the air temperature warmest? When is it coolest?

Lesson 2

The Water Cycle

Look and Wonder

Rain falls from the sky. Where does the water for rain come from?

Explore

Inquiry Activity

How do raindrops form?

Purpose

Find out how raindrops form in the atmosphere.

Procedure

1. Fill a jar one-fourth full with warm water.
2. Stretch plastic wrap over the top of the jar. Use a rubber band to hold the plastic wrap in place. Place a marble in the center of the plastic wrap.

Step 2

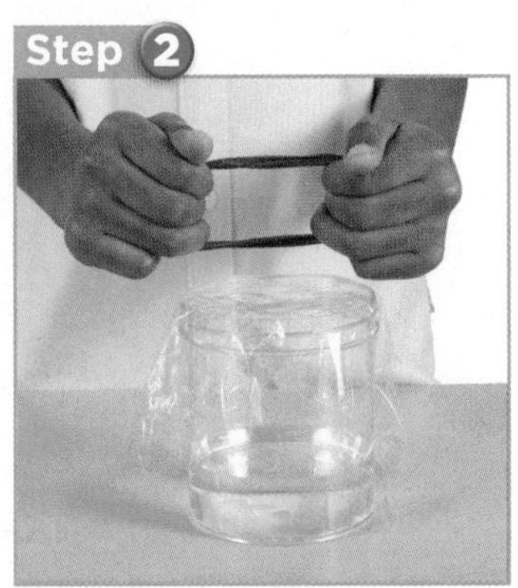

3. **Make a Model** Place a few ice cubes on the top of the plastic wrap to cool the air above the water. The warm water represents a lake. The air above it represents the atmosphere.
4. **Infer** Observe the bottom of the plastic wrap for several minutes. What forms there? Where did it come from?

Draw Conclusions

5. **Infer** Where does the water that forms raindrops come from?

Explore More

Experiment What would happen if you used cold water instead of warm water? Try it.

Materials

clear plastic jar

warm water

plastic wrap

rubber band

marble

ice cubes

Step 3

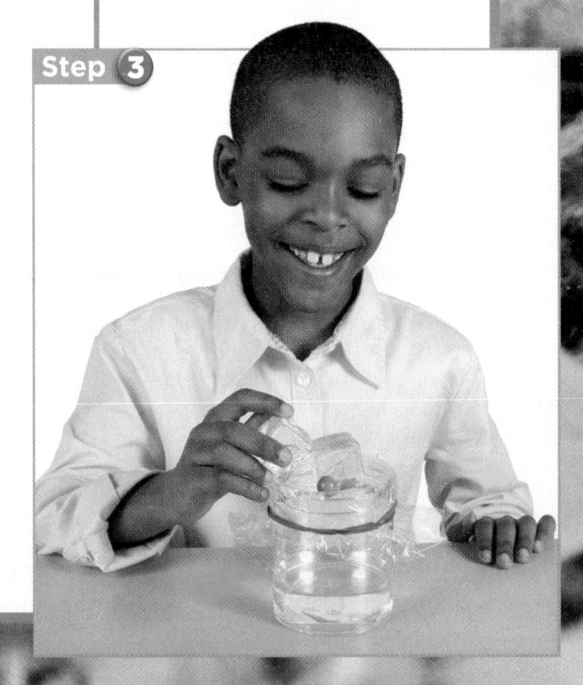

Read and Learn

Essential Question
Where does water go?

Vocabulary

Reading Skill 
Compare and Contrast

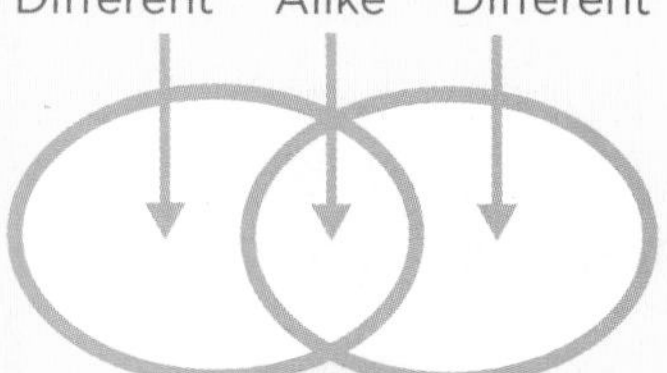

Technology 
e-Glossary, e-Review, and animations online at www.macmillanmh.com

Stratus clouds look like layered blankets.

What are clouds?

Is it cloudy today? A **cloud** is a collection of tiny water drops or ice crystals in the air. There are many different kinds of clouds. Precipitation falls from clouds. However, not all clouds bring precipitation.

Stratus Clouds

Stratus (STRA•tus) clouds are low, flat layers of clouds. They can cover most of the sky. They can be gray or white. Some kinds of stratus clouds bring rain or snow.

FACT Falling raindrops are shaped like the top part of a hamburger bun.

Cirrus Clouds

Cirrus (SIHR•us) clouds are thin, wispy white clouds that form high above the ground. They are usually seen during fair weather. If you see these clouds, precipitation might fall in a day or so.

Cumulus Clouds

Cumulus (KYEW•myuh•lus) clouds are white, puffy clouds with flat bottoms. You usually see them during fair weather. If they become dark though, they might bring thunderstorms.

▲ Cirrus clouds are sometimes called "mares' tails."

 **Quick Check**

Compare and Contrast How do cirrus clouds and cumulus clouds differ?

Critical Thinking There are cirrus clouds in the sky today. What can you predict about tomorrow's weather?

Cumulus is a Latin word meaning "pile" or "heap." ▼

Fog is a stratus cloud that forms near the ground.

How do clouds form?

If you have ever walked in fog, you know that fog feels moist. *Fog* is a stratus cloud that forms near the ground. Like other clouds, fog is made of tiny drops of water.

Evaporation

The water in fog and other clouds comes from water on Earth's surface. The water for clouds even comes from small rain puddles. When the Sun shines on a puddle, the water seems to disappear, but it does not. It *evaporates,* or changes to a gas. This changing of a liquid to a gas is called **evaporation** (ih•va•puh•RAY•shun). Water in the gas form is called **water vapor**. You can not see water vapor, but it is in the air all around you.

The water in a puddle changes to water vapor when it evaporates. ▼

Quick Lab

Cloud in a Jar

1. **Make a Model** Fill a clear jar half full with warm water. Place a metal tray of ice cubes on top of the jar. Wait for about one minute.
2. **Observe** Darken the room. Shine a flashlight into the jar. What do you see? What is it made of?
3. **Infer** Where do clouds come from?

Condensation

Have you ever seen water on the inside of a window? The water forms when water vapor in air touches the cooler window. The water vapor *condenses*, or changes into liquid water, on the glass. The changing of a gas to a liquid is called **condensation** (kahn•den•SAY•shun).

Condensation forms clouds the same way. Water vapor rises into the air and cools. Then it condenses, and the water collects around dust specks in the air. This forms clouds.

▲ Water drops formed on the inside of this window when water vapor in air cooled and condensed.

Quick Check

Compare and Contrast How are condensation and evaporation different?

Critical Thinking Is there evaporation of water from soil? Explain your answer.

What is the water cycle?

Water moves from Earth's surface into the atmosphere and back again. If it didn't, Earth's surface would soon run out of water! During the **water cycle**, water moves between Earth's surface and the atmosphere.

There would be no water cycle without the Sun. The Sun's energy heats water and causes it to evaporate. Water vapor condenses and forms clouds. The water falls back to Earth as precipitation.

Water that falls might seep into the ground and become groundwater. It might flow over the ground. Water flows downhill. It enters bodies of water. Some water becomes water vapor. The process begins again.

The Water Cycle

Water Condenses
Water vapor rises and cools. The water vapor changes into liquid water drops. The drops form clouds.

Water Evaporates
The Sun's energy heats water in lakes, rivers, streams, oceans, and on land. The water changes from a liquid into water vapor, a gas.

Quick Check

Compare and Contrast How is water in the water cycle different after evaporation?

Critical Thinking Why is the water cycle called a cycle?

Water Falls
When enough water has condensed in clouds, the water falls to Earth as precipitation. It can fall as rain, snow, sleet, or hail.

Water Flows
Some precipitation flows over Earth's surface. It flows downhill into lakes, rivers, and oceans. Precipitation can also soak into the ground. Groundwater flows underground through rock.

Read a Diagram

What happens after precipitation falls?

Clue: Follow the arrows and read the captions.

Science in Motion Watch the water cycle at www.macmillanmh.com

What are some kinds of severe weather?

Most of the time, the water cycle brings gentle rains and snowfalls. If you have ever seen a thunderstorm, however, you know weather can be severe. A *thunderstorm* is a storm with thunder, lightning, heavy rains, and strong winds. It can produce hail too. A thunderstorm is just one kind of severe weather.

Tornadoes

A **tornado** (tor•NAY•doh) is a powerful storm with rotating winds that forms over land. It looks like a big, tall funnel. A tornado damages most things in its path.

A tornado's winds move in a circle. They move at 160 kilometers per hour (about 100 miles per hour) or more.

FACT The United States has many more tornadoes than hurricanes every year.

This is what a hurricane looks like from space.

▲ A hurricane's winds move in a circle. They can blow at 119 km/h (74 mph) or more.

Hurricanes

A **hurricane** (HUR•ih•kayn) is a large storm with strong winds and heavy rain. It forms over the ocean.

When a hurricane moves over land, its winds and rain damage property. Trees are blown over. Hurricanes can also cause flooding.

Blizzards

A **blizzard** is a storm with lots of snow, cold temperatures, and strong winds. Blizzards bury plants, cars, and buildings under snow.

▲ The strong winds of a blizzard blow snow. This makes it hard to see.

Quick Check

Compare and Contrast How are hurricanes, thunderstorms, and blizzards alike?

Critical Thinking How might a thunderstorm affect living things?

How can you stay safe in severe weather?

People can be harmed during severe weather. There are things you can do to stay safe.

During a thunderstorm, do not stand under a tree. Do not use phones, computers, or other electrical devices. Stay inside a sturdy building.

During a blizzard, stay inside a warm building. If you must leave, be sure to wear warm clothing.

If a hurricane or tornado is on the way, stay inside. Stay away from doors and windows. During a tornado, go to the basement. If you can not get to a basement, lie flat in a low place.

Quick Check

Compare and Contrast How can you stay safe during hurricanes, tornadoes, and thunderstorms?

Critical Thinking What might happen if you stay outside during a blizzard?

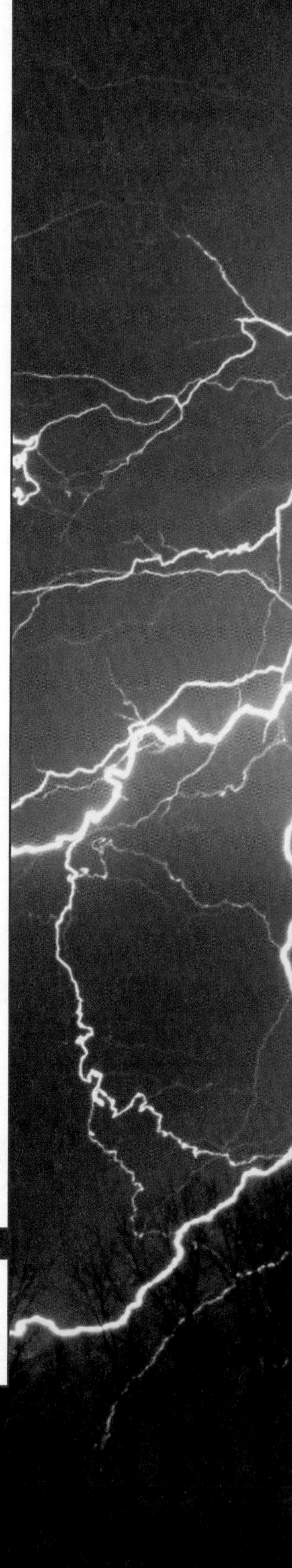

Never stand under a tree in a thunderstorm. Lightning tends to strike tall objects. ▶

Lesson Review

Visual Summary

There are different kinds of **clouds.** Each cloud brings different types of weather.

Clouds form when water vapor in the air condenses.

The **water cycle** describes how water moves between Earth's surface and the atmosphere.

Make a FOLDABLES® Study Guide

Make a trifold book. Use it to summarize what you learned about the water cycle.

Main Ideas	What I learned...	Sketches
Types of Clouds		
How Clouds Form		
The Water Cycle		

Think, Talk, and Write

1. **Vocabulary** What do you call water in the atmosphere that is a gas?

2. **Compare and Contrast** How are evaporation and condensation alike? How are they different?

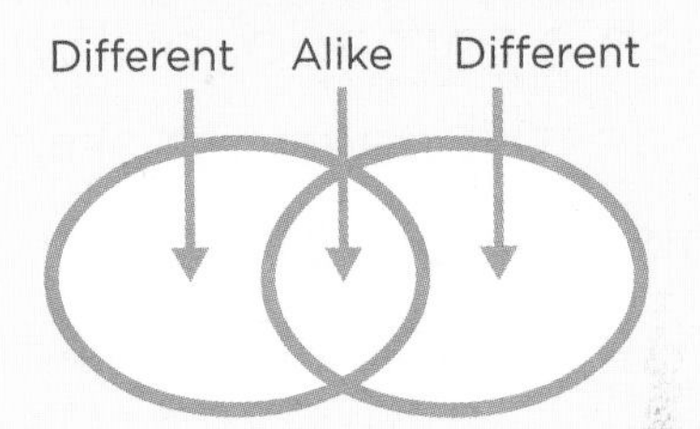

3. **Critical Thinking** What are two steps in the water cycle that you have observed?

4. **Test Prep** **Which kind of severe weather is not likely to happen in summer?**
   - **A** a thunderstorm
   - **B** a blizzard
   - **C** a hurricane
   - **D** a tornado

5. **Essential Question** Where does water go?

Writing Link

Write a Report
Some areas of the United States have more hurricanes than others. Do research to find out which regions have the most hurricanes. Explain why in a report.

Health Link

Promoting Safety
Choose one kind of severe weather. Write a public safety announcement. Tell people how to stay safe in this kind of severe weather. Read your announcement to the class.

Tracking Twisters

When a tornado, or twister, touches down, it can destroy almost anything in its path. For this reason, scientists gather information about tornadoes to help predict where they may happen.

First, scientists observe and measure weather to see if conditions are right for a tornado to form. Tornadoes occur when warm, moist air near the ground mixes with cool, dry air above it and rises rapidly.

Doppler radar is used to track storms. Radar works by sending out radio waves from an antenna. Objects in the air, such as raindrops, bounce the waves back to the antenna. Doppler radar can track the direction and speed of a moving object, such as a tornado or other storm.

People called storm chasers get a close-up look at tornadoes from planes or cars. The information they gather is used to warn communities about tornadoes before they strike.

Doppler radar is used to locate places where tornadoes are likely to occur.

The information gathered by storm chasers helps to warn communities about the size and direction of a tornado.

Predict

When you predict,

- you use what you know to tell what you think might happen in the future.

Write About It

Predict What if there were no storm chasers? What if there were no technology to warn people of tornadoes? Write about what might happen.

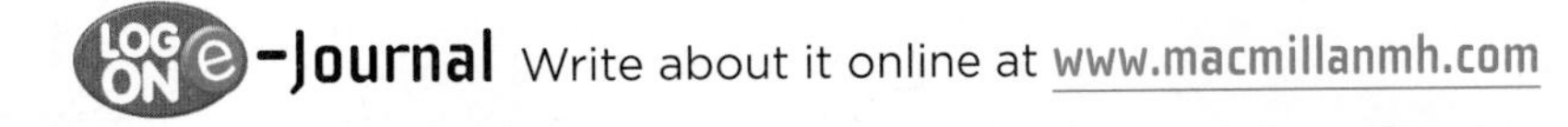

Lesson 3

Climate and Seasons

Look and Wonder

An orchid grows best in a warm, wet place. Could it grow well where you live?

Explore

Inquiry Activity

How do temperature and precipitation patterns compare?

Purpose

Find out how temperature and precipitation compare between two cities.

Procedure

1. Study the data in these tables. What do they show?

Average Air Temperature (in °F)												
City	Jan.	Feb.	Mar.	Apr.	May	June	July	Aug.	Sept.	Oct.	Nov.	Dec.
A	21	25	37	49	59	69	73	72	64	53	40	27
B	78	78	78	78	81	82	82	84	84	84	81	80

Average Precipitation (in inches)												
City	Jan.	Feb.	Mar.	Apr.	May	June	July	Aug.	Sept.	Oct.	Nov.	Dec.
A	2	1	3	4	3	4	4	4	4	2	3	2
B	10	10	14	15	10	6	10	10	9	9	15	12

2. **Use Numbers** What are the highest and lowest temperatures for each city? Which city's temperature changes the most during the year? How much precipitation does each city get? Use a calculator.

Draw Conclusions

3. **Interpret Data** How do the temperature and precipitation patterns for these cities compare?
4. **Infer** Which city is better for growing orchids? Why?

Explore More

Interpret Data What are the monthly air temperatures and precipitation like where you live? How could you find out?

Read and Learn

Essential Question

How do weather patterns change?

Vocabulary

climate, p. 304

sphere, p. 305

axis, p. 305

seasons, p. 308

Reading Skill

Summarize

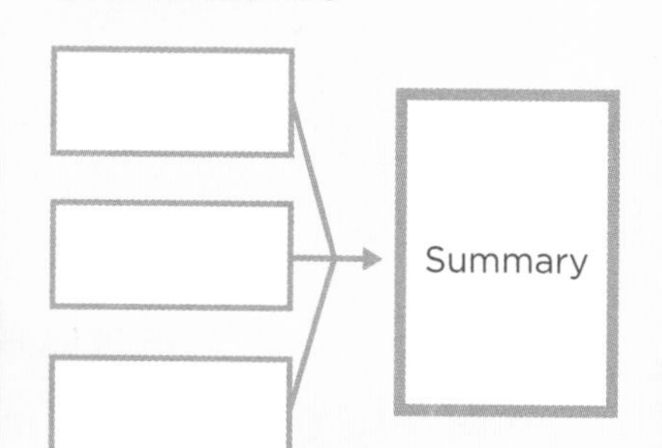

Technology LOG ON

e-Glossary and e-Review online at www.macmillanmh.com

What is climate?

The weather might change every day where you live, but the climate does not. **Climate** is the pattern of weather at a certain place over a long time. A climate is described by its average temperature and precipitation.

Some climates are hot and dry. Some climates are hot and wet. Some climates are cold and dry. Other climates vary. They might have hot and cold temperatures and wet and dry periods at different times of the year.

Different Climates

The Sonoran Desert is hot and dry.

Chicago can be hot or cold, dry or wet.

Antarctica is cold and dry.

Rio de Janeiro is hot and wet.

The Sun's Rays

Read a Diagram

Which place has a warmer climate—A or B? Why?

Clue: Find where the Sun's rays strike Earth nearly straight on.

A place's climate depends on where it is on Earth. Earth is shaped like a **sphere** (SFEER), or ball. Earth spins around its axis (AK•sus). An **axis** is a line through the center of a spinning object. Earth's axis is not straight up and down. It is slightly tilted.

The Sun's rays strike Earth differently at different places because of Earth's shape and tilt. The Sun's rays strike some places on Earth nearly straight on. These places get the most energy and have warmer climates. The Sun's rays strike other places on Earth at a slant. These places get less energy because the Sun's rays are more spread out. These places have colder climates.

Quick Check

Summarize Why are some climates warmer than others?

Critical Thinking How are weather and climate related?

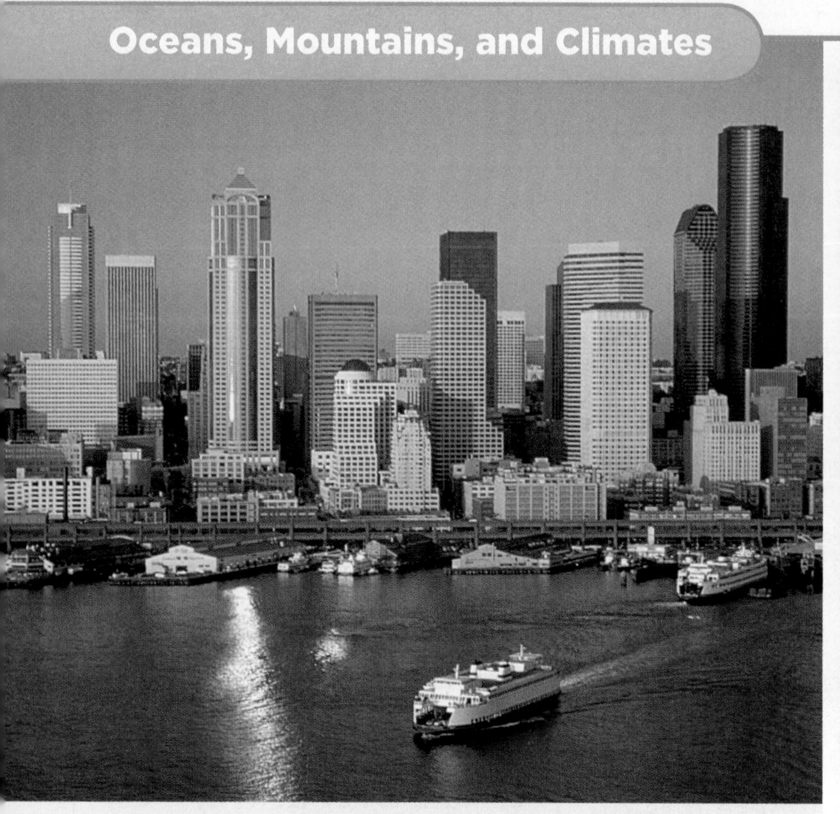

Seattle, Washington, is near the ocean. It has milder temperatures and more rain than places farther from the ocean.

Breckenridge, Colorado, is high in the Colorado Rockies. It has cool temperatures.

What affects climate?

You just learned how the Sun affects climate. Being near an ocean or large lake affects climate too. Oceans keep air temperatures of nearby land from becoming too hot or cold. Places near the ocean have milder climates than those much farther from a coast.

How high a place is also affects its climate. Air temperature gets colder as you go higher in the atmosphere. Places in mountains tend to have colder air temperatures and climates than low areas.

Mountains also affect how wet a climate is. One side of a mountain might be wet while the opposite side might be dry.

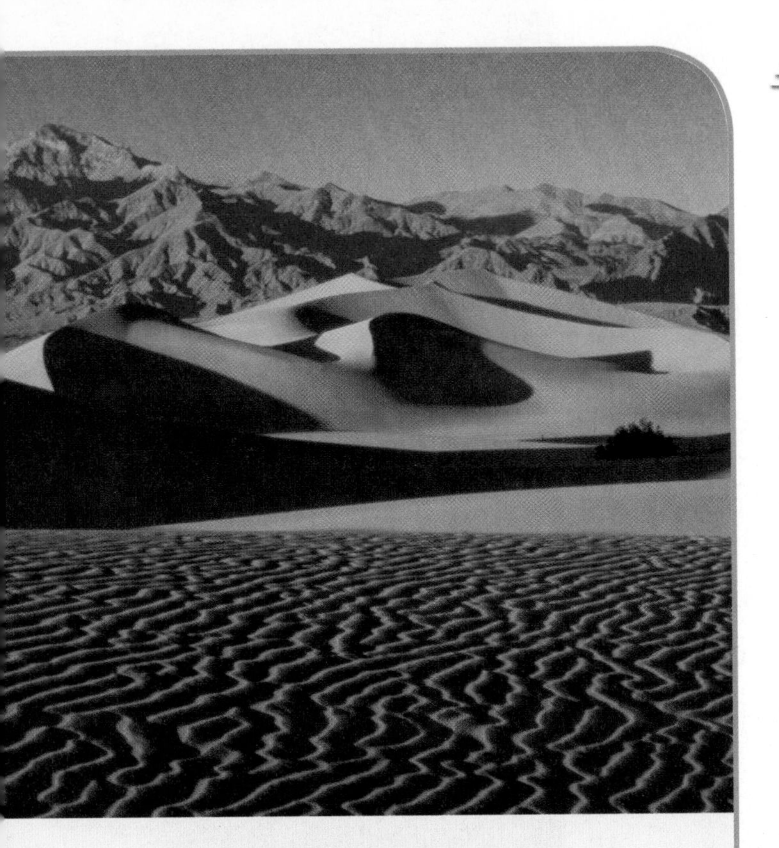

Death Valley, California, is a dry place. The mountains keep moist air from reaching here.

Moist air from the ocean moves toward mountains along the coast. The mountains force the air upward. The rising air cools and forms clouds. Rain or snow might fall. This causes places on the ocean side of mountains to have a wet climate.

Air that blows over mountains is dry. This is because the air has lost its moisture on the ocean side. This causes places on the opposite side to have a dry climate.

Quick Lab

Compare Climates

1. **Make a Model** Label one sheet of paper *City A* and another *City B.* Use a flashlight to model the Sun. Hold it about 6 cm above the paper for *City A.* Shine it straight down. Have a partner use a pencil to trace along the edge of the light.

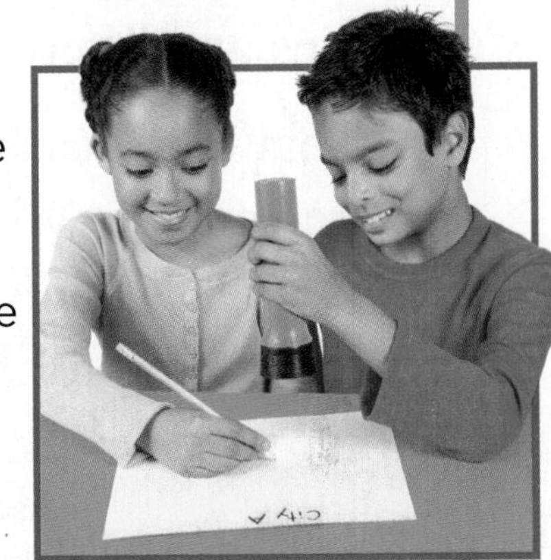

2. **Make a Model** Repeat step 1 for *City B.* This time, tilt the flashlight as you shine it on the paper.
3. **Interpret Data** Over which city is the shape of light larger? Over which city is the Sun's energy more spread out?
4. **Infer** Which city would have a colder climate?

Quick Check

Summarize How do mountains affect climate?

Critical Thinking What might happen to the climate of a city if a large lake nearby evaporated?

What are seasons?

Every year there are four seasons—winter, spring, summer, and fall. **Seasons** are times of the year with different weather patterns. Winter is the coldest season. Summer is the warmest season. You will find out why seasons are different in the next chapter.

What each season is like depends on where you live. Winter in Arizona is different from winter in Wisconsin. Winter is still the coldest season for both places. However, winter temperatures are very different for each place.

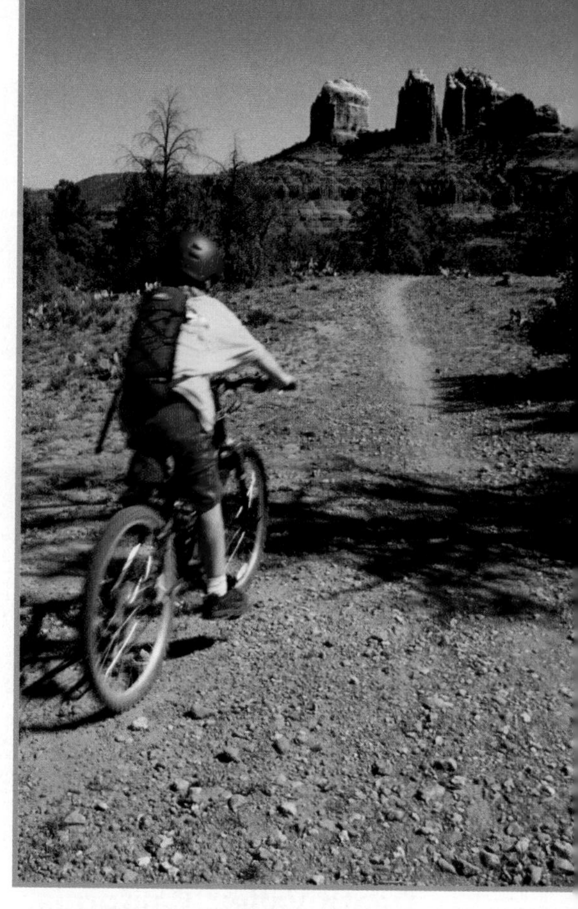

▲ In Sedona, Arizona, winter is not very cold. Snow rarely falls.

Quick Check

Summarize What are seasons?

Critical Thinking Why is winter in Wisconsin colder than winter in Arizona? **Hint** Find each on a globe.

In Bayfield, Wisconsin, winter is cold and snowy.

FACT Animals can not predict what seasons will be like.

Lesson Review

Visual Summary

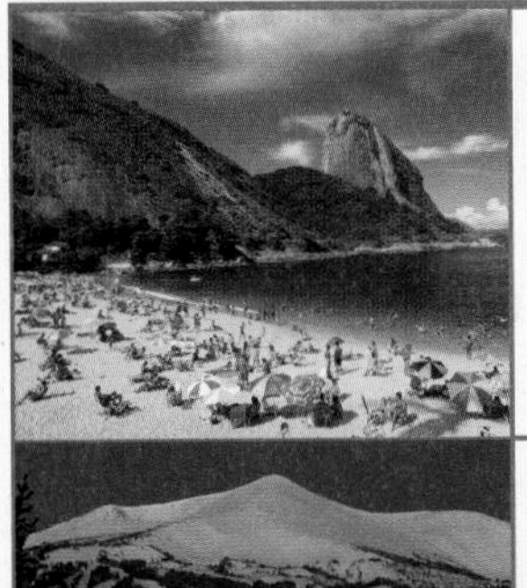

Climate is the pattern of weather at a certain place over a long time.

Oceans, large lakes, mountains, and how high a place is can all **affect climate.**

Seasons are different from place to place.

Make a FOLDABLES® Study Guide

Make a layered-look book. Use it to summarize what you learned about climates and seasons.

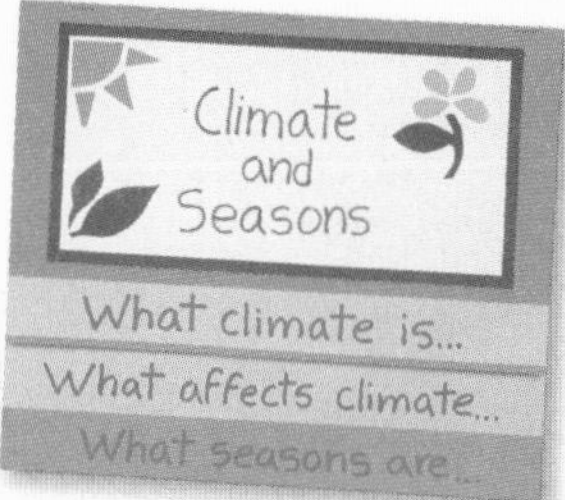

Think, Talk, and Write

1. **Vocabulary** What are times of the year with different weather patterns called?

2. **Summarize** In what ways do climates differ?

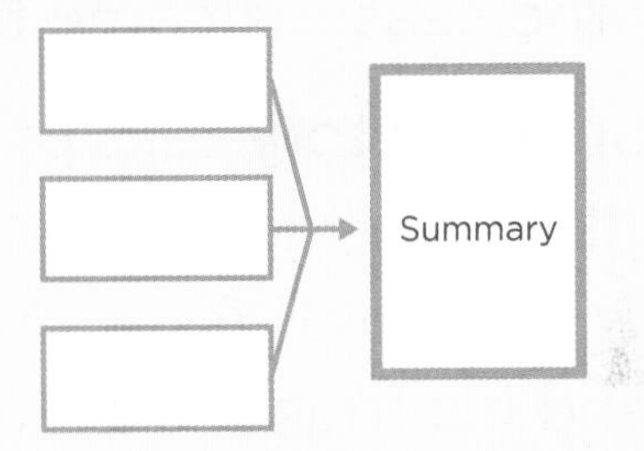

3. **Critical Thinking** How would you describe the climate where you live?

4. **Test Prep** **Which word best describes the ocean side of a mountain?**
   - **A** hot
   - **B** cold
   - **C** dry
   - **D** wet

5. **Essential Question** How do weather patterns change?

Math Link

Solve a Problem

There are four seasons in a year. Each season is about the same number of months. How many months long is each season? Explain how you found your answer.

Music Link

Find a Season Song

Think of a song that reminds you of a climate or season. Write the words in the song that are related to the climate or season.

A Season Myth

A myth is a story. Many myths tell how or why things happen in nature. For example, some myths tell how the seasons came to be.

One Seneca story tells of Old Man Winter who makes the world icy and cold. Young Man Spring and his friend, the Fawn, come to send Old Man Winter away. Fawn represents the warm south wind. Together, they melt Old Man Winter. Spring, flowers, and animals return to the land once more.

Fictional Narrative

A good story

- has a beginning, middle, and end;
- has a plot with a problem that needs to be solved;
- has characters who do things and a setting where the events take place;
- often has dialogue.

Write About It

Fictional Narrative Write your own myth about how a season came to be. Use an animal as a character in the story. Use all the parts of a good story when writing.

Average Temperature

Weather reports give the high and low temperatures for each day. You can use this information to find the average temperature. The table below shows the high and low temperatures for three cities on the same early spring day.

City	High Temperature	Low Temperature
Detroit, Michigan	10°C	2°C
Charleston, South Carolina	13°C	3°C
Phoenix, Arizona	28°C	14°C

Find an Average

- First, add all the numbers.
- Then, divide by the number of numbers you added.

To find the average temperature in Charleston:

$13 + 3 = 16$

$16 \div 2 = 8$

The average temperature was 8°C.

Solve It

What was the average temperature in Detroit? What was the average temperature in Phoenix?

spring in Charleston, S.C.

CHAPTER 7 Review

Visual Summary

Lesson 1 Weather is what the air is like at a certain time and place. Weather is described by air temperature, precipitation, wind, and air pressure.

Lesson 2 In the water cycle, water moves between the atmosphere and Earth's surface.

Lesson 3 Different places on Earth have different climates and seasons.

Make a FOLDABLES® Study Guide

Glue your lesson study guides to a piece of paper as shown. Use your study guide to review what you have learned in this chapter.

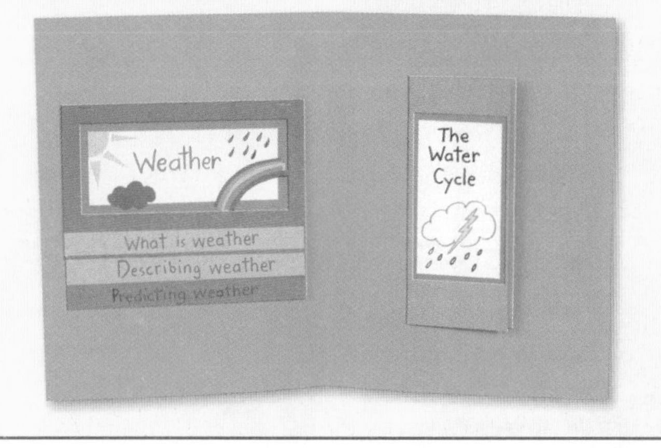

Vocabulary

DOK I

Fill each blank with the best term from the list.

atmosphere, p. 280	**precipitation,** p. 282
climate, p. 304	**seasons,** p. 308
clouds, p. 290	**temperature,** p. 280
condensation, p. 293	**water cycle,** p. 294
evaporation, p. 292	**weather,** p. 280

1. Winter, spring, summer, and fall are ________.

2. What the air is like at a certain time and place is known as ________.

3. Water that falls to the ground from the atmosphere is ________.

4. The changing of a liquid to a gas is called ________.

5. Gas changes to a liquid in the process of ________.

6. The movement of water between the Earth's surface and the atmosphere is called the ________.

7. The pattern of weather in a certain place over a long time is known as ________.

8. The blanket of gases that surrounds Earth makes up the ________.

9. Stratus, cirrus, and cumulus are kinds of ________.

10. A thermometer is a tool that measures ________.

LOG ON e-Glossary Words and definitions online at www.macmillanmh.com

Skills and Concepts

DOK 2–3

Answer each of the following.

11. Predict You see low, dark clouds that look like layered blankets. It is a warm day. What do you think the weather will soon be like? Explain your answer.

12. Fictional Narrative Suppose you are a tiny drop of water in a puddle. Write a story about what happens to you as you travel through the water cycle.

13. Interpret Data Look at the chart. City A and City B are on opposite sides of a mountain near an ocean. Which city is on the side of the mountain that faces the ocean? How do you know?

City	Average Rainfall
City A	74 cm
City B	31 cm

14. Critical Thinking Where do you think the coldest places on Earth are located? How do you know?

15. Critical Thinking People take water from ponds and lakes. Why do ponds and lakes not dry up?

16. Make a Model Design a rain gauge. Explain how it works.

17. What type of cloud is shown below? What is the weather probably like?

18. True or False *A barometer measures temperature.* Is this statement true or false? Explain.

19. The chart below shows the weather conditions on four different days. On which day is it most likely to snow?

Day	Cloudiness	Temperature (°F)
1	no clouds	30
2	cloudy	29
3	partly cloudy	38
4	cloudy	55

A Day 1
B Day 2
C Day 3
D Day 4

The Big Idea

20. How does the weather where you live change throughout the year?

Performance Assessment

DOK 1

Weather in Our Lives

Make a poster about a weather event that you remember, such as a hurricane or a thunderstorm.

- How would you describe the weather for that day? Was it frightening? Did the weather cause damage? Did your home lose electricity?
- Illustrate your weather event on poster board. Next to your drawing, write several sentences to describe how the weather event affected you.

Test Preparation

1 Which tool is used to measure air pressure?

A
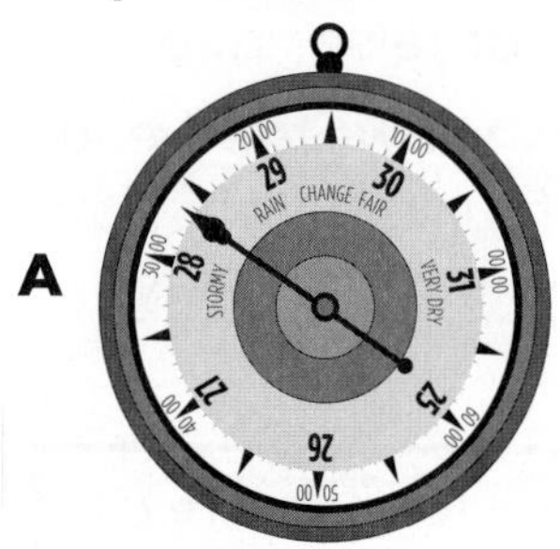

B
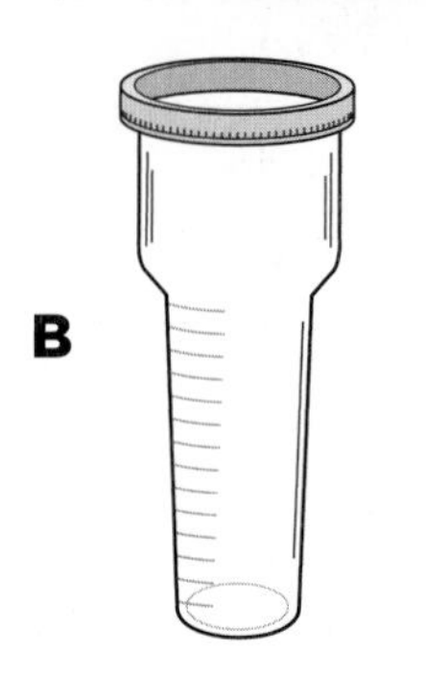

C
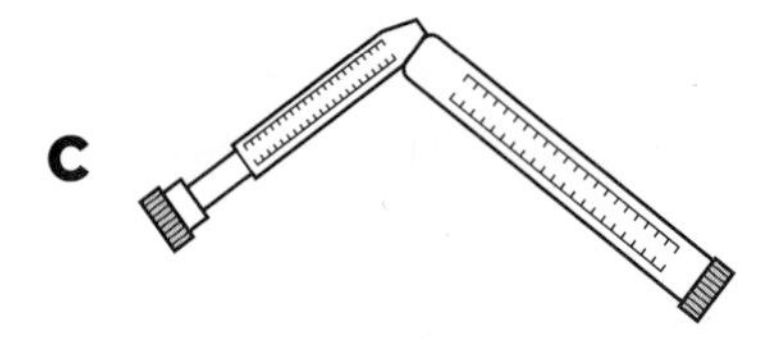

D
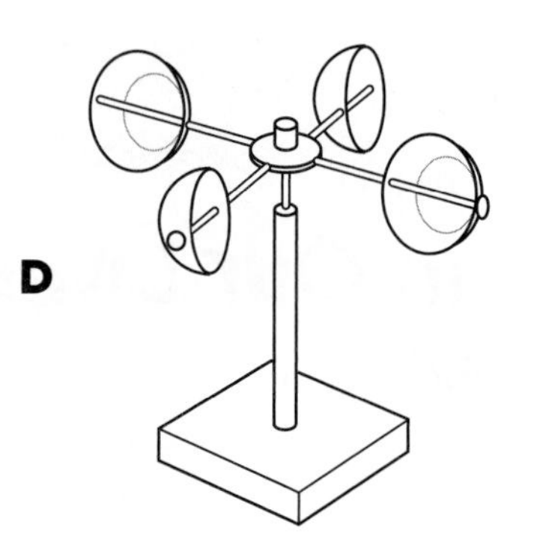

DOK I

2 Which tool measures wind speed?

A wind vane

B barometer

C thermometer

D anemometer

DOK I

3 After it rains, some water soaks into the soil and becomes

A water vapor.

B groundwater.

C salt water.

D sleet.

DOK I

4 Look at the clouds below.

If these clouds become darker, which type of weather can you predict?

A fair

B wet

C dry

D foggy

DOK 2

5 What the air is like at a certain time and place describes

A air pressure.

B atmosphere.

C weather.

D temperature.

DOK I

6 **Look at the chart below.**

City	Average January Temperature	Average Snowfall
Atlanta, GA	43°F	2.6 in.
Buffalo, NY	24°F	93.6 in.
Lexington, KY	32°F	16.1 in.
New Orleans, LA	53°F	0.2 in.

Which city most likely has the warmest climate?

A Atlanta, GA

B Buffalo, NY

C Lexington, KY

D New Orleans, LA

DOK 2

7 **Some trees lose their leaves to save water during cold, dry winters.**

In which location would you most likely find a tree that loses its leaves?

A Sonoran desert

B Chicago

C Antarctica

D Rio de Janeiro

DOK 2

Use the illustration of the water cycle to answer questions 8–10.

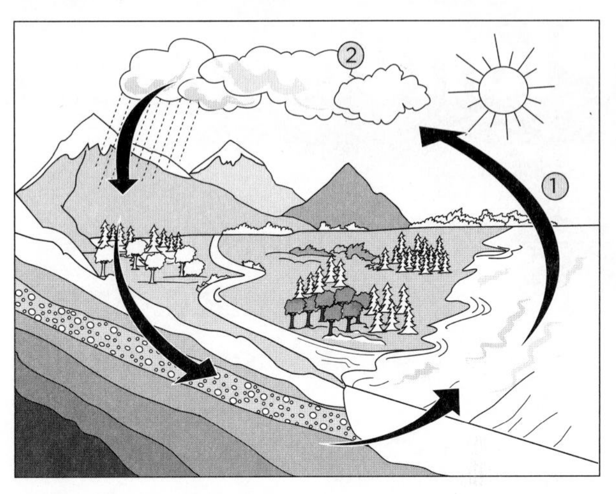

8 **Describe how water moves through the water cycle. Use the words *evaporation, condensation,* and *precipitation* in your answer.**

DOK 2

9 **Compare and contrast steps 1 and 2.**

DOK 3

10 **As the Sun rises in the sky, how will evaporation be affected?**

DOK 2

Check Your Understanding

Question	Review	Question	Review
1	pp. 282–283	6	pp. 304–305
2	pp. 282–283	7	pp. 304–305
3	pp. 294–295	8	pp. 292–295
4	pp. 290–291	9	pp. 294–295
5	pp. 280–281	10	p. 294

CHAPTER 8

Planets, Moons, and Stars

What objects move through space, and how do they move?

Essential Questions

Lesson 1

How do the Sun and Earth interact?

Lesson 2

What can we learn about the Moon?

Lesson 3

How can Earth be compared to the other objects in the solar system?

Lesson 4

What can we see in the night sky?

Saturn and one of its moons, Tethys

Big Idea Vocabulary

rotate to spin (p. 318)

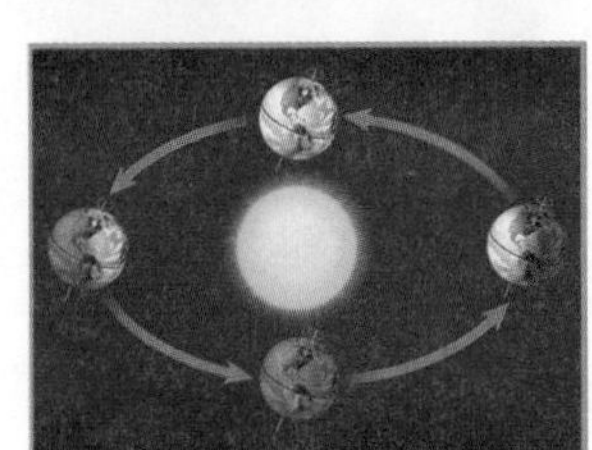

revolve to move around another object (p. 320)

phase each shape of the Moon we see (p. 328)

solar system a system made up of a star and the objects that move around it (p. 338)

planet a large body of rock or gas with a nearly round shape that revolves around a star (p. 338)

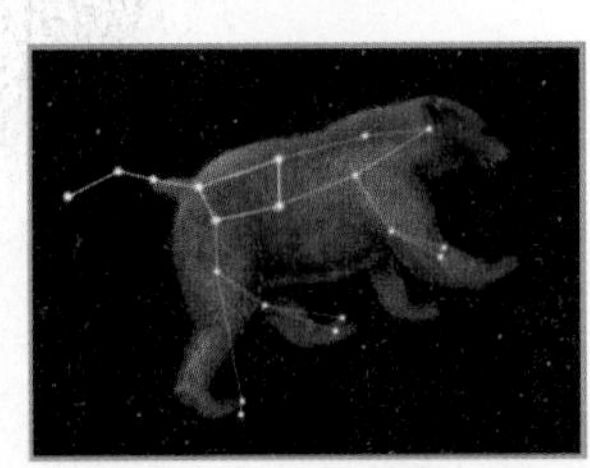

constellation a group of stars that seem to form a picture (p. 349)

Visit www.macmillanmh.com for online resources.

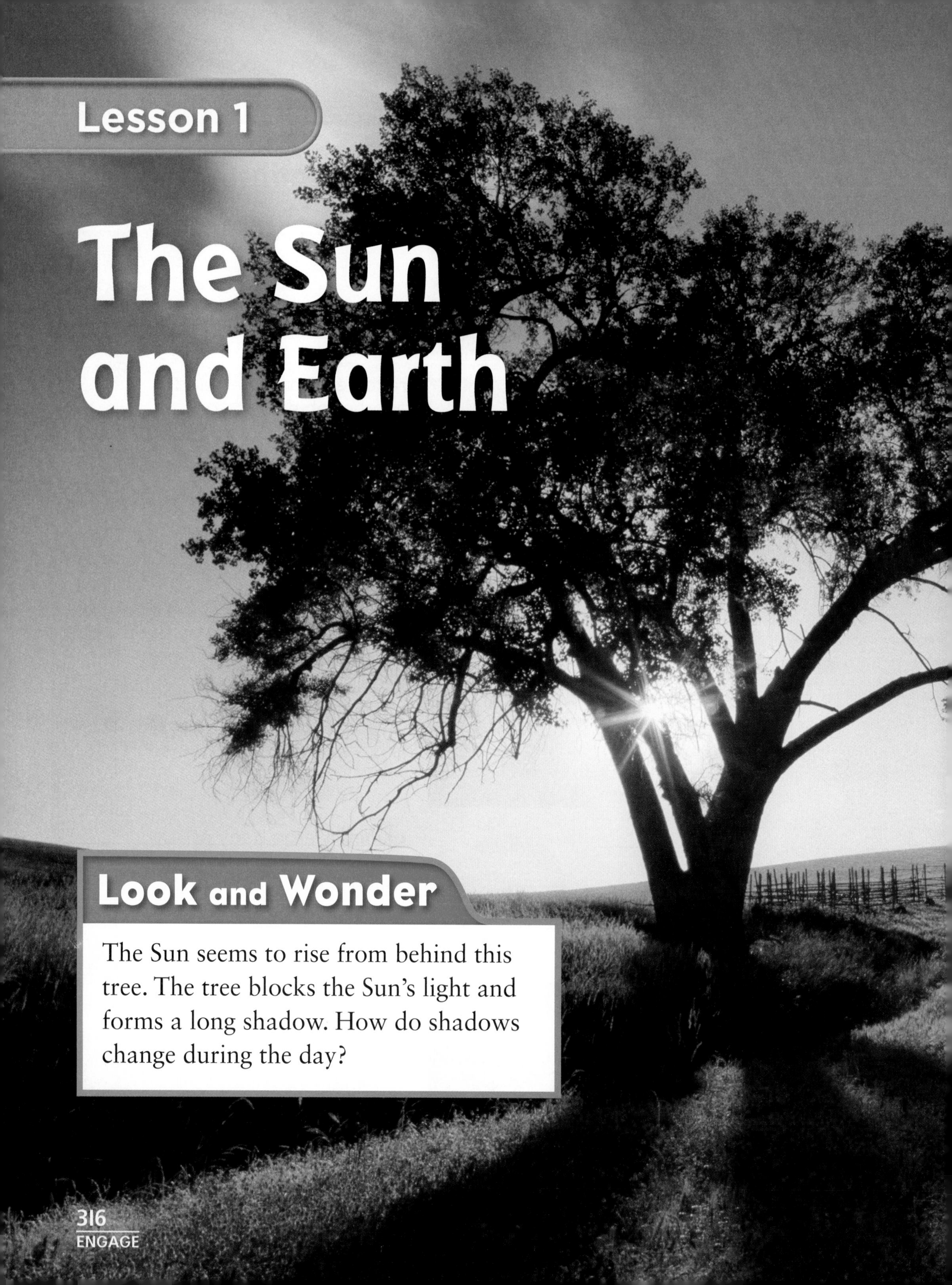

Lesson 1

The Sun and Earth

Look and Wonder

The Sun seems to rise from behind this tree. The tree blocks the Sun's light and forms a long shadow. How do shadows change during the day?

Explore

Inquiry Activity

How do shadows change throughout the day?

Form a Hypothesis

How does the Sun's location affect the length of shadows cast on the ground? How does it affect their position? Write a hypothesis.

Test Your Hypothesis

1. Work in pairs outside on a sunny morning. Use chalk to mark an *X* on the pavement.
2. Have your partner stand on the X. Trace your partner's shadow.
3. **Measure** Use a measuring tape to find the length of the shadow. Observe the location of the Sun and the shadow. Record the data in a table.
   △ **Be Careful.** Never look directly at the Sun.
4. **Observe** Repeat steps 2 and 3 at midday and in the afternoon. How did the Sun's location and the shadow change?

Draw Conclusions

5. **Interpret Data** When was the shadow shortest? When was the shadow longest? How did the Sun's position change?
6. **Infer** What caused the shadow to change position and length?

Explore More

Experiment How can you find out in which month the Sun's position appears highest in the sky? Lowest?

Materials

chalk

measuring tape

Step 2

Step 3

Time	Length and Location of Shadow	Location of Sun
Morning		
Midday		
Afternoon		

Read and Learn

Essential Question
How do the Sun and Earth interact?

Vocabulary
rotate, p. 318
axis, p. 319
revolve, p. 320
orbit, p. 320
star, p. 322

Reading Skill 
Sequence

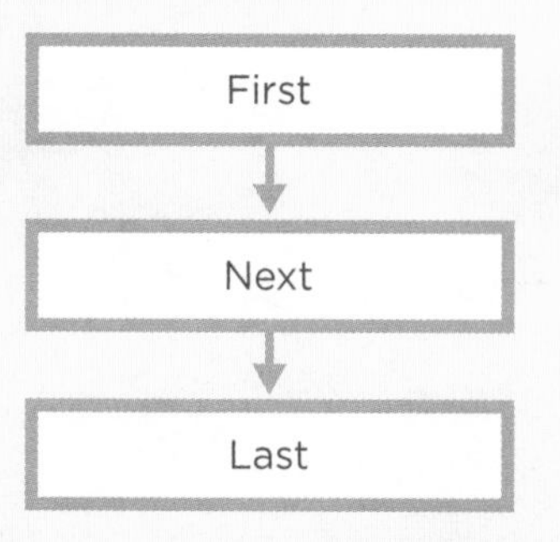

Technology LOG ON
e-Glossary, e-Review, and animations online at www.macmillanmh.com

Why is there day and night?

As you read this page, you are moving through space. All the other things on Earth are moving too. Earth **rotates** (ROH•tayts), or spins, like a giant top in space.

As Earth rotates, we see the Sun in a different place in the sky. At sunrise, the Sun is low in the eastern sky. The Sun's light casts long shadows. At midday, the Sun is high in the sky. Shadows are short then. At sunset, the Sun is low in the western sky, and shadows are long again. The Sun seems to move in the sky, but it does not. Earth does all the moving!

The Sun's Position in the Sky

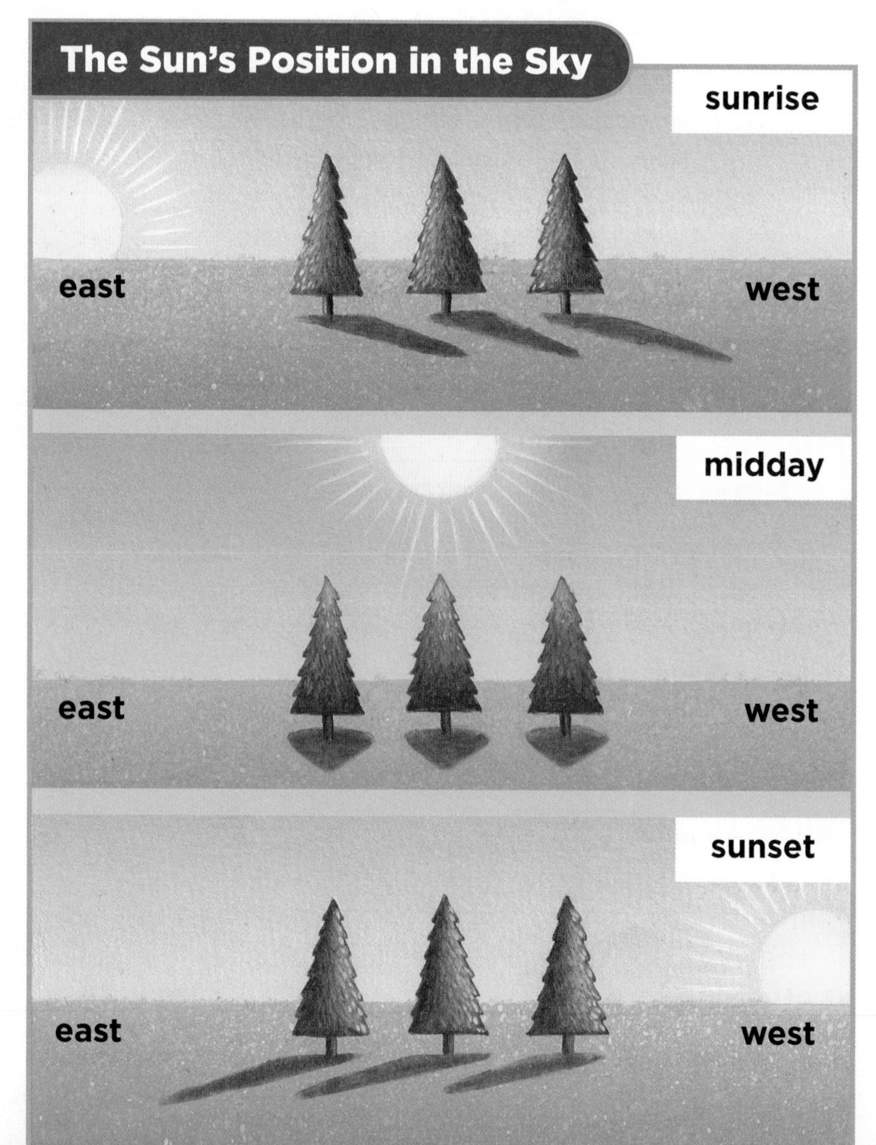

The Sun seems to rise in the east and set in the west. Looking at these pictures, you are facing south. ▶

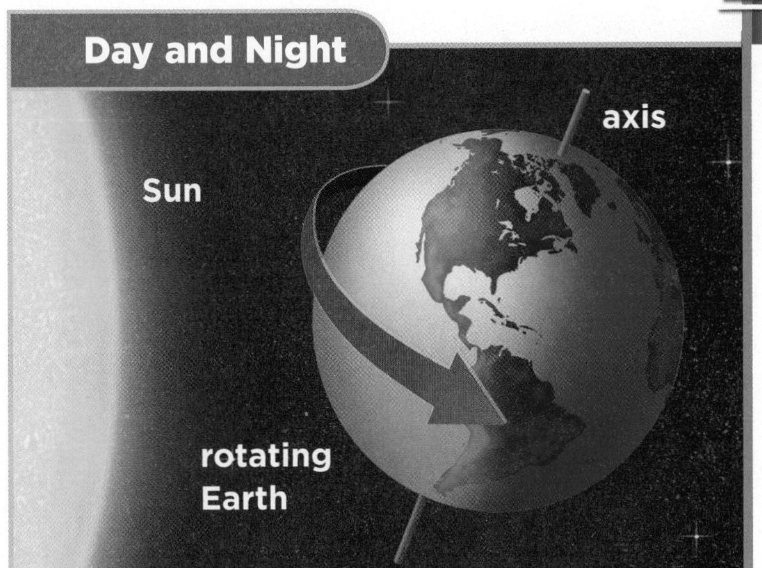

▲ As Earth rotates on its imaginary axis, we have day and night.

Earth's rotation causes day and night. As Earth rotates, one side of Earth faces the Sun. This side has daytime. At the same time, the other side of Earth faces away from the Sun. This side has nighttime. Earth keeps spinning. Day becomes night for one side of Earth. Night becomes day for the other side of Earth. This happens over and over again.

It takes 24 hours for Earth to make one complete turn on its axis (AK•sus). An **axis** is a line through the center of a spinning object. One complete turn is one *day*. We think of a day as daytime hours. However, a day is made up of daytime and nighttime.

Quick Lab

Model Earth's Rotation

1. **Make a Model** Carefully push a pencil through a foam ball. The ball represents Earth. The pencil represents Earth's axis. Press a paper clip into the side of the ball. The paper clip represents you.
2. **Observe** Shine a light on the paper clip. The light represents the Sun. On the model, where is it day?
3. **Observe** Rotate the pencil so the paper clip faces away from the light. Where is it night?
4. **Interpret Data** What does this model help to show?

Quick Check

Sequence How does the Sun appear to move across the sky?

Critical Thinking What might happen if Earth did not rotate?

Why are there seasons?

Earth rotates on its axis. It also revolves (ri•VOLVZ) around the Sun. To **revolve** means to move around another object. Earth revolves around the Sun in a regular path called an **orbit** (OR•but). It takes Earth about 365 days to make one trip around the Sun. We call this one *year*.

Seasons change as Earth revolves because of Earth's tilted axis. Earth's axis is not straight up and down compared with its orbit. For part of Earth's orbit, the *Northern Hemisphere,* or upper half of Earth, is tilted toward the Sun. The Northern Hemisphere has summer. At the same time, the *Southern Hemisphere,* or lower half of Earth, is tilted away from the Sun. The Southern Hemisphere has winter.

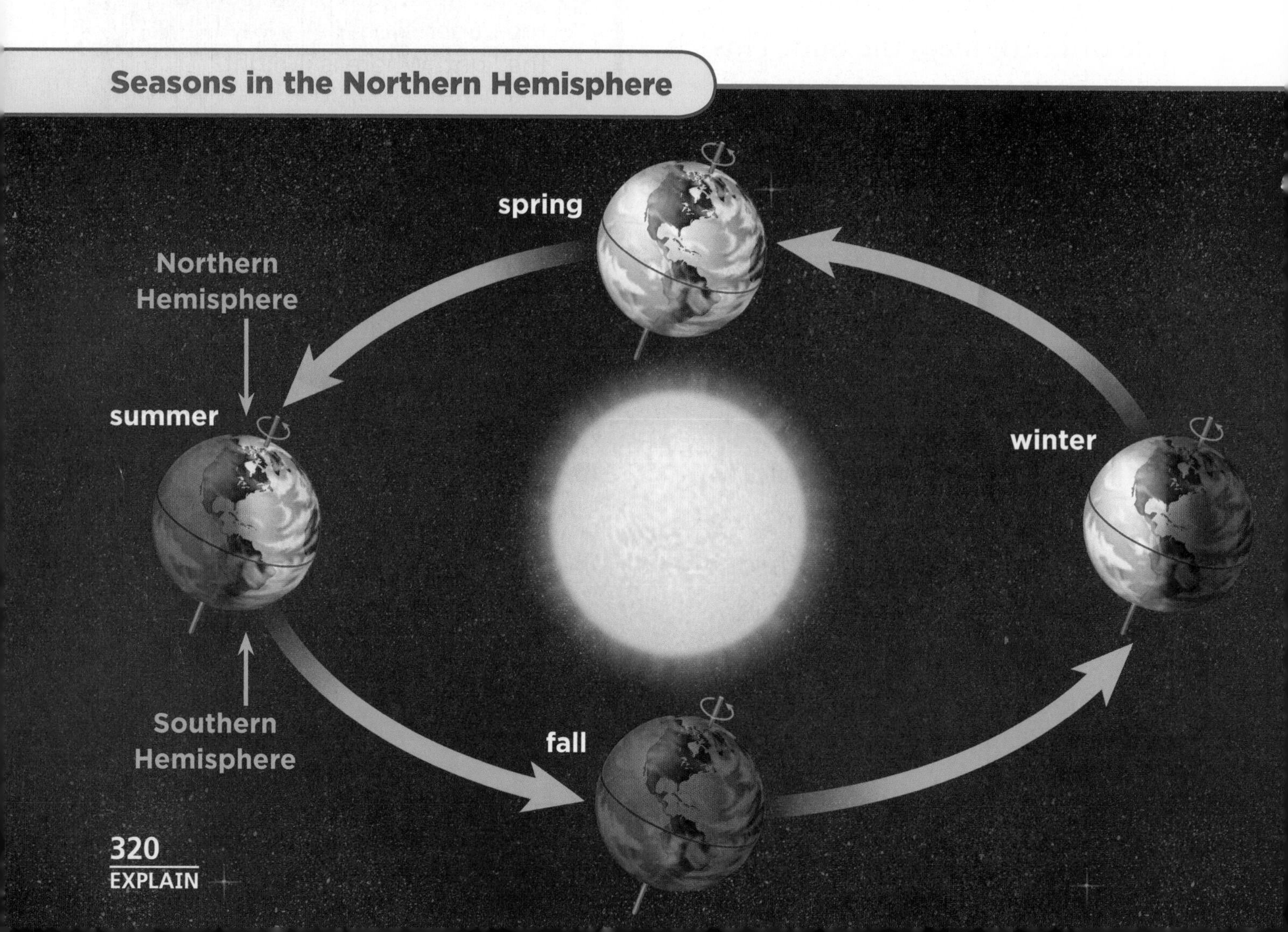

At other times, the Northern Hemisphere is tilted away from the Sun and has winter. At the same time, the Southern Hemisphere is tilted toward the Sun and has summer. The diagram can help you see this.

The Northern Hemisphere gets the most light and heat from the Sun in summer. Days are longer. Temperatures are warmer than in any other season. As Earth continues its orbit, the days get shorter and cooler. Fall begins.

The Northern Hemisphere gets the least light and heat from the Sun in winter. Days are shorter. Temperatures are cooler than in any other season. As Earth continues its orbit, the days get longer and warmer. Spring begins.

Read a Diagram

What is the same about Earth's axis as it moves around the Sun?

Clue: Compare the drawings of Earth's axis for each of the four seasons.

Science in Motion Watch how seasons change at www.macmillanmh.com

Quick Check

Sequence Which season comes before fall?

Critical Thinking How would seasons change if Earth moved more slowly around the Sun?

▲ Like many living things, this tree changes through the seasons.

What is the Sun like?

The Sun is a star. A **star** is a ball of hot, glowing gases. The temperature inside the Sun is about 59 million degrees Fahrenheit (15 million degrees Celsius). This is hot enough to melt Earth!

The Sun is about 150 million kilometers (93 million miles) away from Earth. If you could travel there by car on a highway, it would take more than 160 years. Even though the Sun is that far away, it provides Earth with light and heat. Life on Earth could not exist without the Sun. Plants could not grow. Earth would freeze.

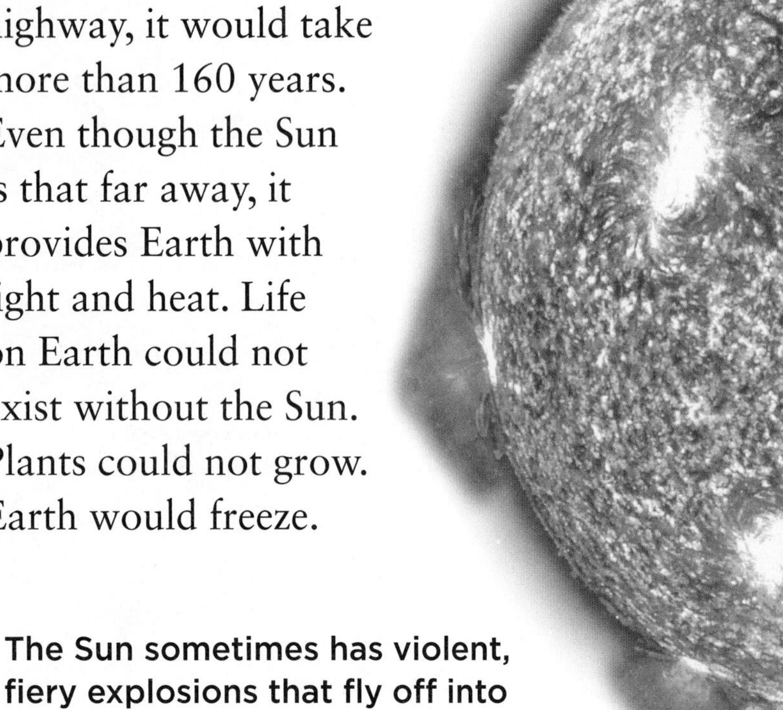

The Sun sometimes has violent, fiery explosions that fly off into space. They are called solar flares. ▶

Quick Check

Sequence What would happen to Earth if the Sun's light were blocked for many years?

Critical Thinking Do you think that people could travel to the Sun? Explain your answer.

FACT It would take 109 Earths to fit across the Sun.

Lesson Review

Visual Summary

Earth rotates on its axis. This causes **day and night.**

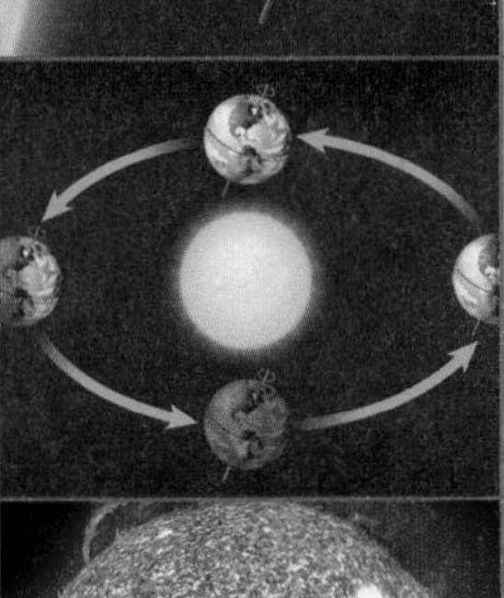

Seasons change because Earth revolves around the Sun and is tilted on its axis.

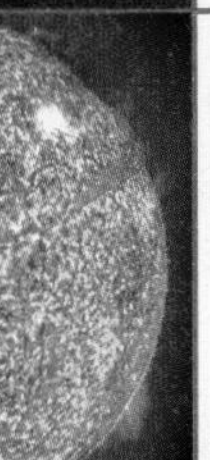

The Sun is a star that supplies Earth with light and heat.

Make a FOLDABLES Study Guide

Make a trifold book. Use it to summarize what you learned about the Sun and Earth.

Think, Talk, and Write

1. **Vocabulary** What is a star? Give an example.

2. **Sequence** How do seasons change in the Northern Hemisphere?

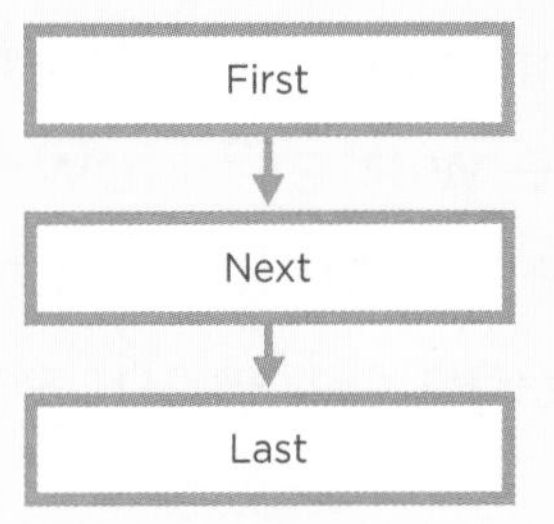

3. **Critical Thinking** How would the seasons be different if Earth were not tilted on its axis?

4. **Test Prep** **Why are there seasons? Choose the best answer.**
   - **A** Earth rotates on its axis and revolves around the Sun.
   - **B** Earth has a tilted axis and revolves around the Sun.
   - **C** The Sun revolves around Earth.
   - **D** Earth's distance from the Sun changes.

5. **Essential Question** How do the Sun and Earth interact?

Math Link

Use a Calendar
Use a calendar to find out when the next season begins. What is the date? How many days from now is that?

Health Link

Sun Safety
Find out how sunlight can benefit human health. How can it be harmful to human health? How can people avoid the harmful effects of sunlight? Write a report.

LOG ON e-Review Summaries and quizzes online at www.macmillanmh.com

Seasons Where You Live

You have learned about seasons. Now think about what the seasons are like in your state. Do seasons have different kinds of weather? Does it snow in the winter? How hot is it in the summer? Think about the things you do during different seasons.

Personal Narrative

A good personal narrative

- tells a story from the writer's personal experience;
- expresses the writer's feelings;
- tells events in an order that makes sense.

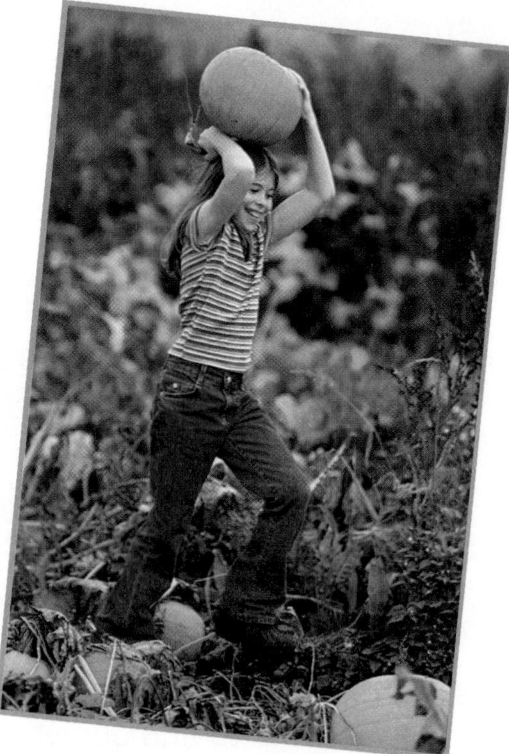

Write About It

Personal Narrative Choose a season. Tell a true story about something you did during that season. Explain why you still remember the event. How did it make you feel? Describe what the weather was like.

Write about it online at www.macmillanmh.com

Converting Hours to Minutes

Daytime is longer in summer than it is in winter. Look at the table below. It shows the number of daylight hours for summer and winter in Boston, Massachusetts.

Convert Hours

- Multiply the number of hours by the number of minutes in one hour. There are 60 minutes in one hour.

Example:
$5 \text{ h} \times 60 \text{ min/h} = 300 \text{ min}$

Longest and Shortest Days in Boston

Date	Approximate Hours of Daylight
June 21	15
December 21	10

Solve It

How long is Boston's longest day in minutes? How long is the shortest day in minutes? What is the difference in minutes between the longest day and the shortest day?

Lesson 2

The Moon and Earth

St. Louis, Missouri

Look and Wonder

The Moon does not look the same today as it did yesterday. How will it look in one week? Does the Moon's shape really change?

Explore

Inquiry Activity

How does the Moon's shape seem to change?

Make a Prediction

How will the Moon seem to change during one school week? Make a prediction.

Test Your Prediction

1. **Observe** Work with an adult. Look at the Moon each night for one school week. You can also use research materials to find out what the Moon looks like each night.

2. **Communicate** Record what the Moon looks like each night. Also record where the Moon is in the sky and the time that you observed it.

Draw Conclusions

3. **Interpret Data** How did the Moon's shape seem to change each night? How does its location seem to change?

4. **Predict** What will the Moon's shape look like for the next few nights?

Explore More

Experiment How does the Moon's shape and position in the sky seem to change over several months? Make a plan to find out.

Step 1

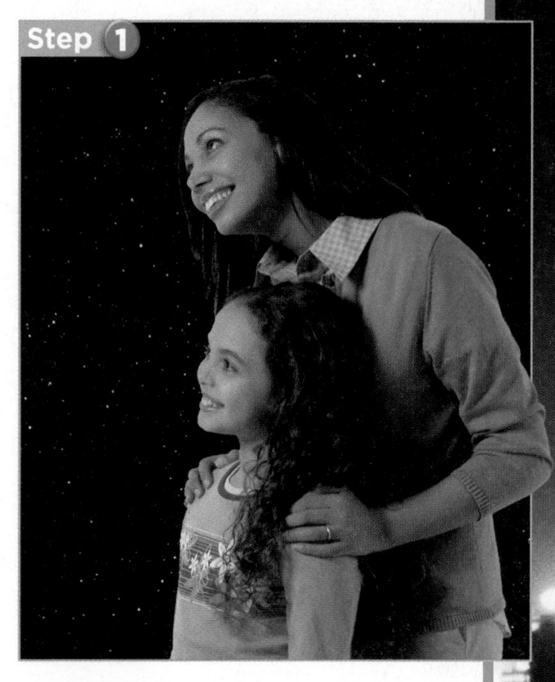

Step 2

Day	Observation
Monday	
Tuesday	
Wednesday	
Thursday	
Friday	

Read and Learn

Essential Question
What can we learn about the Moon?

Vocabulary

phase, p. 328

crater, p. 332

Reading Skill

Cause and Effect

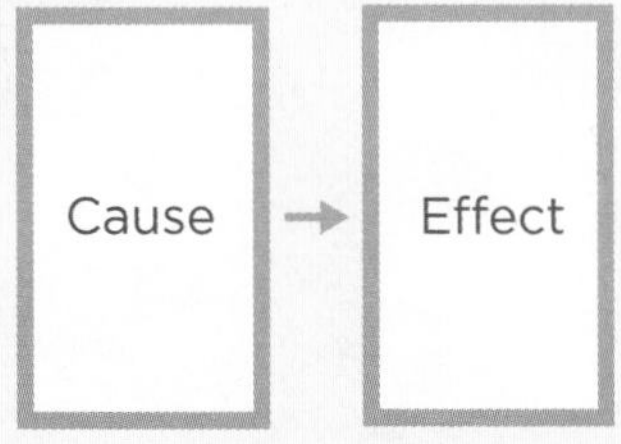

Technology LOG ON
e-Glossary and e-Review online at www.macmillanmh.com

What are the phases of the Moon?

On many nights you can see the Moon. The Moon is a *sphere*, or ball, of rock. It is Earth's closest neighbor in space. The Moon is a satellite of Earth. A *satellite* is anything that revolves around another larger object in space. The Moon revolves around Earth, much like Earth revolves around the Sun.

Each day the Moon seems to change shape. Sometimes you can see only a small part of the Moon. At other times the Moon is a big, bright circle. Still other times you can not see the Moon at all. Each shape of the Moon we see is called a **phase** (FAYZ). The photos below show the eight main phases of the Moon. Have you seen each phase of the Moon?

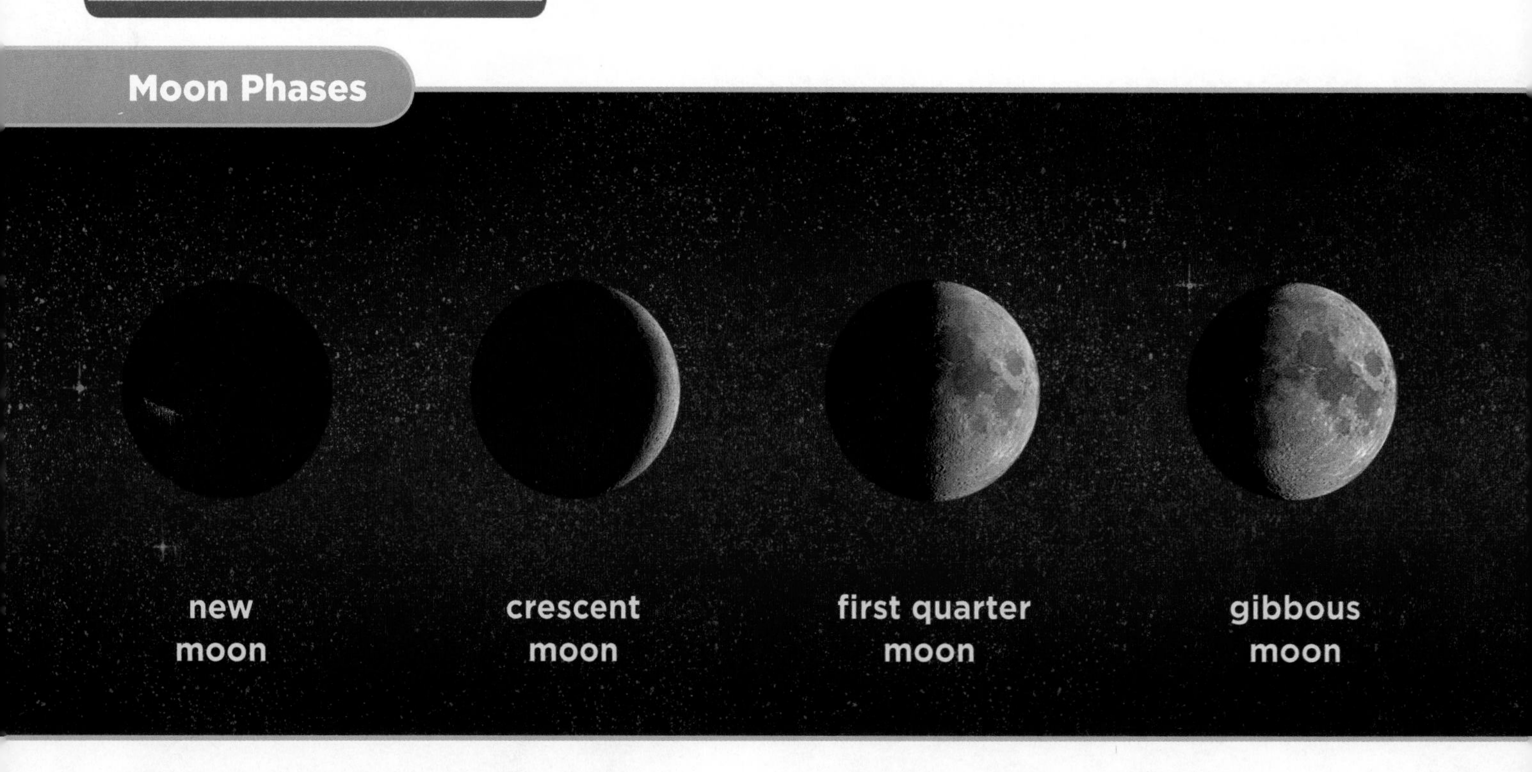

Like the Sun, the Moon seems to rise, move across the sky, and set each day. This happens because of Earth's rotation. As Earth rotates on its axis, the Moon's position in the sky moves from east to west.

The Moon seems to glow in the night sky. However, the Moon does not make its own light. The Moon appears to shine because sunlight *reflects,* or bounces, off the Moon.

▲ Sometimes we can see the Moon during the daytime.

Quick Check

Cause and Effect Why does the Moon seem to rise and set each day?

Critical Thinking How could you predict what the shape of the Moon will be next week?

The Moon's Orbit

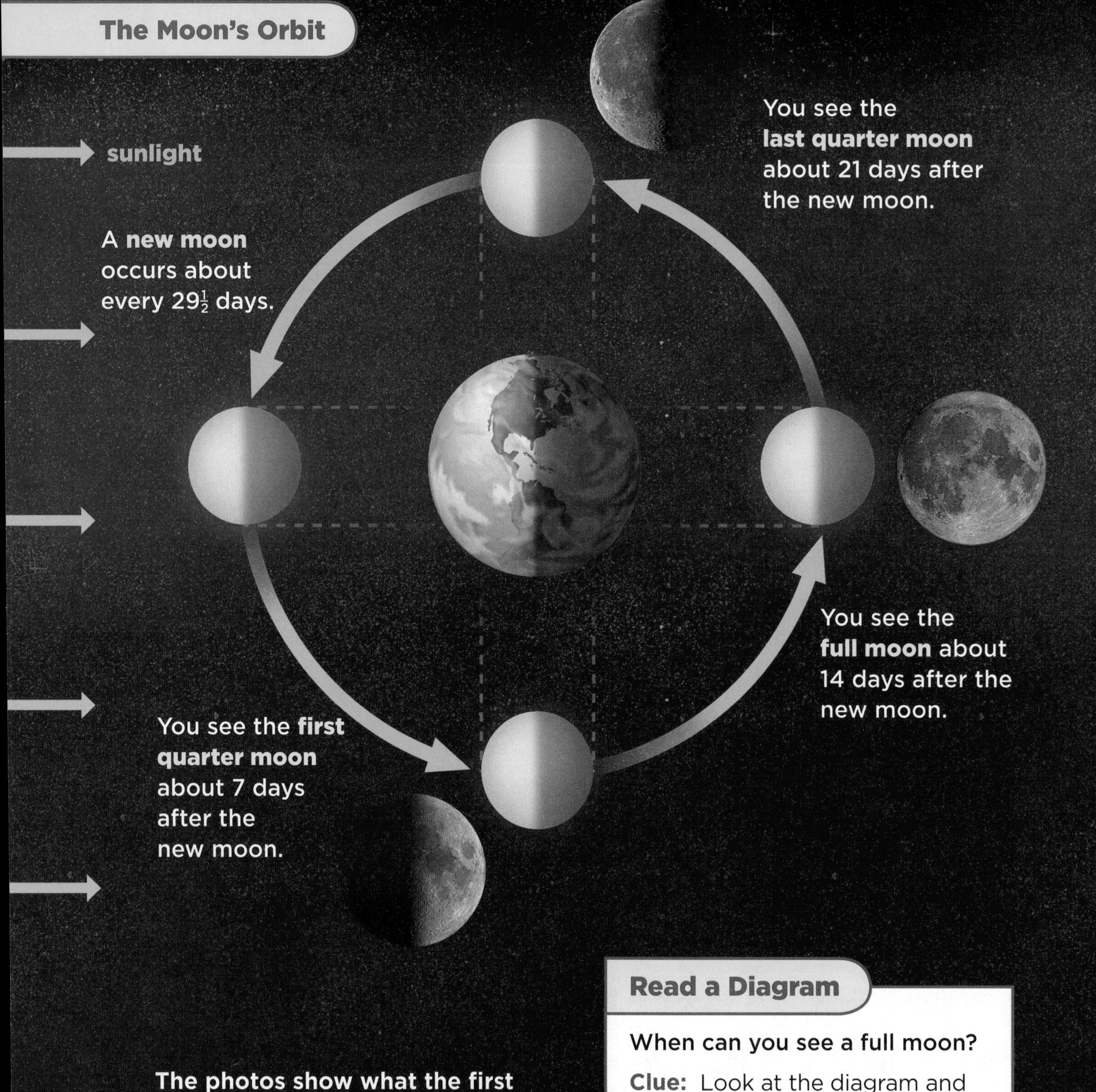

The photos show what the first quarter moon, full moon, and last quarter moon look like from Earth.

Read a Diagram

When can you see a full moon?

Clue: Look at the diagram and read the captions.

Why does the Moon's shape seem to change?

The Moon might seem to change shape, but it is always a sphere. It only appears to change shape because you see different amounts of its lit side. The Sun shines on and lights half of the Moon's surface. The other half faces away from the Sun. It is in darkness. As the Moon revolves around Earth, different amounts of the Moon's lit half face Earth. The lit parts are the phases you see.

When the Moon is between Earth and the Sun, there is a new moon. You can not see the Moon at all. The Moon's lit side faces away from Earth. As the Moon continues to revolve, you see more and more of the Moon's lit side. When you see the entire lit half, there is a full moon. As the Moon continues to revolve, you begin to see less and less of the lit side. This cycle happens about every $29\frac{1}{2}$ days.

Quick Lab

Make a Flip Book

1. Draw what a moon phase looks like on the right side of an index card. Label the phase. Do this for each of the eight phases.

2. Put the cards in order. Staple them together on the left side.
3. Flip the pages with your thumb. How does the Moon's shape seem to change?

Quick Check

Cause and Effect What causes the Moon's phases to change?

Critical Thinking What would the Moon look like if it did not revolve around Earth?

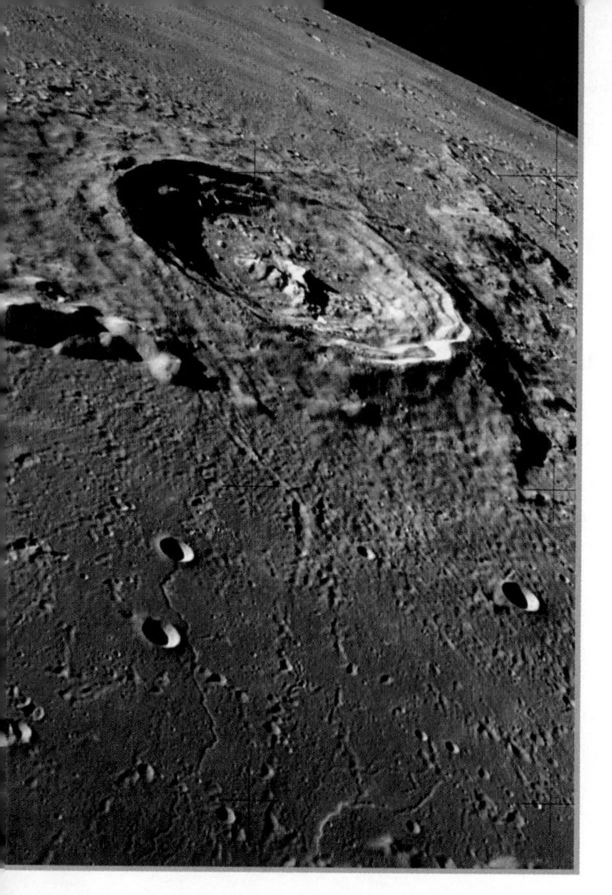

▲ Craters of all sizes are found on the Moon's surface.

What is it like on the Moon?

Both the Moon and Earth are spheres. They both rotate. They both revolve. They both get light from the Sun. They are both made of rock. In other ways, though, they are very different.

Earth is about four times larger than the Moon. Unlike Earth, the Moon has no atmosphere. It has no liquid water. There are no living things on the Moon.

The Moon's surface is also different from Earth's. Most of its surface is covered with craters (KRAY•turz). A **crater** is a hollow area, or pit, in the ground. Craters formed on the Moon when chunks of rock from space crashed into the Moon.

Quick Check

Cause and Effect What caused craters to form on the Moon?

Critical Thinking Why are there no living things on the Moon?

This is what Earth looks like from the Moon. ▼

FACT The Moon's temperatures are colder and warmer than Earth's.

Lesson Review

Visual Summary

The Moon seems to change shape each night. These shapes are called phases.

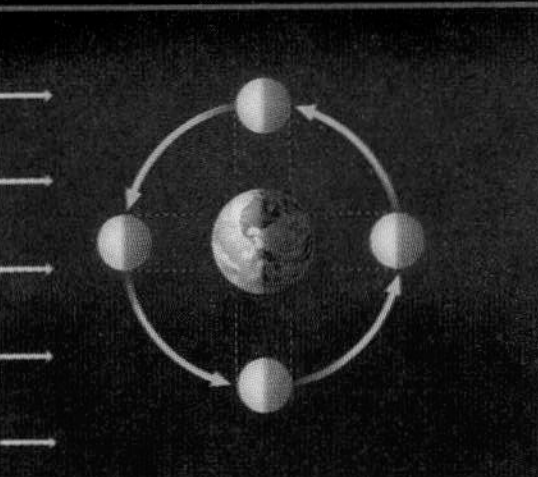

The Moon revolves around Earth. This causes the different phases of the Moon.

In many ways, **the Moon and Earth are** different.

Make a FOLDABLES® Study Guide

Make a trifold book. Use it to summarize what you learned about the Moon and Earth.

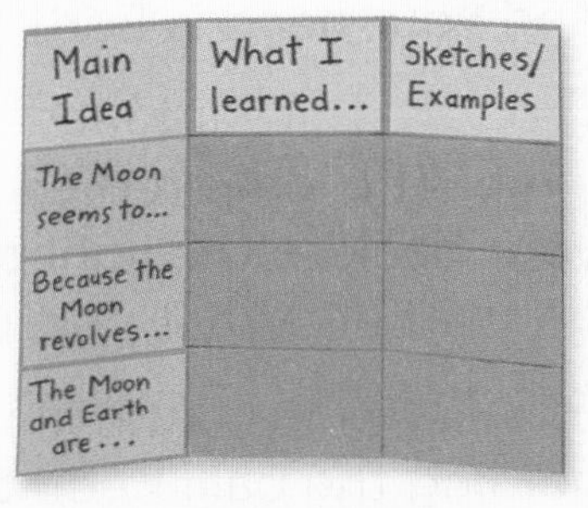

Think, Talk, and Write

1. **Vocabulary** What is a crater?

2. **Cause and Effect** What causes the Moon to rise and set?

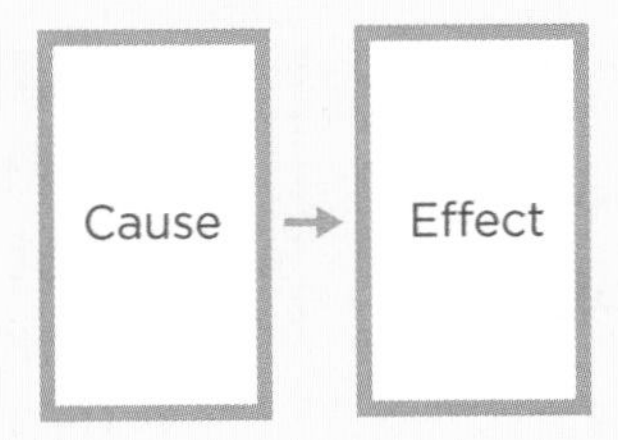

3. **Critical Thinking** Why can you not see the new moon phase?

4. **Test Prep** **How are Earth and the Moon similar?**
   - **A** Both have an atmosphere.
   - **B** Both make their own light.
   - **C** Both have living things.
   - **D** Both are spheres.

5. **Essential Question** What can we learn about the Moon?

Math Link

Solve a Problem
Objects weigh about six times more on Earth than they would on the Moon. If a moon rock weighs 9 pounds on the Moon, about what would it weigh on Earth?

Social Studies Link

Do Research
Some astronauts have actually walked on the Moon. Do research to find out about one of these astronauts. Share your findings with the class.

Be a Scientist

Materials

lamp

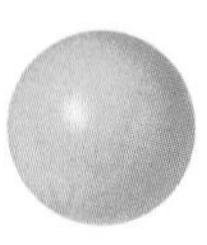

ball

Structured Inquiry

Why does the Moon's shape appear to change?

Form a Hypothesis

The Moon completes one orbit around Earth about every $29\frac{1}{2}$ days. How does the Moon's position in space affect how we see the Moon? When do we see a full moon? When do we see a last quarter moon? Write a hypothesis.

Step 2

Test Your Hypothesis

1. **Make a Model** Hold a ball and stretch out your arm in front of you. Position your arm so the ball is a little higher than your head. The ball is the Moon, and your head is Earth.
2. **Observe** Your teacher will turn on a lamp. The lamp represents the Sun. Turn your back to the light so the light shines on the ball. What part is lit up?
3. **Experiment** Turn in place while holding the ball. Keep the ball in front of you. Notice the change in light and shadow on the ball.
4. **Communicate** Record your observations by drawing or describing what you saw as you turned.
5. Repeat the experiment several times. Are your observations the same with each trial?

Draw Conclusions

6. **Observe** Where is the ball when it is lit like a full moon? Where is the ball when it looks like a last quarter moon?
7. **Infer** Why does the Moon's shape appear to change? How do you know?

Guided Inquiry

How does the Moon's position change?

Form a Hypothesis

How does the Moon's position change in the night sky over the course of one month? Write a hypothesis.

Test Your Hypothesis

Design a plan to test your hypothesis. Remember your plan should test only one variable—the Moon's position in the sky. Decide on the materials you will need. Then write the steps you plan to follow.

Draw Conclusions

Did your results support your hypothesis? Why or why not? Share your results with your classmates.

Open Inquiry

What other questions do you have about the Moon? Talk with your classmates about questions you have. How might you find the answers to your questions?

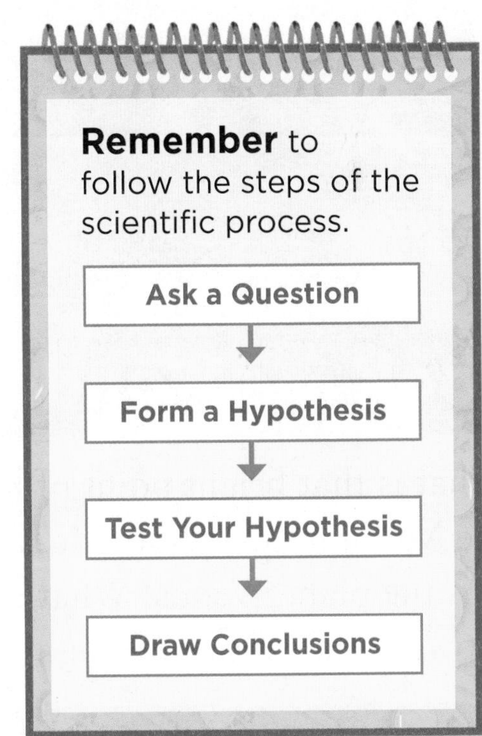

Lesson 3

The Planets

Look and Wonder

What is that bright point of light near the Moon in this photo? It is not a star. It is the planet Venus! What are planets? How do they move through space?

Explore

Inquiry Activity

How do the planets move through space?

Purpose

Model how the positions of the planets change.

Procedure

1. Work in a large room or other area. Put a chair in the center of the room. Tape the card labeled Sun to the chair. Tape a line from the chair to a wall.
2. Form two groups. Each student in the first group will take a card and line up along the tape in order: Mercury, Venus, Earth, Mars, Jupiter, Saturn, Uranus, and Neptune.
3. **Make a Model** Students in the first group model how the planets move by walking in a circle around the Sun. Take steps of the same size. Count steps together. Stop counting when you reach the tape again. Students in the second group repeat steps 2–3.
4. **Use Numbers** Do all the planets complete one trip around the Sun in the same number of steps?

Draw Conclusions

5. What was different about the orbits?
6. **Infer** How do planets move through space?

Explore More

Predict What planets can you see in the night sky where you live? Make a plan to find out.

Materials

masking tape

8 planet cards and 1 Sun card

Step 1

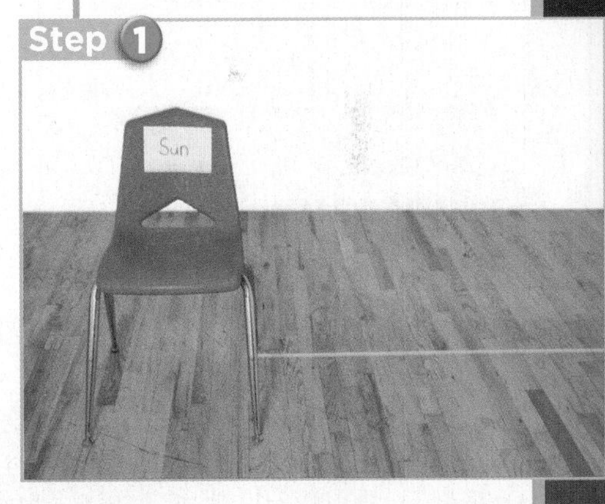

Step 2

Read and Learn

Essential Question

How can Earth be compared to the other objects in the solar system?

Vocabulary

solar system, p. 338

planet, p. 338

telescope, p. 342

space probe, p. 342

Reading Skill

Classify

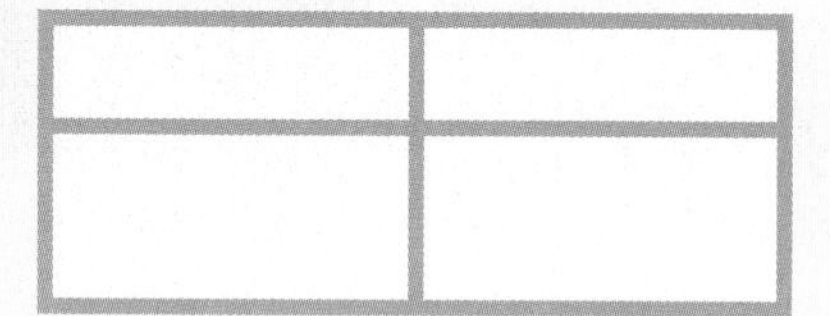

Technology LOG ON e

e-Glossary, e-Review, and animations online at www.macmillanmh.com

What is our solar system?

The Sun, Earth, and the Moon are part of a larger system called a solar system (SOH•lur SIS•tum). A **solar system** is made up of a star and the objects that move around it.

Earth is one of eight planets (PLA•nuts) that move around the Sun in our solar system. A **planet** is a large body of rock or gas with a nearly round shape that revolves around a star. Some planets in our solar system are smaller than Earth. Some are larger. Many have one or more moons. All rotate. All revolve in a nearly circular orbit.

It takes Earth about 365 days to complete one orbit. It takes Mercury only 88 Earth days to travel its orbit. Neptune takes 165 Earth years. The farther away a planet is from the Sun, the longer it takes to complete one orbit.

Our Solar System

Sun

Mercury

Venus

Earth

Mars

FACT Scientists have found over 100 planets that revolve around other stars.

You can sometimes see Mercury, Venus, Mars, Jupiter, or Saturn in the night sky. These planets look like stars from Earth. However, like the Moon, planets do not make their own light. Planets appear to shine because sunlight reflects, or bounces, off them. Although these planets are larger than the Moon, they look tiny. The Moon seems large because it is close to Earth. The planets are farther away, so they look much smaller than the Moon.

Quick Check

Classify Name an object in our solar system that is not a planet.

Critical Thinking How would the Sun look from Mercury?

Quick Lab

Sizing Up Planets

1. **Measure** Work with a partner. Hold a marble about 30 centimeters away from you.
2. **Measure** Have a partner hold a tennis ball about 5 meters away from you.
3. **Observe** Which object seems larger? Why? Which object really is larger?
4. **Infer** How can larger planets look smaller to us than smaller planets?

Read a Diagram

Where is Earth in our solar system?

Clue: Diagrams can show the positions of things.

LOG ON *Science in Motion* Watch the planets at www.macmillanmh.com

Jupiter

Neptune

Uranus

Saturn

What are the inner and outer planets?

The four planets closest to the Sun are Mercury, Venus, Earth, and Mars. They are called the *inner planets*. The inner planets are among the smallest planets in our solar system. They are solid and made up of rocky material. They are warmer than the other planets because they are closer to the Sun.

Mercury is the planet closest to the Sun. It is rocky and full of craters.

The surface of **Venus** is covered with thick clouds that trap heat. This makes it the hottest planet.

Saturn is known for its beautiful rings made up of bits of ice and rock that revolve around the planet.

Jupiter is the largest planet. It now has two red spots. The larger spot, shown here, is known as the Great Red Spot.

The planets farthest from the Sun are Jupiter, Saturn, Uranus, and Neptune. These are the *outer planets*. They are called "gas giants," since they are largely made up of gases. They might have solid cores, but scientists do not know yet.

Quick Check

Classify Is Saturn an inner planet or an outer planet?

Critical Thinking If Mercury is closer to the Sun than Venus, why is Venus hotter than Mercury?

◀ **Earth** is the only planet known to have oxygen, liquid water, and living things.

◀ **Mars,** the Red Planet, has reddish-brown soil and polar ice caps that contain frozen water.

◀ **Uranus** is called the "sideways planet" because it rotates on its side.

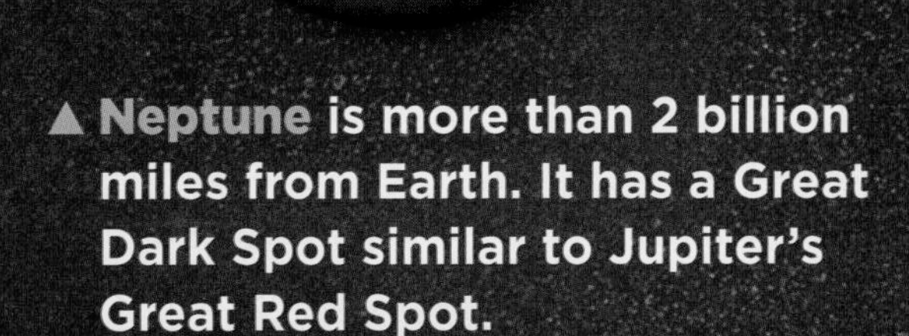

▲ **Neptune** is more than 2 billion miles from Earth. It has a Great Dark Spot similar to Jupiter's Great Red Spot.

How can we view the planets?

Are there other planets? What is Neptune like? We can not use only our eyes to find out. Special tools are needed. A **telescope** (TEL•uh•skohp) is a tool used to make objects far away appear closer and larger. A telescope works by gathering light with mirrors and lenses (LENZ•uz). A *lens* is a clear material that helps us see objects in more detail.

Another tool used is a space probe. A **space probe** is a machine that leaves Earth and travels through space. Some space probes land on planets. Others take pictures as they fly by planets and other objects in space. In 2006 the space probe *New Horizon* was launched. It will travel far beyond Neptune to find out about other space objects.

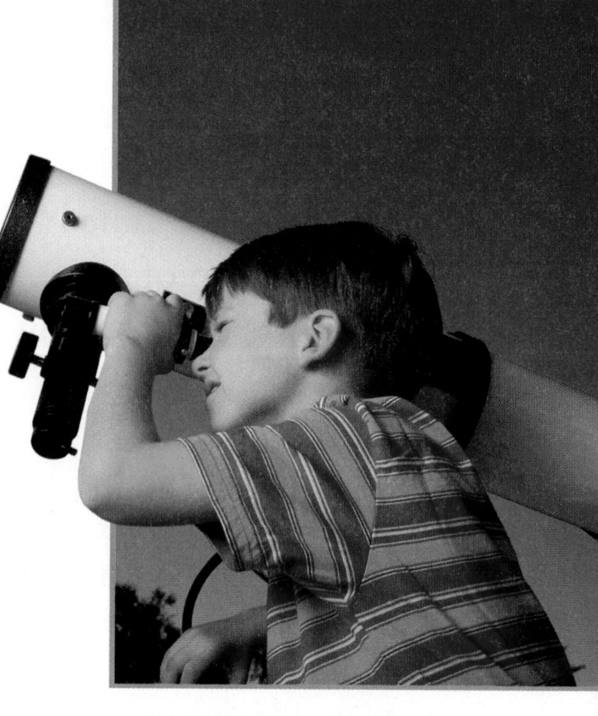

▲ Telescopes help us see planets and other objects in space.

Quick Check

Classify What kind of tool is the *New Horizon* spacecraft?

Critical Thinking How could you see the Moon's craters better?

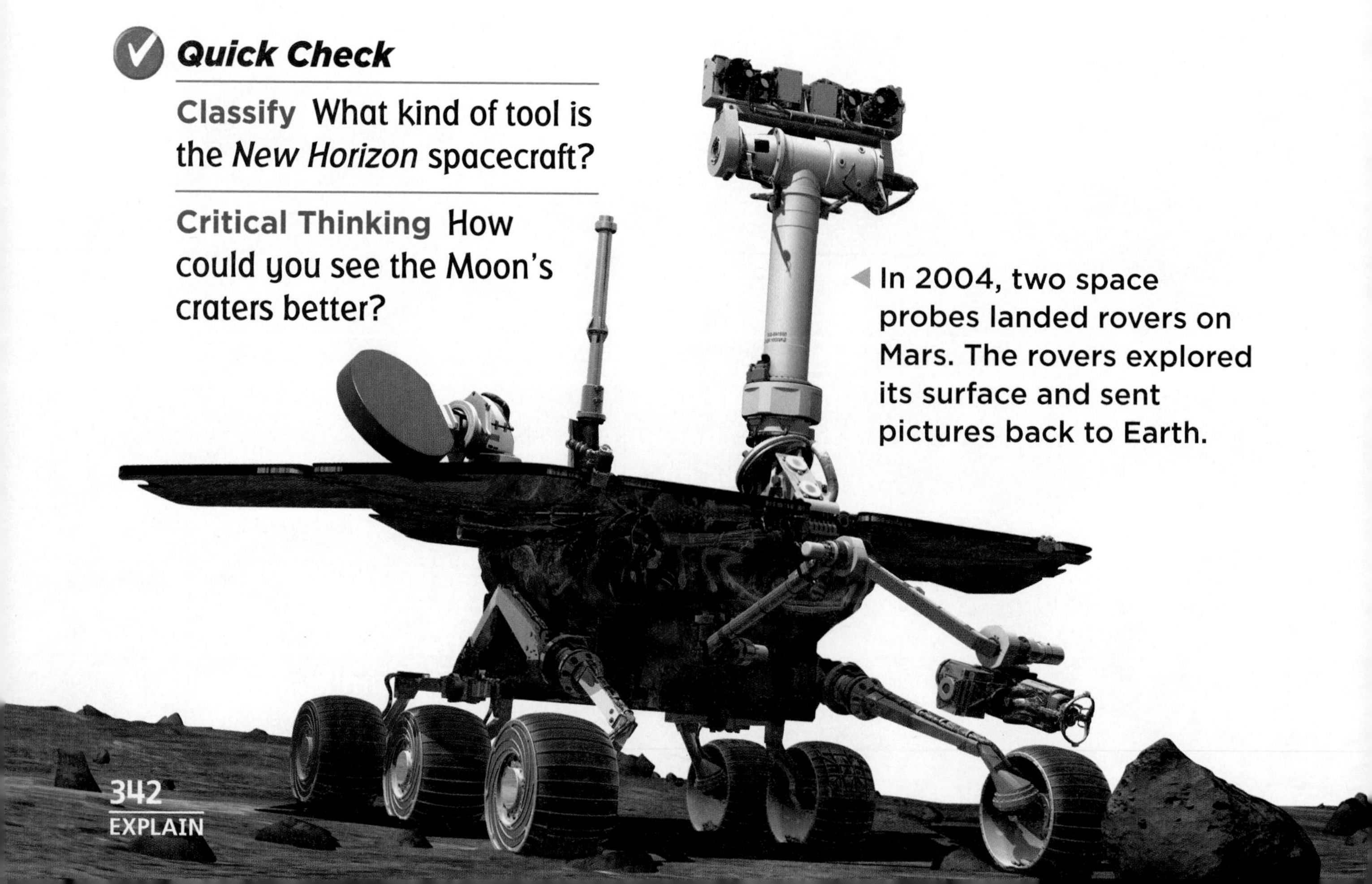

◄ In 2004, two space probes landed rovers on Mars. The rovers explored its surface and sent pictures back to Earth.

Lesson Review

Visual Summary

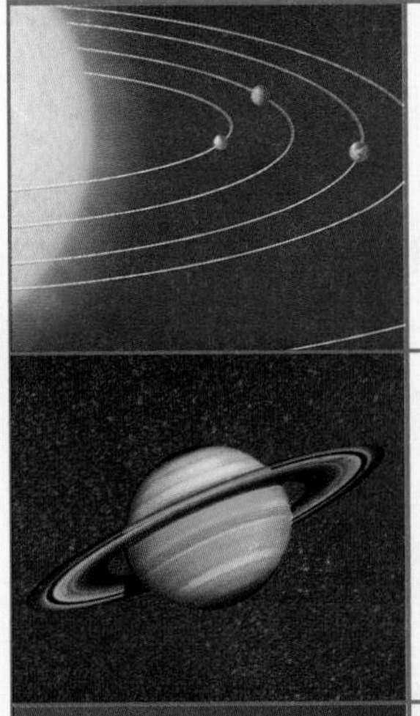

Our solar system is made up of the Sun, the planets, and other objects that revolve around the Sun.

Our solar system has four inner planets and four outer planets.

Telescopes and space probes are used to observe the planets and other objects in space.

Make a FOLDABLES® Study Guide

Make a trifold book. Use it to summarize what you learned about Earth's solar system.

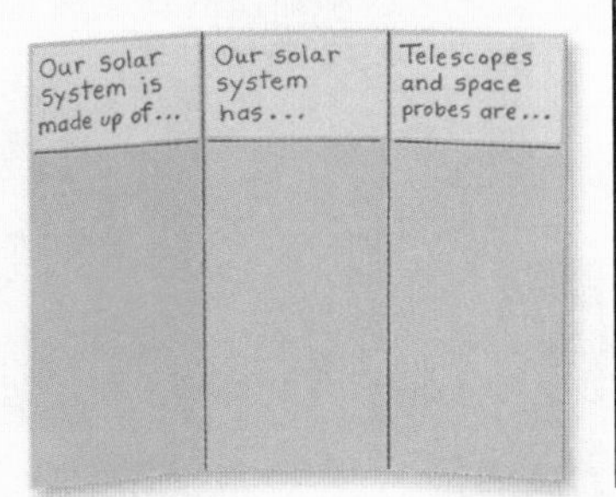

Think, Talk, and Write

1. **Vocabulary** What is a planet? Name five planets.

2. **Classify** Is Venus a planet or a star? Explain your answer.

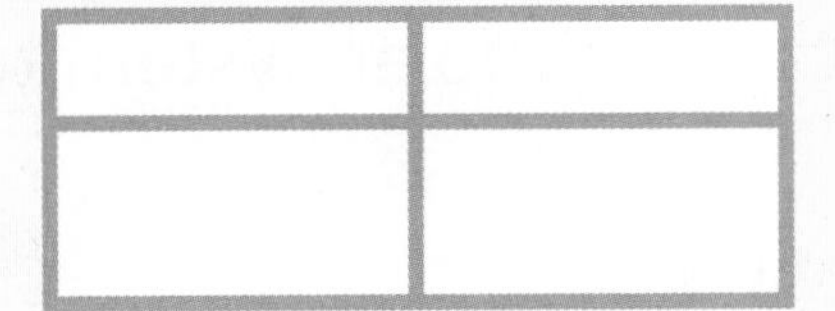

3. **Critical Thinking** What is the same about all of the planets in our solar system?

4. **Test Prep** **Which tool would best show details of Saturn?**
   - **A** binoculars
   - **B** a microscope
   - **C** a hand lens
   - **D** a telescope

5. **Essential Question** How can Earth be compared to the other objects in the solar system?

Math Link

Solve a Problem

In 2006 Saturn had 28 times as many known moons as Mars. Mars had two moons. How many moons did Saturn have?

Art Link

Planet Picture

Suppose you could visit another planet. If you could stand on its surface, what would you see? Draw or paint a picture. Write the name of the planet on your picture.

Focus on Skills

Inquiry Skill: Observe

You know that Earth is only one of the planets in our solar system. How do scientists learn about other planets? How do they learn about other objects, such as comets, that orbit the Sun? They **observe** the Sun, planets, moons, and other objects in our solar system to learn more about them.

▶ Learn It

When you **observe**, you use one or more of your senses to learn about an object or event. Remember, your senses are sight, hearing, smell, taste, and touch. Scientists often use tools, such as binoculars, microscopes, and telescopes, to make their observations. They use their observations to draw conclusions about objects and events.

△ **Be Careful.** It can be dangerous to observe things by tasting. You should not taste things in school unless your teacher tells you it is safe.

▶ Try It

You can **observe** things too. Look at the detail of this comet. **Observe** its color and shape. Look for unique features that help you identify what it is. What detail helps you know this is a comet and not a planet?

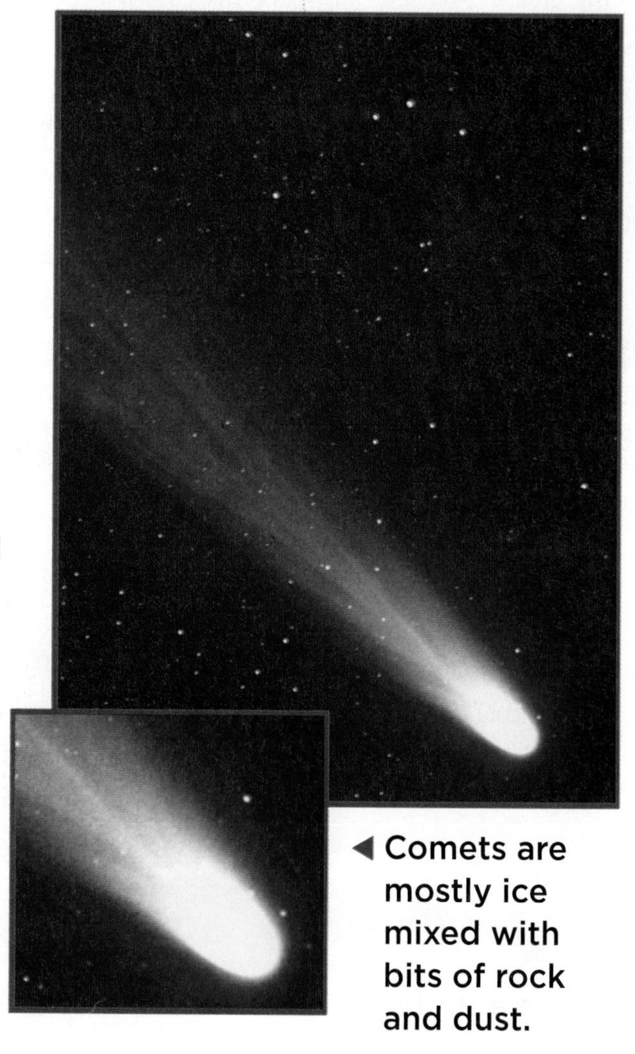

◀ **Comets are mostly ice mixed with bits of rock and dust.**

▶ Apply It

The photos below show details of planets and other objects in our solar system. **Observe** each photo carefully. Use your observations to identify what each object is. Which details helped you to identify the objects?

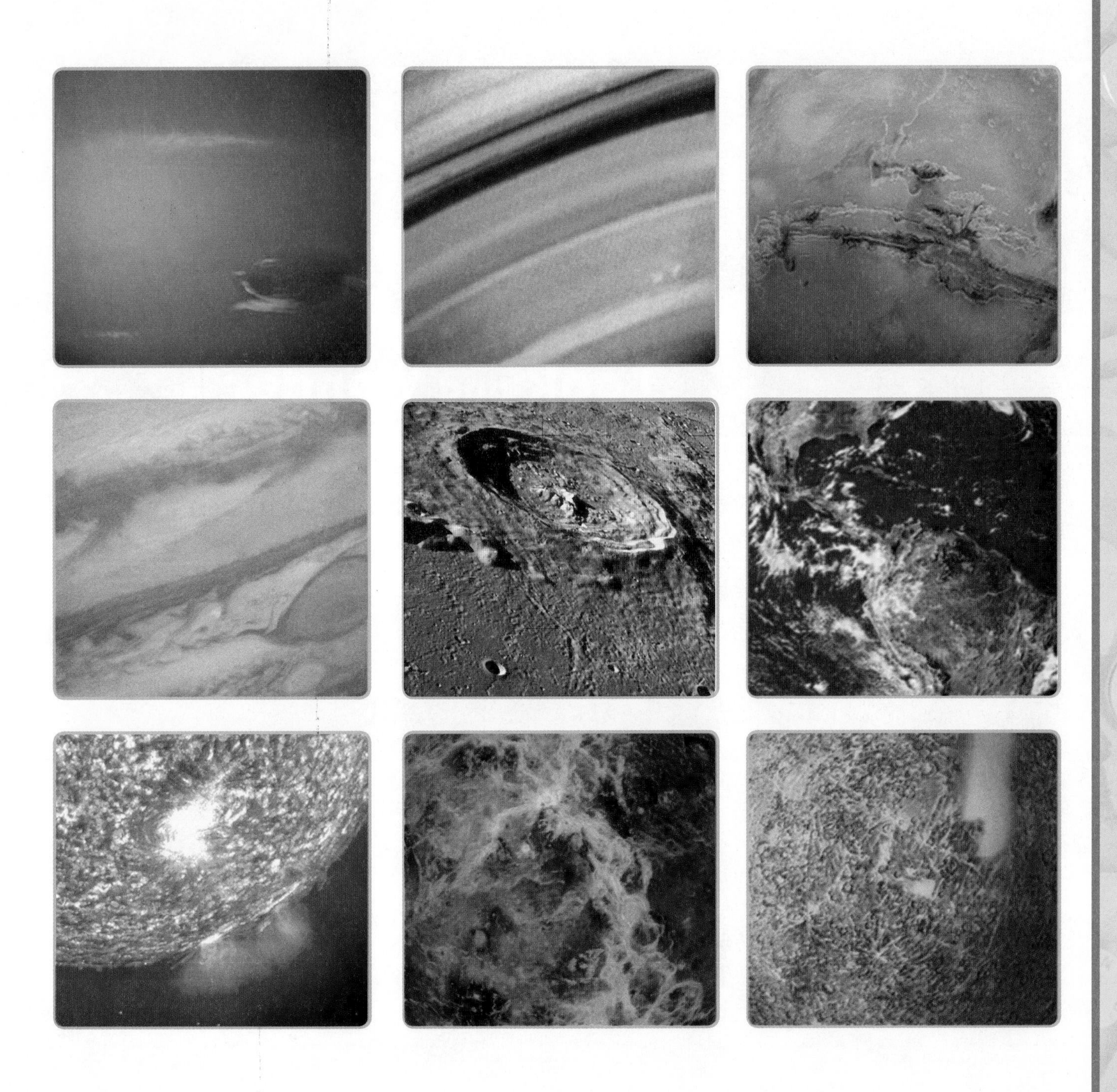

Lesson 4

The Stars

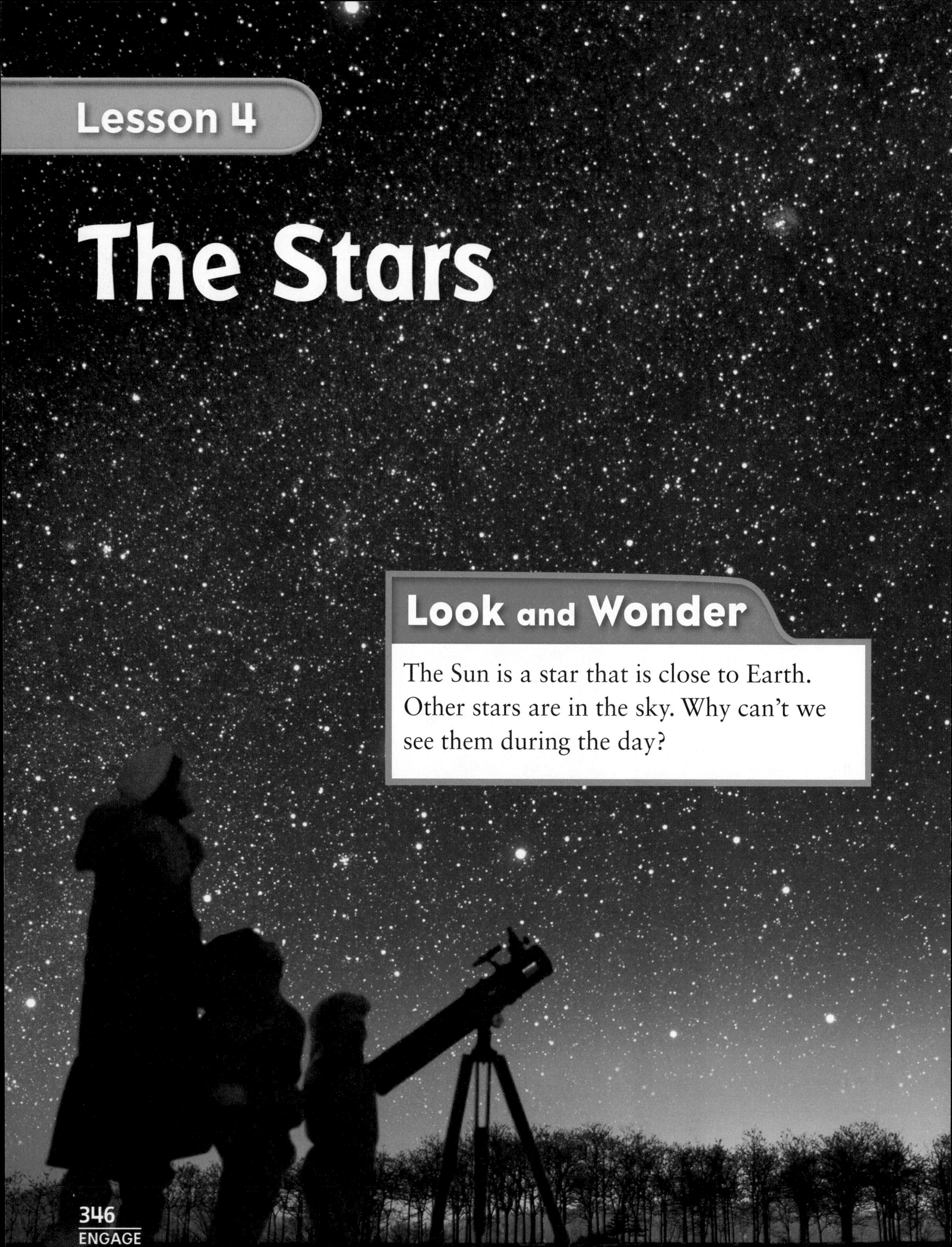

Look and Wonder

The Sun is a star that is close to Earth. Other stars are in the sky. Why can't we see them during the day?

Explore

Inquiry Activity

Why do we see stars at night?

Purpose

We want to understand why we do not see stars in the daytime sky.

Procedure

1. Draw a 3-cm dot with white chalk or crayon on black paper.
2. Draw a 3-cm dot with white chalk or crayon on white paper.
3. **Measure** Have your partner hold both papers 2 m away from you.
4. **Observe** Which dot is easier to see?

Draw Conclusions

5. **Infer** Why do you think one dot was easier to see than the other?
6. **Infer** Suppose the dots on the papers were stars. Why do you think you see stars only at night?

Explore More

Experiment Design an experiment to show how we sometimes see the Moon during the day.

Materials

white chalk or white crayon

black paper

white paper

measuring tape

Step 1

Step 3

Read and Learn

Essential Question
What can we see in the night sky?

Vocabulary
constellation, p. 349

Reading Skill
Summarize

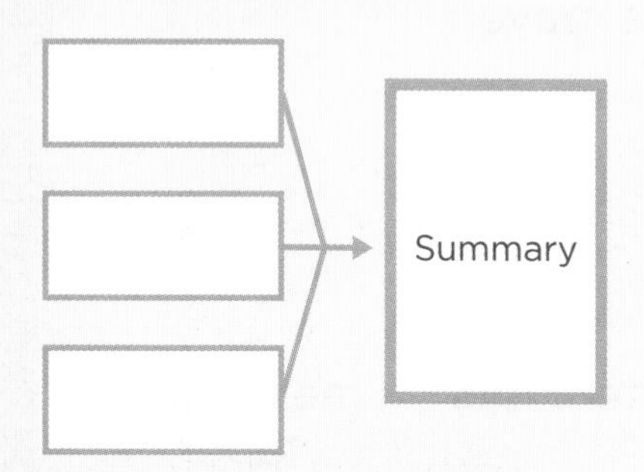

Technology

e-Glossary and e-Review online at www.macmillanmh.com

What are stars?

Stars are glowing balls of hot gases that give off light and heat. The Sun is the only star in our solar system. It is the closest star to Earth.

There are more stars in the sky than anyone can easily count. Stars are always in the sky. We can not see stars in daytime because the Sun's light prevents us from seeing them.

The Sun seems bigger and brighter than other stars because it is closer to Earth. The Sun is only a medium-size star, however. It is much smaller and younger than many other stars. Other stars seem tiny and faint because they are so far away.

Size is only one way stars differ. Some stars are brighter than others. Some are hotter. Blue stars are hottest. Red stars are coolest.

Sizing Up Stars

Stars are classified by size, color, and brightness.

medium star

small star

large star

giant star

Constellations

Long ago, people gazed at the stars. They thought groups of stars formed pictures in the night sky. We can still see those star groups today. A **constellation** (kahn•stuh•LAY•shun) is a group of stars that seem to form a picture.

Quick Check

Summarize What are some ways that stars differ?

Critical Thinking How is the Sun different from all other stars?

The Big Dipper is shown by the bright yellow lines. It is part of the constellation Ursa Major, or the Great Bear. ▼

Quick Lab

Make a Constellation

1. Find out about a constellation. Use chalk to draw it on a piece of black paper. Use a pencil point to poke holes in the pattern. △ **Be Careful.**
2. **Observe** Trade papers with a partner. In a darkened room, hold your partner's paper out at arm's length toward a lamp or flashlight. What does the star pattern look like?

3. **Communicate** Tell your partner the name of your constellation.

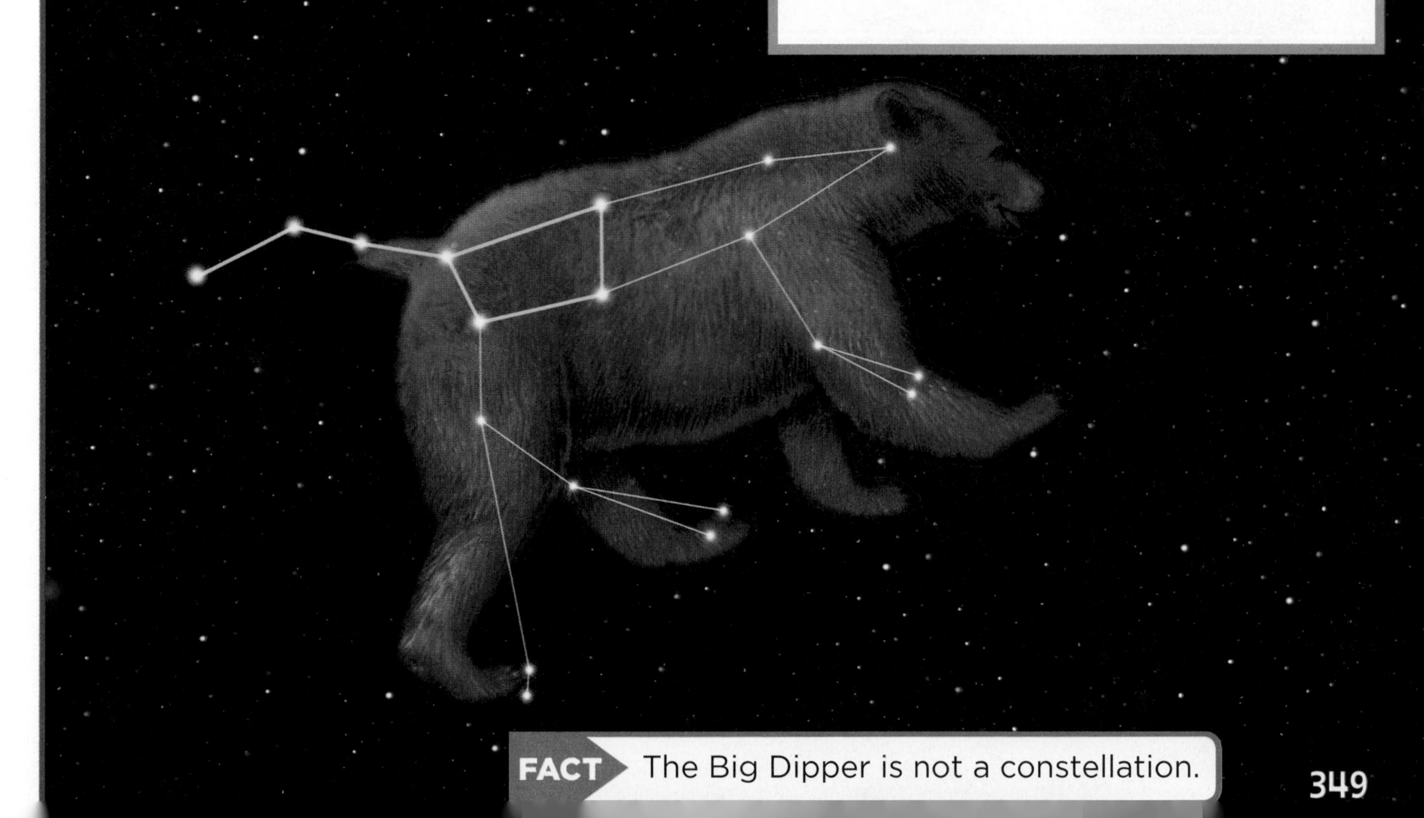

FACT The Big Dipper is not a constellation.

Why do we see different stars during different seasons?

Each season you can see different stars. As Earth revolves around the Sun, the dark part of Earth faces different directions. This movement gives us a different view of the nighttime sky.

Find the constellation Orion below. In winter you can see Orion in the night sky. In summer Earth has moved to the other side of the Sun. When you look at the night sky in summer, you are facing away from Orion. Orion is in the daytime sky and can not be seen.

Stars and Seasons

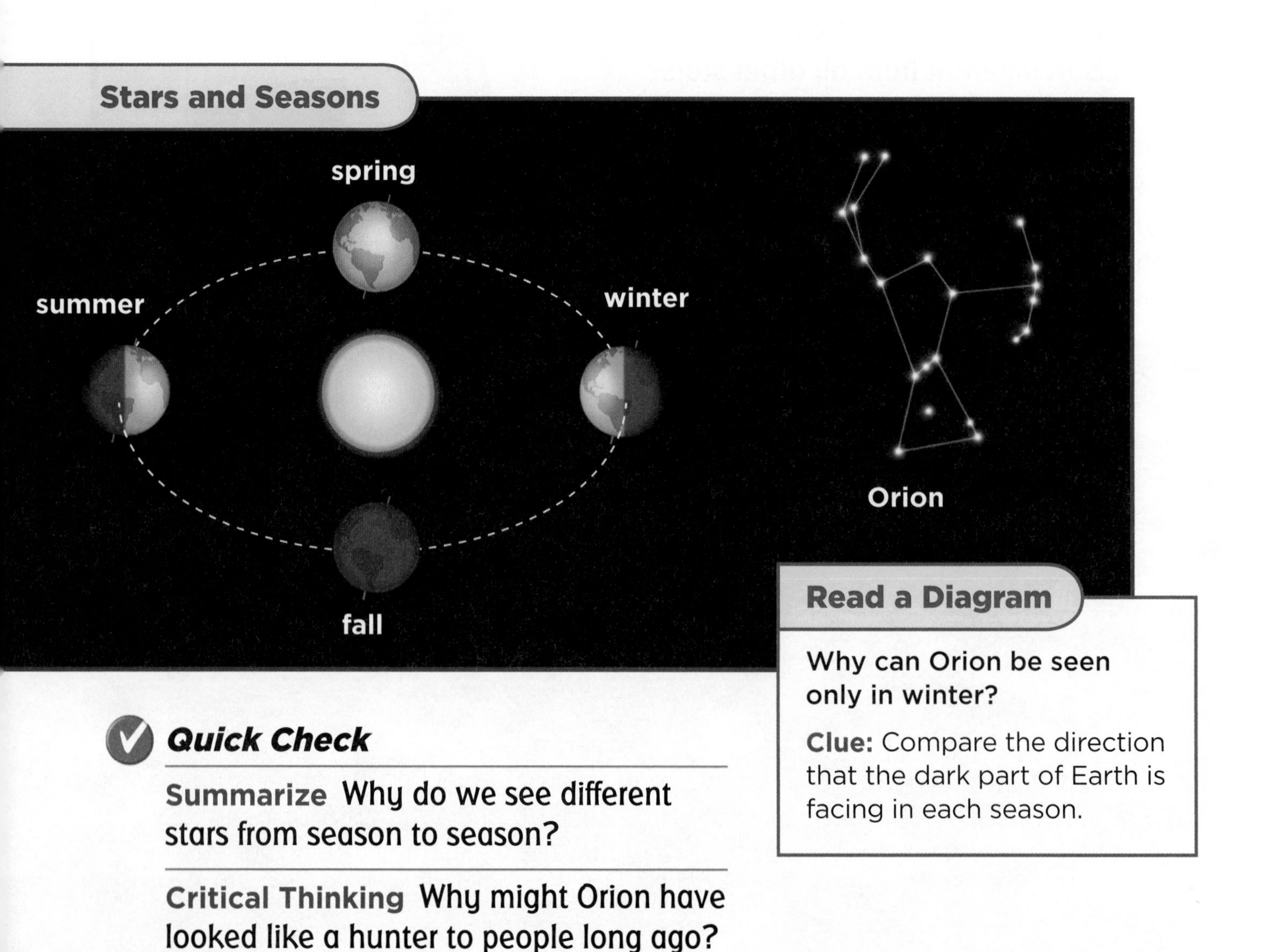

Read a Diagram

Why can Orion be seen only in winter?

Clue: Compare the direction that the dark part of Earth is facing in each season.

Quick Check

Summarize Why do we see different stars from season to season?

Critical Thinking Why might Orion have looked like a hunter to people long ago?

Lesson Review

Visual Summary

Stars differ by size, brightness, and color.

Constellations are **groups of stars that seem to form** pictures in the night sky.

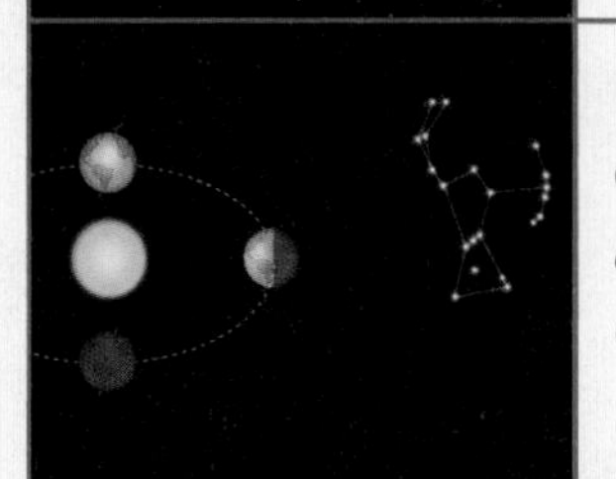

Different constellations are seen during different seasons.

Make a FOLDABLES® Study Guide

Make a trifold book. Use it to summarize what you learned about stars.

Think, Talk, and Write

1. **Vocabulary** What is a constellation?

2. **Summarize** Why do stars much larger than the Sun look much smaller than the Sun from Earth?

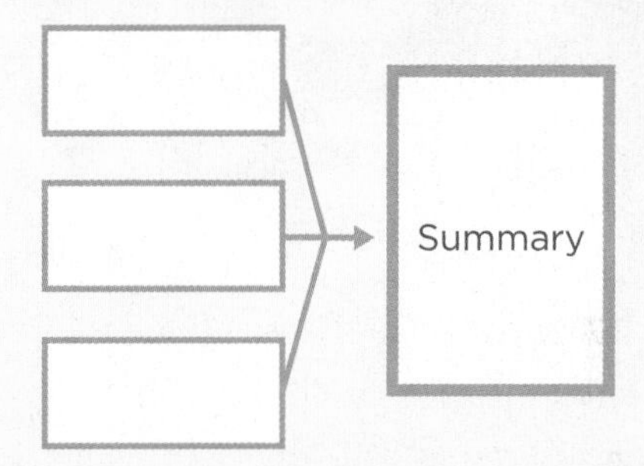

3. **Critical Thinking** Do you think that you can see all the stars there are? Explain your answer.

4. **Test Prep How would you describe the Sun?**
   A a small star
   B a medium star
   C a giant star
   D a blue star

5. **Essential Question** What can we see in the night sky?

Writing Link

Write a Report
Long ago, people thought the constellation Cassiopeia looked like a queen. Research this constellation or another one. Then write about it.

Music Link

Sing a Star Song
Learn the words to the song "Twinkle, Twinkle, Little Star." Write a verse of your own about stars to go with the tune. Read or sing the new words to the class.

LOG ON e-Review Summaries and quizzes online at www.macmillanmh.com

Meet Orsola De Marco

Do you ever wonder about the stars? Orsola De Marco does. She is an astrophysicist at the American Museum of Natural History in New York. An astrophysicist is a scientist who studies stars. Orsola studies stars that are found together in pairs. As far as we know, our Sun is a star that stands alone. Most stars in the universe have a partner. They are called binary stars.

These binary stars orbit each other at a very close distance. Scientists think that one star is being absorbed by the other. The butterfly-shaped "wings" are probably caused by gases from the surface of the central star.

Summarize

A good summary

- states the main idea;
- gives the most important details;
- is brief;
- is told in your own words.

Of course, Orsola can not go to the stars to learn about them. Instead, she travels to Arizona, Hawaii, and Chile to use large telescopes. She gazes billions of miles into space to get a good look at binary stars. She watches how the stars affect each other. When a star gets old, it becomes larger. If there is another star nearby, it might get eaten up, or absorbed, by the old star. No one is sure what will happen after that. Orsola is working to find out.

Write About It

Summarize Read the article with a partner. List the most important information in a chart. Then use the chart to summarize the article. Remember to start with a main-idea sentence and to keep your summary brief.

LOG ON e-Journal Write about it online at www.macmillanmh.com

Connect to

at www.macmillanmh.com

CHAPTER 8 Review

Visual Summary

Lesson 1 Earth rotates and revolves around the Sun. Earth's movement causes day and night and the seasons.

Lesson 2 The Moon's shape and position in the sky seem to change.

Lesson 3 Earth is one of eight planets that revolve around the Sun.

Lesson 4 Stars differ in several ways. Different stars are seen during different seasons.

Make a FOLDABLES® Study Guide

Glue your lesson study guides to a piece of paper as shown. Use your study guide to review what you have learned in this chapter.

Vocabulary

DOK I

Fill each blank with the best term from the list.

axis, p. 319	**rotates,** p. 318
constellation, p. 349	**solar system,** p. 338
craters, p. 332	**space probe,** p. 342
orbit, p. 320	**star,** p. 322
phase, p. 328	**telescope,** p. 342

1. Earth revolves around the Sun in a regular path called an ______.

2. A line through the center of a spinning object is an ______.

3. A group of stars that seems to form a picture is called a ______.

4. Faraway objects look closer when seen through a ______.

5. A star and the objects that orbit it make up a ______.

6. Every 24 hours Earth ______.

7. A machine that travels through space from Earth is a ______.

8. Each shape of the Moon seen from Earth is known as a ______.

9. The Sun is a _____.

10. The hollow areas and pits on the Moon's surface are _____.

Skills and Concepts

DOK 2–3

Answer each of the following.

11. **Summarize** What do the planets in our solar system have in common? How do they differ?

12. **Personal Narrative** Write a journal entry about what you observe in the night sky. How does what you observe make you feel?

13. **Observe** What can you observe about the Moon by only looking at it? What other details might you be able to observe using a telescope?

14. **Critical Thinking** How might life on Earth change if Earth were farther away from the Sun? What if Earth were much closer?

15. **Critical Thinking** If stars are always in the sky, why can we not see them during the day?

16. **Predict** How would Earth's climates be different if Earth's axis were straight up and down instead of tilted?

17. Observe the Moon's shape below. How much time must pass before the new moon phase occurs?

18. **True or False** *The Moon changes shape as it revolves around Earth.* Is this statement true or false? Explain.

19. Why is it warmer in summer than in winter?

A There is more cloud cover in winter to block the Sun's rays.

B The Sun's rays strike Earth's surface more directly in summer.

C The Sun's rays strike Earth's surface more directly in winter.

D Earth is closer to the Sun in summer than in winter.

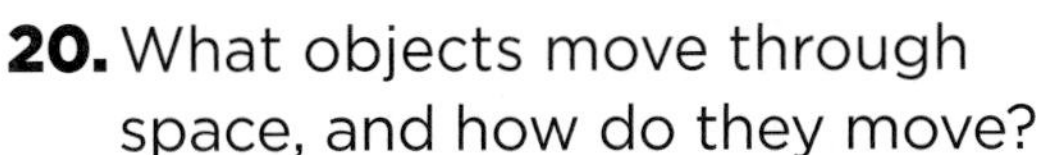

20. What objects move through space, and how do they move?

Performance Assessment

DOK 2

Space Exploration

- Make a time line of what you think were important events in space exploration.
- Research some important events in space exploration. Mark on your time line each event and the year that each occurred.
- Which event on your time line do you think was most important? Why? Write about it.

Test Preparation

1 Which picture explains the Moon's phases?

A

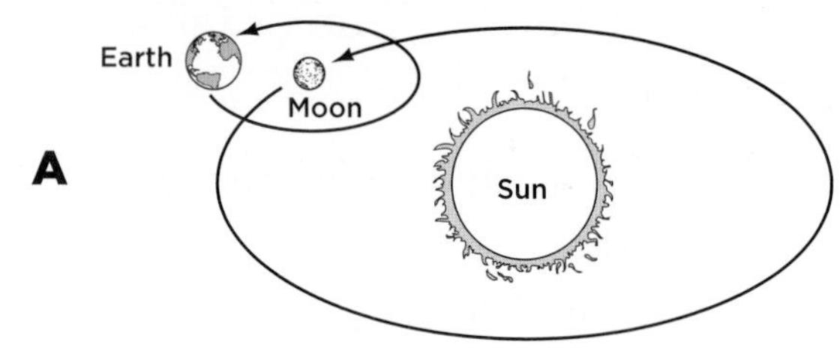

B

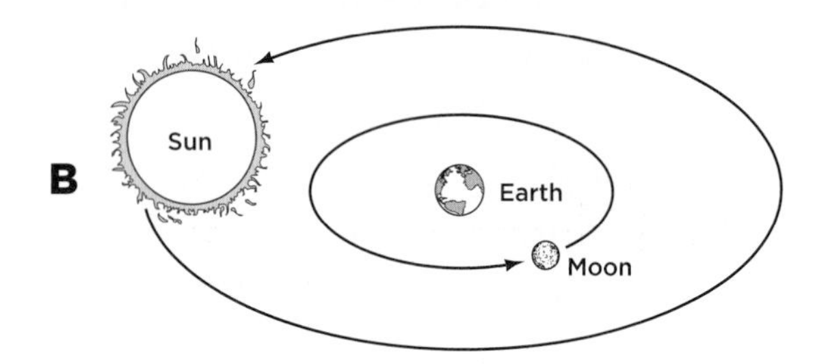

C

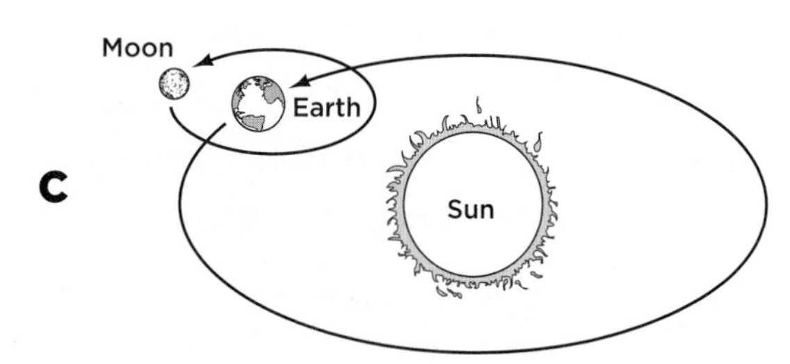

D

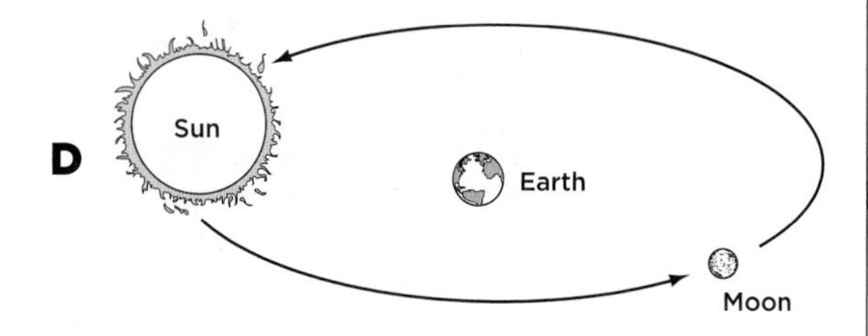

DOK 2

2 Which part of the solar system does Earth move around?

A the Moon

B the Sun

C other planets

D other moons

DOK 1

3 Look at the diagram below.

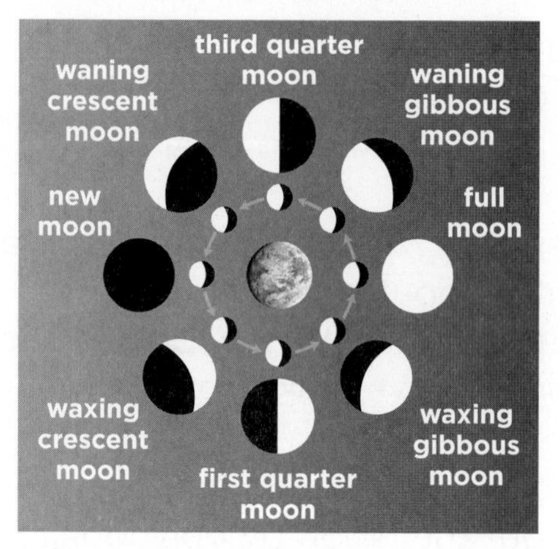

If the new moon occurs on a given day, when will the next full moon occur?

A one day later

B one week later

C two weeks later

D two months later

DOK 1

4 How does a telescope make it easier to observe the Moon's features?

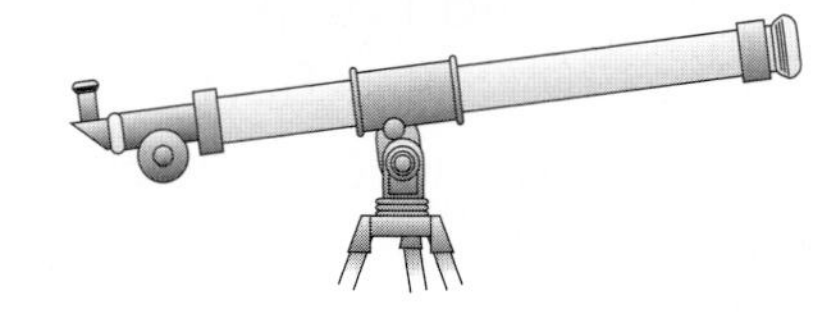

A It makes objects that are far away look closer.

B It makes objects that are far away look darker.

C It makes objects that are far away look smaller.

D It makes objects that are far away look brighter.

DOK 1

5 Look at the picture of the moon phase below.

Which moon phase occurs right before this phase?

A gibbous moon

B full moon

C crescent moon

D quarter moon

DOK 2

6 Why do we see different stars each season?

A We see different constellations as they revolve around Earth.

B The dark part of Earth faces different directions at different times.

C Constellations revolve around the Sun and go to different parts of the sky.

D Other planets pass in front of Earth and block our view of the stars during certain times.

DOK I

7 How are the inner planets alike?

A They are all made of gas.

B They are all the same size.

C They are all made of rocky material.

D They are all cold.

DOK I

8 The diagram below shows Earth and the Sun.

Use this diagram to explain what causes day and night.

DOK 2

Then contrast this with what causes seasons.

DOK 2

Check Your Understanding

Question	Review	Question	Review
1	pp. 328–331	5	pp. 328–329
2	pp. 338–339	6	p. 350
3	p. 330	7	pp. 340–341
4	p. 342	8	pp. 318–321

Meteorologist

Not all meteorologists work on TV. Some work as aviation meteorologists. They predict weather for airlines. Airline pilots need to know what weather will be like so people can travel safely.

Other meteorologists gather data at weather stations in rain forests or at the North Pole and South Pole. The information they gather helps other scientists learn more about Earth's climate.

To be a meteorologist, you need to study weather in college. You need good computer skills. You will also need to learn to use special weather tools and how to read radar scans and satellite photos.

▲ Meteorologists use technology to collect data about weather.

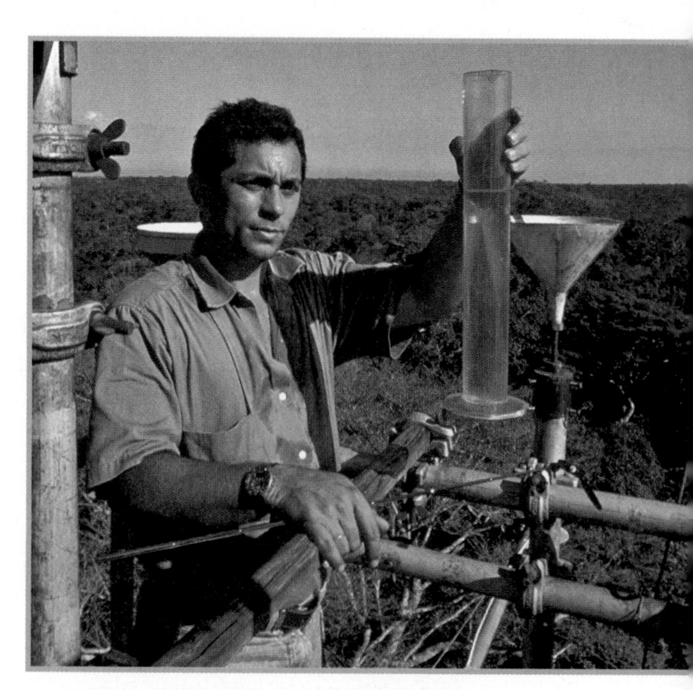

▲ This meteorologist checks rainfall in the Amazon rain forest of Brazil.

Here are some other Earth science careers:

- weather observer
- climatologist
- astronaut
- astronomer

Matter

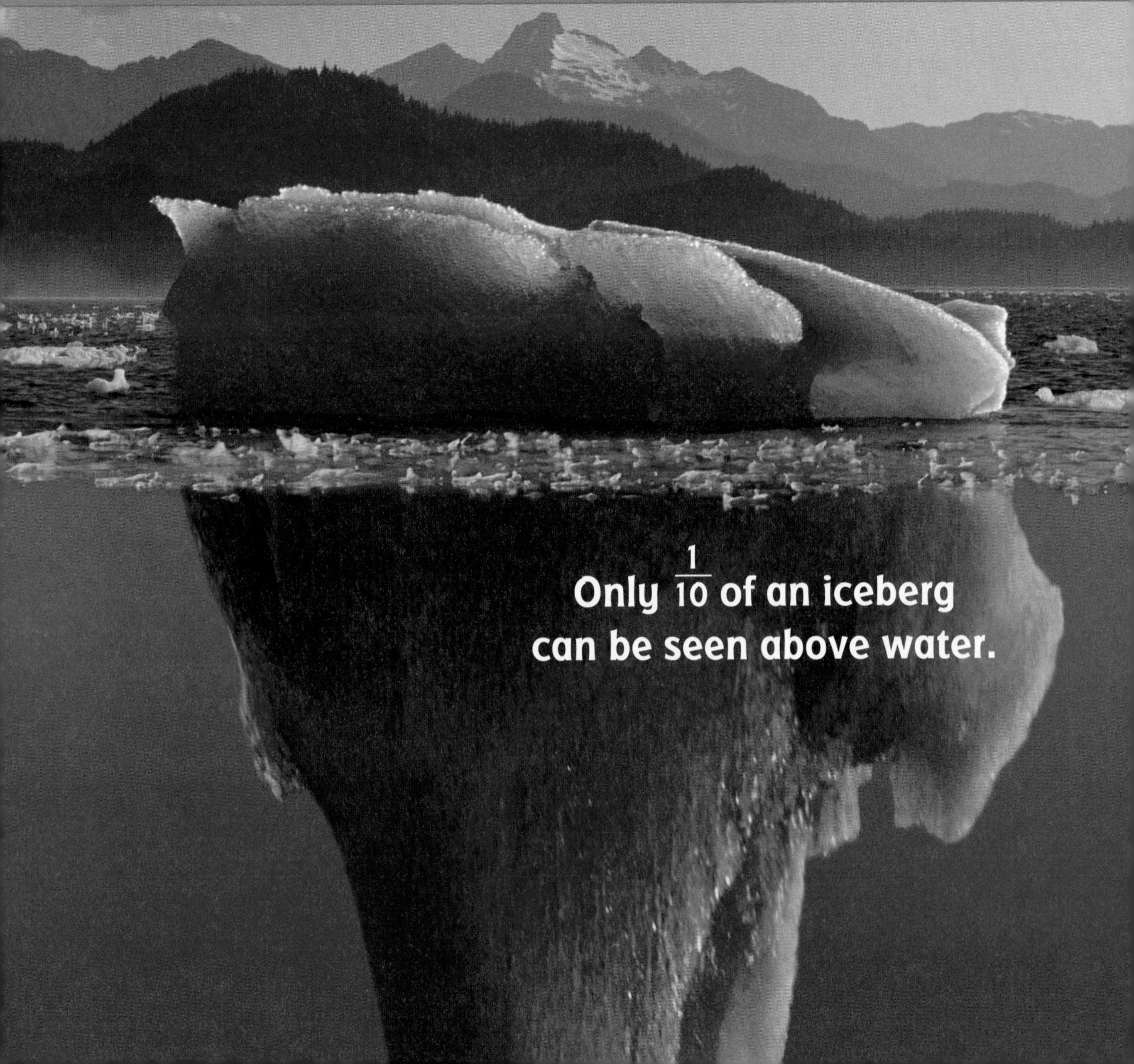

Only $\frac{1}{10}$ of an iceberg can be seen above water.

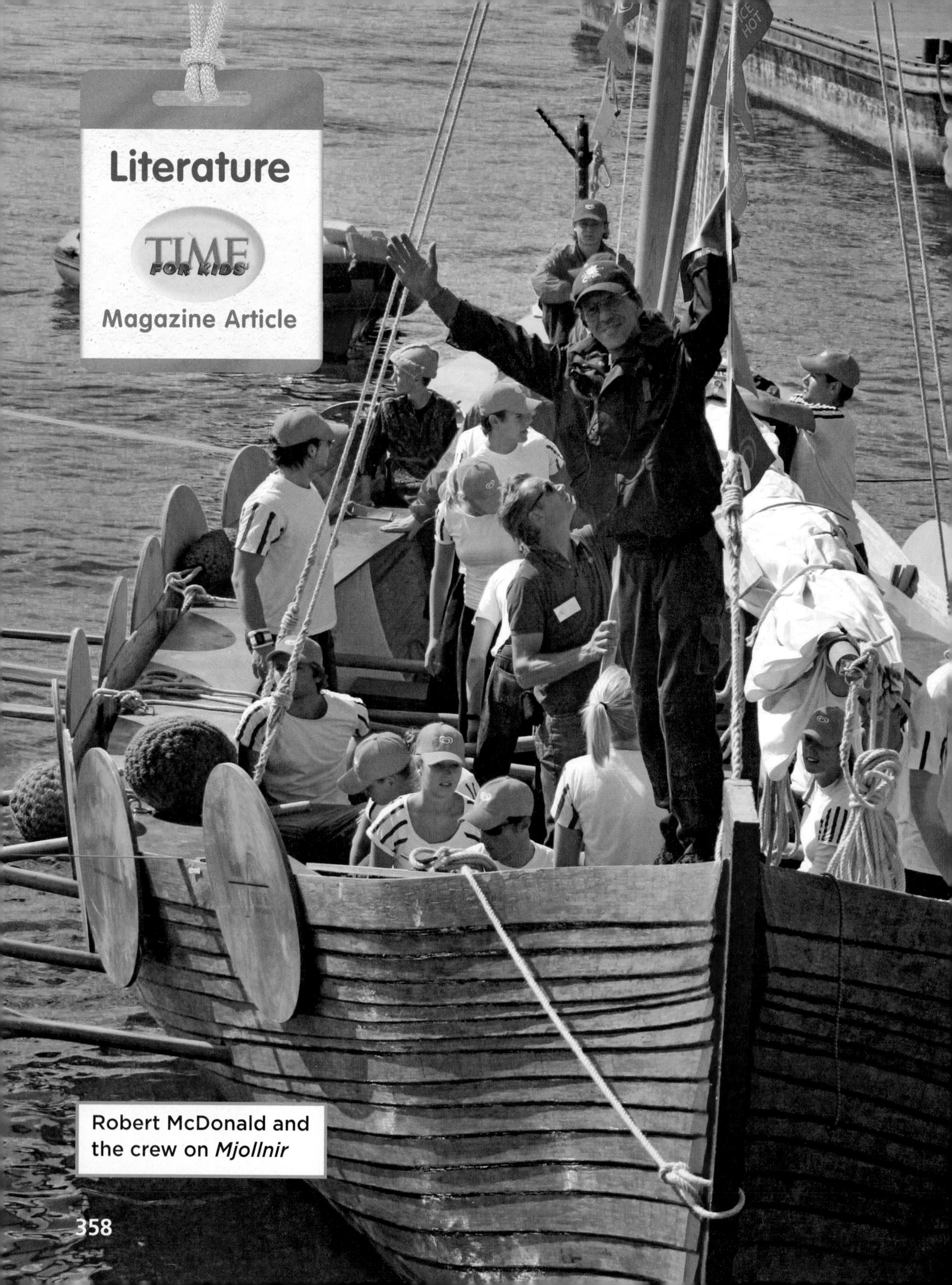

Robert McDonald and the crew on *Mjollnir*

The Good Ship Popsicle Stick

from *Time for Kids*

September 2, 2005

On August 16, former Hollywood stuntman Robert McDonald performed a record-breaking stunt. He floated a ship made of wooden ice-cream sticks on a river that flows through the city of Amsterdam, in the Netherlands. And the boat didn't sink!

McDonald built the 50-foot-long replica of a Viking ship with 15 million ice-cream sticks and more than two tons of glue. The 13-ton cruiser is named *Mjollnir* (MIHL•nur), after the hammer of Thor, the Viking god of thunder.

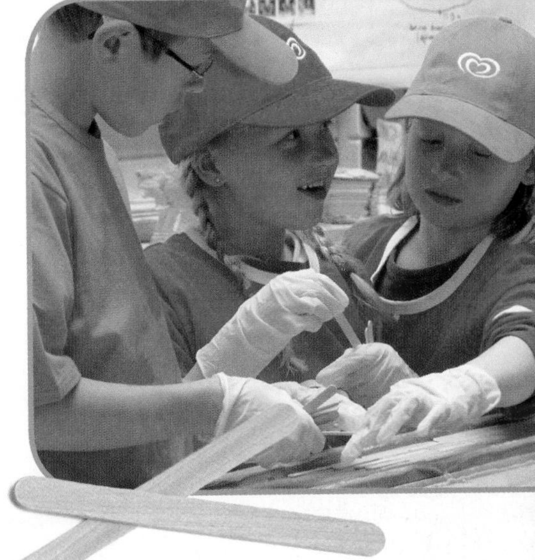

Captain Rob is the president of the Sea Heart Ship Foundation. The group's goal is to spread fun to kids in hospitals around the world. "I have a dream to show children they can do anything," he says. "If they can dream it, they can do it."

Write About It

Response to Literature This article is about a ship made from ice-cream sticks. What words are used to describe the ship? Choose an object around you. Then use words to tell about it.

CHAPTER 9

Observing Matter

What are some ways you can describe matter?

Essential Questions

Lesson 1
What are all objects made of?

Lesson 2
How can you compare different kinds of matter?

Lesson 3
What are the states of matter?

 Castle Geyser in Yellowstone National Park

Big Idea Vocabulary

matter anything that takes up space and has mass (p. 364)

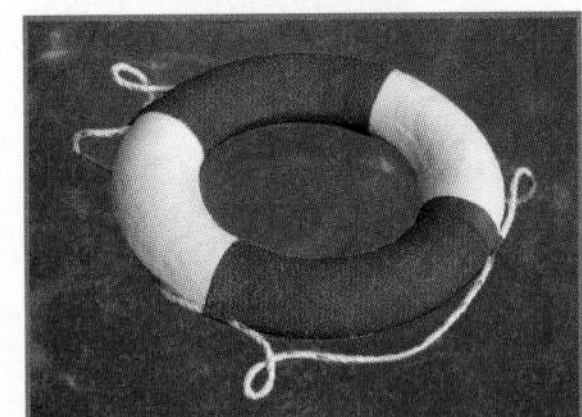

property a characteristic of something (p. 365)

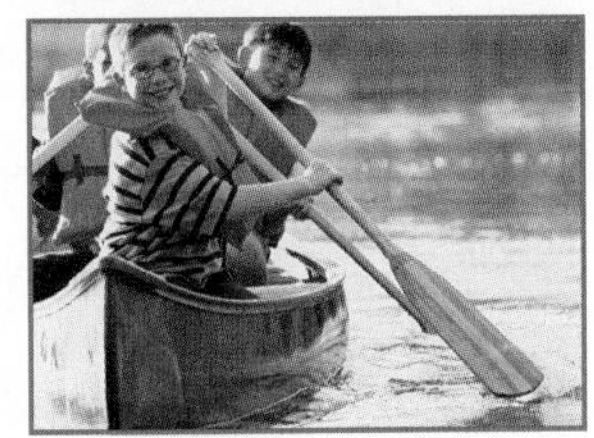

state of matter a form of matter, such as solid, liquid, or gas (p. 384)

solid matter that has a certain volume and shape (p. 384)

liquid matter that has a certain volume but a shape that can change (p. 386)

gas matter that does not have a certain volume or shape (p. 387)

Visit www.macmillanmh.com for online resources.

Lesson 1

Properties of Matter

Look and Wonder

How can you tell different objects from one another? Objects can have different colors, shapes, and sizes. Each feels a bit different. How can you describe the objects in this photograph?

Explore

Inquiry Activity

How do you describe objects?

Purpose

Explore ways to describe objects.

Procedure

1. **Observe** Select a "mystery object" in your classroom. Observe the object. What color is it? How does it feel? What is the object's shape and size?
2. **Communicate** Record your observations in a word web like the one shown. Label each line with a word that describes your mystery object. Leave the circle blank.
3. **Infer** Trade webs with a partner. Think about the descriptive words on your partner's web. What classroom object do the words describe? Label the circle with the name of your partner's mystery object.

Draw Conclusions

4. Were you able to guess your partner's mystery object? Was your classmate able to guess your mystery object?
5. What helped you most in figuring out your partner's object?

Explore More

Experiment How might your web be different if you were blindfolded and could only touch the mystery object? Try it to find out.

Materials

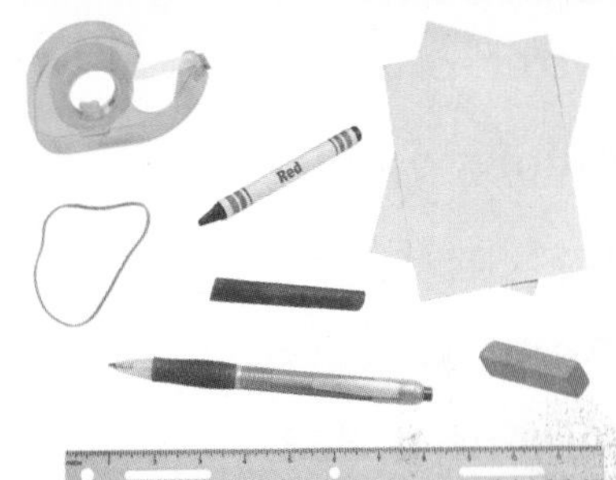

classroom objects

hand lens

Step 1

Step 2

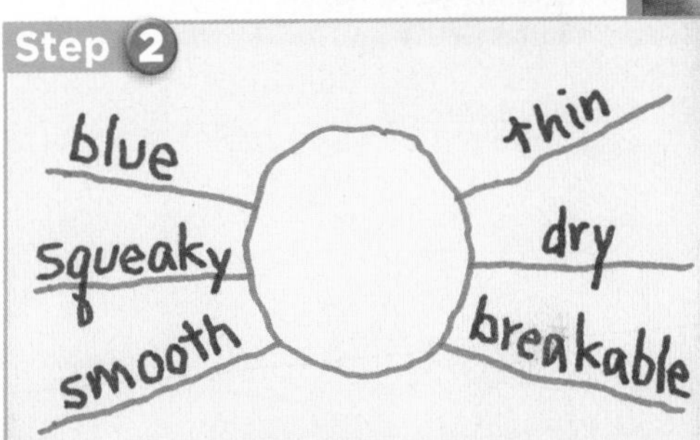

Read and Learn

Essential Question
What are all objects made of?

Vocabulary
matter, p. 364
volume, p. 365
mass, p. 365
property, p. 365
element, p. 368

Reading Skill
Main Idea and Details

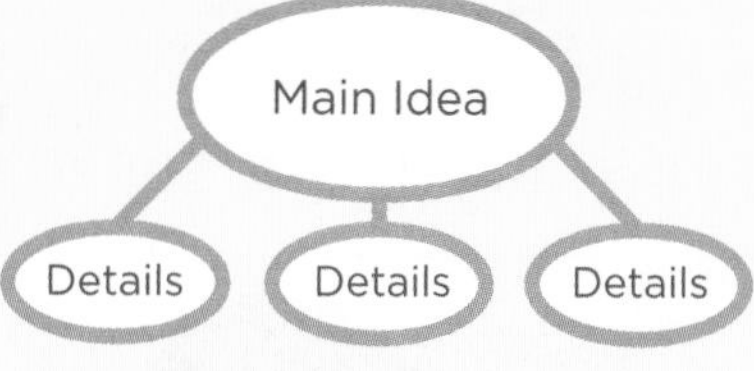

Technology LOG ON
e-Glossary and e-Review online at www.macmillanmh.com

What is matter?

Explore the area around you. Can you find things with different colors, sizes, and shapes? Things differ in the way they look, feel, sound, and smell. All the things around you are alike in one way, however. All are made of matter (MA•tur).

Matter is anything that takes up space. You are matter. This book is matter. Even the air you breathe is matter. All of these things take up space.

What can you see, hear, and touch at the beach? ▼

Volume

Volume (VOL•yewm) describes how much space an object takes up. It tells how big or small an object is. This beach ball takes up more space than this bowling ball. The beach ball has more volume.

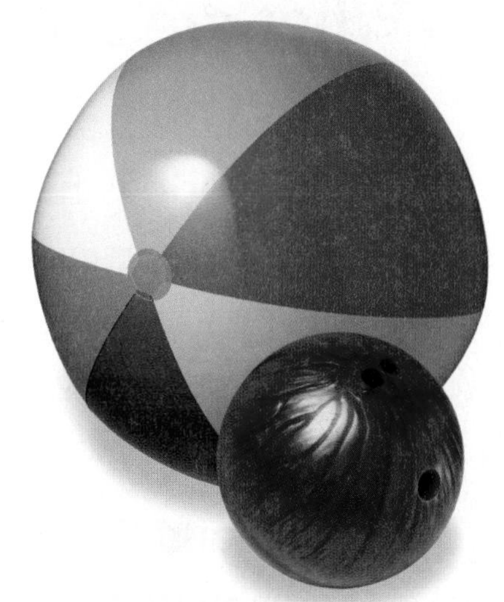

▲ This beach ball has more volume but less mass than this bowling ball.

Mass

All objects have mass. **Mass** is a measure of the amount of matter in an object. An object with a large mass feels heavy. An object with a small mass feels light. This bowling ball feels heavier than this beach ball. This is because the bowling ball contains more matter. The bowling ball has more mass.

Volume and mass are properties (PRAH•pur•teez) of matter. A **property** is a characteristic of something. The way an object looks, tastes, smells, sounds, and feels are other properties that you can observe.

Properties of a Pineapple

Property	Description
color(s)	brown, green
shape	round and spiky
feel	rough
taste	sweet

Read a Table

How does a pineapple taste?

Clue: Headings help you find information.

Quick Check

Main Idea and Details What are two properties of all types of matter?

Critical Thinking Why is sound not matter?

What are some properties of matter?

The world is full of many kinds of matter. We use properties to tell them apart. An object might be hot or cold. It could feel smooth or rough, wet or dry. Here are some properties that help us describe and identify matter.

Sinking and Floating

Some matter sinks in water. Some matter floats. For example, a rock sinks in water and an apple floats. Metal objects usually sink, while wooden objects often float. Objects sink or float because of their mass and volume. Objects with a lot of mass and little volume tend to sink. Objects with little mass and a lot of volume tend to float.

A life preserver floats on water. ▼

An anchor sinks in water.

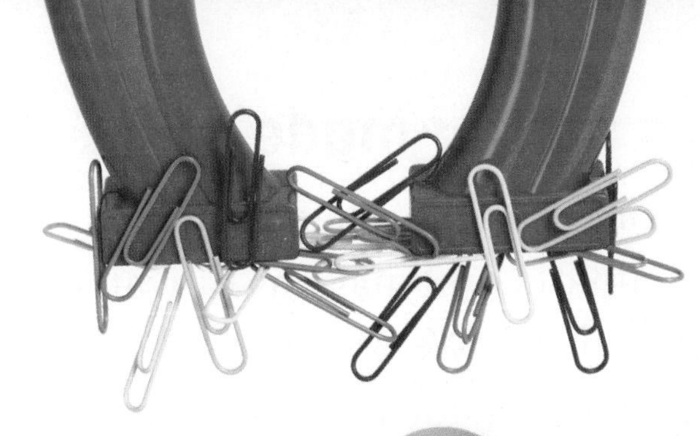

Magnetism

Magnets have a special property. Magnets pull on, or *attract,* certain metals, such as iron. They do not attract wood, plastic, or water. Put a magnet near an object made of iron. What happens? The magnet pulls on the object, and then the object "sticks" to the magnet.

Conducting Heat

Some matter *conducts* heat. This means that some kinds of matter let heat move through them easily. For example, heat moves easily through metals such as iron and copper. This is why these materials make good cooking pots. Heat moves from the stove through the metal pot. The pot gets warm. Wood does not heat up quickly. This is why wood is good to use for pot handles and cooking spoons.

Quick Lab

Classify Matter

1. Look at ten objects.
2. **Communicate** List the properties of each object in a table like the one shown.
3. **Classify** Sort the objects into groups that have similar properties. Give each group a name that describes how its items are alike.
4. **Interpret Data** Did some of the objects in one group have the same properties as objects in another group? How did you decide how to classify each object?
5. **Communicate** Is there more than one way to classify these objects? Explain your answer.

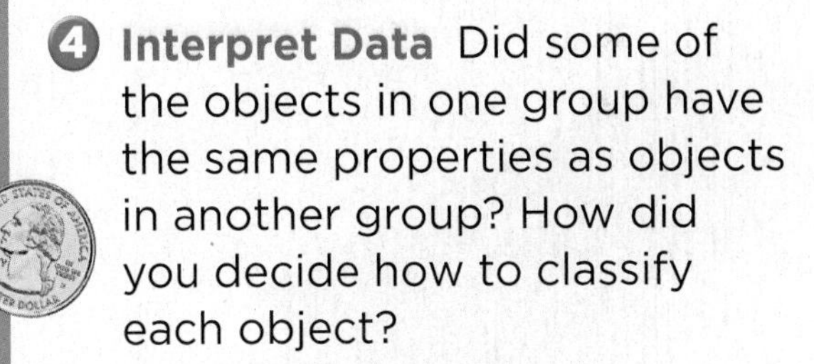

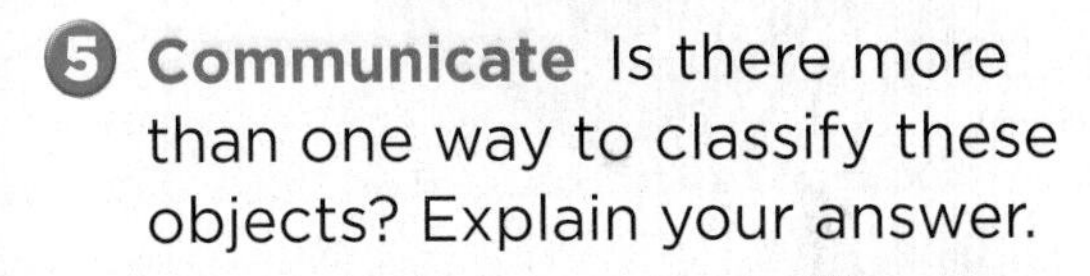

Object	Properties

Quick Check

Main Idea and Details Name three properties of matter.

Critical Thinking What properties of plastic make it useful as a bowl but not as a cooking pan?

FACT Only some metals are attracted to magnets.

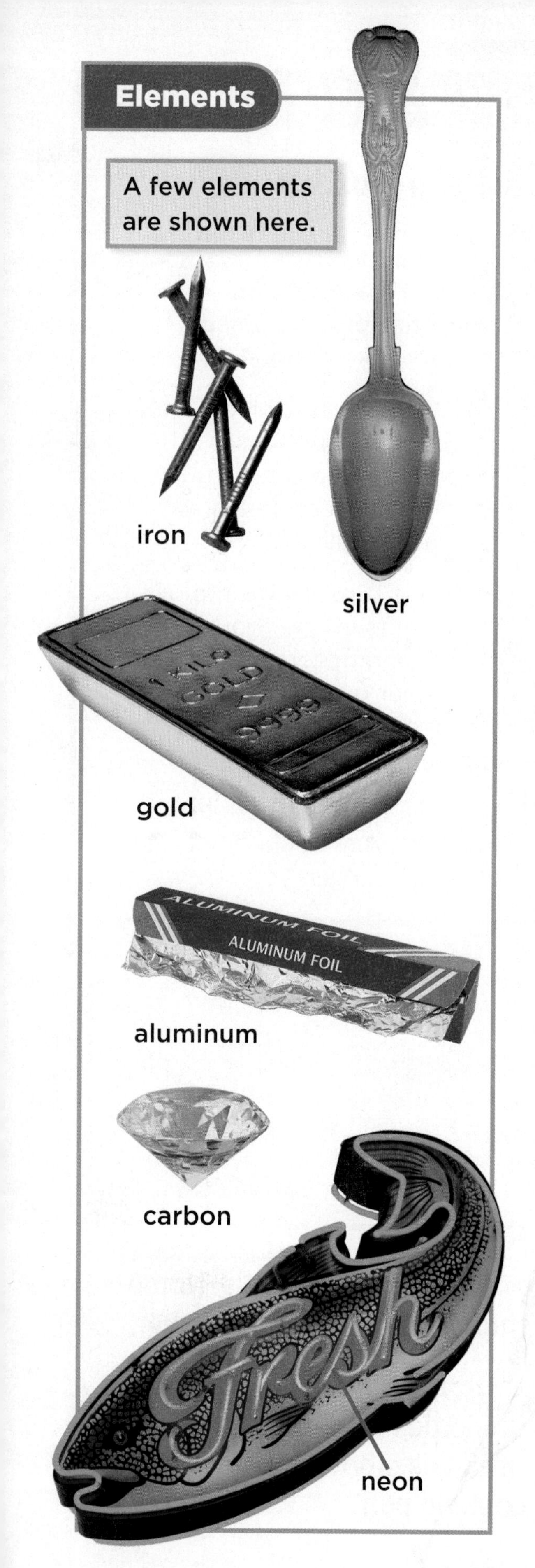

What is matter made of?

People once thought that all matter was made up of combinations of water, air, earth, and fire. We now know that all matter is made up of elements (E•luh•munts). **Elements** are the building blocks of matter. There are more than 100 different elements. They make up all the matter in the world.

Some matter is made of mostly one element. An iron nail contains mostly the element iron. Aluminum (uh•LEW•muh•num) foil contains mostly the element aluminum.

Most matter on Earth is made of more than one element. Water is made up of the elements hydrogen and oxygen. Sugar is made up of hydrogen, oxygen, and a third element called carbon. Elements join in different ways and in different amounts to form everything in our world.

Quick Check

Main Idea and Details Why are elements called the building blocks of matter?

Critical Thinking How is an iron nail different from water?

Lesson Review

Visual Summary

Matter is anything that has volume and mass.

Matter has properties that can describe and identify it.

Matter is made of elements.

Make a FOLDABLES® Study Guide

Make a trifold book. Use it to summarize what you learned about matter and its properties.

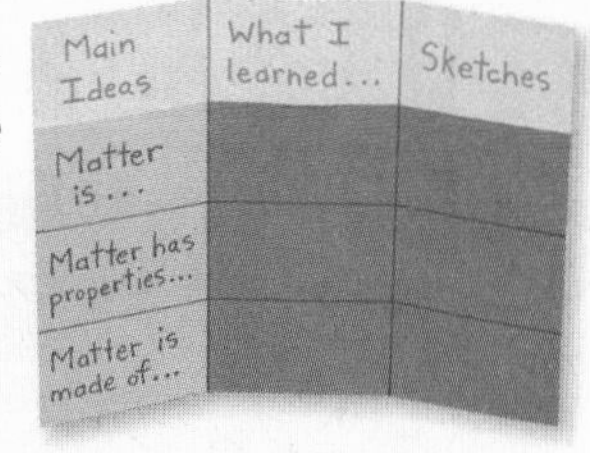

Think, Talk, and Write

1. **Vocabulary** What is matter?
2. **Main Idea and Details** Choose two objects. List all the properties you can to describe each one.

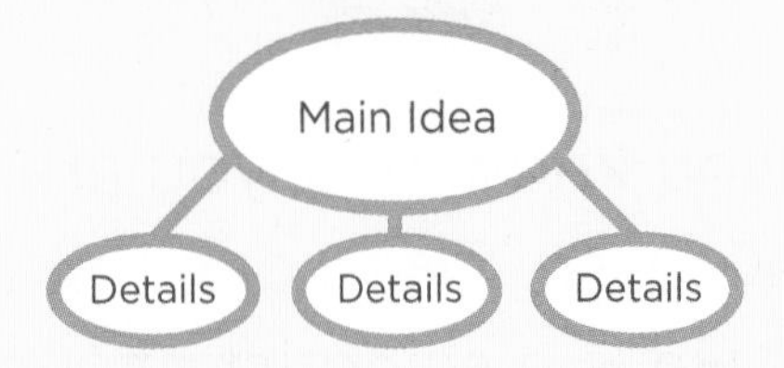

3. **Critical Thinking** What property of glass makes it a good material for windows?
4. **Test Prep** **What are the building blocks of matter?**
   A liquids
   B elements
   C wood
   D water
5. **Essential Question** What are all objects made of?

Writing Link

Writing That Describes
Suppose you brought your favorite toy to school and lost it. Write a notice to hang on your classroom bulletin board. What properties of the toy will you describe?

Math Link

Make a Table
Collect five small objects. Predict which ones will sink and which ones will float. Then put the objects in a tank of water. Record your results in a table.

Meet Neil deGrasse TYSON

Did you know that you are "star dust"? Neil deGrasse Tyson can tell you what that means. He is a scientist at the American Museum of Natural History in New York. He studies how the universe works.

Your body is full of hydrogen, carbon, and many other elements. All these elements were first formed in stars a long time ago. How did these elements make their way from the stars to your body?

Most elements form inside the dense and fiery centers of stars. Hydrogen combines to form all of the other elements in these conditions. Throughout their lives, stars scatter elements into space. Over millions of years, these elements combine to form new stars, planets, or even living things like you!

Neil is an astrophysicist. An astrophysicist is a scientist who studies how the universe works. ▲

Connect to

at www.macmillanmh.com

Main Idea and Details

- A main idea tells what the article is about.
- Details, such as facts and examples, support the main idea.

A nebula is a cloud of gas and star dust in space. The Horsehead Nebula shown here gets its name from its horselike shape.

Write About It

Main Idea and Details Read the article with a partner. What is the main idea? What details add to the main idea? Fill in a main-idea chart. Then write a few sentences to explain the main idea.

LOG ON e-Journal Write about it online at www.macmillanmh.com

Lesson 2

Measuring Matter

Look and Wonder

This diver is measuring the length of a grass plant on the seafloor. Why is it important to know how to measure matter?

Explore

Inquiry Activity

How can you measure length?

Make a Prediction

How wide is your classroom? Make a prediction.

Test Your Prediction

1. **Measure** Work with a partner. Stand with your back against one wall. Slowly walk across the room, placing one foot in front of the other. The heel of your front foot should meet the toe of your back foot. Your partner will count the number of steps it takes to cross the room.
2. Trade roles with your partner and repeat step 1.
3. **Communicate** Compare your data with the class's data. Make a table listing the data for the entire class.

Draw Conclusions

4. **Interpret Data** What is the highest measurement? What is the lowest measurement? Did anyone get the same measurement?
5. **Infer** Why were there different measurements? Why is it useful to use measuring tools, such as a ruler?

Explore More

Measure Scientists use the metric system to measure matter. Predict how wide your classroom is in meters and centimeters. Then use a metric ruler to measure the width of your classroom. How do your measurements compare with your predictions?

Read and Learn

Essential Question

How can you compare different kinds of matter?

Vocabulary

metric system, p. 374

pan balance, p. 376

gravity, p. 378

weight, p. 378

Reading Skill

Summarize

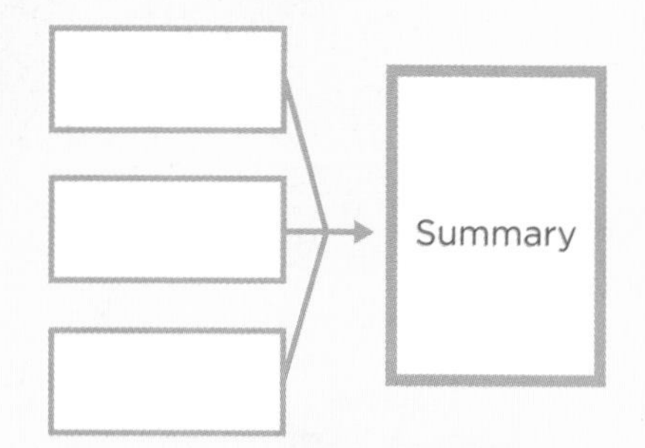

Technology 

e-Glossary and e-Review online at www.macmillanmh.com

How is matter measured?

Many properties of matter can be observed or measured with tools. You can look closely at an object with a hand lens. You can measure its length and width with a ruler. You can use a thermometer to measure its temperature.

Measuring is a way to compare sizes or amounts. People use tools marked with standard units to measure matter. A *standard unit* is a unit of measurement that people agree to use, such as feet or miles. A common system of standard units is the **metric system** (ME•trik SIS•tum). Scientists use the metric system.

Length

You measure length to find out how long something is. You have probably used rulers to measure how tall you are. In the metric system, length is measured in units called meters.

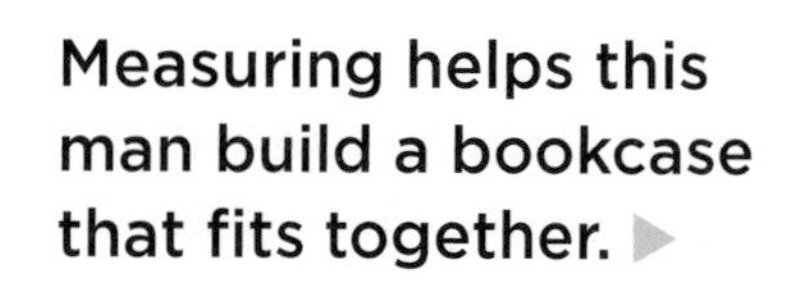

Measuring helps this man build a bookcase that fits together. ▶

Measuring the Volume of a Solid

Read a Photo

How can you measure the volume of this rock?

Clue: Look how the water level changes.

Volume

Volume describes how much space an object takes up. You probably have used measuring cups to measure the volume of liquids. You can also use beakers or graduated cylinders. In the metric system, a liquid's volume is measured in units called liters.

You can measure the volume of a solid too. First, measure some water. Then place a solid object completely under the water. Subtract the original water level from the new water level. The difference is the solid's volume.

▲ The volume of a liquid can be measured using a graduated cylinder, beaker, or measuring cup.

Quick Check

Summarize What are three measurements you could make to describe matter?

Critical Thinking Why is it important to use standard units?

How do we measure mass?

You can use a **pan balance** to measure mass. Remember that mass is a measure of the amount of matter in an object. To find an object's mass, you balance it with objects whose masses you know. First, place the object on one end of a pan balance. Then add the known masses to the other side until both sides are level. When the two sides are level, you know the mass of the object.

▲ Gram masses can be used to find the mass of an object.

In the metric system, mass is measured in grams. A gram is close to the amount of mass in two small paper clips. A kilogram is the same as 1,000 grams.

Objects with the same volume do not always have the same mass. A marble is about the same size as a piece of popcorn. However, the marble has a greater mass. How is that possible?

This pan balance measures mass. ▼

Matter is made up of tiny particles. In some objects the particles are close together. In other objects they are farther apart. The particles inside a marble are packed together more tightly than those inside a piece of popcorn. A marble has more particles than a piece of popcorn. It has more mass.

Quick Check

Summarize How can you measure mass using a pan balance?

Critical Thinking How could you measure the mass of a liquid with a balance?

Quick Lab

Measure Mass and Volume

1. **Predict** Look at a toy car, golf ball, and marble. Predict which object has the most mass. Which has the greatest volume?
2. **Measure** Find the mass of each object. List the objects from most mass to least mass.
3. **Measure** Fill a measuring cup with 250 milliliters of water. Add one object at a time to the measuring cup. Record the water level for each object.
4. **Interpret Data** List the objects from greatest to least volume.
5. **Interpret Data** Which object has the most mass? Which object has the greatest volume? How did the results compare with your prediction?

◀ The bag of marbles has more mass than the bag of popcorn.

FACT Air has mass.

How are mass and weight different?

What happens when you leap into the air? Do you float away? No, you come back to the ground. This happens because of gravity (GRA•vuh•tee). **Gravity** is a pulling force that holds you on Earth. Gravity keeps you and everything on Earth from floating into space.

You can measure how much Earth's gravity pulls on you. This measurement is your weight (WAYT). **Weight** is a measure of the pull of gravity on you. Weight can be measured with a spring scale.

Weight is different from mass. If you visited the Moon, your mass would stay the same. The matter inside you would not change. However, your weight would change. This is because the pull of the Moon's gravity is weaker than the pull of Earth's gravity. Your weight on the Moon would be less than your weight on Earth.

▲ Spring scales are used to measure weight.

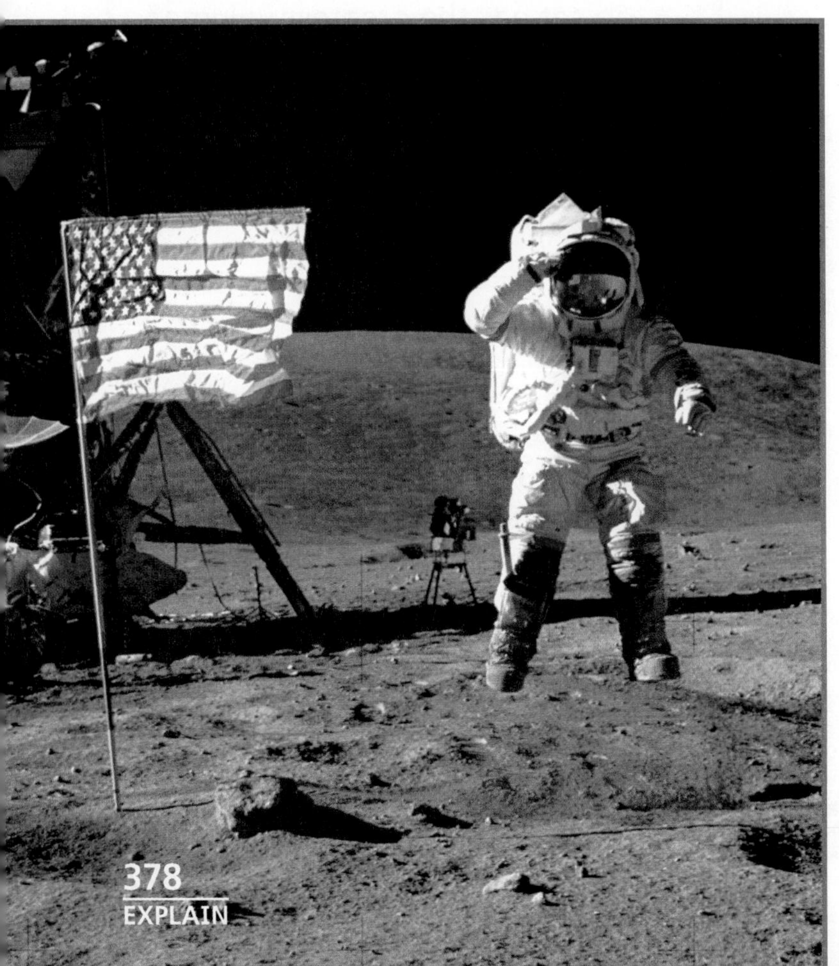

◀ The pull of gravity is weaker on the Moon than on Earth.

Quick Check

Summarize How is weight different from mass?

Critical Thinking Do you think you could jump higher on the Moon? Explain.

Lesson Review

Visual Summary

Properties of matter, such as **length** and **volume,** can be measured and observed with tools.

Mass can be measured with a pan balance.

Mass stays the same. The **weight** of an object depends on the force of gravity.

Make a FOLDABLES® Study Guide

Make a layered-look book. Use it to summarize what you learned about measuring matter.

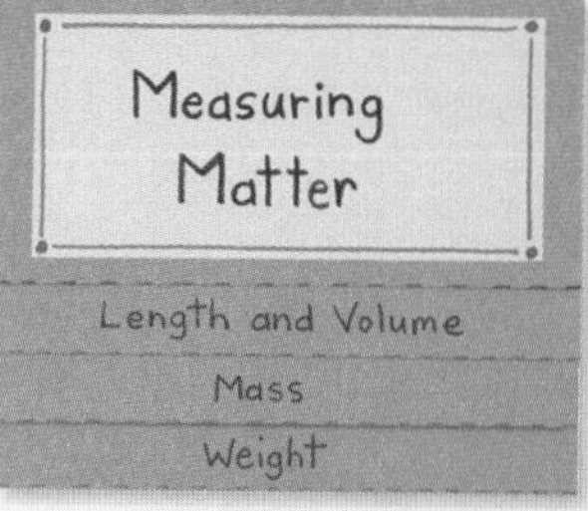

Think, Talk, and Write

1. **Vocabulary** What is gravity?

2. **Summarize** Does a large object always have a lot of mass? Explain your answer.

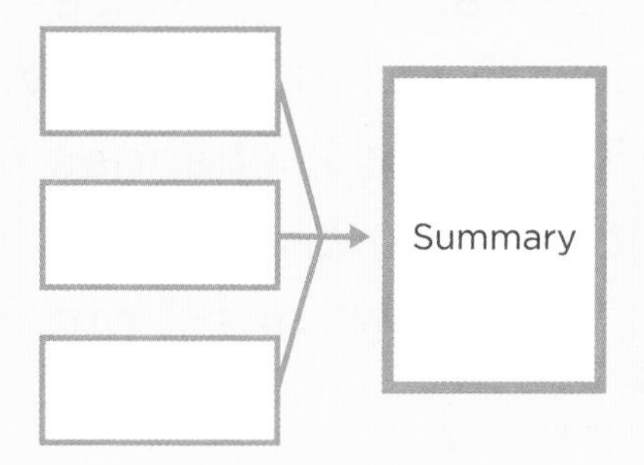

3. **Critical Thinking** Why is it important to measure accurately?

4. **Test Prep** Which tool would you use to measure weight?
   - **A** thermometer
   - **B** hand lens
   - **C** spring scale
   - **D** ruler

5. **Essential Question** How can you compare different kinds of matter?

Math Link

Metric Measurements
What tool would you use to measure the length of a pencil in centimeters? Use this tool to measure the length of four objects. List them in order from shortest to longest.

Social Studies Link

Do Research
About 5,000 years ago, people began using standard weights. Some early weights were the shekel (SHE•kul) and the mina (MI•nuh). Find out more about early systems of measurement.

LOG ON e-Review Summaries and quizzes online at www.macmillanmh.com

Inquiry Skill: Measure

You have learned that matter is anything that takes up space and has mass. Water is matter that is important to life on Earth. It is found on Earth as solid ice and liquid water. It is even found in the air. What happens to water's mass as it changes from a chunk of solid ice to liquid water? Scientists **measure** things to answer questions like this.

measuring cup

▶ Learn It

When you **measure**, you find such things as the mass, volume, length, or temperature of an object. You can also **measure** distances and time. Scientists use many tools to **measure** things. Some of these tools are shown on this page. Scientists use measurements to describe and compare objects or events.

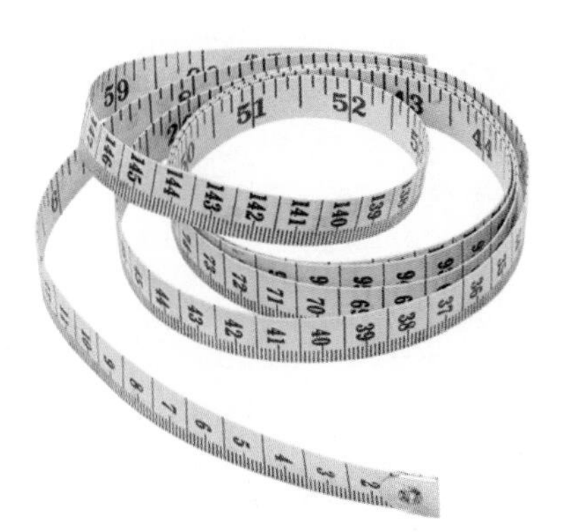

tape measure

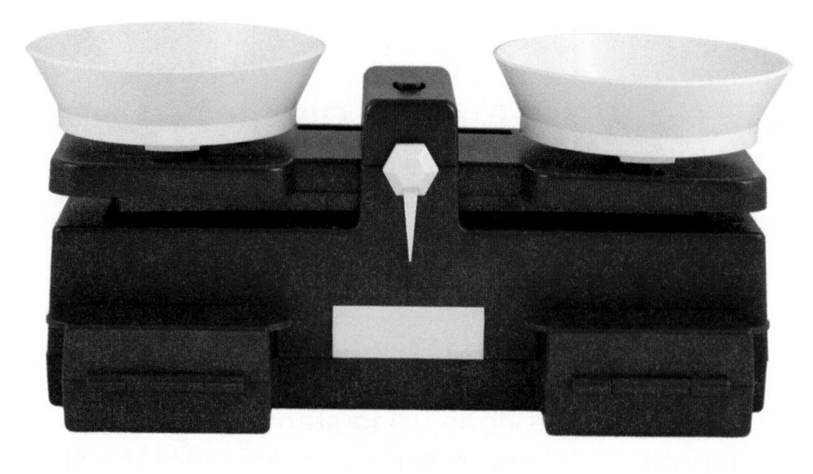

pan balance

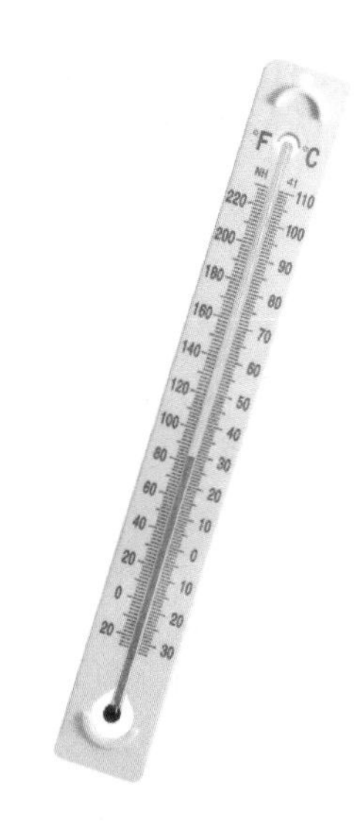

thermometer

▶ Try It

You know that scientists **measure** things to answer questions. You can measure too. Answer this question: *Do ice cubes have the same mass after they melt?*

1. To start, place several ice cubes in a cup. Then cover the cup with plastic wrap so the water stays inside the cup.
2. Measure mass by placing the cup on one end of a pan balance. Add masses to the other side of the pan balance until both sides are level. Record the mass on a chart.

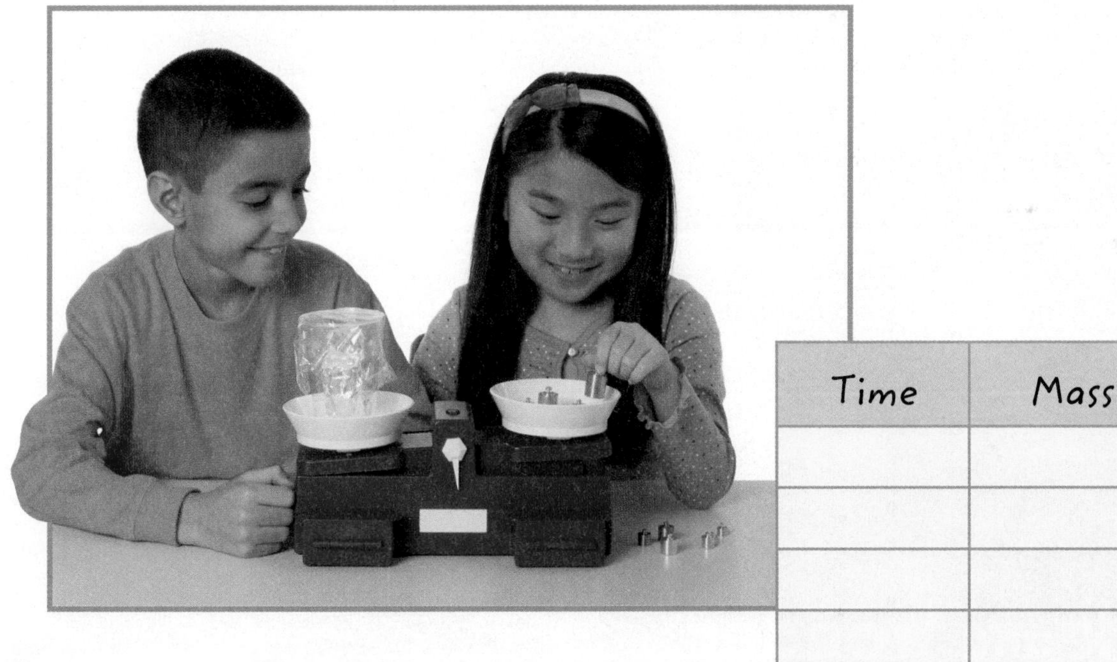

Time	Mass

3. Measure the mass every 30 minutes until the ice is completely melted.
4. Now use your measurements to answer the question. Do ice cubes have the same mass after they melt?

▶ Apply It

Now **measure** to answer this question: *Does ice cream have the same mass after it melts?* How do you know?

Lesson 3

Solids, Liquids, and Gases

Look and Wonder

This person is soaring through the air. What do you think you would notice on the ground below? How would you describe the land and water?

Explore

Inquiry Activity

How are solids different from liquids?

Make a Prediction

How do you know if something is solid? How do you know when something is a liquid? Make a prediction.

Test Your Prediction

1. **Observe** Touch the block. Does it feel more like a solid or more like a liquid? Why?
2. **Experiment** Put the block into the beaker. Record your observations.
3. **Experiment** Use the spoon to stir the block. What happens? Record your observations. Empty the beaker.
4. Repeat steps 1–3. Instead of the block, use the water, salt, hand soap, and clay. Test each object one at a time.

Draw Conclusions

5. Which objects did not change shape? Which objects were easy to stir?
6. **Classify** Which objects are solids? Which are liquids?
7. Explain how solids are different from liquids.

Explore More

Experiment What would happen if you put each object in the freezer? What would happen if you put each object in a warm place? Form a hypothesis and test it.

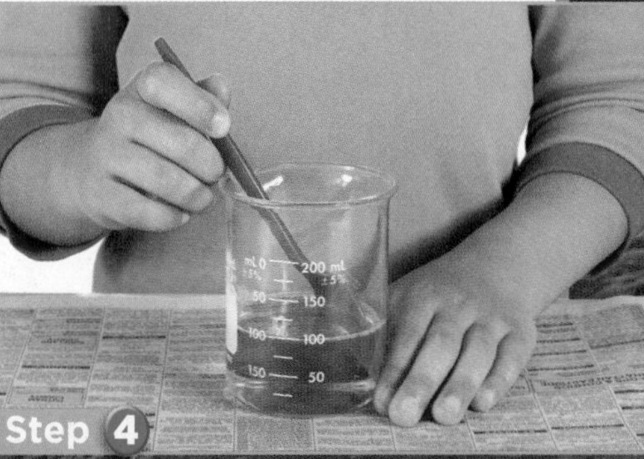

Read and Learn

Essential Question
What are the states of matter?

Vocabulary
states of matter, p. 384
solid, p. 384
liquid, p. 386
gas, p. 387

Reading Skill ✓
Classify

Technology 
e-Glossary, e-Review, and animations online at www.macmillanmh.com

What are three forms of matter?

Matter comes in many forms. Look at the picture below. The canoe is a solid. The river is made of water, a liquid. The air is made of gases. Solids, liquids, and gases are three forms of matter. Scientists call these forms **states of matter**. Each state of matter has certain properties.

Solids

Most of the things you notice around you are solids (SAH•lidz). A **solid** is matter that takes up a definite, or certain, amount of space. A solid has its own shape. This book is a solid. Pencils, desks, and pillows are solids too. Solids have a definite shape and volume. If you put a pencil into a jar or a box, its shape and size stay the same.

How are these boys using three states of matter?

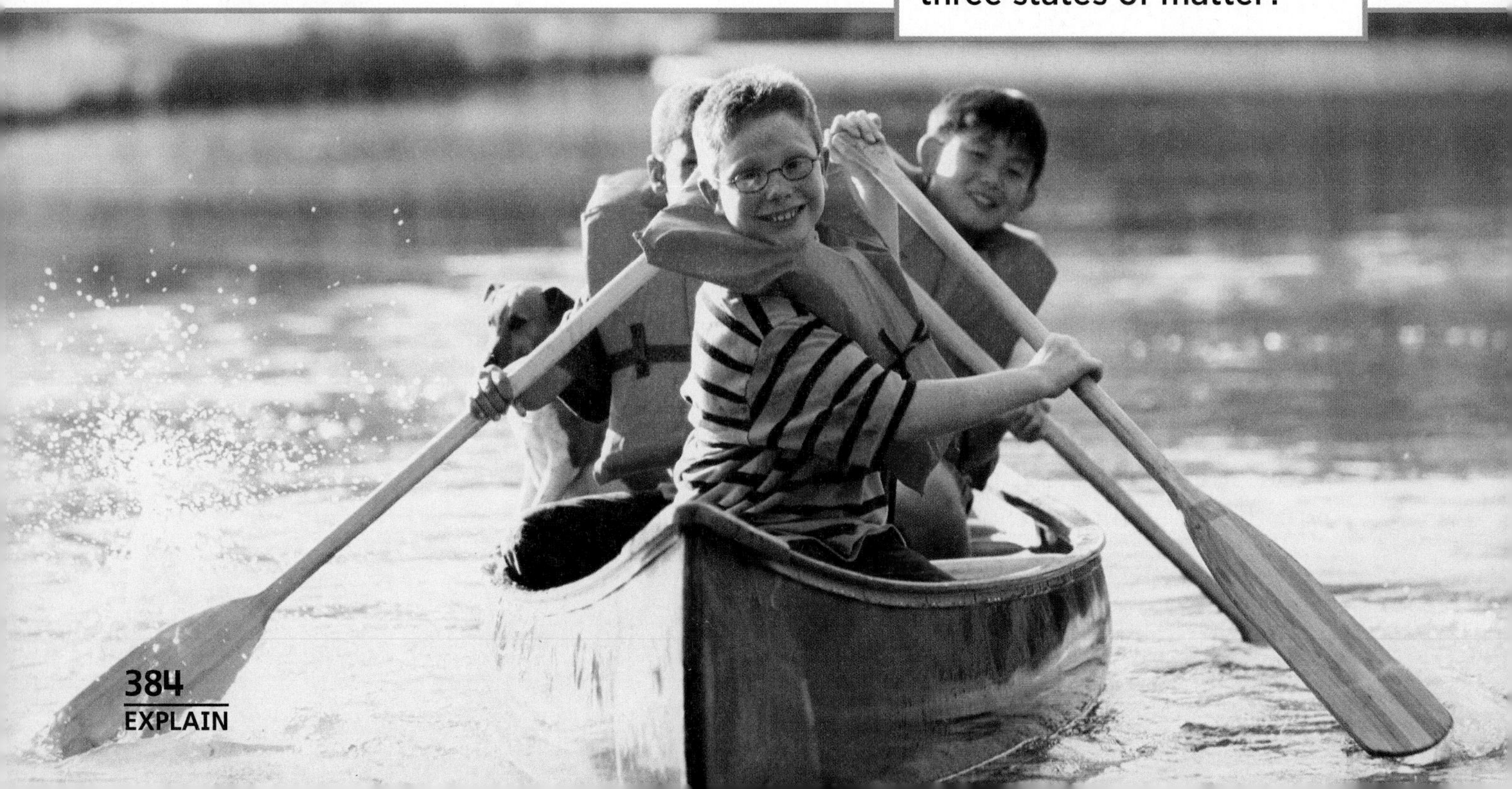

Remember that matter is made up of tiny particles. These particles are too small to see. In a solid these particles are packed closely together. They do not have a lot of room to move around. This helps the solid keep its shape.

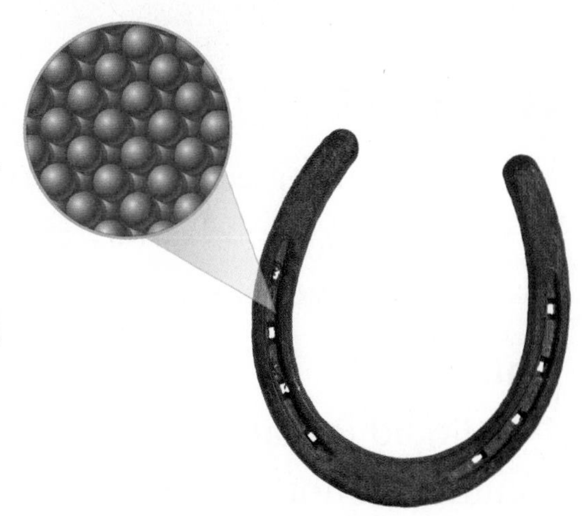

▲ The particles in this solid horseshoe can not move much.

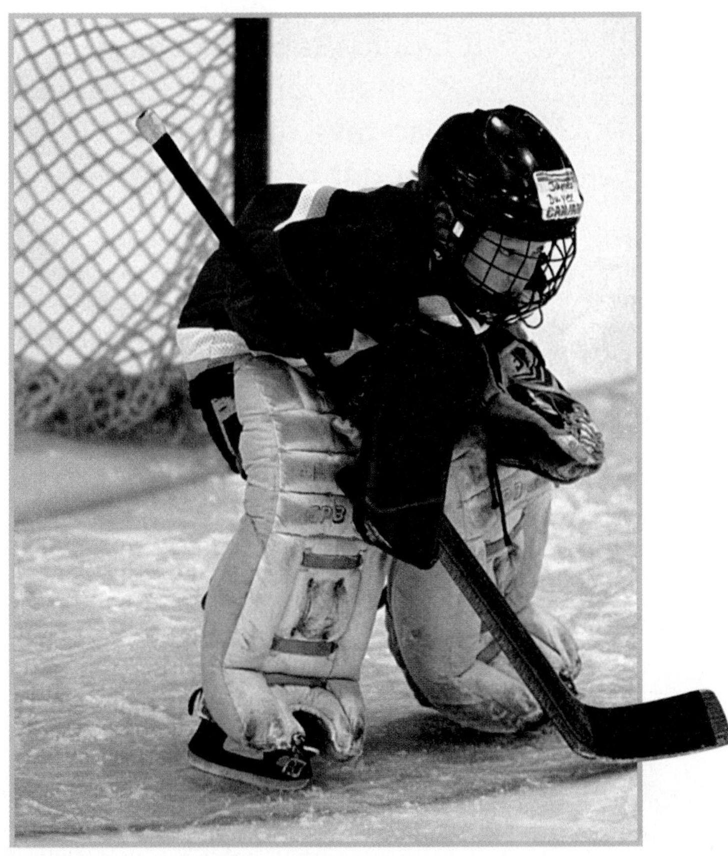

▲ Solids can be hard or soft. This goalie's helmet is hard, but his leg pads are soft.

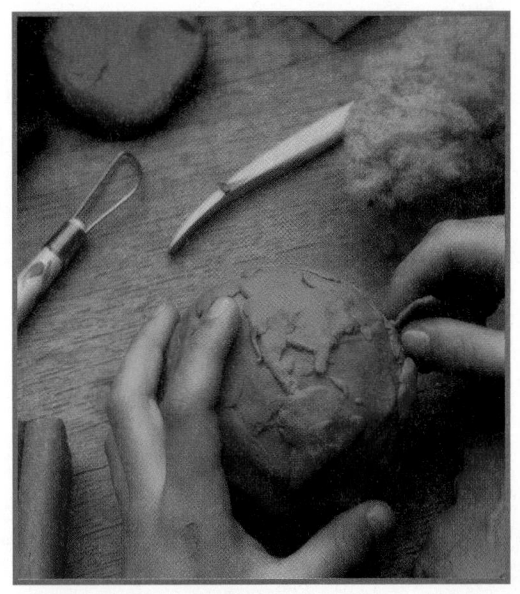

▲ Even though you can change the shape of clay, it is still a solid.

Quick Check

Classify What are three solids you use every day?

Critical Thinking A rubber band can change its shape when it is stretched. Do you think a rubber band is a solid or a liquid? Explain your answer.

What are liquids and gases?

Liquids and gases are two other states of matter. Like solids, they take up space and have mass.

▲ Liquids take the shape of their containers. Liquids also take up a definite amount of space inside their containers.

Liquids

A **liquid** is matter that has a definite volume but not a definite shape. A liquid takes the shape of its container. Water, shampoo, and milk are some liquids. When milk is inside a carton, it takes the shape of the carton. When you pour milk into a glass, it takes the shape of the glass. If you spill the milk, it will spread out over the floor. If you were able to mop up the milk and put it back into the carton, it would still be the same amount of milk. The volume of the milk stays the same. Only its shape changes.

Liquid Particles

◀ The particles in a liquid are able to slide past one another. That is why liquids can change shape.

Read a Diagram

How would you describe the particles in a liquid?

Clue: Illustrations can help show things that are hard to see.

Science in Motion Watch how the particles of a liquid move at www.macmillanmh.com

Gases

You can not always see gases, but they are all around you. A **gas** is matter that has no definite shape or volume. A gas takes the shape and volume of its container.

Think about balloons being blown up with a helium tank. Helium is a gas. When it is in the tank, it has a small volume. It has the shape of the tank. When the gas is used to fill balloons, it spreads out. It then has a much greater volume. It also changes shape. It takes the shape of the balloons.

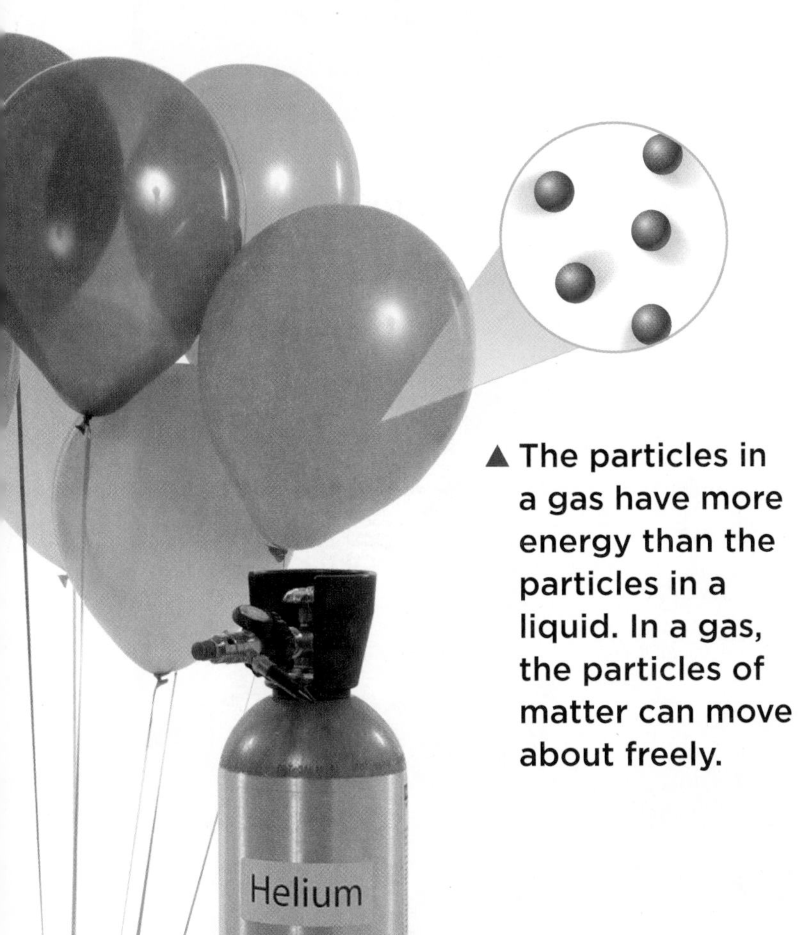

▲ The particles in a gas have more energy than the particles in a liquid. In a gas, the particles of matter can move about freely.

Quick Lab

Compare Solids, Liquids, and Gases

gas

1. Blow into an empty bag. Then quickly seal the bag.
2. Fill a second bag with water and seal this bag. Put a rock in a third bag and seal it.

liquid

3. **Observe** Each bag contains matter in a different state. How does each bag look and feel? Record your observations.

solid

4. **Observe** Open each bag. What happens?
   △ **Be Careful.** Hold the bag filled with water over a container. Then open it.
5. **Communicate** Describe the properties of a solid, a liquid, and a gas. Tell how these three states of matter are different from one another.

Quick Check

Classify List three liquids you drink every day.

Critical Thinking Suppose a balloon filled with helium bursts. What would happen to the gas?

How do you use all the states of matter?

Solids, liquids, and gases are all around you. You use them in many ways. Many of the foods you eat are solids. Your body needs water, a liquid. You need oxygen, a gas from the air. Oxygen helps you get the energy you need from the food you eat.

You use the states of matter in other ways too. You can find three states of matter on a bicycle, for example. Many parts of the bicycle are made of solids. The handlebars, seat, and the rubber of the tires are solids. The tires are filled with air, a gas. The oil on the bicycle chain is a liquid.

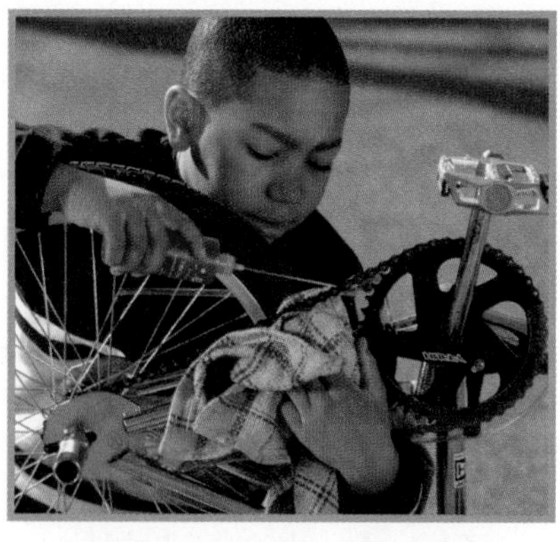

▲ Oil, a liquid, helps a bicycle chain move smoothly.

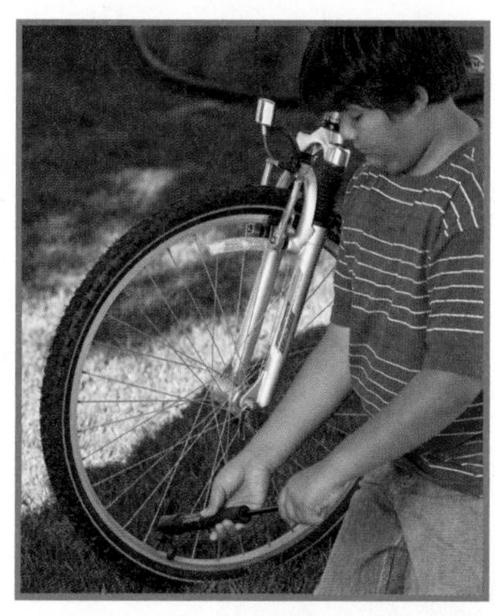

▲ You pump air into the tires to inflate them.

Quick Check

Classify What are three states of matter found on a bicycle?

Critical Thinking How do you use the different states of matter?

The bicycle frame is solid. It has to be solid to keep the bicycle together. ▶

Lesson Review

Visual Summary

	A **solid** is matter that has a certain volume and shape.
	A **liquid** is matter that has a certain volume but not a certain shape.
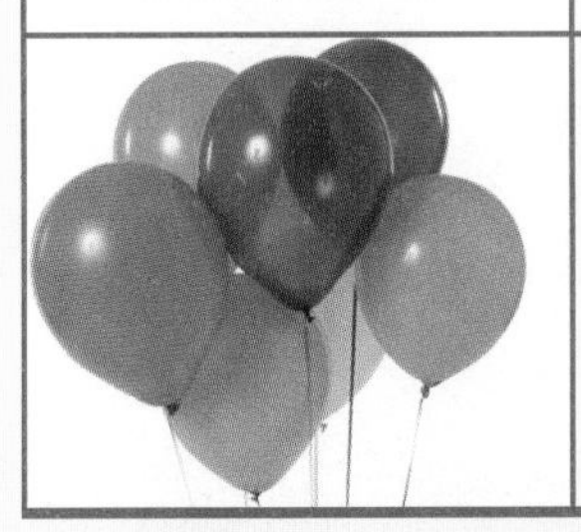	A **gas** is matter that does not have a certain volume or shape.

Make a FOLDABLES® Study Guide

Make a layered-look book. Use it to summarize what you learned about states of matter.

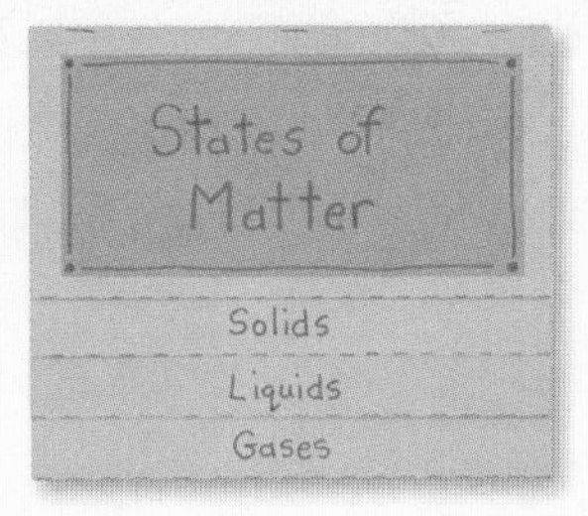

Think, Talk, and Write

1. **Vocabulary** What is matter that does not have a certain shape or size?

2. **Classify** What kind of matter is this book? What kind of matter is water? What kind of matter is air?

3. **Critical Thinking** Compare solids, liquids, and gases. How are they alike? How do they differ?

4. **Test Prep** **Matter that spreads out to fill its container is a**
   - **A** gas.
   - **B** liquid.
   - **C** mass.
   - **D** solid.

5. **Essential Question** What are the states of matter?

Math Link

Solve a Problem
A helium tank can inflate 126 large balloons. It can inflate three times as many small balloons. How many small balloons can the tank inflate?

Art Link

Make a Poster
Draw diagrams that show the differences among solids, liquids, and gases. Write a brief explanation of each diagram.

LOG ON e-Review Summaries and quizzes online at www.macmillanmh.com

Describe Matter

You can describe matter in many ways. How would you describe a pizza to someone who has never seen one? How does it look? How does it smell? These are some of the pizza's observable properties. How big is the pizza? What is its mass? These are some of its measurable properties. Is it a solid or a liquid? This is its state of matter.

Descriptive Writing

A good description

- includes describing words to tell how something looks, sounds, feels, smells or tastes;
- uses details to create a picture for the reader;
- groups together details in an order that makes sense.

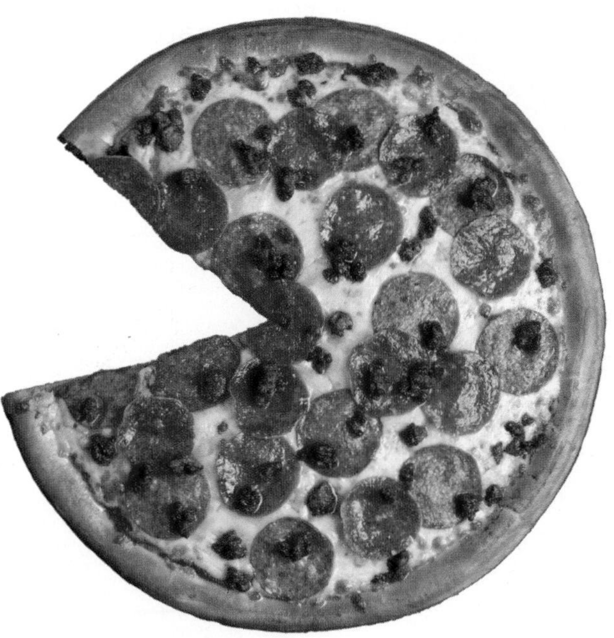

Write About It

Descriptive Writing Think of an object you use every day, such as your book bag. How would you describe it to someone who has never seen it before? Use the object's properties to write a description of the object.

Measuring Perimeter

Solids come in many shapes and sizes. They can be round like a ball or square like a brick. They can be huge like a skyscraper or tiny like a grain of sand. You can measure the distance around a solid. The distance around a solid object is called the *perimeter*.

Find the Perimeter

▶ To find the perimeter of an object, add the lengths of all of its sides.

6 + 2 + 6 +2 = 16

This rectangle's perimeter is 16.

6

2 2

6

Solve It

Find the perimeter of the red square. Find the perimeter of the blue triangle. How can you find the perimeter of the entire house? Try it.

CHAPTER 9 Review

Visual Summary

Lesson 1 Matter is anything that has volume and mass. You can use properties to describe and identify matter.

Lesson 2 Matter can be measured using tools that record standard units.

Lesson 3 Solids, liquids, and gases are three forms of matter.

Make a FOLDABLES® Study Guide

Glue your lesson study guides to a piece of paper as shown. Use your study guide to review what you have learned in this chapter.

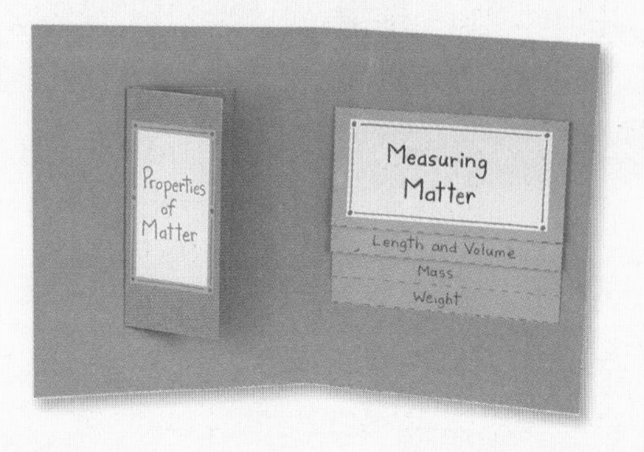

Vocabulary

DOK I

Fill each blank with the best term from the list.

elements, p. 368	**matter,** p. 364
gas, p. 387	**metric system,** p. 374
gravity, p. 378	**properties,** p. 365
liquid, p. 386	**solid,** p. 384
mass, p. 365	**volume,** p. 365

1. Matter with no certain shape or volume is a ________.
2. The amount of space an object takes up is its ________.
3. Scientists make measurements using the ________.
4. If matter has a certain volume but not a certain shape, it is in a ________ state.
5. The pulling force that holds you on Earth is called ________.
6. Matter with a certain shape and volume is a ________.
7. The amount of matter in an object is its ________.
8. All matter is made up of ________.
9. Size and color are examples of ________.
10. Anything that has mass and volume is ________.

LOG ON e-Glossary Words and definitions online at www.macmillanmh.com

Skills and Concepts

DOK 2–3

Answer each of the following.

11. **Summarize** Name three properties of an object that you can measure using the metric system. What standard units would you use for each?

12. **Descriptive Writing** Write a brief description of a solid, a liquid, and a gas. Include a diagram with your description.

13. **Measure** What steps do you take to measure the mass of an object with a pan balance?

14. **Critical Thinking** Where can you find the three states of matter in a car?

15. **Critical Thinking** Hannah has a rock, a measuring cup, and some water. How can she measure the volume of a rock?

16. **Infer** How could you find out what is inside a brown paper bag without looking inside?

17. What properties might the two objects shown below have in common? How do you think their properties might be different?

aluminum

gold

18. **True or False** *Air can not be weighed.* Is this statement true or false? Explain.

19. In the metric system the volume of a liquid is measured in

    A liters. **C** centimeters.

    B inches. **D** meters.

20. What are some ways you can describe matter?

Performance Assessment

DOK 3

What Is It Made Of?

- Make a book about some of the matter that surrounds you every day—the clothes you wear. Choose some of your favorite clothes. Describe their physical properties.

- Put a picture or drawing of each piece of clothing on a page in your book. Include a description of the properties of the clothing next to each item.
- Choose two pieces of clothing. Use their properties to describe their similarities and differences.

Test Preparation

1 Look at the objects below.

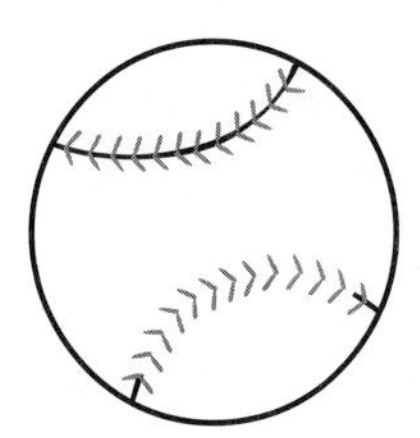

Which property do the two balls have in common?

A length

B volume

C shape

D color

DOK 1

2 The mass of an object refers to

A how large it is.

B the amount of matter it has.

C whether it will float.

D how much volume it has.

DOK 1

3 Helium in a balloon is in which state of matter?

A air

B gas

C solid

D liquid

DOK 1

4 A magnet can attract certain types of objects.

Which object is attracted to a magnet?

A wood

B iron

C plastic

D water

DOK 1

5 Why does a life preserver float in water?

A A life preserver is large in size.

B A life preserver has an equal amount of mass and volume.

C A life preserver has a small mass and a large volume.

D A life preserver has a large mass and a small volume.

DOK 2

6 Which statement about matter is correct?

A Matter is too small to be seen without a microscope.

B Matter comes from the Sun.

C All matter can be seen.

D All matter takes up space.

DOK 2

7 Look at the picture of a pan balance.

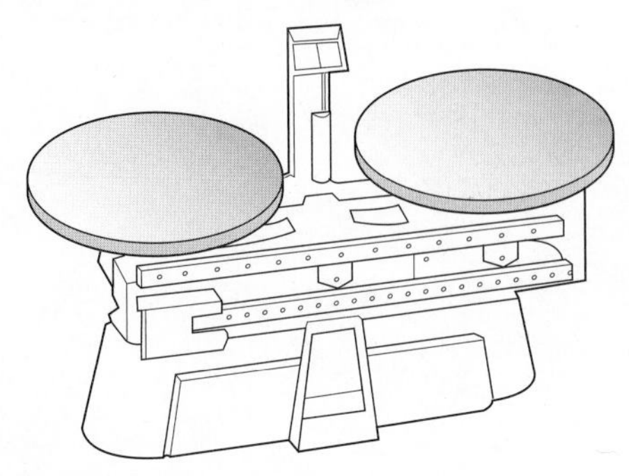

Which property of matter does the pan balance measure?

A mass

B weight

C volume

D magnetism

DOK I

8 Look at the picture of a spring scale.

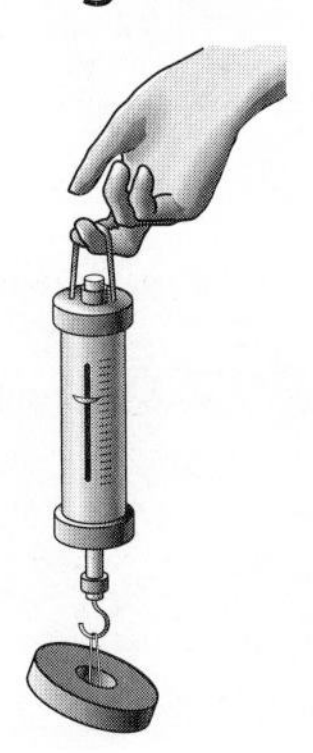

Which property of matter does the spring scale measure?

A mass

B weight

C volume

D magnetism

DOK I

9 Make a table like the one shown below.

Property	Metric Unit
length	
mass	
volume	

Fill in the table with the correct metric units.

DOK I

10 Make a table like the one shown below.

State of Matter	Definite Volume	Definite Shape
solid		
liquid		
gas		

Place an *X* in the correct column for each state of matter.

DOK I

Compare and contrast the spacing and movement of the particles in each state of matter. You can use a drawing to help explain your answer.

DOK 2

Check Your Understanding

Question	Review	Question	Review
1	pp. 364–365	6	pp. 364–365
2	p. 365	7	pp. 376–377
3	p. 387	8	p. 378
4	p. 367	9	pp. 374–376
5	p. 366	10	pp. 384–387

CHAPTER 10

Changes in Matter

In what ways can matter change?

Essential Questions

Lesson 1
How can matter change states?

Lesson 2
What happens when matter goes through a physical change?

Lesson 3
What happens when matter goes through a chemical change?

rusty shipwreck, Straits of Magellan, Chile

Big Idea Vocabulary

melt to change from a solid to a liquid (p. 398)

boil when a heated liquid changes into a gas (p. 399)

physical change a change in the way matter looks (p. 408)

mixture different kinds of matter mixed together (p. 410)

solution when one or more kinds of matter are mixed evenly into another kind of matter (p. 411)

chemical change a change that causes different kinds of matter to form (p. 418)

Visit www.macmillanmh.com for online resources.

Lesson 1

Changes of State

Look and Wonder

A winter storm can bring snow and ice. What happens to snow on a warm, sunny day? What causes this change?

Explore

Inquiry Activity

What happens when ice is heated?

Make a Prediction

How does ice change as it is heated? Make a prediction.

Test Your Prediction

1. **Measure** Place a thermometer in a cup of ice. Measure the temperature of the ice. Record the temperature in a table like the one shown.
2. Place the cup in a warm place, such as on a sunny windowsill.
3. **Measure** Stir the ice and measure its temperature every ten minutes for the next hour. Record the temperature in the table.
4. Describe how the ice changes.

Draw Conclusions

5. **Communicate** How did the ice change as it was heated? Was your prediction correct?
6. **Infer** What happened to the temperature of the water as the ice melted? At what temperature does ice melt?

Explore More

Predict What will happen to the water as it continues to sit in the warm place after the ice has melted? Test your prediction to find out.

Materials

thermometer

plastic cup of ice

spoon

Step 1

Time	Temperature

Read and Learn

Essential Question
How can matter change states?

Vocabulary
melt, p. 398
boil, p. 399
evaporate, p. 399
condense, p. 400
freeze, p. 401

Reading Skill 
Predict

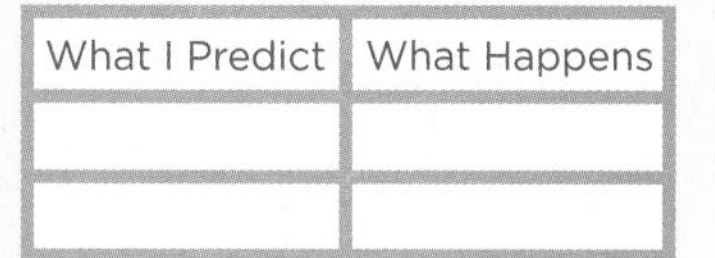

What I Predict	What Happens

Technology
e-Glossary, e-Review, and animations online at www.macmillanmh.com

What happens when matter is heated?

When matter is heated, it gains energy. Its temperature rises. At certain temperatures, matter will change state.

Changing from a Solid to a Liquid

If you heat most solids to a high enough temperature, they will melt. To **melt** is to change from a solid to a liquid. Different kinds of matter melt at different temperatures. Ice melts at a lower temperature than rocks. Ice melts at 32°F (0°C). Rocks melt at over 1,100°F (593°C)!

Remember that all matter is made up of tiny particles. In solids these particles are held closely together. When a solid is heated and gains energy, its particles begin to move away from one another. They flow around one another and are no longer held tightly together. This causes the solid to lose its shape. It becomes a liquid.

◀ The lava flowing from this volcano is rock that melted deep beneath Earth's surface.

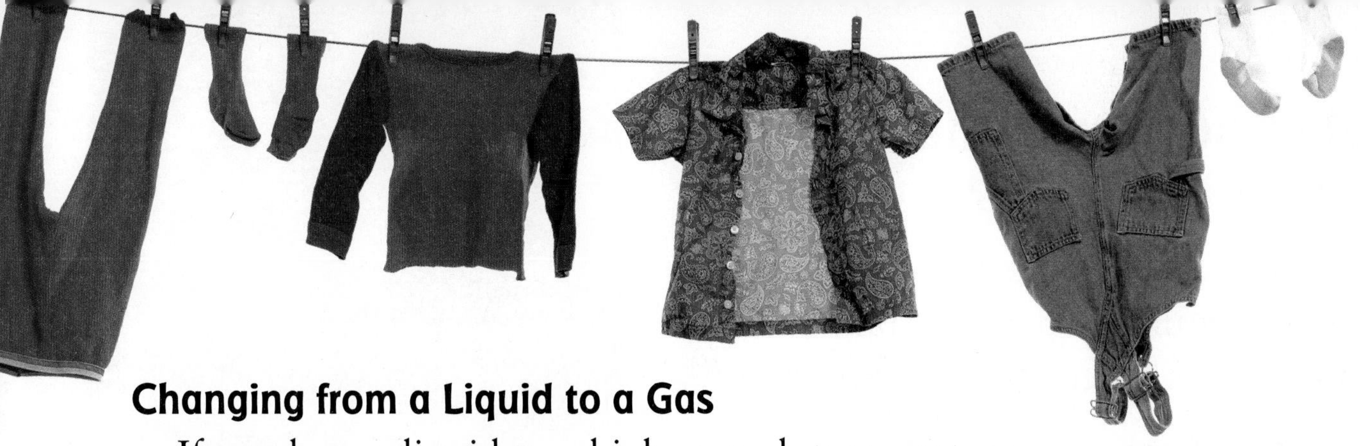

▲ These wet clothes will be dry when the water finishes evaporating.

Changing from a Liquid to a Gas

If you heat a liquid to a high enough temperature, it will **boil**. When the liquid boils, it changes from a liquid to a gas. Energy from heat causes the particles in a liquid to move faster. They spread apart. The liquid turns into a gas. Bubbles form.

Liquids can also **evaporate**, or change into a gas without boiling. When wet clothes are placed in the Sun, the water in the clothes evaporates. The Sun heats water droplets in the clothes. The water turns slowly into a gas, and the clothes dry. The gas state of water is called *water vapor.* You can not see water vapor, but it is part of the air.

Heating Water

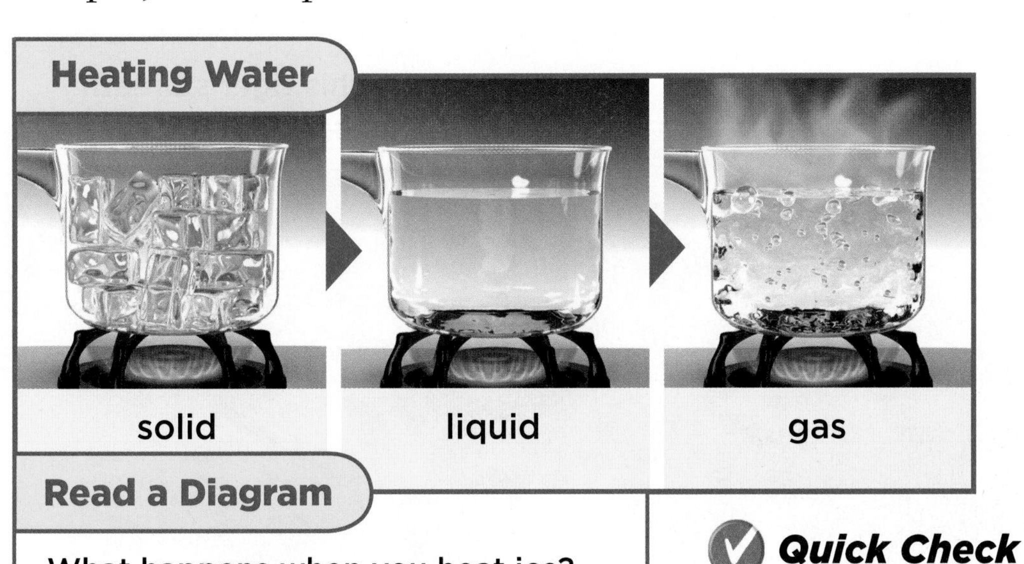

Read a Diagram

What happens when you heat ice?

Clue: Arrows show a sequence.

LOG ON *Science in Motion* Watch how matter changes at www.macmillanmh.com

Quick Check

Predict What will happen to cheese when it is heated?

Critical Thinking How does a blow dryer get your hair dry?

What happens when matter is cooled?

When matter is cooled, it loses energy. Its temperature drops. At certain temperatures, matter will change state.

Changing from a Gas to a Liquid

If you cool a gas to the right temperature, it will condense (kun•DENS). To **condense** is to change from a gas to a liquid. For example, on cool mornings, small droplets of water called dew can appear on grass and windows. This happens when water vapor in the air touches cool objects and loses energy. Particles of water vapor come closer together. They change into drops of liquid water.

▲ Dew forms when water vapor in the air cools and condenses.

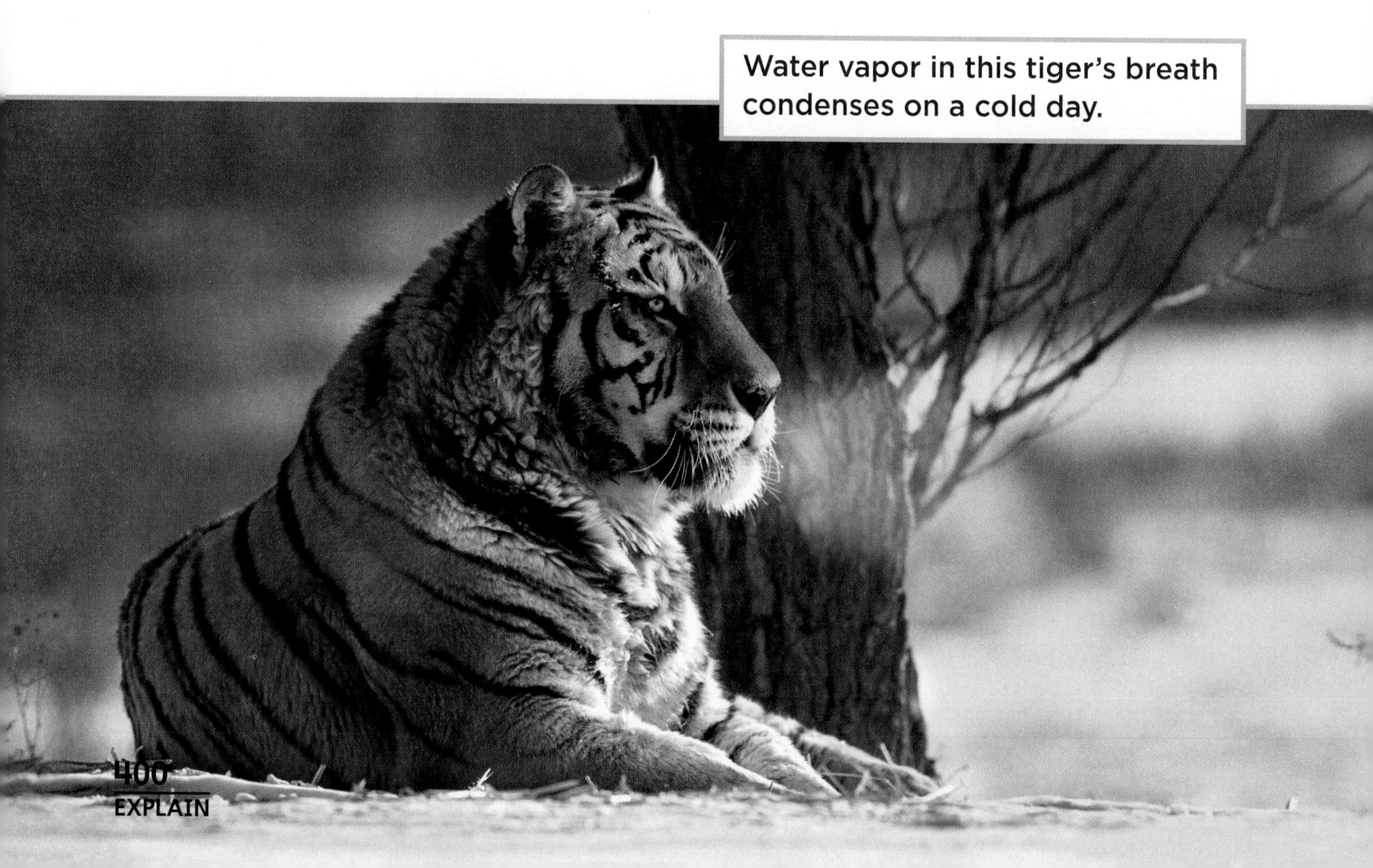

Water vapor in this tiger's breath condenses on a cold day.

Changing from a Liquid to a Solid

If you cool a liquid to the right temperature, it will freeze. To **freeze** is to change from a liquid to a solid. The particles in the liquid lose energy and move slower and closer together. They get locked into position and form a solid. For example, when you put liquid water into the freezer, it loses energy. It cools to a certain temperature and turns into ice.

Quick Lab

Condense Water Vapor

1. **Observe** Feel an empty plastic cup. Does it feel wet or dry? Does it feel hot or cold? Record your observations.
2. Fill your cup with ice cubes. Next add cold water to the cup.
3. **Observe** Feel your cup again. Does the cup feel wet or dry? Does the cup feel hot or cold? Record your observations.
4. **Observe** Look at your cup after five minutes. What do you notice about the outside of the cup? Is it wet or dry?
5. **Infer** Where did the water on the cup come from?

When juice is cooled enough, it will freeze and become a solid.

Quick Check

Predict What will happen to water vapor when it is cooled?

Critical Thinking How could you make an ice pop?

How is water different from other kinds of matter?

Most kinds of matter shrink as they freeze. Their particles get packed more closely together. They take up less space. Yet water gets larger when it freezes.

As water freezes, its particles move around. They make a special pattern. Empty spaces form between the particles. The frozen water takes up more space than the liquid water. This is why freezing a glass of water cracks the glass.

Ice floats in liquid water. This keeps lakes and ponds from freezing from the bottom up. Living things can survive under the ice.

Quick Check

Predict What would happen if you put a plastic bottle filled with liquid water in the freezer? Why does this happen?

Critical Thinking Describe how water changes when it melts.

The particles in ice are more spread out than the particles in liquid water. This is why ice floats.

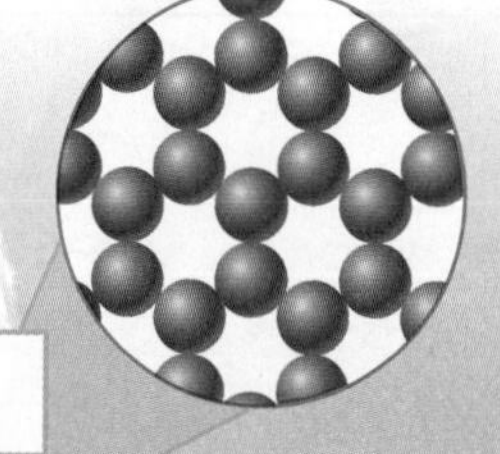

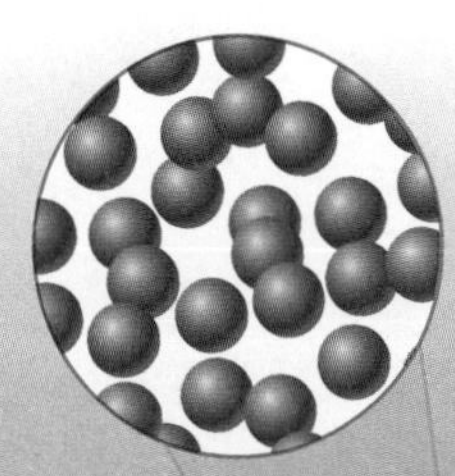

FACT Ice, liquid water, and water vapor are all forms of water.

Lesson Review

Visual Summary

	When most **solids** are heated, they melt into **liquids.** When a **liquid** is heated, it changes into a **gas.**
	When a **gas** cools, it usually condenses into a **liquid.** When a **liquid** cools, it freezes into a **solid.**
	**Water** is a special kind of matter. It gets larger when it freezes.

Make a FOLDABLES® Study Guide

Make a layered-look book. Use it to summarize what you learned about the changes of state for matter.

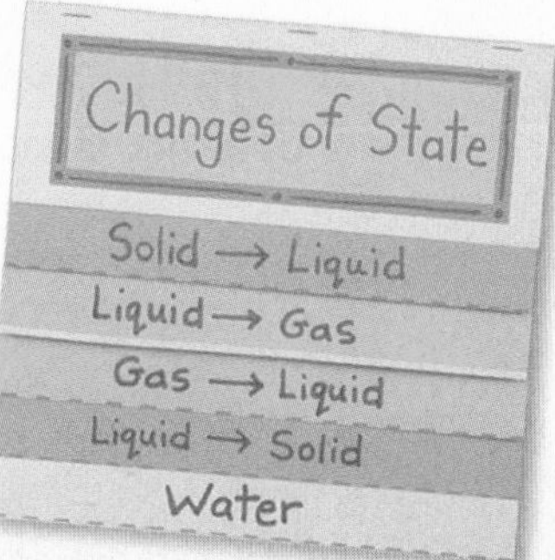

Think, Talk, and Write

1. **Vocabulary** What happens when a gas condenses?

2. **Predict** After a rainstorm, the Sun comes out and shines brightly. What will happen to puddles of rainwater?

What I Predict	What Happens

3. **Critical Thinking** You see drops of water on the bathroom mirror after a shower. What caused the water drops to form?

4. **Test Prep** **How is water different from other liquids?**
   - **A** Water gets larger as it freezes.
   - **B** Water gets smaller as it freezes.
   - **C** Water stays the same as it freezes.
   - **D** Water never freezes.

5. **Essential Question** How can matter change states?

Writing Link

Write a Story
Describe how your life would be different if liquids changed into solids when heated. For example, it could snow when the temperature outside was very hot.

Math Link

Find the Difference
Ice melts at 32°F. Water boils at 212°F. How many degrees are there between water's melting and boiling temperatures?

LOG ON e-Review Summaries and quizzes online at www.macmillanmh.com

Focus on Skills

Inquiry Skill: Predict

You just learned about how liquids change to solids. Which do you think freezes faster: salt water or freshwater? To find answers to questions like this, scientists first **predict** what they think will happen. Next, they experiment to find out what does happen. Then, they compare their results with their prediction.

▶ Learn It

When you **predict**, you state the possible results of an event or experiment. It is important to record your prediction before you do an experiment. Next, you record your observations as you experiment and record the final results. Then you have enough data to figure out if your prediction was correct.

Ice floats on the salt water of Shoup Bay in Alaska.

▶ Try It

Which do you think freezes faster, salt water or freshwater? **Predict** what will happen when you freeze freshwater and salt water. Write your prediction on a chart like the one shown. Then do an experiment to test your prediction.

Materials **measuring cup, water, 2 plastic cups, salt, measuring spoon**

1. Pour 125 milliliters of water into a plastic cup. Label this cup *Freshwater.*
2. Pour 125 mL of water into another plastic cup. Add 1 tablespoon of salt and stir with a spoon. Label this cup *Salt Water.*
3. Place both cups into the freezer. Check them every 15 minutes. Draw or write your observations.

Now answer these questions. Which freezes faster: freshwater or salt water? Was your prediction correct?

Which Freezes Faster?	
My Predictions	
Observations of freshwater	
Observations of salt water	
Results	

▶ Apply It

Now that you have learned to think like a scientist, make another prediction. Do you **predict** that salt water or freshwater will evaporate faster? Plan an experiment to find out if your prediction is correct.

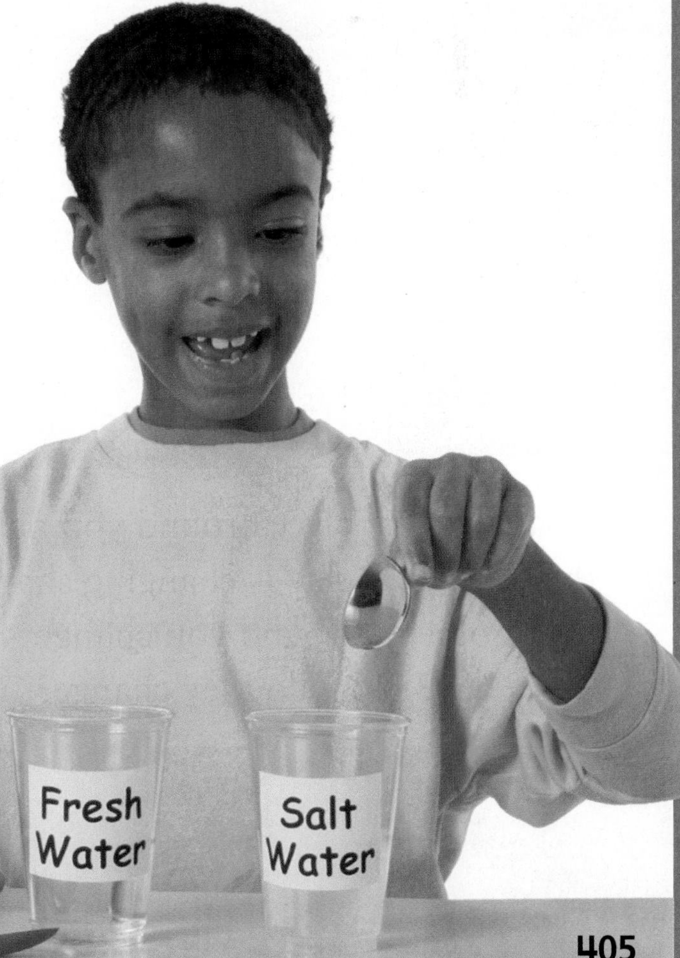

Lesson 2

Physical Changes

Look and Wonder

Changes take place around you all the time. This clay is changing shape. What objects around you change every day? How do they change?

Explore

Inquiry Activity

How can you change matter?

Purpose

Find out some ways you can change matter.

Procedure

1. Make a table like the one shown below.

Object	Change	Properties changed
Paper		
Clay		
Ice cubes		

2. **Observe** Look at each object. What properties does each object have? How can you change each object? Record your plan.
3. **Experiment** Change each object. What properties does each have now? Record the property that has changed. **Be Careful.** Handle scissors carefully.

Draw Conclusions

4. How are the objects different after you made the changes?
5. **Infer** Do you think you changed the kind of matter making up the object? Explain.

Explore More

Experiment What would happen if you added a spoon of salt to a cup of water? How would the salt and water change? How could you remove the salt from the water?

Materials

paper

clay

ice cubes

scissors

Step 3

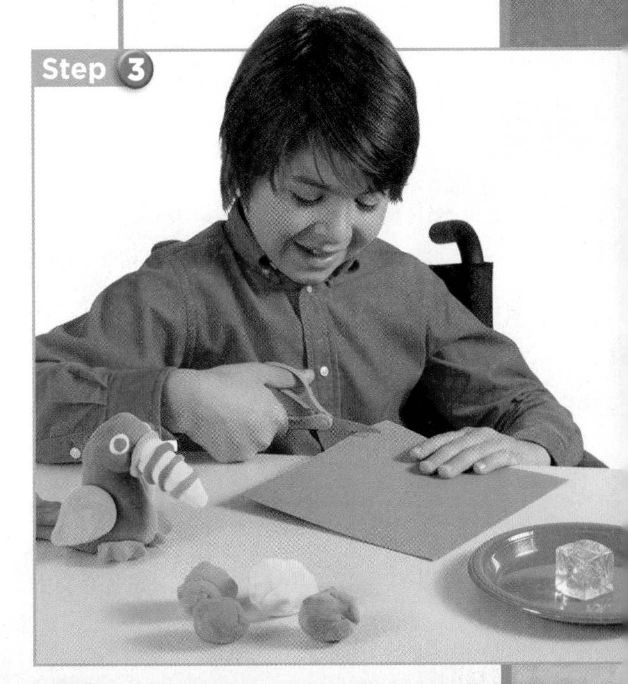

Read and Learn

Essential Question

What happens when matter goes through a physical change?

Vocabulary

physical change, p. 408

mixture, p. 410

solution, p. 411

Reading Skill

Draw Conclusions

Text Clues	Conclusions

Technology 

e-Glossary and e-Review online at www.macmillanmh.com

What are physical changes?

Matter can change. A **physical change** (FIH•zih•kul CHAYNJ) is a change in the way matter looks. Tearing a sheet of paper is a physical change. The size and shape of the paper are different, but the paper is still paper. Matter looks different after a physical change, but it is still made of the same kind of matter.

A change of state is also a physical change. When liquid water freezes, its state changes from a liquid to a solid. The water looks different, but it is still water.

Not all types of matter change in the same way. If you pull on a rubber band, it stretches. When you let go, it returns to its original size. If you pull on a metal spoon, nothing happens. If you pull on a piece of thread, it might break.

Painting an object does not change what the object is made of. ▼

How Steel Changes

Solid steel is melted into a liquid. The melted steel can be shaped to make the frame of a car.

The steel hardens. Now it is a solid. It is combined with other materials to make a car.

The steel is now part of a car. The car is ready to drive on the highway.

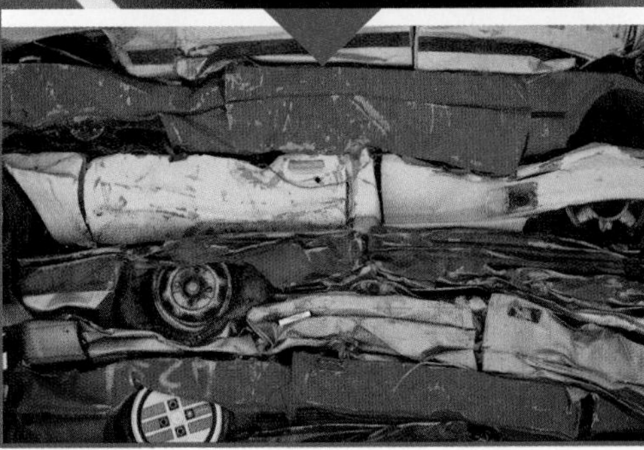

In time, the car is crushed. The steel can be melted and used again to make other steel products.

Read a Chart

What physical changes have happened to the steel?

Clue: Notice how the steel has changed in each photograph. Use the captions to help.

Quick Check

Draw Conclusions Why is a change of state a physical change?

Critical Thinking Make a list of three physical changes you could make to a piece of paper.

What happens when you mix matter?

Another kind of physical change is a mixture (MIHKS•chur). A **mixture** is different kinds of matter mixed together. When you pour milk on your cereal, you are making a mixture. In a mixture the properties of each kind of matter might change. For example, the cereal might get soggy. However, the milk is still milk, and the cereal is still cereal.

▲ What makes up this mixture?

A mixture can be a combination of solids, liquids, and gases. Vegetable soup is a mixture of liquids and solids. Salad dressing can be a mixture of different liquids. Clouds are a mixture of air, dust, and water droplets.

FACT Solutions can be solid.

Solutions

There are many kinds of mixtures. One kind of mixture is a solution (suh•LEW•shun). A **solution** forms when one or more kinds of matter are mixed evenly into another kind of matter.

Salt water is a solution. If you add salt to water, the salt mixes evenly with the water. You can not see the salt, but it is still there. If the water evaporates, the salt will be left behind.

Not all solids form solutions in liquids. Try to mix sand with water. The sand will just sink to the bottom. Some things will not form solutions no matter how long you stir.

Some solutions contain no liquids at all. Air is a solution of different gases. Brass is a solution of several solids, including copper and zinc.

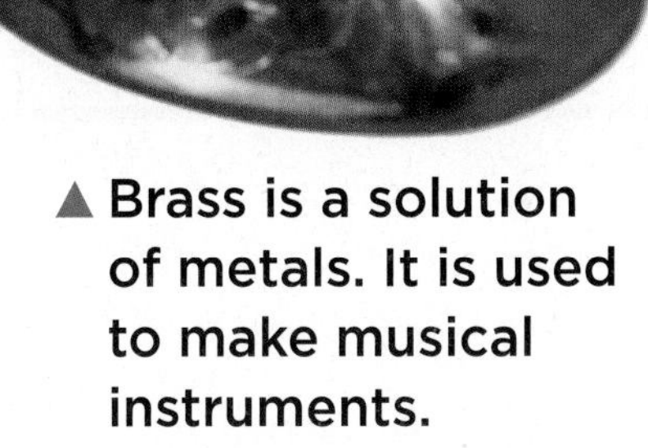

▲ Brass is a solution of metals. It is used to make musical instruments.

Ocean water is a mixture. It contains many different types of matter, including salt, water, and oxygen.

Quick Check

Draw Conclusions Do all kinds of matter form solutions with water? Explain your answer.

Critical Thinking You can not see that salt is in salt water. How do you know it is there?

Quick Lab

Separating Mixtures

1. Mix some sand, marbles, and paper clips together in a bowl.

2. **Experiment** Design an experiment to separate this mixture.
3. **Observe** Were you able to completely separate the mixture? How do you know?
4. **Experiment** How could you separate a mixture of sugar and water?

How can mixtures be separated?

Some properties help you separate mixtures. These properties include size, shape, and color. One way to separate a mixture is to pick out each different type of matter. In a mixture of spaghetti and meatballs, you can pick out the meatballs.

Another way to separate a mixture is by evaporation. Leave a solution of salt and water in a warm place. As the water evaporates, the salt is left behind. The photos below show some other ways to separate mixtures.

Quick Check

Draw Conclusions How can you separate peas from carrots?

Critical Thinking List some ways to separate sand from salt.

Objects that float, such as these cranberries, can be separated from objects that sink. ▼

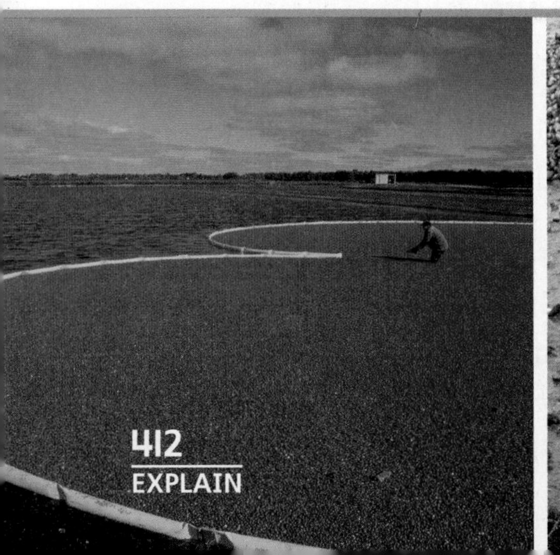

Filters separate mixtures by size. ▼

Magnets separate certain metals from other objects. ▼

Lesson Review

Visual Summary

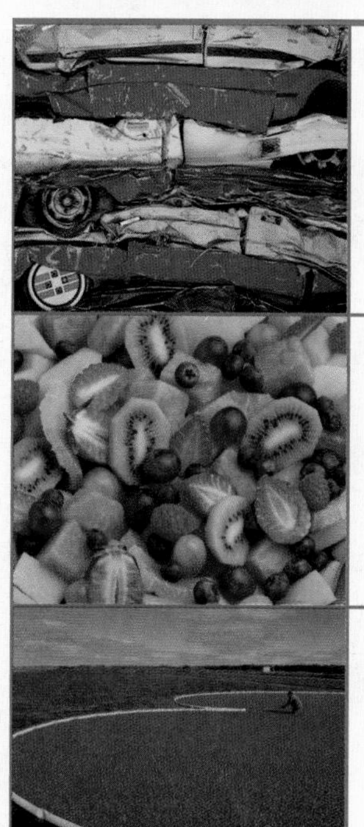

Physical changes cause matter to look different. The kind of matter stays the same.

A **mixture** is a combination of two or more types of matter.

Some properties can help you **separate a mixture.**

Make a FOLDABLES® Study Guide

Make a trifold book. Use it to summarize what you learned about physical changes.

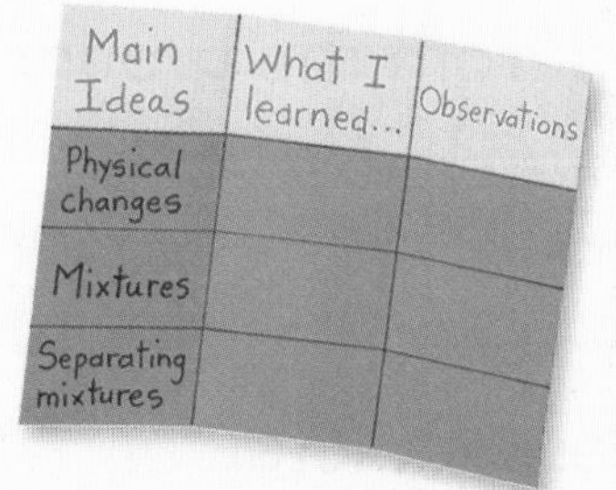

Think, Talk, and Write

1. **Vocabulary** What is a mixture?

2. **Draw Conclusions** A sculptor carves a statue out of a rock. Is this a physical change? How do you know?

Text Clues	Conclusions

3. **Critical Thinking** How could you separate plastic paper clips from metal paper clips?

4. **Test Prep** **Noodles and broth could be separated by**
   - **A** heating in an oven.
   - **B** boiling in a pot.
   - **C** filtering.
   - **D** freezing.

5. **Essential Question** What happens when matter goes through a physical change?

Math Link

Sort and Classify Materials
What materials dissolve in water? Try mixing several materials, such as salt, flour, sugar, soil, and cooking oil, with water. Then classify the materials into groups to show which dissolve and which do not. Make a chart to show your results.

Art Link

Experiment with Color
Cut a circle out of a coffee filter. Use a black marker to draw a spot in the center of the filter. Put the filter on a plate. Add a few drops of water to the spot. Watch what happens. Why do you think this happens? What does this tell you about ink?

LOG ON e-Review Summaries and quizzes online at www.macmillanmh.com

Mining Ores

Did you use something made of metal today? You might have if you ate breakfast with a spoon or rode your bike to school. Both are made of metals.

Metals come from Earth. Metals are found in ores. Ores are rocks that have useful minerals. Some of these minerals contain metals, such as silver or copper. Ores can be found in all kinds of places, from volcanoes to river valleys to mountains.

An ore is usually mined from the ground. Then it is crushed into powder. Magnets, oil, chemicals, and streams of water can be used to separate the minerals from the powdered rock. Later, these minerals are heated to high temperatures to draw out the metals inside them. Once separated, the metals can be mixed with other metals. Then the metals are used to create products, such as the spoon or bicycle you used today.

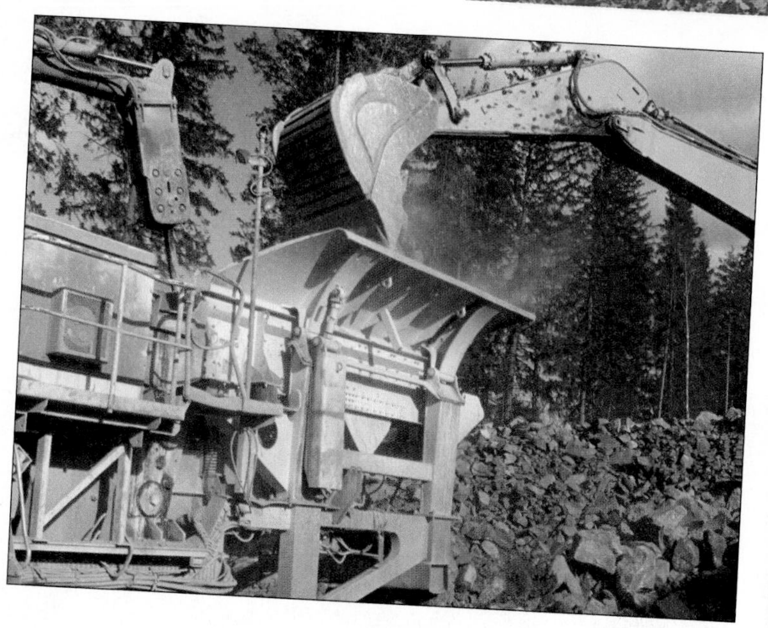

rock crusher

Infer

When you infer,

- you use what you already know;
- you use facts in the article;
- you form new ideas.

Write About It

Infer Read the article with a partner. Use what you know and what you read in the article to answer this question. *Why do you think it is important for people to recycle metals?* Write a paragraph to share your ideas.

-Journal Write about it online at www.macmillanmh.com

Connect to

at www.macmillanmh.com

Lesson 3

Chemical Changes

Look and Wonder

Have you ever baked a cake? Why doesn't a cake taste like the ingredients it is made from? What happens to the ingredients to make them taste different?

Explore

Inquiry Activity

How can matter change?

Make a Prediction

How do flour and baking soda change when each is mixed with vinegar? Make a prediction.

Test Your Prediction

⚠ **Be Careful.** Wear goggles.

1. **Observe** List the properties of the vinegar, flour, and baking soda.
2. **Measure** Use a funnel to put 2 tablespoons of flour in one balloon. Add 50 mL of vinegar to a plastic bottle.
3. **Experiment** Carefully, put the balloon over the bottle's opening without letting any flour fall into the bottle. After you attach the balloon, raise it up so the flour goes into the bottle. Record your observations.
4. Repeat steps 2 and 3 using the second balloon and baking soda instead of flour.

Draw Conclusions

5. Did your results match your prediction? Explain your answer.
6. **Infer** What do you think caused the differences in the balloons?

Explore More

Experiment What might happen to the balloon if you add 2 tablespoons of baking soda and 50 mL of water to a container? Try it and find out.

Materials

vinegar

flour

baking soda

goggles

funnel

measuring cup and spoons

2 balloons

2 plastic bottles

Step 3

Read and Learn

Essential Question
What happens when matter goes through a chemical change?

Vocabulary
chemical change, p. 418

Reading Skill
Infer

Clues	What I Know	What I Infer

Technology
e-Glossary and e-Review online at www.macmillanmh.com

What are chemical changes?

You might have seen an apple turn brown or a burning log change into ash and smoke. Both are examples of a chemical change (KE•mih•kul CHAYNJ). A **chemical change** is a change that causes different kinds of matter to form. The properties of the new matter are different from those of the original materials.

Chemical changes happen every day. Your body uses chemical changes to break down the food you eat. Green plants use the Sun's energy to change carbon dioxide and water into food and oxygen. Cooking also uses chemical changes. Cake batter changes when you bake it. You know that it has changed because it feels and tastes different.

A Chemical Change

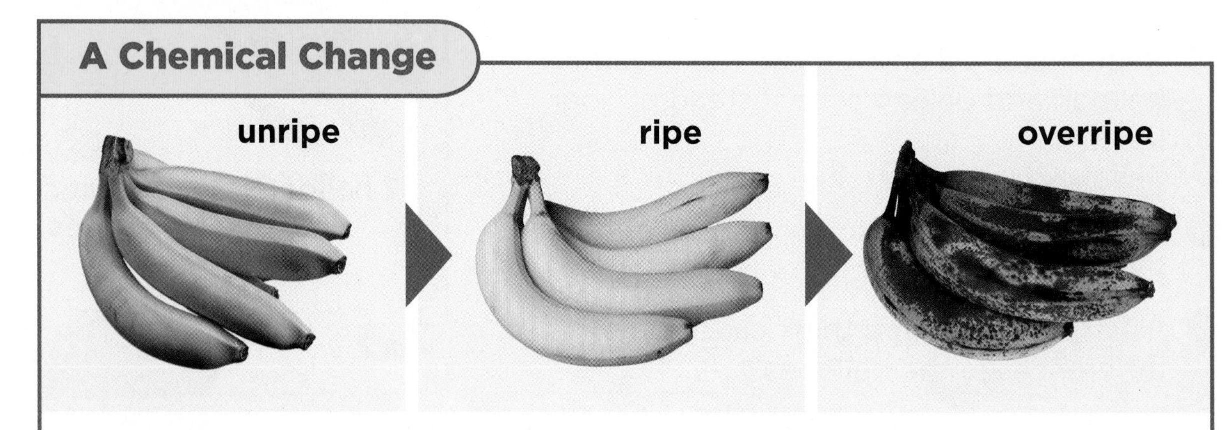

A chemical change happens when fruit ripens. As bananas ripen, they change color. They also become softer and sweeter.

Read a Diagram

How have the bananas in these photographs changed?

Clue: Compare the three photographs to find differences.

Not all chemical changes are useful. A car made from iron might rust. Rust is very different from iron. It is weaker and it peels. Food spoils as a result of chemical changes. Materials in food can break down and form new materials. When this happens, food might change color or smell bad.

Quick Check

Infer Is it a physical or chemical change when milk spoils? Why?

Critical Thinking How are chemical changes important to living things?

Quick Lab

A Chemical Change

1. **Observe** Look closely at some pennies. Make a list of their properties.
2. Place 1 teaspoon of salt in a bowl. Add 150 mL of vinegar. Stir until the salt dissolves.
3. **Experiment** Dip a penny halfway into the liquid. Slowly count to 20 as you hold the coin there. Then remove the penny. Compare the half you held with the half that was in the liquid.
4. **Infer** What caused the change in appearance?

Water and oxygen caused the iron in this train to rust. ▼

What are the signs of a chemical change?

Sometimes a chemical change happens when different materials are put together. Certain signs can show that a chemical change has happened. Here are a few.

Light and Heat

A burning log changes into carbon dioxide gas and ash. As the log burns, it releases light and heat. The light and heat are signs of a chemical change.

▲ Heat and light are two signs of a chemical change.

Formation of Gas

The formation of a gas can be a sign of a chemical change. When you add baking soda to vinegar, carbon dioxide gas forms. As this gas escapes from the liquid, bubbles form.

▲ These bubbles tell you that a chemical change is occurring.

Color Change

Sometimes a color change shows that a chemical change has happened. The Statue of Liberty used to be the same color as a penny. It turned green as a result of a chemical change.

The Statue of Liberty got its green color from a chemical change. ▼

 ***Quick Check***

Infer Is a burning match a physical or chemical change? How do you know?

Critical Thinking Is sugar dissolving in water a physical or chemical change? Explain.

Lesson Review

Visual Summary

	**Chemical changes** cause different kinds of matter to form.
	You observe chemical changes every day.
	Light and heat, formation of a gas, and a color change are **signs of a chemical change.**

Make a FOLDABLES® Study Guide

Make a trifold book. Use it to summarize what you learned about chemical changes.

Main Ideas	What I learned...	Observations
Chemical changes are ...		
Signs of a chemical change are...		

Think, Talk, and Write

1. **Vocabulary** What is a chemical change? Give an example.

2. **Infer** Two clear liquids are combined. Bubbles form. What kind of change might have happened? Explain.

Clues	What I Know	What I Infer

3. **Critical Thinking** Mrs. Hall wiped a discolored pot with a special cleaner. The pot returned to its original color. What happened?

4. **Test Prep Which is a chemical change to a piece of paper?**
   - **A** folding
   - **B** cutting
   - **C** tearing
   - **D** burning

5. **Essential Question** What happens when matter goes through a chemical change?

Math Link

Solve a Problem

A log takes one hour to burn down into ash. A banana turns brown and mushy in four days. How many hours did the longer chemical change take?

Social Studies Link

Conduct Research

Bread is made differently in other countries. Different ingredients make different chemical changes. Research how bread is made in other countries.

LOG ON e-Review Summaries and quizzes online at www.macmillanmh.com

Be a Scientist

Materials

chalk

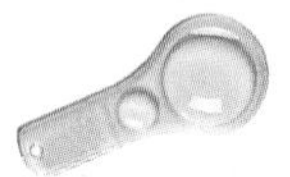

hand lens

black construction paper

vinegar

dropper

Structured Inquiry

How can physical and chemical changes affect matter?

Form a Hypothesis

How will breaking chalk change the chalk? How will adding vinegar to the chalk change it? Write a hypothesis.

Test Your Hypothesis

1. **Observe** Break a piece of chalk in half. Use a hand lens to look at the broken end of the chalk. Record your observations. Is this a chemical or physical change?

2. **Experiment** Rub one of the chalk pieces on a piece of black paper. Using the hand lens, look at the chalk on the paper. Record your observations. Is this a chemical or physical change?

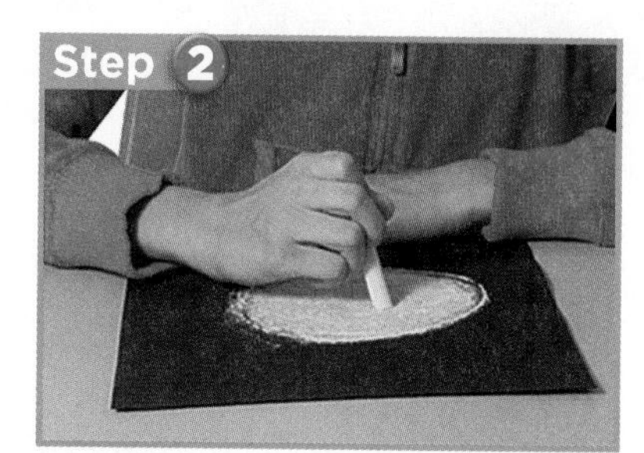

3. **Experiment** Use a dropper to add one drop of vinegar to the chalk on the black paper. Record your observations. Is this a chemical or physical change?

Draw Conclusions

4. **Interpret Data** What did you observe? Which changes were physical changes? Was there a chemical change?

5 **Infer** Describe what happened to the chalk when you added the vinegar. What caused this to happen?

6 **Communicate** Use your observations to write your own definitions of chemical and physical change.

Guided Inquiry

What are the signs of a chemical change?

Form a Hypothesis

How can you tell a chemical change has happened? Write a hypothesis.

Test Your Hypothesis

Design an experiment to investigate chemical changes. Use the materials shown. Write the steps you plan to follow. Record your results and observations.

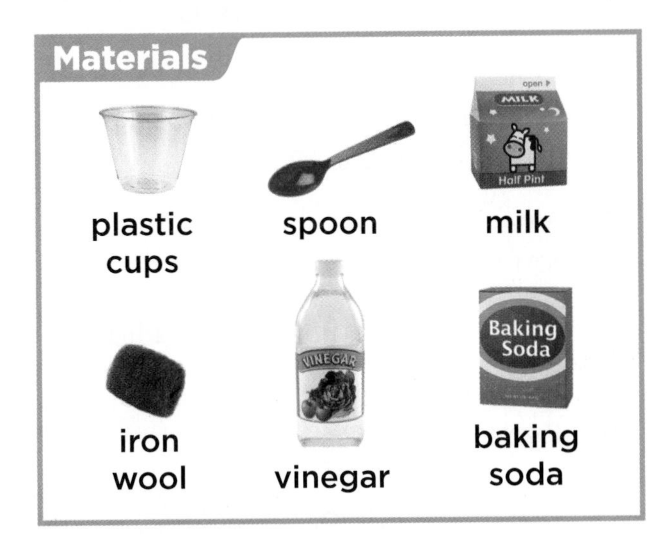

Draw Conclusions

What changes did you observe? Did your experiment support your hypothesis? Why or why not?

Open Inquiry

What else would you like to know about physical and chemical changes? Think of a question to investigate. For example, how does iron rust? Design an experiment to answer your question.

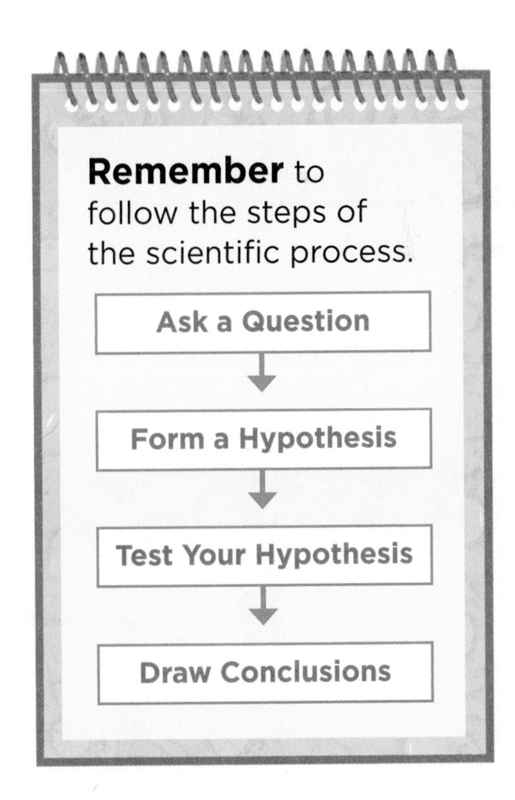

CHAPTER 10 Review

Visual Summary

Lesson 1 Adding or removing heat can cause matter to change state.

Lesson 2 Matter looks different after a physical change, but it is still the same kind of matter.

Lesson 3 Chemical changes cause different kinds of matter to form.

Make a FOLDABLES® Study Guide

Glue your lesson study guides to a piece of paper as shown. Use your study guide to review what you have learned in this chapter.

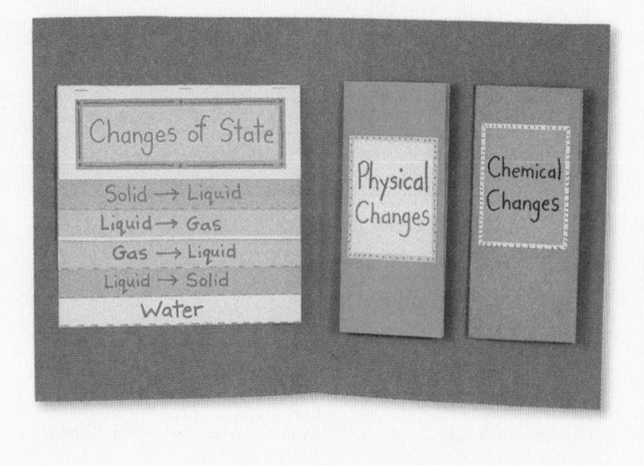

Vocabulary

DOK I

Fill each blank with the best term from the list.

boil, p. 398
chemical change, p. 418
condense, p. 400
evaporate, p. 399
freeze, p. 401
mixture, p. 410
melt, p. 398
physical change, p. 408
solution, p. 411
water vapor, p. 399

1. When you stir spaghetti and meatballs together, you make a ________.

2. Tearing a sheet of paper is a ________.

3. If you ________ a liquid, it becomes a solid.

4. A change that causes different kinds of matter to form is a ________.

5. To change from a liquid to a gas slowly is to ________.

6. When you mix salt with water, you make a ________.

7. To change from a solid to a liquid is to ________.

8. If you cool a gas to the right temperature, it will ________, or turn into a liquid.

9. To change from a liquid to a gas is to ________.

10. The gas state of water is called ________.

LOG ON e-Glossary Words and definitions online at www.macmillanmh.com

Skills and Concepts

DOK 2–3

Answer each of the following.

11. Infer What kind of change occurs when you toast bread? What kind of change occurs when butter melts on a piece of toast? Explain your answer.

12. Expository Writing Describe what happens to water as it freezes.

13. Predict It is a warm, sunny day. You leave a bar of chocolate on the windowsill. How do you think it will change? Can you change it back?

14. Critical Thinking You add sugar to a glass of lemonade and stir it. You can not see the sugar anymore. The lemonade tastes sweet now. What kind of mixture is this? How do you know?

15. Infer Two clear liquids are mixed in a flask. The liquid inside the flask turns orange. Is this most likely a physical or chemical change? Explain.

16. Study the photograph below. In which two states of matter is water shown? Describe how they are different.

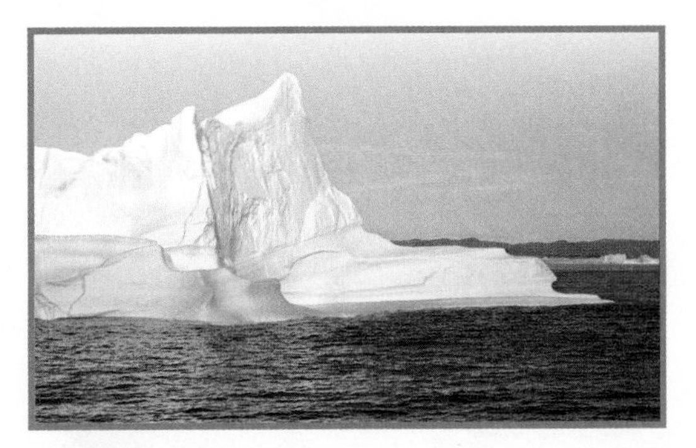

17. True or False *A brass tuba is a solution.* Is this statement true or false? Explain.

18. True or False *A glass of apple juice is a solution.* Is this statement true or false? Explain.

19. Which of the following best describes what happens when a log burns?

A A chemical change is taking place.

B The logs are becoming a liquid.

C The logs are getting bigger.

D The smoke is boiling.

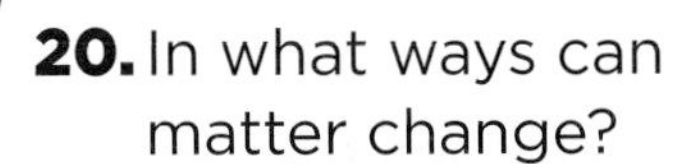

20. In what ways can matter change?

Performance Assessment

DOK 3

Act It Out!

- With a partner, act out one important term or idea from this chapter. For example, you could choose a term such as *melt.* You can use props but may not speak during your skit.
- Present your skit to the class. Then let other students guess the term.
- What information about your term or idea did you show? How did you show it?
- What details helped you guess other pairs' terms and ideas?

Test Preparation

1 This chart shows the melting points of four substances.

Substance	Melting Temperature (in °F)
iron	2,795
silver	1,780
potassium	146
bromine	19

Which substance will most likely be a liquid at room temperature?

A iron

B silver

C potassium

D bromine

DOK 2

2 Look at these clothes that are hanging outside to dry.

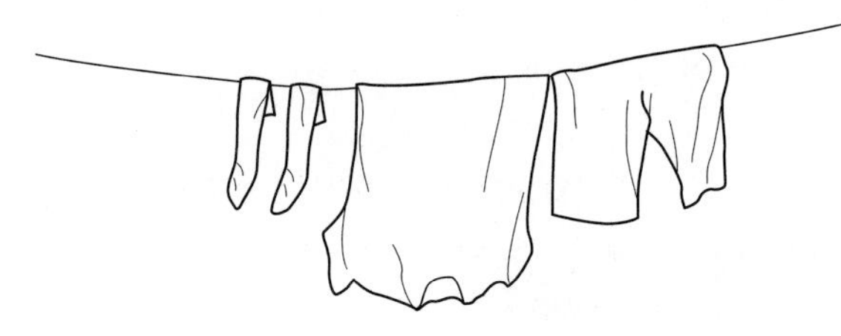

How will these clothes become dry?

A melting

B boiling

C evaporation

D condensation

DOK 1

3 Which is most likely a solution?

A glass of grape juice

B cup of vegetable soup

C plate of fruit salad

D bowl of cereal with milk

DOK 1

4 How do most types of matter change as they are heated?

A They lose mass.

B They gain mass.

C They take up less space.

D They take up more space.

DOK 1

5 What happens when water vapor cools?

A A gas becomes a solid.

B A gas becomes a liquid.

C The water vapor freezes.

D The water vapor evaporates.

DOK 1

6 Which is the best example of a chemical change?

A cutting a carrot

B tossing a salad

C melting ice

D rusting metal

DOK 1

7 The picture below shows a tool used to filter flour.

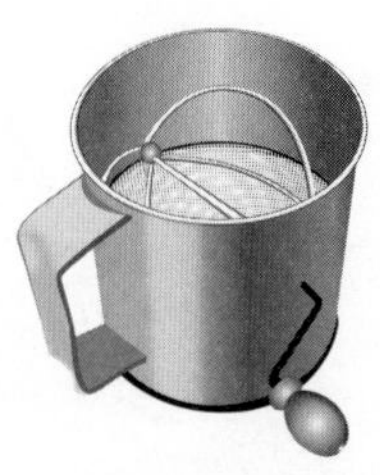

flour sieve

Which mixture could this tool most likely separate?

A a mixture of white flour and wheat flour

B a mixture of white flour and powdered sugar

C a mixture of powdered sugar and rice

D a mixture of wheat flour and powdered sugar

DOK 2

8 What process is shown in the picture below?

A melting

B boiling

C precipitation

D condensation

DOK 1

9 A substance that gets larger when it freezes is

A air.

B oil.

C vinegar.

D water.

DOK 1

10 Look at the ingredients shown below.

Describe one physical change that can occur when using these ingredients.

DOK 1

Describe one chemical change that can occur when using these ingredients.

DOK 1

Check Your Understanding

Question	Review	Question	Review
1	p. 398	6	pp. 418–419
2	p. 399	7	p. 412
3	pp. 410–411	8	p. 399
4	p. 398	9	p. 402
5	pp. 400–401	10	pp. 408–409, 418–419

Environmental Chemist

Do you like helping keep plants and animals healthy? Are you concerned with keeping the environment clean? If so, then you might like to be an environmental chemist.

An environmental chemist is a type of scientist. These scientists help keep the water, land, and air free of pollution. Pollution can hurt plants, animals, and people. Environmental chemists protect living things by helping clean up pollution. Environmental chemists also show people how to reduce pollution.

To become an environmental chemist, you should begin learning about the environment where you live. Start a recycling program in your home or at school to reduce waste. You could also join a group that helps protect the environment.

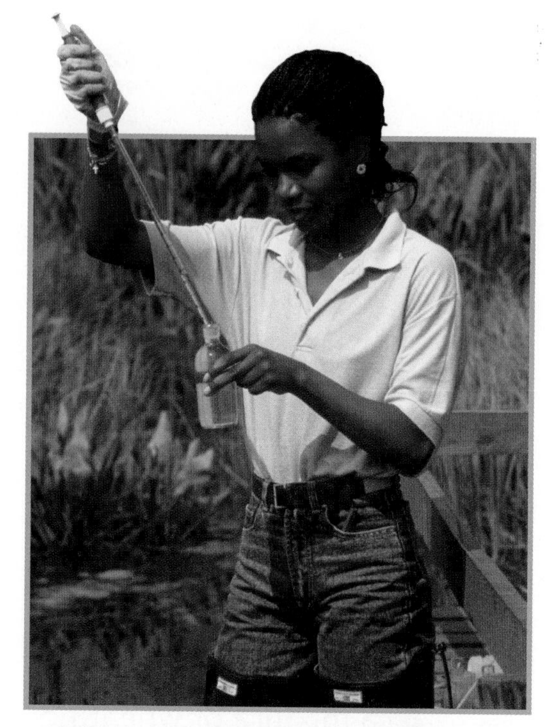

▲ This scientist is collecting data on water pollution.

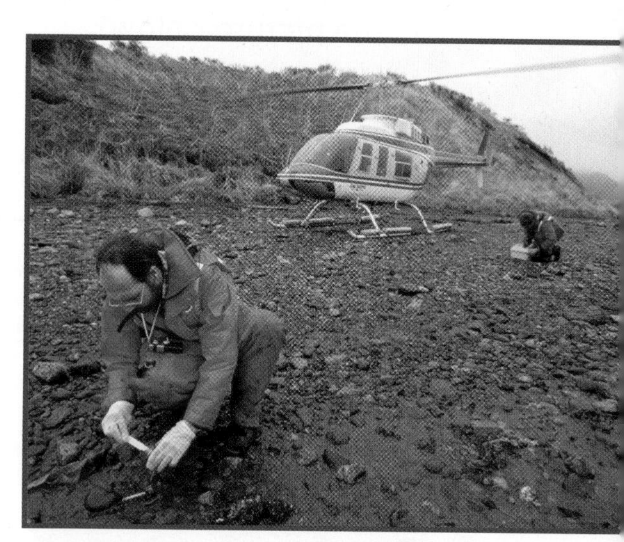

▲ This chemist is testing to see how much oil is left on a beach after a spill.

Here are some other physical science careers:

- carpenter
- lab technician
- chemical engineer
- pharmacist

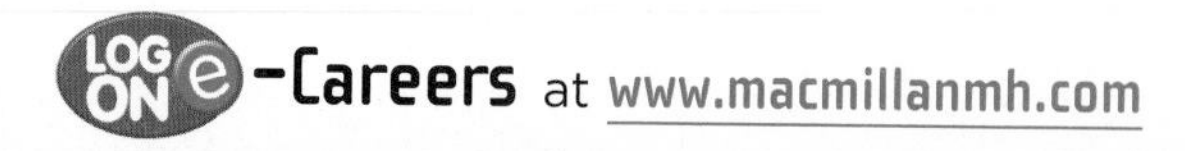

e-Careers at www.macmillanmh.com

Forces and Energy

The Ferris wheel was designed by George W. Ferris in 1893.

Ferris wheel at the Del Mar Fair in California

Literature
Poem

Jump Rope

by Rebecca Kai Dotlich

Swings
up,
whirls around,
brushes ground
beneath quick feet.
Sweeps walks,
slap, slip,
double Dutch,
scissor skip.
Flip, flap,
LOOPS around,
slip, slap,
swoops down.
Slides and swirls,
twirls and twists,
song for a jump rope
sounds like this:
Tiger leap,
spider spin,
your turn next,
jump on in!

Write About It

Response to Literature This poet uses rhythm and rhyme to describe how a jump rope moves. How else do things move on the playground? Write a poem about another movement game.

e-Journal Write about it online at www.macmillanmh.com

CHAPTER 11

Forces and Motion

What makes something move?

Essential Questions

Lesson 1

How can you tell if something is moving?

Lesson 2

How do forces change motion?

Lesson 3

How do we do work?

Lesson 4

How can a simple machine reduce force?

Grand prix race car

Big Idea Vocabulary

position the location of an object (p. 434)

motion a change in position (p. 436)

force a push or pull; a force can make an object move (p. 444)

work what is done when a force moves an object or changes an object's motion (p. 454)

energy the ability to do work; energy is what makes motion possible (p. 456)

simple machine a machine with few or no moving parts (p. 465)

Visit www.macmillanmh.com for online resources.

Lesson 1

Position and Motion

Look and Wonder

Snowboarding is like skateboarding on snow. How does this snowboarder's position change as she travels down the mountain?

Explore

Inquiry Activity

How can you describe an object's position?

Purpose

Find out ways to describe a block's position.

Procedure

1. Sit opposite a partner at a table. Prop up a notebook between the two of you.
2. One partner, "the builder," uses the blocks to make a building. Make sure the other partner, "the copier," can not see the building.
3. **Communicate** The builder tells the copier how to make the same building. Make a list of the words you use.
4. **Observe** Remove the notebook. Are the buildings the same? Switch roles and try the activity again.

Draw Conclusions

5. What words did you use to describe your building?
6. **Infer** Could you describe the position of each block without comparing it to other blocks around it?

Materials

notebook

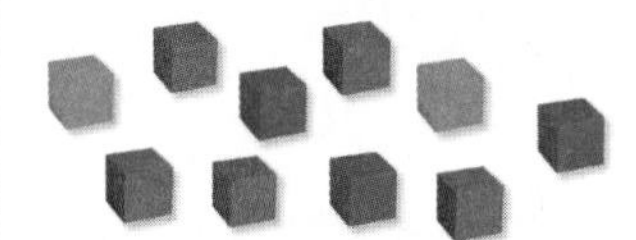

two sets of 10 colored blocks

Explore More

Communicate
How could you direct someone from your home to your school?

Read and Learn

Essential Question
How can you tell if something is moving?

Vocabulary
position, p. 434
distance, p. 435
motion, p. 436
speed, p. 438

Reading Skill
Compare and Contrast

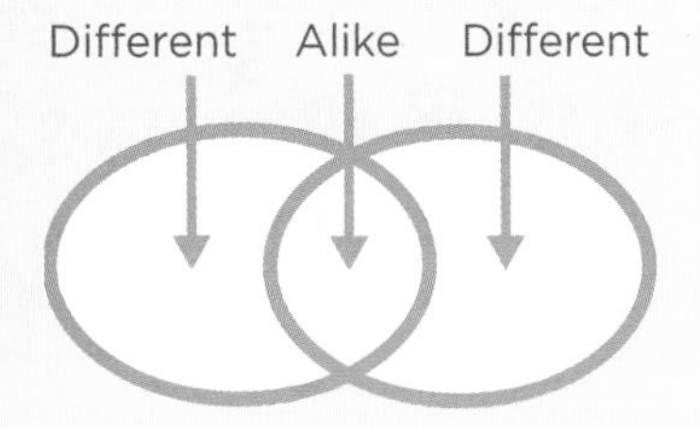

Technology 

e-Glossary and e-Review online at www.macmillanmh.com

How can you describe position?

In the picture below, where is the boy in the red shirt? He is next to the girl in the pink shirt. He is under the girl wearing the blue overalls. When you describe where something is, you describe its position (puh•ZI•shun). **Position** is the location of an object.

You can describe something's position by comparing it to the position of other things. Words such as *over, under, left, right, on top of, beneath,* and *next to* give clues about position. You could say that a mouse is under a table or that a cat is on top of a shelf. When we describe the position of something, we compare it with objects around it.

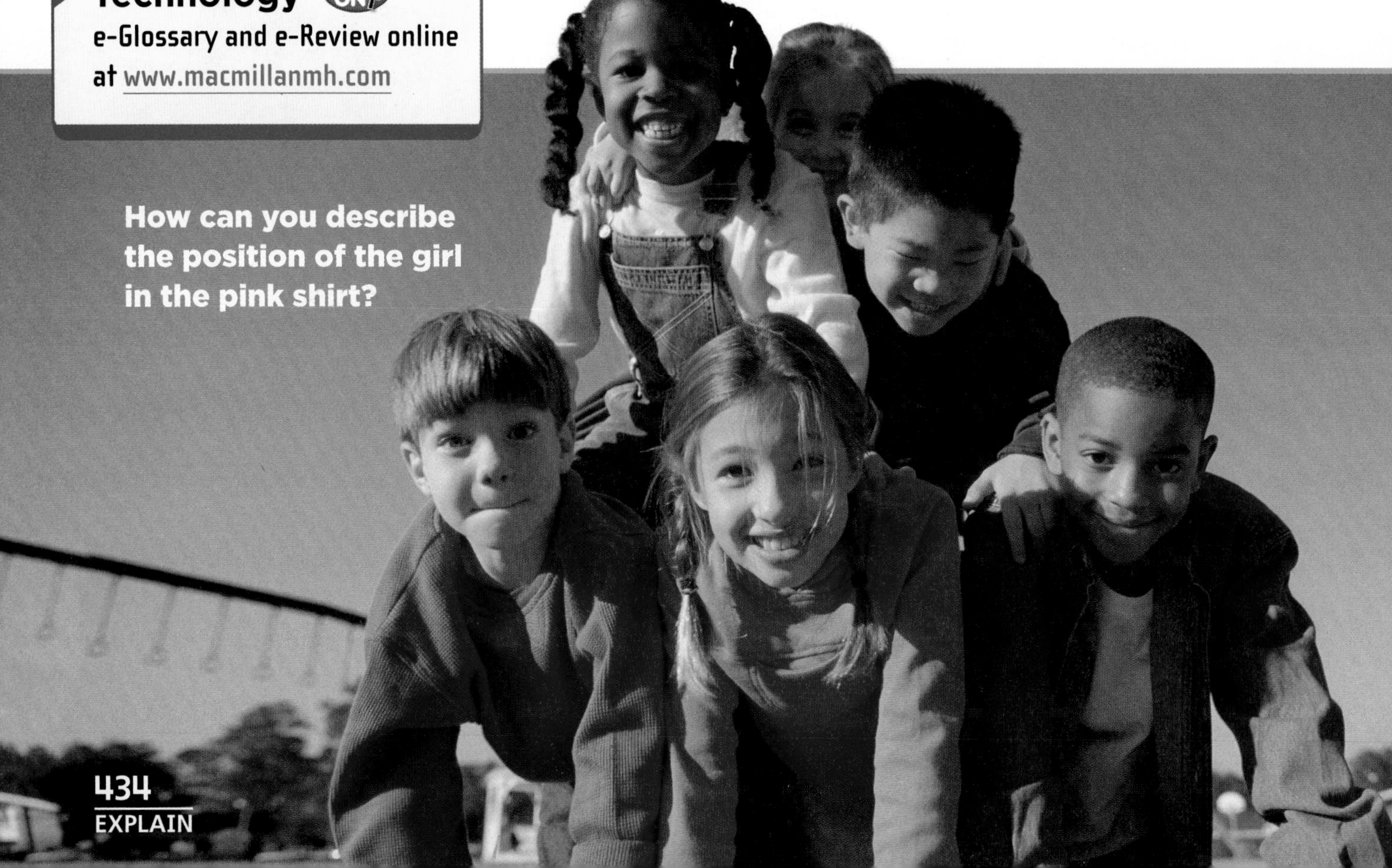

How can you describe the position of the girl in the pink shirt?

Distance

You can also describe something's position by measuring its distance (DIS•tuns) from other objects. **Distance** is the amount of space between two objects or places. Distance can be measured in inches, yards, or miles. In the metric system, distance is often measured in centimeters, meters, or kilometers. You can use a ruler or meter stick to measure distances. The distance between the two toys shown below is 10 centimeters.

 ***Quick Check***

Compare and Contrast What must you compare an object to in order to describe its position?

Critical Thinking Use position words to describe the location of your classroom.

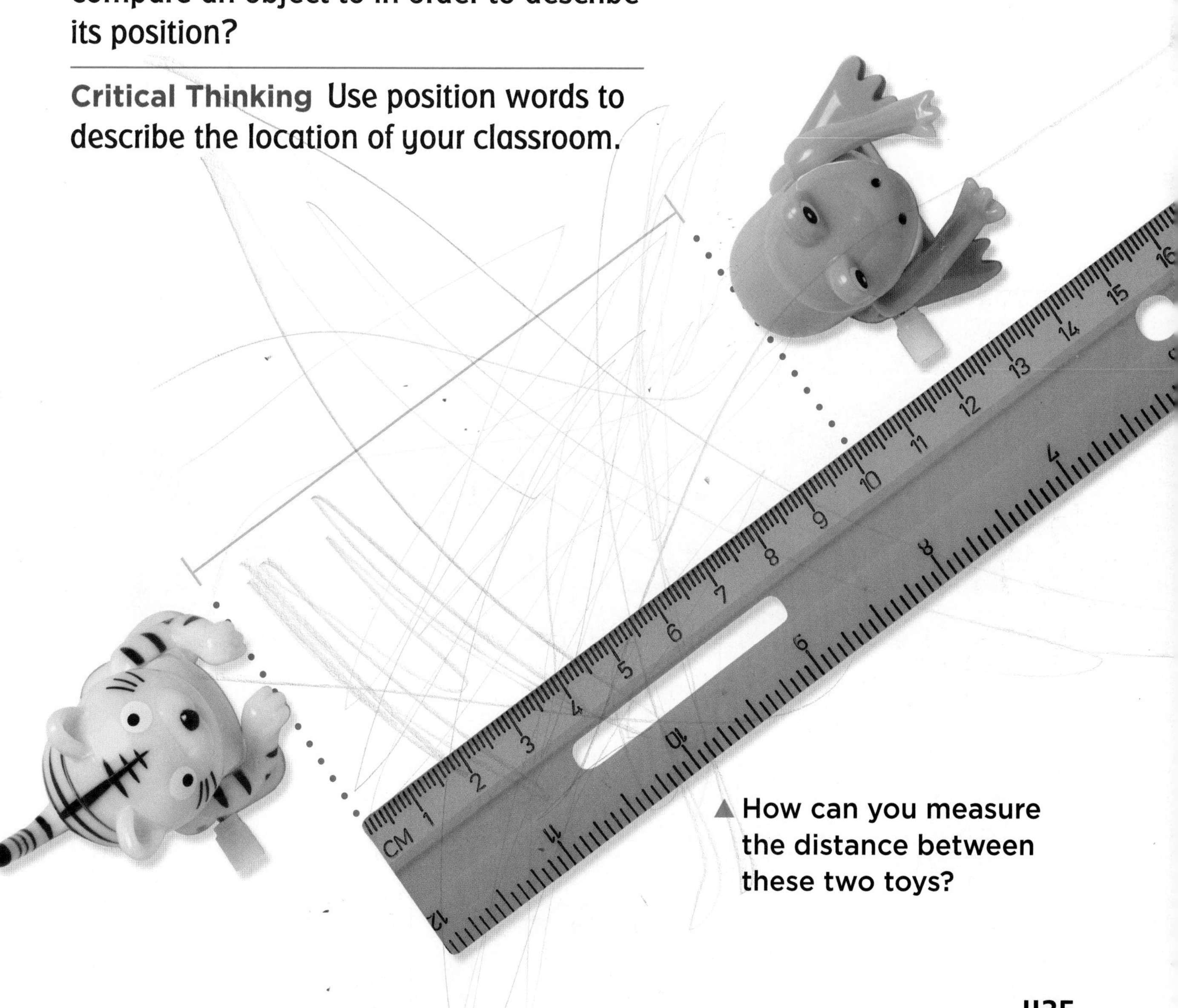

▲ How can you measure the distance between these two toys?

What is motion?

Look at the pictures of the mouse below. In the first box, the mouse is on a rock. In the second box, it is between the two rocks. What happened to the mouse? It moved. You know that the mouse moved because its position changed. While an object is changing position, it is in motion (MOH•shun). **Motion** is a change in position.

▲ A swing moves back and forth.

Objects can move in different ways. Look at the chart on the next page. The runner moves forward in a straight line. The figure skater spins round and round on the ice. The snowboarder moves down the hill in a zigzag. A zigzag is a path with short, sharp turns from one side to another. The skateboarder moves back and forth in the pipe. Straight line, round and round, zigzag, and back and forth are types of motion.

▼ How can you tell that the mouse has moved?

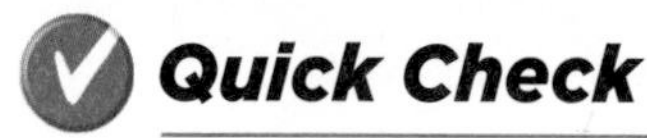

Compare and Contrast How are zigzag and back and forth motions similar?

Critical Thinking List some objects that move round and round.

Types of Motion

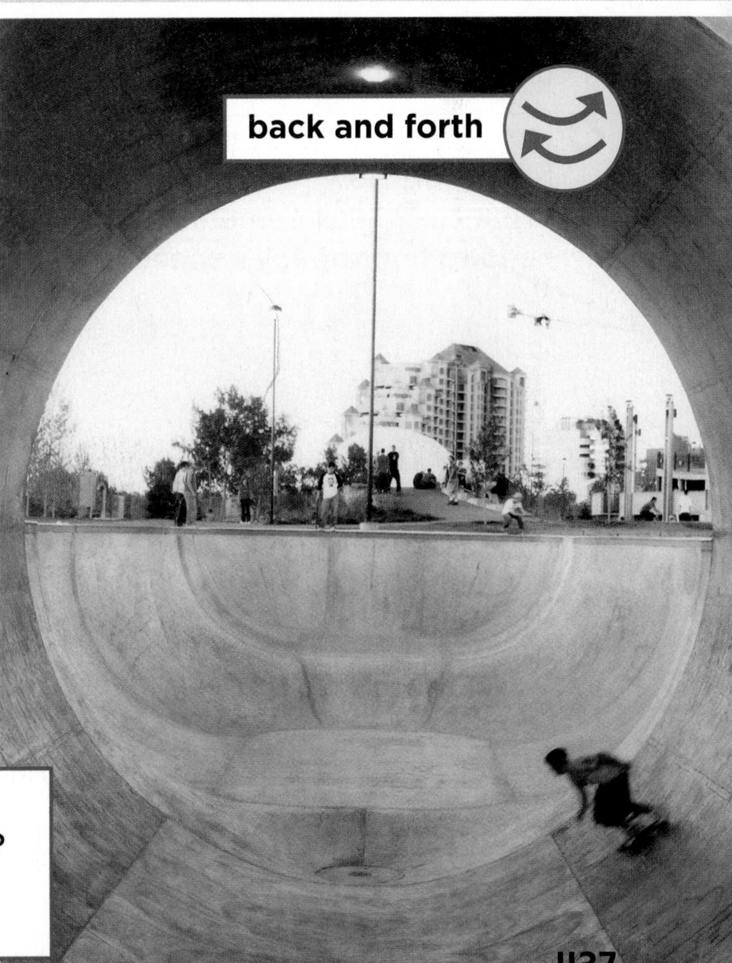

Read a Chart

What are some ways objects can move?

Clue: Arrows can show directions.

Quick Lab

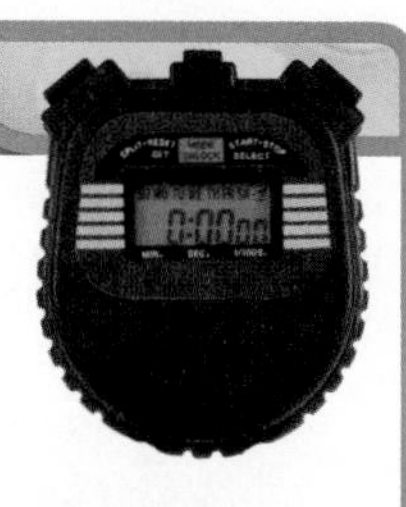

Measure Speed

1. Set up a racetrack as shown below.

2. **Measure** Wind up a wind-up toy. Place it at the starting line and let it go. Have a partner use a stopwatch to time the toy's trip. Measure how far the toy travels. Record your measurements.
3. **Communicate** Make a drawing to show how the toy moved.
4. **Use Numbers** How far did the toy travel? How fast did it travel? What two measurements do you need to find the toy's speed?

What is speed?

Some things move faster than others. A cheetah moves faster than a snail. **Speed** describes how quickly an object moves. An object's speed tells how far it will move in a certain amount of time.

You can measure the speed of an object. You need to know how far the object traveled. You also need to know how much time it took for the object to travel that distance. If a car traveled 50 kilometers per hour, its speed was 50 km/h.

Quick Check

Compare and Contrast Which is faster: a plane or car? Explain.

Critical Thinking A red car moves faster than a green car. Both move for three seconds. Which car moves farther? Why?

Slow-moving objects take longer to travel a distance than fast-moving objects.

Lesson Review

Visual Summary

Position is the location of an object.

Motion is a change in an object's position. Objects can move in different ways.

Speed describes how quickly an object moves.

Make a FOLDABLES® Study Guide

Make a three-tab book. Use it to summarize what you learned about position and motion.

Think, Talk, and Write

1. **Vocabulary** What is the position of an object?

2. **Compare and Contrast** How is zigzag motion like back and forth motion? How are they different?

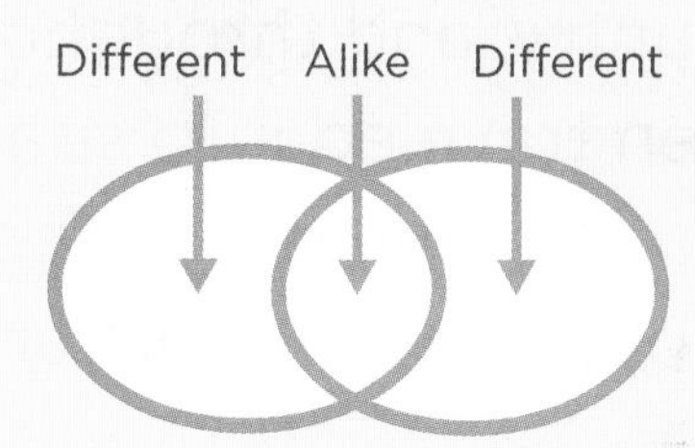

3. **Critical Thinking** Suppose you rode a bike at 10 km/h for 3 hours. How far would you travel?

4. **Test Prep** **Which tool measures distance?**
   - **A** stopwatch
   - **B** thermometer
   - **C** pan balance
   - **D** meter stick

5. **Essential Question** How can you tell if something is moving?

Writing Link

Write a Description
Hold a ball in your hand. Drop it. How does the ball move? Then toss the ball to a friend. How does the ball move? Describe the different ways the ball moves.

Math Link

Make a Graph
Use research materials to find the speed of five objects. Organize this information into a chart. Then make a bar graph. Is it easier to compare data using a chart or a bar graph? Explain your answer.

LOG ON e-Review Summaries and quizzes online at www.macmillanmh.com

Travel through Time

People have always wanted to travel. They found ways to travel within their state, across the country, and throughout the world. People have even traveled into space. The time line below shows some of the first machines that helped people travel to distant places.

1804

In England, Richard Trevithick built the first steam engine for a train. The steam engine helped people travel great distances. It also helped them get to their destinations more quickly.

1884

In Germany, Karl Friedrich Benz built the first car to run on gasoline. It worked similarly to the cars you see on the road today. However, his car had only three wheels!

Connect to

at www.macmillanmh.com

1903

Wilbur and Orville Wright constructed the first motorized airplane that flew and landed safely. Their airplane's engine ran on gasoline. It flew for 12 seconds over 36 meters (120 feet).

1961

Russian astronaut Yuri Gagarin was the first person in space. His spaceship had special engines. They produced a force that was stronger than the pull of Earth's gravity. These engines helped the spaceship leave Earth's surface and orbit the planet.

Problem and Solution

A problem and solution

- gives a problem;
- tells how to solve the problem.

Write About It

Problem and Solution How have machines helped people learn about distant places? Read the article again. Then write about ways machines have helped people solve problems.

e-Journal Write about it online at www.macmillanmh.com

Lesson 2

Forces

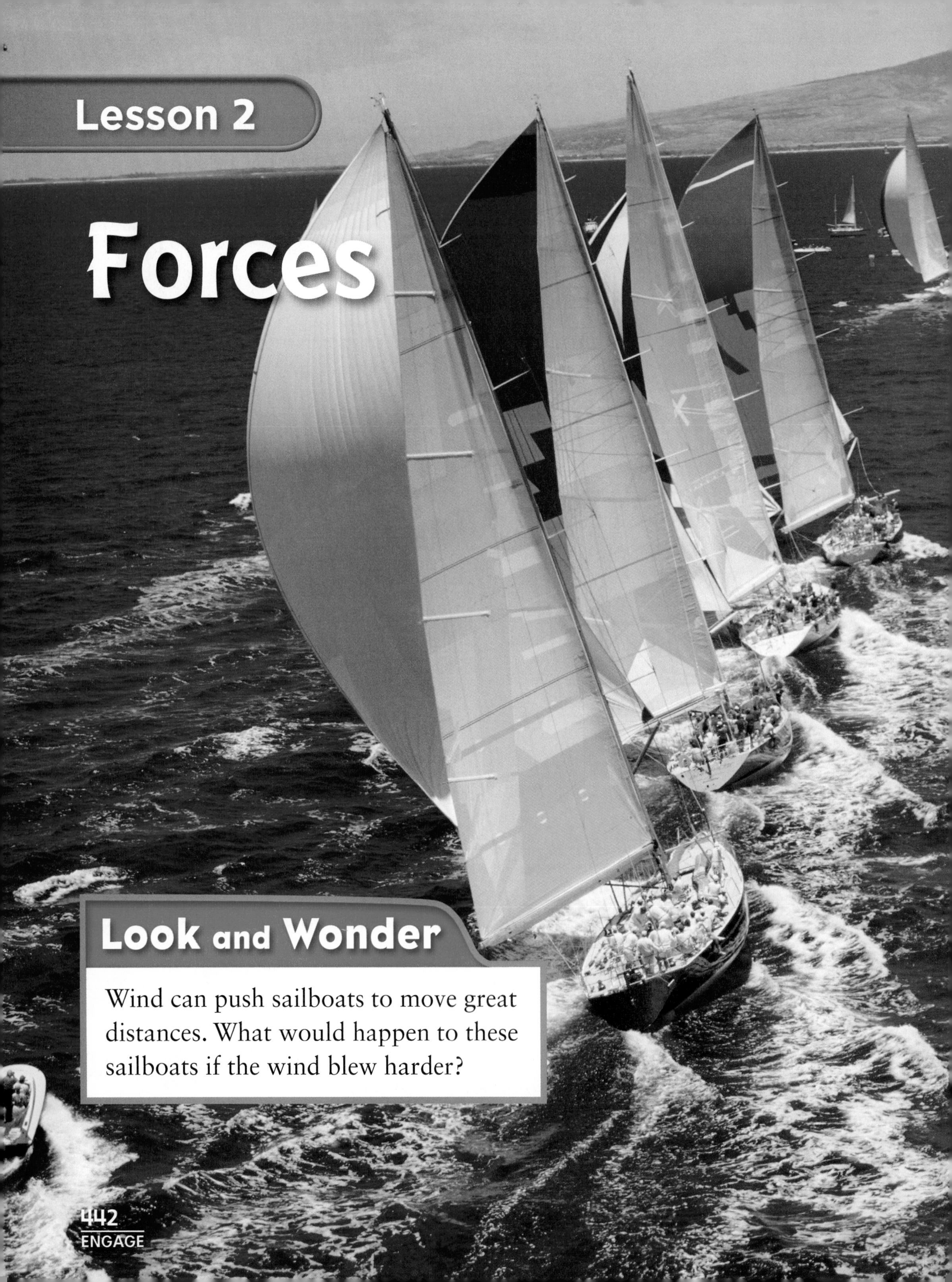

Look and Wonder

Wind can push sailboats to move great distances. What would happen to these sailboats if the wind blew harder?

Explore

Inquiry Activity

How can pushes affect the way objects move?

Form a Hypothesis

What will happen to an object if you increase the force you use to push it? Write a hypothesis. Start with "If I push an object with more force, then..."

Test Your Hypothesis

1. Stack three books on the floor. Then lean a piece of cardboard against the top book to make a ramp. Tape down the edge along the floor.
2. **Observe** Place a toy car at the bottom of the ramp. Hold a tennis ball at the top of the ramp. Then let the ball go so it pushes the toy car. What happens?
3. **Measure** Find out how far the car travels.
4. **Use Variables** Add three more books to the stack. The ball pushes the car with more force when you increase the height of the ramp. Repeat steps 2 and 3.

Draw Conclusions

5. **Infer** What caused the car to move?
6. **Interpret Data** When did the car travel farther?
7. **Infer** How does the amount of force you use to push an object affect how far the object travels?

Explore More

Experiment What would happen if you added a weight to the toy car and repeated the activity?

Materials

- 6 books
- cardboard
- masking tape
- toy car
- tennis ball
- ruler

Step 1

Step 2

Read and Learn

Essential Question

How do forces change motion?

Vocabulary

- **force,** p. 444
- **magnet,** p. 446
- **gravity,** p. 447
- **weight,** p. 447
- **friction,** p. 448

Reading Skill

Cause and Effect

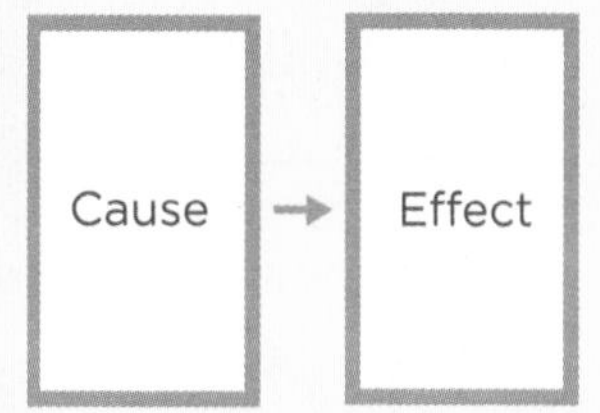

Technology

e-Glossary and e-Review online at www.macmillanmh.com

What are forces?

Objects do not move by themselves. You have to apply a force (FORS) to make them start moving. A **force** is a push or a pull. You use forces to move things all the time. When you pull on a door handle or push a wagon, you apply a force to make something move.

Forces can be large or small. The force a crane uses to lift a truck is huge. The force your hand uses to lift a feather is tiny. It takes more force to move heavy objects than light objects. Forces also affect an object's speed. The more force you use, the faster an object will move.

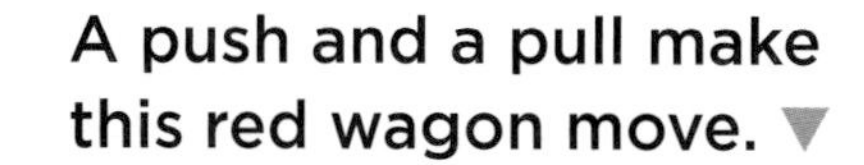

A push and a pull make this red wagon move. ▼

pull

push

Changes in Motion

Forces can change the motion of objects. They can make objects start moving, speed up, slow down, or stop moving. They can make objects change direction too.

Forces can change a soccer ball's motion. A goalie applies a force to throw the ball to a teammate. The ball starts to move. The teammate applies another force when he kicks the ball. The ball changes direction. Each time a force is applied, the motion of the ball changes. When a goalie catches the ball, the ball's motion stops.

A change in an object's motion is the result of all the forces that are acting on the object. Think of the game tug-of-war. When both sides pull equally on the rope, the forces are balanced. Nothing moves. If one side pulls harder, the forces become unbalanced. Now the rope moves. Both sides move as well.

Cause and Effect How can forces affect an object's motion?

Critical Thinking What happens when you kick a moving ball?

Changes in Motion

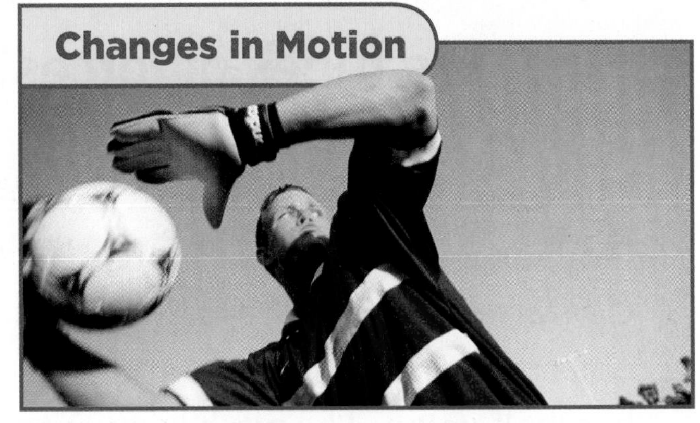

① The goalie throws the ball to start its motion.

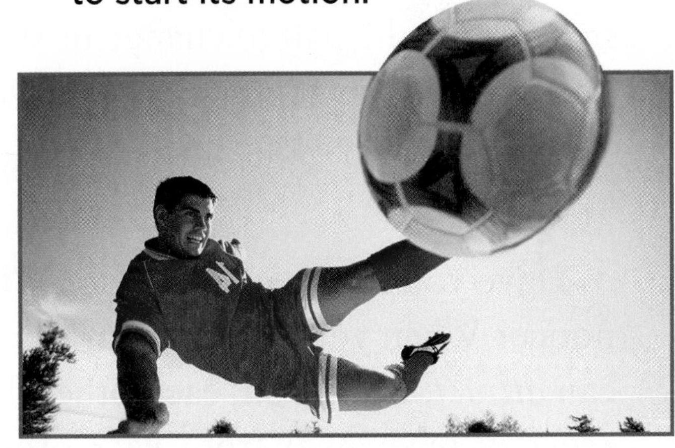

② This player kicks the ball, changing its speed and direction.

③ The goalie catches the soccer ball, stopping its motion.

Read a Photo

How have forces changed the motion of this soccer ball?

Clue: Captions give information.

What are types of forces?

There are many types of forces. The forces you are probably most familiar with are contact forces. *Contact forces* happen between objects that touch. Think about a baseball game. The pitcher must touch the ball to throw it to home plate. A bat must touch the ball to change its direction. Some forces can act on an object without touching the object. Magnetism and gravity are examples.

▲ When the bat hits the ball, the ball changes direction.

Magnetism

Have you ever used magnets? What did you notice? When you bring two magnets together, they can *attract,* or pull on, each other. They can also *repel,* or push away from, each other. Magnets can attract or repel each other without touching. The force that causes this to happen is called *magnetic force*. A **magnet** is any object with a magnetic force.

Magnets can attract or repel each other. They can also attract things made of certain metals like iron. They can not attract things made of wood, glass, plastic, or rubber. Magnets can attract or repel objects through solids, liquids, or gases.

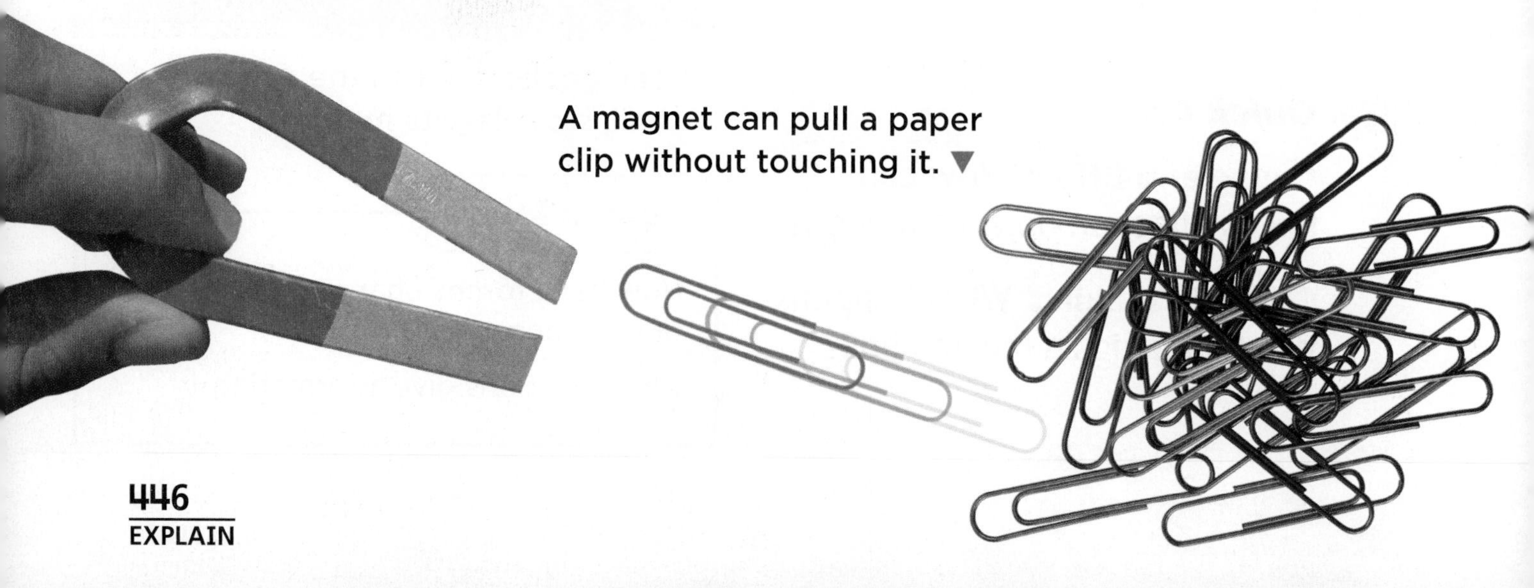

A magnet can pull a paper clip without touching it. ▼

Gravity

You can not see gravity, but it is what keeps you on Earth. **Gravity** is a pulling force between two objects, such as you and Earth. Gravity pulls objects together. When you jump up, Earth's gravity pulls you down. Gravity pulls through solids, liquids, or gases.

How much gravity does it take to keep you on Earth? The answer is your weight (WAYT). An object's **weight** is a measure of the pull of gravity on it. The more mass an object has, the more gravity pulls on it.

Quick Check

Cause and Effect What effect does gravity have on objects?

Critical Thinking How can you pick up metal paper clips without touching them?

Quick Lab

Observe Gravity

1. **Predict** Does gravity act the same on all objects? Would it act the same on two plastic bottles that have the same volume but different masses?
2. Hold an empty plastic bottle in one hand. Hold an identical bottle full of water in the other hand. Hold each bottle away from your body.

3. **Observe** Describe what you feel. Is each bottle pulled toward Earth with the same force?
4. **Infer** Is the amount of gravity on the two bottles the same? How could you tell?

Gravity is pulling these skydivers to Earth.

▲ This water slide is smooth and has little friction.

Friction between the brake pad and the bike rim stops the bike. ▼

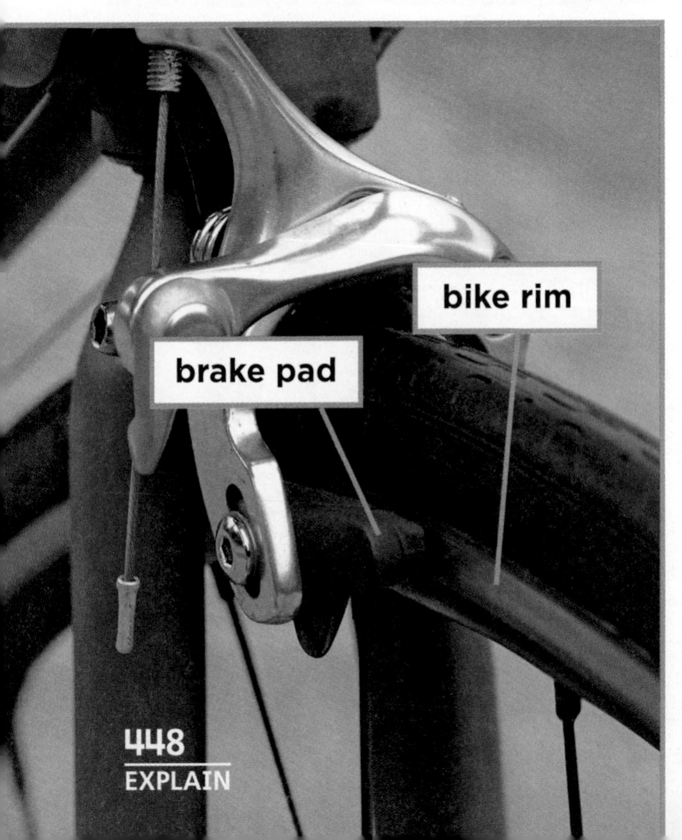

What is friction?

A block slides on the floor. It then slows down and stops. Why does this happen? A force called friction (FRIHK•shun) is acting on the block. **Friction** is a force that occurs when one object rubs against another. It pushes against moving objects and causes them to slow down.

Different surfaces produce different amounts of friction. Rough surfaces, such as sandpaper, usually produce a lot of friction. Smooth surfaces, such as ice, usually produce less friction.

People use slippery things to reduce friction. Oil is often put on moving parts of machines to reduce friction. People use rough or sticky things to increase friction. The brakes on a bike use rubber pads to increase friction. When you squeeze the brake handles, the brake pads press against the rim of the wheel. The friction between the pads and rim cause the bike to stop.

Quick Check

Cause and Effect What happens when you squeeze a hand brake on a bicycle?

Critical Thinking How can you tell that friction is a force?

Lesson Review

Visual Summary

A force is a push or a pull. Forces can change the motion of objects.

Contact, magnetism, and gravity are different types of forces.

Friction is a force that occurs when one object rubs against another.

Make a FOLDABLES® Study Guide

Make a trifold book. Use it to summarize what you learned about forces.

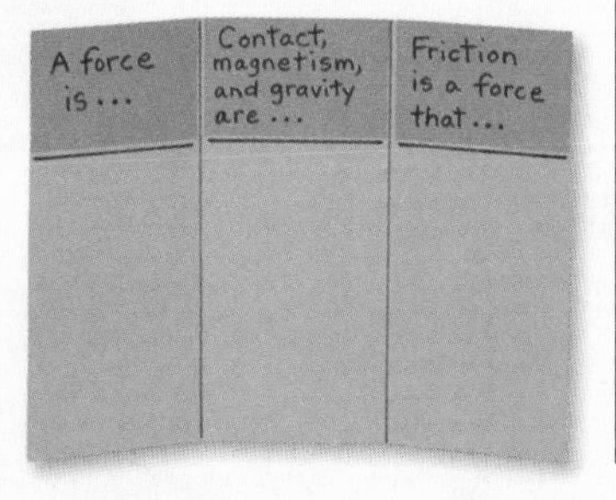

Think, Talk, and Write

1. **Vocabulary** What is friction? Define it and give an example.

2. **Cause and Effect** You are swinging on the playground. What force causes you to slow down as you go up?

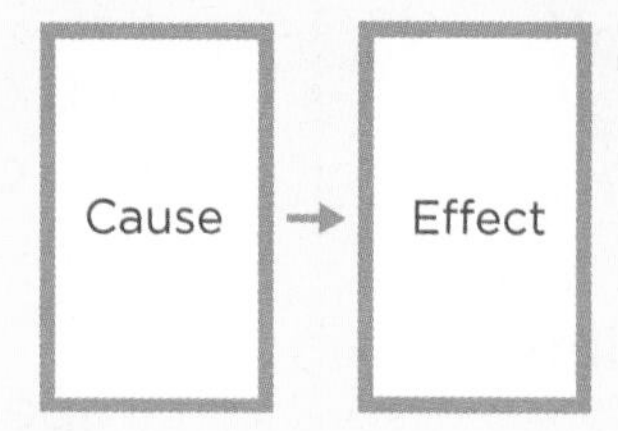

3. **Critical Thinking** How can friction help keep you safe?

4. **Test Prep** **Which is an example of a contact force?**
   - **A** a magnet attracting a paper clip
   - **B** two magnets repelling each other
   - **C** a bat hitting a ball
   - **D** gravity pulling on a leaf

5. **Essential Question** How do forces change motion?

Math Link

Order Numbers
Weigh five objects on a spring scale. Measure their weight in newtons, the unit of force in the metric system. Organize your data in a bar graph from least weight to greatest weight.

Health Link

Use Your Muscles
You use your muscles when you push or pull things. Find out about some of the muscles in your body. What do your muscles do? How do your muscles help you move?

Be a Scientist

Materials

magnet

paper clips

ruler

Structured Inquiry

How does distance affect the pull of a magnet on metal objects?

Form a Hypothesis

You know that some metal objects, such as paper clips, are attracted to magnets. What happens when you change the distance between a magnet and a pile of paper clips? How does this affect the magnet's pull on the paper clips? Write your answer in the form "If you move a magnet closer to a pile of paper clips, then..."

Test Your Hypothesis

1. Gather a pile of paper clips on your desk. Stand up a ruler near the paper clips.
2. **Experiment** Hold a magnet as shown below. Slowly lower the magnet until it is only 1 centimeter above the pile.

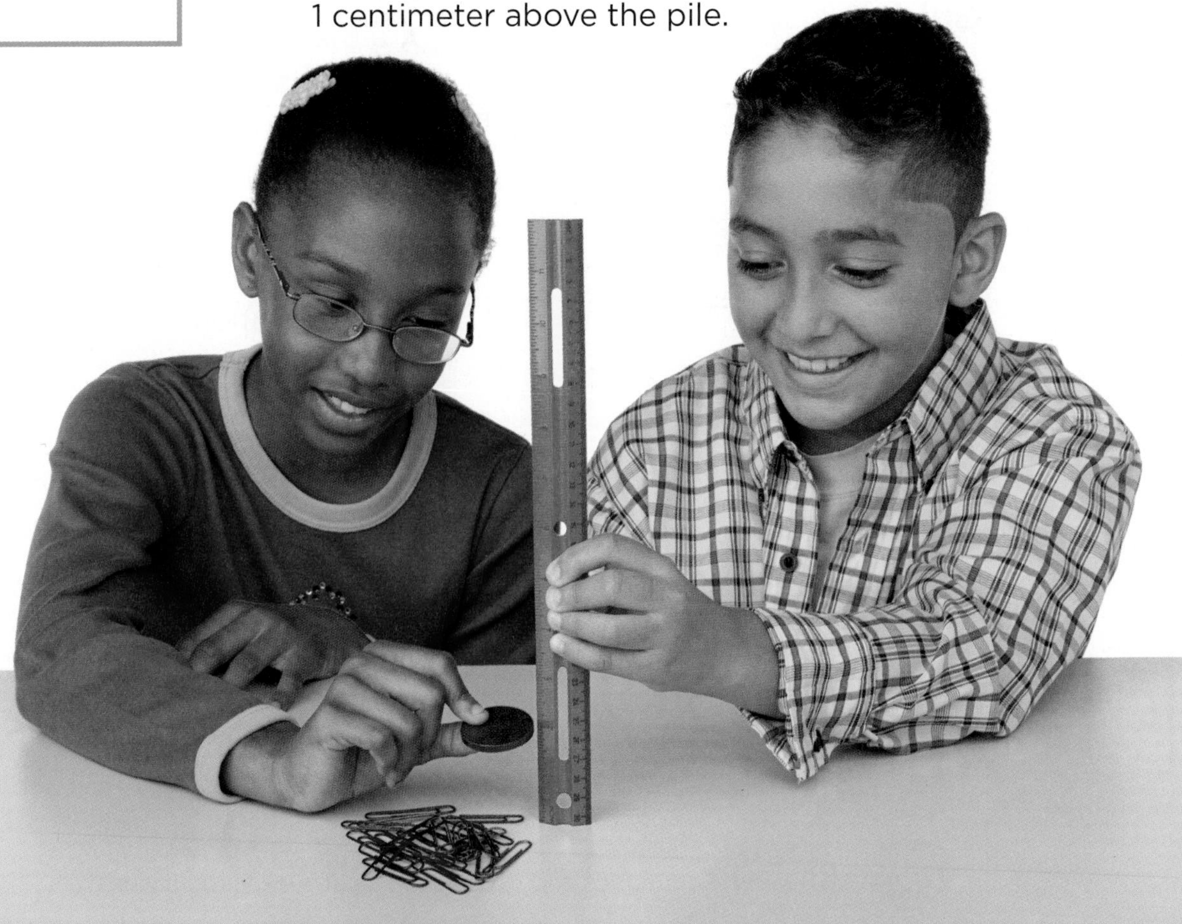

3. **Measure** Move the magnet away from the pile. Remove the paper clips and count how many stuck to the magnet. Record the data in a table.
4. Repeat steps 1–3, holding the magnet 2 cm and 3 cm away from the pile of paper clips. Record your data.

Step 3

Distance	Number of Paper Clips
1 cm	
2 cm	
3 cm	

Draw Conclusions

5. **Use Numbers** At what distance did the magnet pick up the most paper clips?
6. **Interpret Data** Does a magnet's pull on objects get greater or smaller as the magnet moves away from the objects?

Guided Inquiry

Can magnetic force pass through an object?

Form a Hypothesis

Can a magnetic force pass through different objects, such as wood, plastic, paper, or foil? Write a hypothesis.

Test Your Hypothesis

Design a plan to test your hypothesis. List the materials you will use. Write down the steps you plan to follow.

Draw Conclusions

Did any of the objects block magnetic force? Did any of the objects make the magnetic force stronger or weaker? Share your results with your classmates.

Open Inquiry

What other questions do you have about magnets? For example, what common objects are attracted to magnets? Design an experiment to find out.

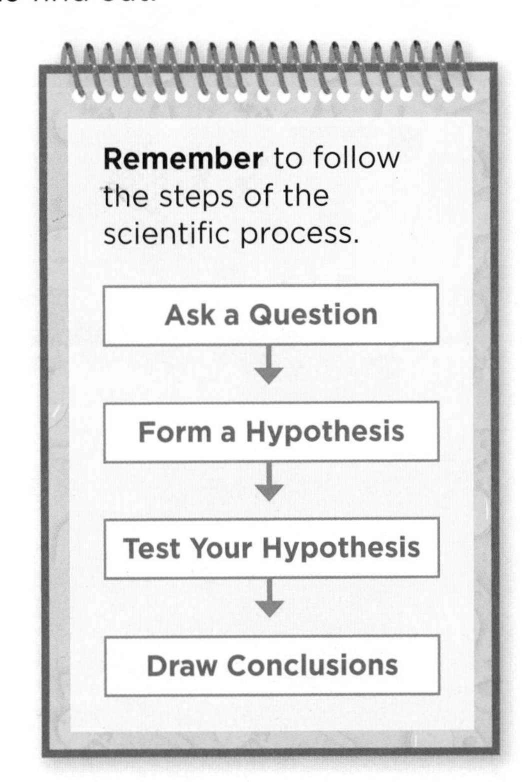

Lesson 3

Work and Energy

Look and Wonder

This tugboat is pulling a large container ship to the dock. Is this tugboat doing work? Why or why not?

Explore

Inquiry Activity

What is work?

book

chair

Make a Prediction

How do you know when work is being done? Make a prediction.

Test Your Prediction

1. Make a table like the one shown below. Perform each action listed in the table.

Actions	Is It Work?	Why or Why Not?
pick up a book		
think about a problem		
slide a chair		
press feet against floor		
push against wall		

2. **Classify** Decide whether each action was work. Ask yourself whether you got something done.

Draw Conclusions

3. **Communicate** Explain why you classified each action the way you did. Record this information in the table.
4. **Infer** What do you think work is?

Explore More

Experiment Perform other actions at home. Try to find actions where you do different amounts of work.

Read and Learn

Essential Question
How do we do work?

Vocabulary
work, p. 454
energy, p. 456
kinetic energy, p. 456
potential energy, p. 456

Reading Skill 
Summarize

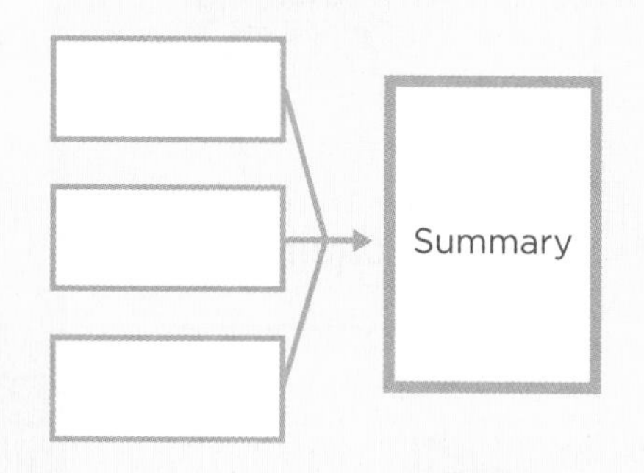

Technology LOG ON
e-Glossary and e-Review online at www.macmillanmh.com

What is work?

Do you know what work is? You might say that you do work every day at school. In science work has a special meaning. **Work** is done when a force moves an object or changes an object's motion. This means that picking up a book is work. A force changes the book's motion. Work is done when a book falls to the floor. Gravity changes the book's motion. Gravity does the work. Pushing on a wall is not work. No matter how hard you push, the wall does not move.

Work can be easy or hard. Picking up a small pebble is work. Lifting a large boulder is too. In both examples a force is used to move an object.

Quick Check

Summarize How can you tell whether an action is work?

Critical Thinking Can play be work? Why or why not?

When you paint at an easel, you are doing work. Your hand moves the brush.

Doing Work

Read a Photo

How is work being done in each picture?

Clue: Look for a force and a change in an object's motion in each picture.

What is energy?

Work can not be done without energy (E•nur•jee). **Energy** is the ability to do work. Energy is what makes motion possible. An object needs energy to move. When you do work on an object, you give it that energy.

Kinds of Energy

When you throw a paper airplane, you do work. You give the airplane energy. The airplane starts to move. Energy of motion is called **kinetic energy** (kuh•NE•tik E•nur•jee). All moving objects—roller coasters, cars, even people—have kinetic energy.

When you pull a sled to the top of a hill, you do work. You give the sled potential energy (puh•TEN•chul E•nur•jee). **Potential energy** is stored energy that is ready to be used. As the sled moves down the hill, its potential energy changes into kinetic energy.

How do you get energy to move, live, and grow? You get most of your energy from food. Food is a source of stored energy. Gasoline, wood, and food all have stored energy that is ready to be used.

Quick Check

Summarize What can energy do?

Critical Thinking When does a roller coaster have the most potential energy?

Quick Lab

Using Energy

1. You get energy to move and play from the foods you eat. Food is a source of stored energy. The table below shows how much stored energy is in some of the foods we eat.

Food	Calories of Energy
1 cup of apple juice	120
1 slice of wheat bread	75
1 slice of turkey	30
1 slice of cheese	60
1 cup of lettuce	7

2. **Use Numbers** Use the table to plan a meal. How many calories are in your meal?

3. **Use Numbers** Choose an activity from the table below. How long can you do that activity before you have used up all the stored energy from your meal?

4. **Use Numbers** Choose another activity and repeat step 3. Which activity uses the most energy?

Activity	Calories Used in 30 Minutes
biking (slow)	100
jogging	160
listening to music	17

Energy from the ball makes the pins move.

How can energy change?

Energy can move from one object to another. When you roll a bowling ball, you transfer energy from your body to the ball. When the ball hits the pins, it transfers energy to the pins. The pins move.

Energy can also change form. Rub your hands together. What do you notice? They get warmer. Your moving hands have energy. As friction slows your hands down, some of that energy is changed into heat.

Quick Check

Summarize How can energy change?

Critical Thinking Why does a bowling ball slow down when it hits a pin?

FACT Energy does not get used up. It changes form.

Lesson Review

Visual Summary

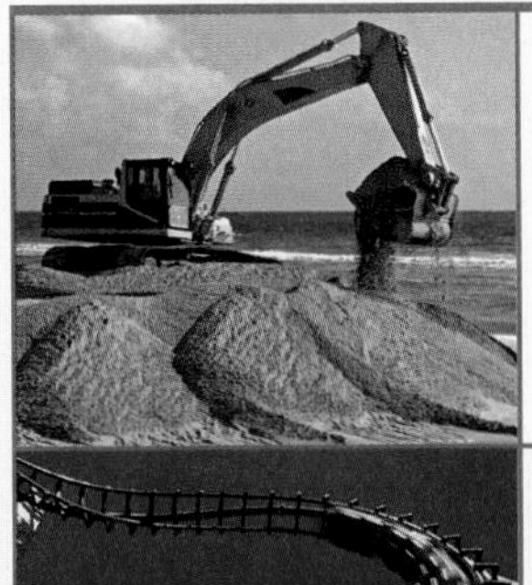

Work is done when a force moves an object or changes an object's motion.

Energy is the ability to do work.

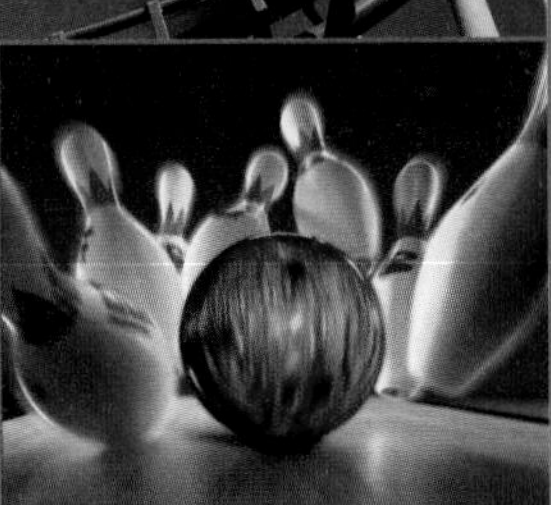

Energy can move from one object to another. Energy can also change form.

Make a FOLDABLES® Study Guide

Make a three-tab book. Use it to summarize what you learned about work and energy.

Think, Talk, and Write

1. **Vocabulary** What is work? Give two examples.

2. **Summarize** A soccer ball is at your feet. You kick the ball, and it travels across the field. Use the terms *work* and *energy* to describe what happens.

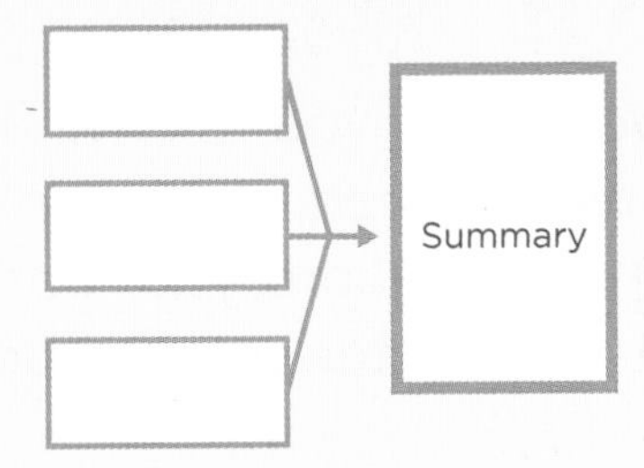

3. **Critical Thinking** How is an apple like gasoline in a car?

4. **Test Prep Which is an example of work being done?**
   - **A** studying for a test
   - **B** picking up a feather
   - **C** holding a heavy box over your head
   - **D** pushing on a wall

5. **Essential Question** How do we do work?

Writing Link

Explanatory Writing

A rock on a hill has potential energy. What happens to this energy as the rock rolls down the hill? Write about it. Then make a drawing to illustrate your writing.

Art Link

Make a Collage

Cut pictures from magazines of objects with kinetic energy. Paste your pictures onto a poster. Write how you think each object got its energy.

LOG ON e-Review Summaries and quizzes online at www.macmillanmh.com

Focus on Skills

Inquiry Skill: Infer

When you do an experiment, you are trying to answer a question. Sometimes you can answer a question from the data you collect. Other times, you must **infer** the answer using facts you know.

▶ Learn It

When you **infer**, you form an idea based on observations and facts. As you make observations, it is important to record your data. The more data you collect, the better you will be able to **infer.**

▶ Try It

Can running water do work? To answer this question, make a water wheel. Then observe what happens to it under running water. Use your observations and what you know about work to **infer** whether water can do work.

Materials **plastic plate, ruler, scissors, pencil, thread, paper clip, tape, faucet**

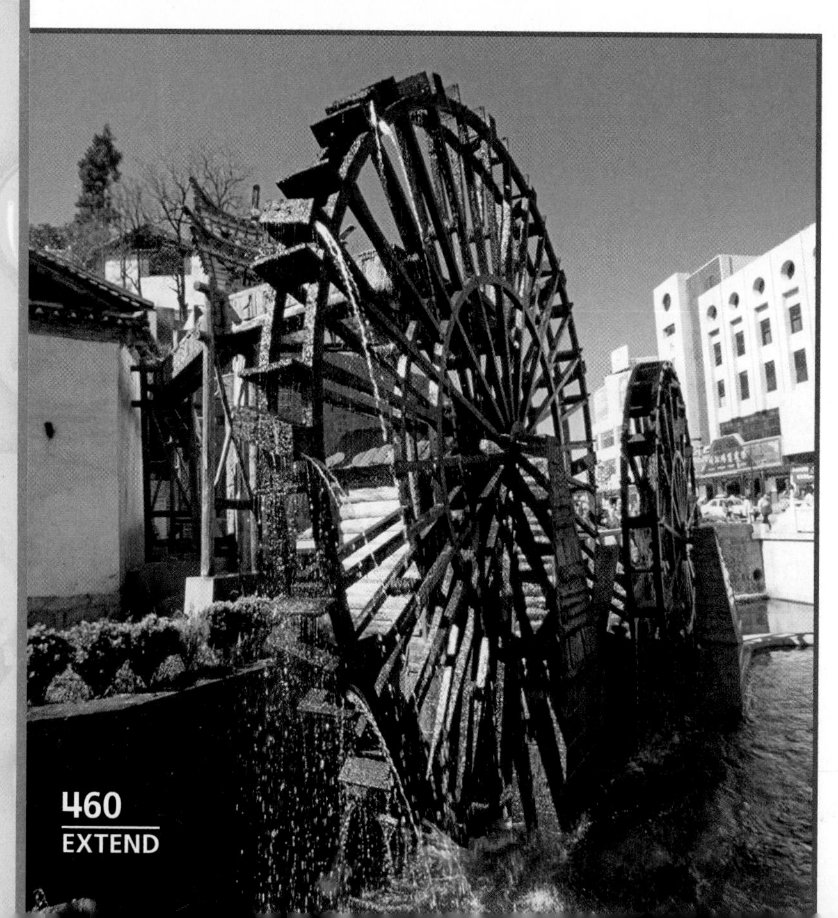

◀ **A water wheel is a machine that uses the energy of moving water to power mills and factories.**

1. Cut four 3-cm slits into a plastic plate. Then bend the slits to create a pinwheel.
2. Gently push a pencil through the center of the plastic plate. △ **Be Careful.** Point the pencil away from your body. Ask an adult for help.
3. Tie one end of a piece of thread to a paper clip. Tape the other end to the pencil, near the hole in the plate.
4. Turn on the faucet until a little water flows out.
5. Rest the pencil across the palms of your hands. Then hold the edge of the plate 2 cm under the water. Record your observations.
6. Repeat with a larger stream of water. Record what you observe.

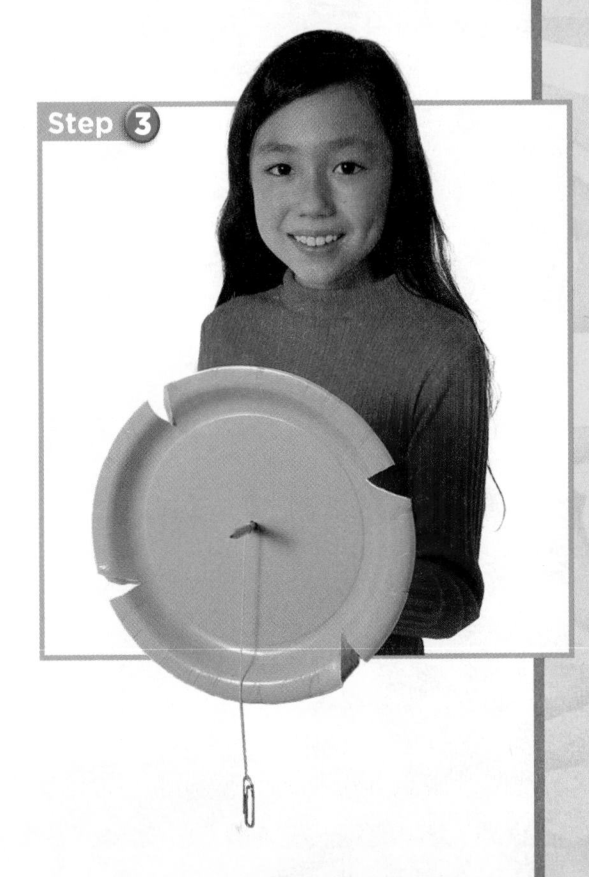
Step 3

Now use observations and facts you know to answer these questions.

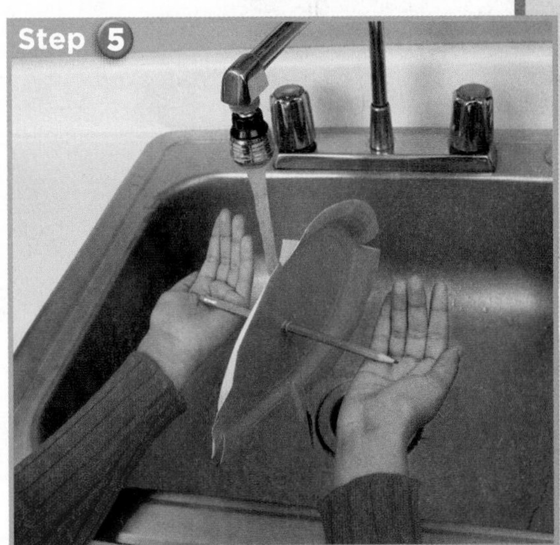
Step 5

- What makes the water wheel move?
- Does using more water give the wheel more energy? How can you tell?
- Can running water do work? Explain your answer.

Apply It

You have learned to **infer** the answer to a question from the data you collect and the facts you know. Now you can **infer** answers to new questions. For example, can wind do work? How might you use your wheel to **infer** the answer?

Lesson 4

Using Simple Machines

Look and Wonder

Machines make our lives easier. How can this wheelbarrow make harvesting a garden easier?

Explore

Inquiry Activity

How can a simple machine help you lift objects?

Form a Hypothesis

Look at the photos of steps 2 and 4. Will moving the ruler's position on the marker change the amount of force needed to lift two blocks? Write a hypothesis.

Test Your Hypothesis

1. Use some clay to stick a marker to the center of a ruler. Then use clay to stick a small cup to each end of the ruler as shown below.
2. **Experiment** Put two large blocks in one cup. Add gram cubes to the other cup. How many cubes does it take to lift the two large blocks?
3. **Use Variables** Change the position of the marker. Move it closer to one end of the ruler.
4. **Experiment** Repeat step 2. How does the marker's new position change your results?

Draw Conclusions

5. **Communicate** How does this simple machine lift objects?
6. **Interpret Data** How does the position of the marker change the number of gram cubes you need to lift the two large blocks?

Explore More

Experiment When are the two blocks lifted higher in the air—when the marker is near the two large blocks or when it is near the gram cubes? Try it to find out.

Read and Learn

Essential Question
How can a simple machine reduce force?

Vocabulary

simple machine, p. 465
lever, p. 466
pulley, p. 467
wheel and axle, p. 467
inclined plane, p. 468
screw, p. 468
wedge, p. 469
compound machine, p. 470

Reading Skill
Problem and Solution

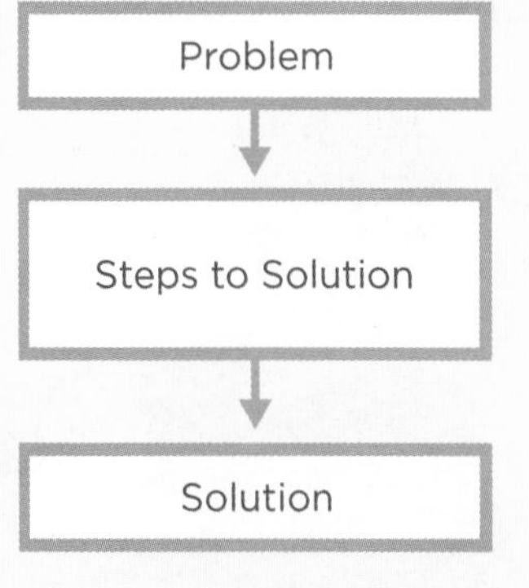

Technology LOG ON
e-Glossary, e-Review, and animations online at www.macmillanmh.com

What are machines?

You use machines every day. You might use a machine to travel to school. You might use a machine to sharpen your pencil. How would you describe a machine? A *machine* is something that makes work easier to do. Machines do not change the amount of work done. They simply change the way you do the work. For example, it is easier to lift and carry a heavy rock with a wheelbarrow than with your hands.

Some machines help you use less force to do work. Other machines change the direction in which you push or pull.

How is this backhoe making work easier? ▼

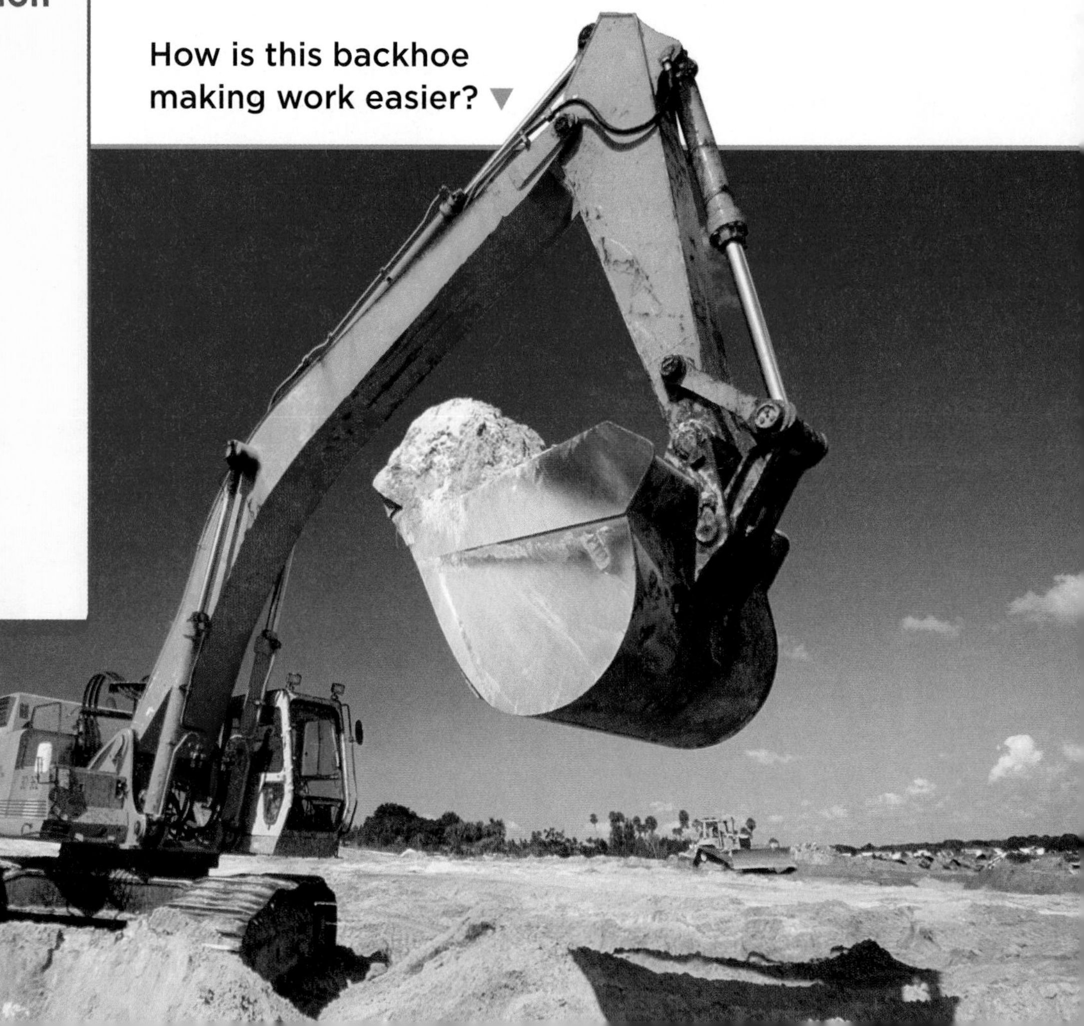

Machines such as cars, giant cranes, and bulldozers have many parts. They help us do a lot of work in a short amount of time. Yet the first machines were simple ones. **Simple machines** are machines with few or no moving parts. There are six types of simple machines. They are the lever, the pulley, the wheel and axle, the inclined plane, the screw, and the wedge.

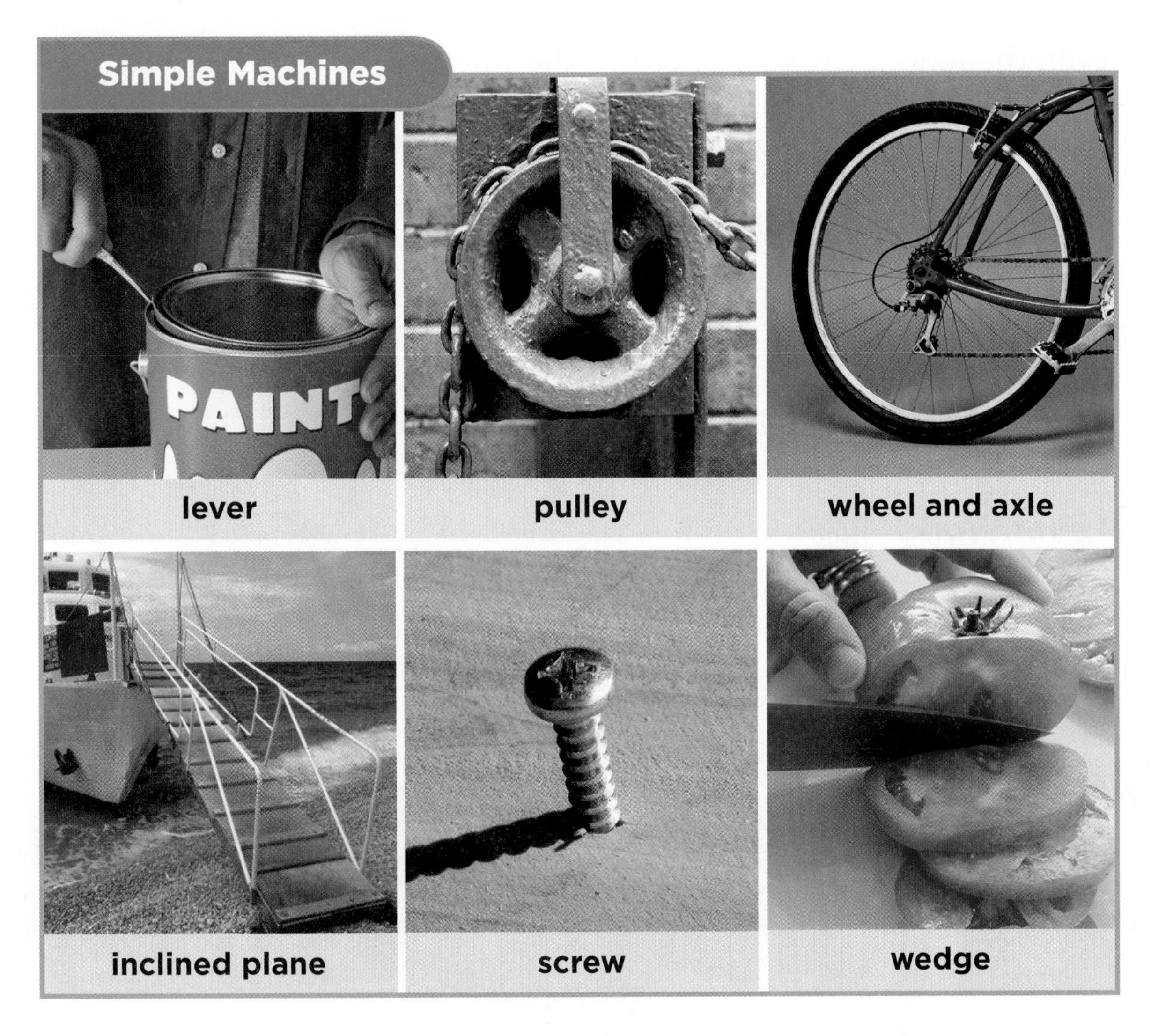

Problem and Solution How do machines help people solve problems?

Critical Thinking What are three simple machines you use every day?

What are levers?

How are a wheelbarrow and a seesaw alike? They are both levers (LE•vurz). A **lever** is a straight bar that moves on a fixed point. The fixed point is the *fulcrum*.

A lever can be used to lift something. The object lifted is called the *load*. In the diagram below, the girl is the load. When the boy presses down on one end of the lever, the load is lifted. The closer the fulcrum is to the load, the less force you need to lift the load.

Levers can make it easier for people to lift objects. They can change how much force you need to move something. They can also change the direction of the force you use. Pressing down on a lever lifts up the load.

How a Lever Works

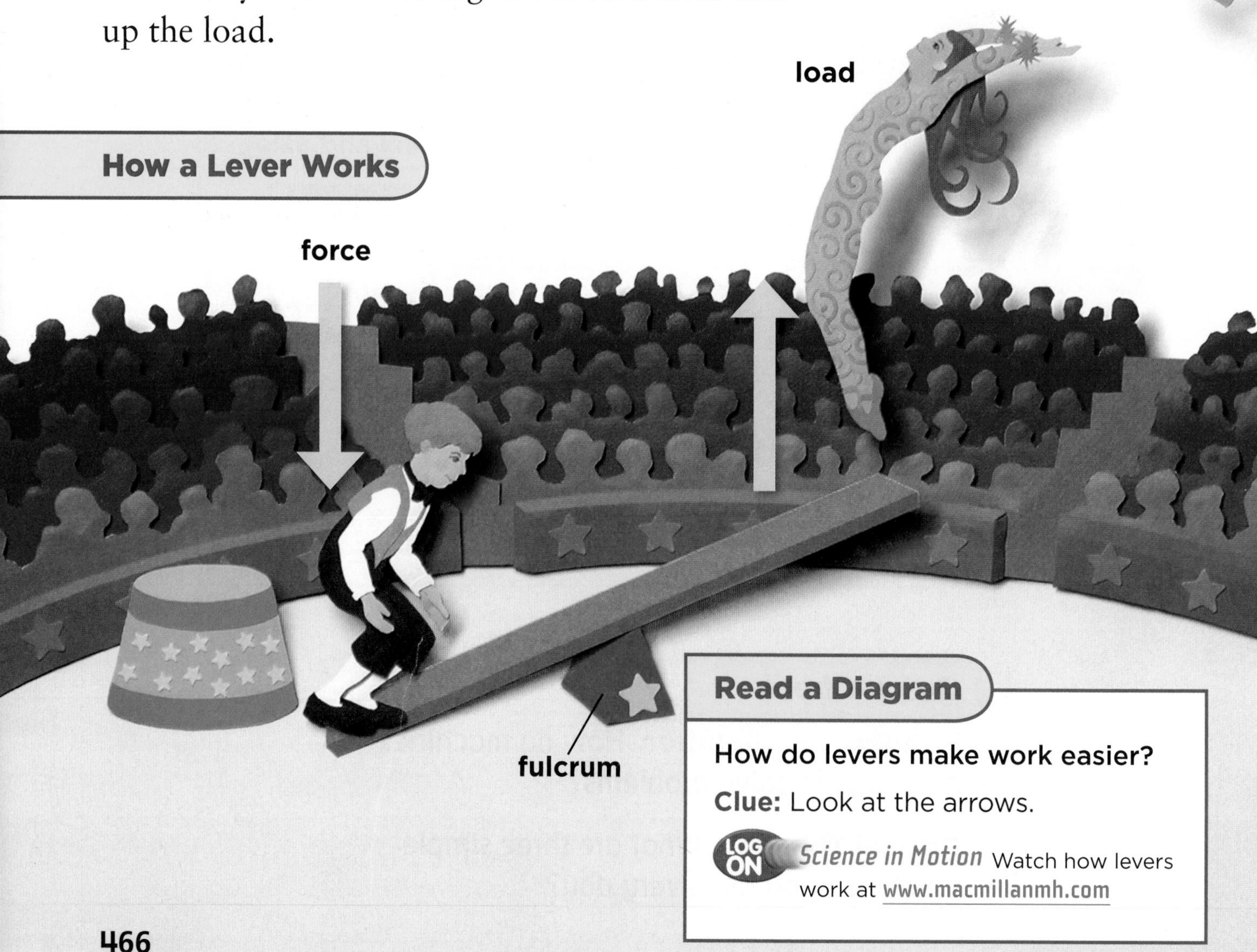

Read a Diagram

How do levers make work easier?

Clue: Look at the arrows.

LOG ON *Science in Motion* Watch how levers work at www.macmillanmh.com

Pulley

A **pulley** (PU•lee) is a special kind of lever. It uses a rope and a wheel to lift an object. When you pull down on one end of the rope, the other end rises up. The pulley shown here makes work easier by changing the direction of the force you use to lift an object.

Wheel and Axle

A **wheel and axle** (AK•sul) is another special lever made up of a wheel that moves around a post. The post is called an axle. Doorknobs and Ferris wheels are wheel and axles.

A wheel and axle can make work easier to do. Try opening a door by turning the knob. Now try it by turning the thin bar behind the knob. Which requires less force? Turning a wheel requires less force than turning an axle.

Quick Check

Problem and Solution How could you move a heavy rock?

Critical Thinking Which simple machine could you use to raise a flag?

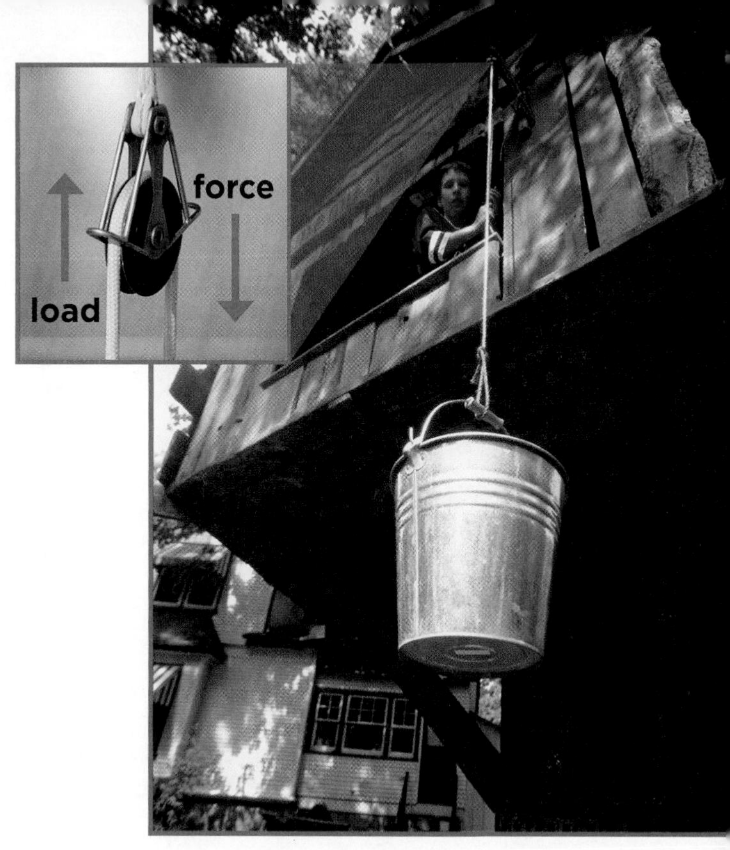

▲ A pulley makes it easier to lift this bucket.

The axle makes a smaller movement. The wheel makes a larger movement. ▼

It takes less force to push a box up a ramp than to lift it straight up.

What are inclined planes?

You have probably seen ramps in buildings such as your school. A ramp is an inclined plane (in•KLINED PLAYN). An **inclined plane** is a simple machine with a flat, slanted surface.

Inclined planes can make work easier to do. They reduce the force you need to move an object. Think about moving a heavy box onto a truck. You could not lift it off the ground to put it in the truck. You could slide it up an inclined plane instead. Sliding a box up an inclined plane requires less force than lifting the box straight up. However, you must push the box a longer distance.

Screw

A **screw** is an inclined plane wrapped into a spiral. It takes less force to turn a screw than to pound a nail. A screw changes a turning force into a downward force.

This machine is called an auger. An auger is a giant screw.

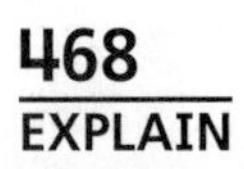

Wedge

If you put two inclined planes back to back, you get a wedge (WEJ). A **wedge** is a simple machine that splits objects apart.

The head of an ax is a wedge. When you swing an ax, the downward force is changed into a sideways force. The sideways force pushes, or splits, the wood apart.

Most cutting tools, such as knives, are wedges. As you press a knife into food, the knife pushes the food apart.

The downward force of the ax changes to the sideways force that splits the log. ▼

Quick Lab

Inclined Planes

1. Make an inclined plane as shown below. Then tie a bag of 25 marbles to a spring scale.
2. **Measure** Lift the spring scale straight up so it is even with the height of the books. Record what the spring scale reads. Measure and record the distance you pulled the marbles.
3. **Measure** Use the spring scale to pull the marbles up the ramp. Record what the spring scale reads. Measure and record the distance you pulled the marbles.

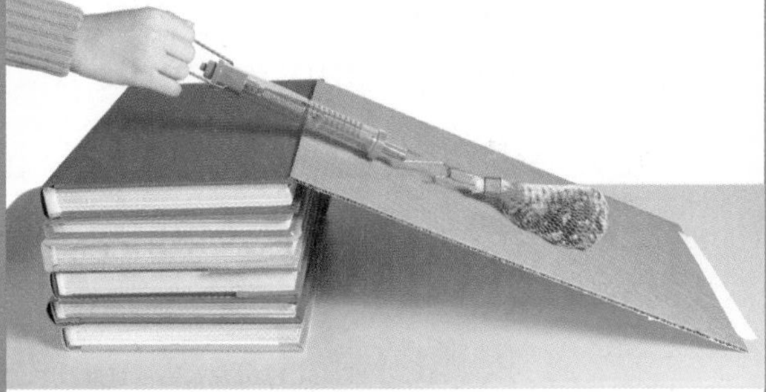

4. **Interpret Data** Which method of moving the marbles required more force? Which method required moving the marbles a greater distance?

Quick Check

Problem and Solution Which simple machine would you use to cut a banana?

Critical Thinking Where have you seen ramps used in your community?

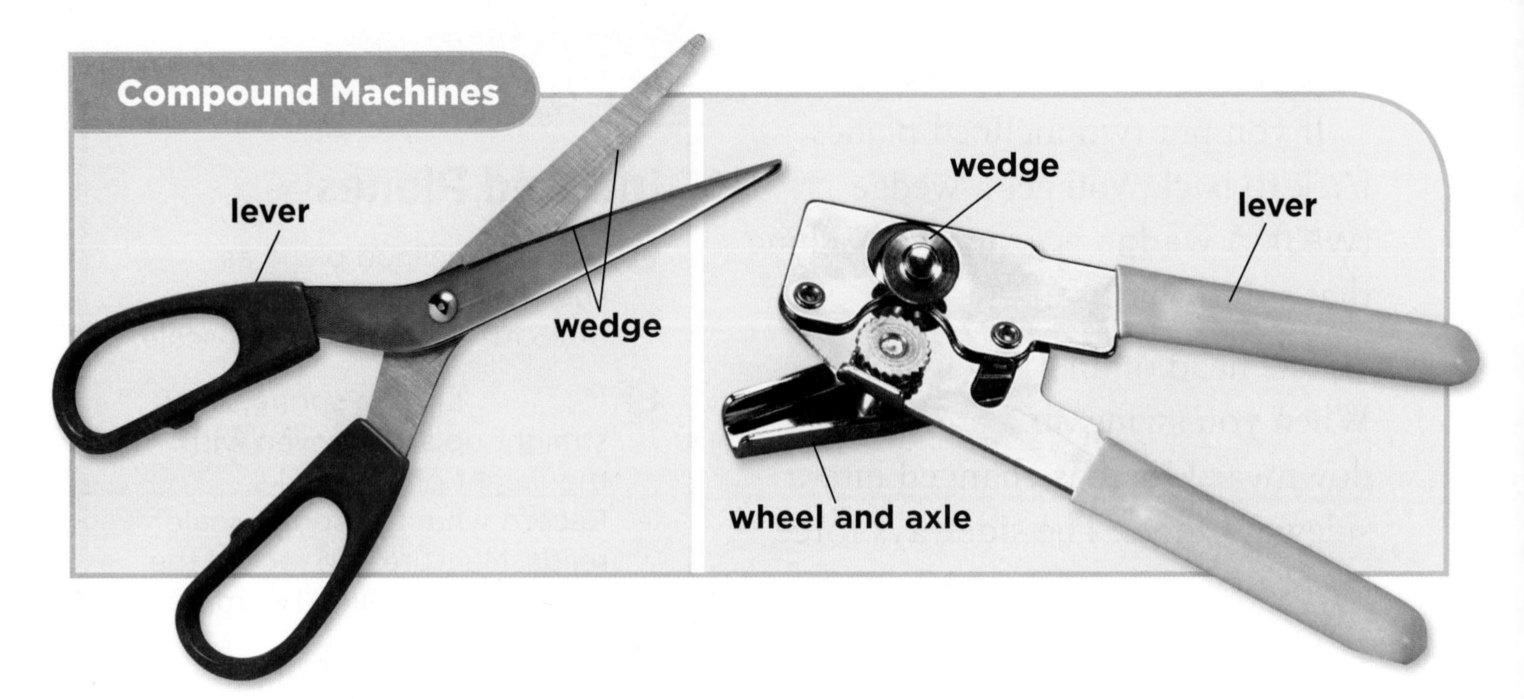

How do machines work together?

Most of the tools you use every day are compound machines. A **compound machine** is two or more simple machines put together.

A pair of scissors is a compound machine. Two wedges and two levers make an excellent cutting tool. The point where they are connected is the fulcrum. When the handles are pushed together, the edges cut through material.

A can opener is also a compound machine. It contains a wedge, a lever, and a wheel and axle acting as one machine.

Quick Check

Problem and Solution What do you get if you put two or more simple machines together?

Critical Thinking Which simple machines are part of a bicycle?

Lesson Review

Visual Summary

	**A machine is** something that helps make work easier to do.
	The lever, wheel and axle, pulley, inclined plane, screw, and wedge are **types of simple machines.**
	**A compound machine is** made of two or more simple machines.

Make a FOLDABLES® Study Guide

Make a trifold book. Use it to summarize what you learned about machines.

Main Idea	What I learned...	Sketches/ Examples
A machine is...		
Types of simple machines...		
A compound machine is...		

Think, Talk, and Write

1. **Vocabulary** What is a simple machine? Describe one.
2. **Problem and Solution** Suppose you were going to build a pyramid that was 10 meters high. How might you build it? What simple machines might you use?

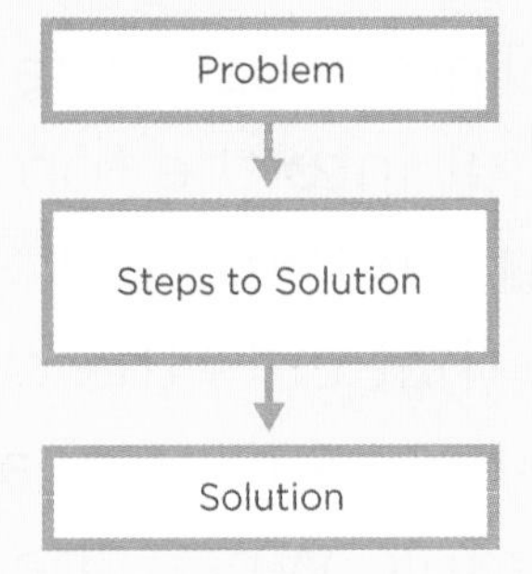

3. **Critical Thinking** How does a woodpecker use its beak as a simple machine?
4. **Test Prep** **Which of the following is a compound machine?**
   A lever
   B inclined plane
   C scissors
   D wheel and axle
5. **Essential Question** How can a simple machine reduce force?

Writing Link

Write a Report

Make a list of the simple machines you find in your neighborhood. For example, you might see a ramp at the library or a seesaw at the playground. Write about how these machines are used.

Social Studies Link

Do Research

The giant pyramids in Egypt were built many years ago. There were no bulldozers, trucks, or tractors then. Use research materials to learn more about how people think these pyramids might have been built.

LOG ON e-Review Summaries and quizzes online at www.macmillanmh.com

A Very Useful Machine

A can opener is a compound machine. It makes opening cans easy. How does it work? First, you attach the cutting wheel to the can's lid. Then, you press the two long handles together. This causes the cutting wheel to cut into the top of the can. Next, you turn the crank. This moves the wheel that cuts the can. The wheel continues to turn as long as you turn the crank. When the top of the can comes off, you can let go of the handles and remove the can opener.

Explanatory Writing

A good explanation

- explains how something works or gives information about how to do something;
- gives details that are easy to follow;
- uses time-order words such as *first*, *next*, and *after.*

Three simple machines are found in a can opener. They are the wedge, the lever, and the wheel and axle.

Write About It

Explanatory Writing Choose another compound machine. Find out how it works. Then write a paragraph that explains how to use it.

e-Journal Write about it online at www.macmillanmh.com

Using Number Patterns

Levers can help people lift heavy loads. Suppose you know you need to use 60 newtons of force to lift a bag of soil with a wheelbarrow. You need two times as much force, 120 N, to lift 2 bags of soil. How much force would you need to lift 3 bags of soil with this wheelbarrow? Find the number patterns in the chart below to solve this problem.

How to Use Patterns

- A number pattern is a series of numbers that follows a rule. To find a pattern, look at how the numbers change.
- To find the next number in this pattern, add 60 to the previous number.

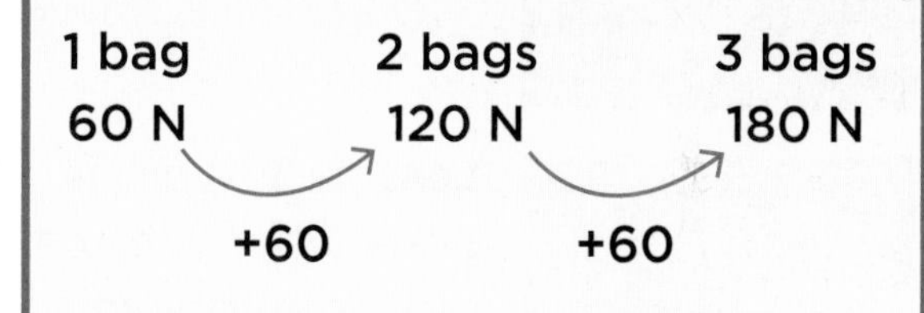

Bags of Soil	Force Needed (in newtons)
1	60
2	120
3	180
4	240

Solve It

How many bags of soil could you lift using this wheelbarrow if you apply 240 N of force? How much force would you need to lift 5 bags of soil? Explain how you found your answer.

CHAPTER 11 Review

Visual Summary

Lesson 1 An object is in motion when its position changes.

Lesson 2 Forces can change an object's motion.

Lesson 3 Work is done when a force moves an object. Energy is the ability to do work.

Lesson 4 Simple machines make work easier to do.

Make a FOLDABLES® Study Guide

Glue your lesson study guides to a piece of paper as shown. Use your study guide to review what you have learned in this chapter.

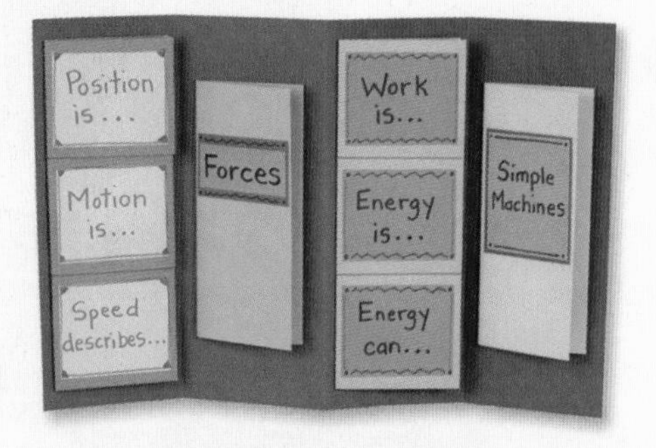

Vocabulary

DOK I

Fill each blank with the best term from the list.

compound machine, p. 470	**lever**, p. 466
energy, p. 456	**magnet**, p. 446
force, p. 444	**motion**, p. 436
friction, p. 448	**speed**, p. 438
inclined plane, p. 468	**wedge**, p. 469

1. An object in ________ is changing its position.

2. The ability to do work is called ________.

3. How quickly an object moves is described by its ________.

4. A ramp is an example of an ________.

5. You can use a ________ to attract things made of iron.

6. A straight bar that moves on a fixed point is a ________.

7. A push or pull is called a ________.

8. A knife acts as a ________ when cutting food.

9. You squeeze the hand brakes on a bike. The force that slows down the bike is ________.

10. A machine made up of two or more simple machines is a ________.

LOG ON e-Glossary Words and definitions online at www.macmillanmh.com

Skills and Concepts

DOK 2–3

Answer each of the following.

11. Problem and Solution A car has just traveled 100 km. What else do you need to know to figure out its average speed?

12. Explanatory Writing When does a roller coaster have the most potential energy? When does it have the most kinetic energy?

13. Infer Would you move faster down a waterslide or a regular slide? Explain your answer.

14. Critical Thinking Cars and people need energy. How is the way they get energy alike?

15. Critical Thinking Four people are pushing very hard on a cardboard box. However, the box is not moving. Explain how this can happen.

16. What simple machines are shown below? How are they the same? How are they different?

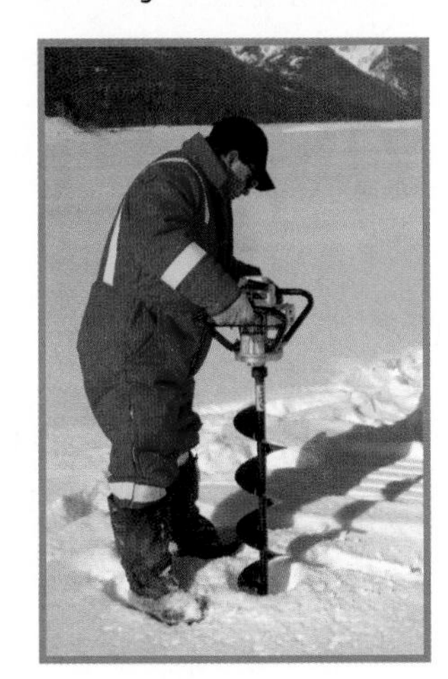

17. True or False *Work always takes a long time.* Is this statement true or false? Explain.

18. True or False *A lever moves back and forth on a fixed point called an axle.* Is this statement true or false? Explain.

19. Which two kinds of simple machines make up a pair of scissors?

A wheels and axles

B pulleys and wedges

C levers and pulleys

D levers and wedges

20. What makes something move?

Performance Assessment

DOK 3

Not-So-Simple Machines

- Machines help make work easier to do. Are there any tasks that you wish could be easier to do? Choose one such task and design an imaginary machine to help you do it.
- Write instructions that tell how the machine works. Draw a simple diagram of the machine.
- List some of the simple machines that make up your imaginary compound machine.

Test Preparation

1 **Which simple machine is used to turn a Ferris wheel?**

A pulley

B wheel and axle

C inclined plane

D wedge

DOK 1

2 **Which has the most potential energy?**

A roller-coaster car at the bottom of a hill

B roller-coaster car halfway up a hill

C roller-coaster car at the top of a hill

D roller-coaster car halfway down a hill

DOK 1

3 **A bowling pin falls over when it is hit by a bowling ball because the ball is transferring**

A heat.

B stored energy.

C energy of motion.

D physical properties.

DOK 1

4 **Which is an example of a contact force?**

A Two magnets repel each other.

B Gravity pulls a ball to the ground.

C A magnet attracts nails to it.

D Brakes stop the wheels of a bicycle.

DOK 2

5 **Which choice describes the position of the triangle in the picture below?**

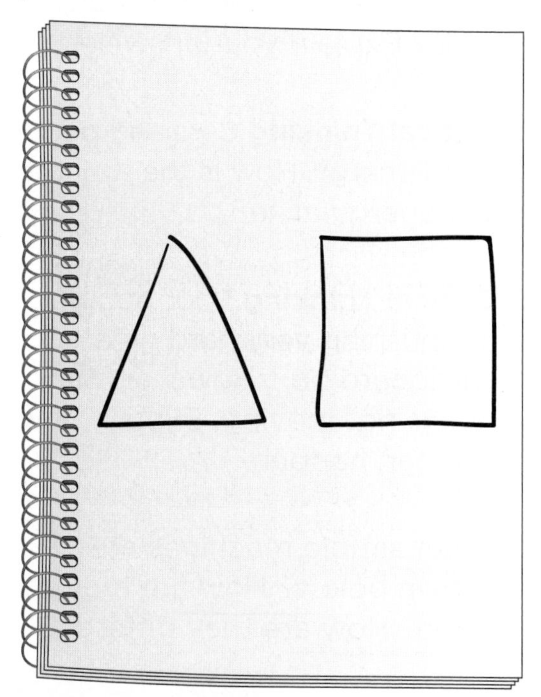

A The triangle is below the square.

B The triangle is above the square.

C The triangle is to the right of the square.

D The triangle is to the left of the square.

DOK 1

6 Which of these could most likely make an object move without touching it?

A 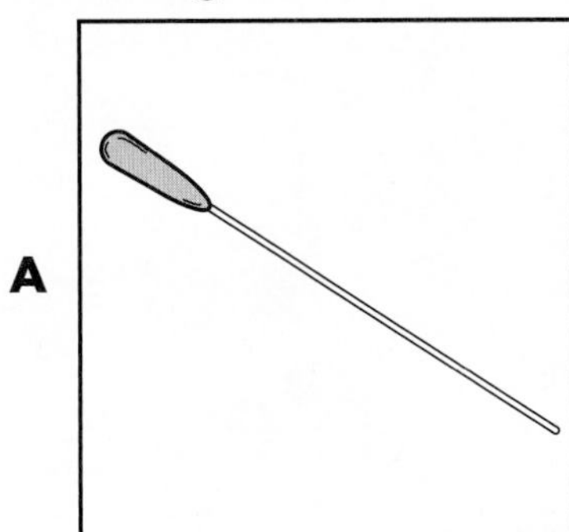

B 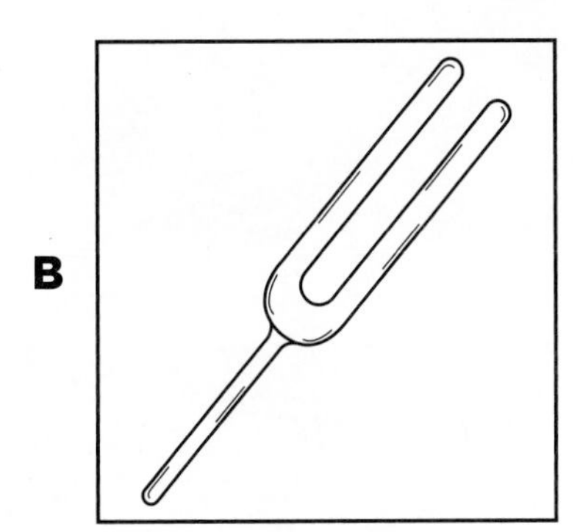

C 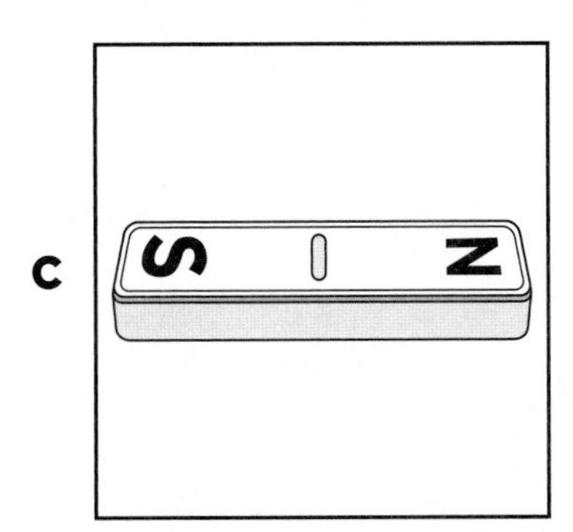

D 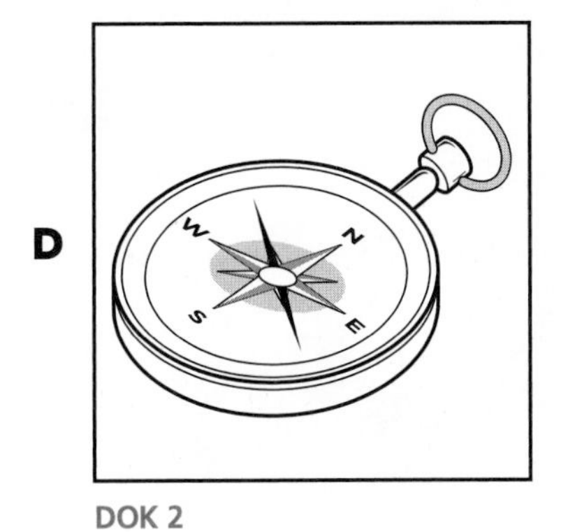

DOK 2

7 The diagram below shows forces acting on a ball. The larger the arrow is, the greater the force is.

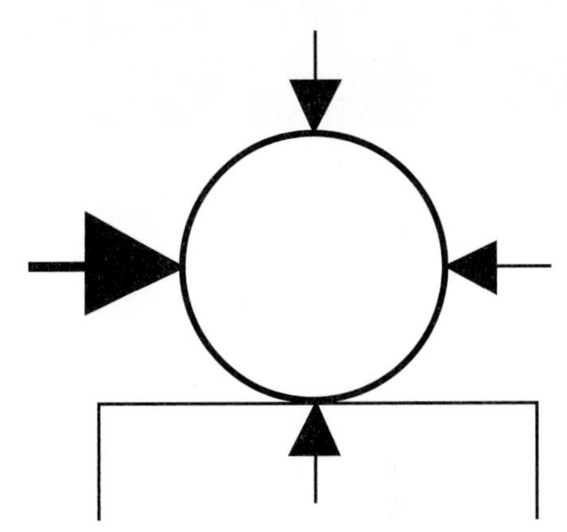

In which direction will the ball move?

A to the left

B to the right

C down

D up

DOK 1

8 Maureen knows the distance a car traveled. What does she need to know to calculate the car's speed?

DOK 1

Check Your Understanding

Question	Review	Question	Review
1	pp. 464–467	5	pp. 434–435
2	p. 456	6	pp. 446–447
3	p. 458	7	pp. 444–445
4	pp. 446–448	8	p. 438

CHAPTER 12

Forms of Energy

What are the main forms of energy, and how are they used?

Essential Questions

Lesson 1
How can you describe heat?

Lesson 2
What is sound?

Lesson 3
How does light allow you to see objects?

Lesson 4
How do you use electricity?

 Beijing, China

Big Idea Vocabulary

heat the flow of energy from a warmer object to a cooler object (p. 480)

temperature a measure of how hot or cold something is (p. 482)

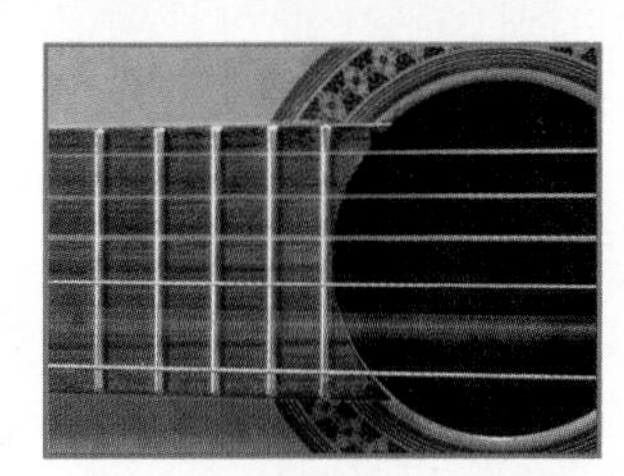

vibrate to move back and forth quickly (p. 490)

sound a form of energy that comes from objects that vibrate (p. 490)

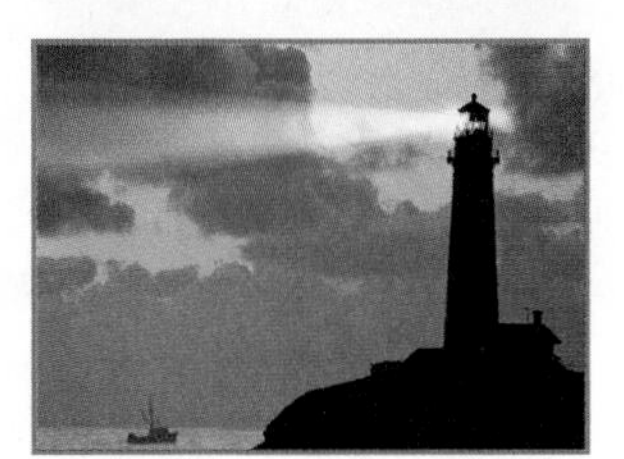

light a form of energy that allows you to see objects (p. 500)

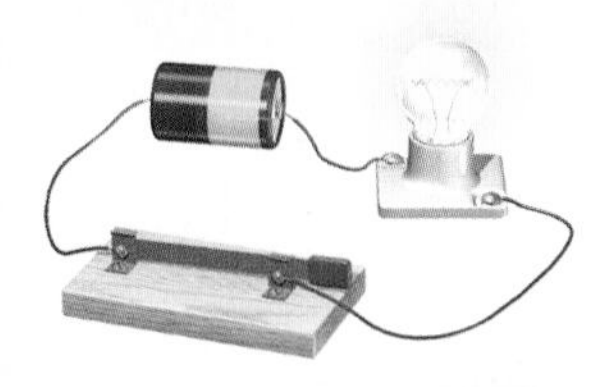

electric current a flow of charged particles (p. 514)

Visit www.macmillanmh.com for online resources.

Lesson 1

Heat

Look and Wonder

Hot air causes these balloons to rise into the sky. What happens to air as it is heated?

Explore

Inquiry Activity

What happens to air when it is heated?

Form a Hypothesis

How does heat affect air? Does it make air get bigger or smaller? Write a hypothesis.

Test Your Hypothesis

1. Use a dropper to place five drops of water along the edge of a bottle's opening. Place a plastic disk on top of the opening. Then put the bottle in a refrigerator for several hours.
2. **Predict** What will happen to the disk if the temperature of the air in the bottle increases?
3. **Observe** Remove the bottle from the refrigerator. Rub your hands together quickly. When your hands feel warm, place them on the bottle. Look at the disk.

Draw Conclusions

4. **Communicate** What happened to the disk? Was your prediction correct?
5. **Infer** Think about what happened to the disk. What happens to air when it is heated?

Explore More

Experiment Place an empty plastic bottle in the refrigerator for several hours. Remove the bottle from the refrigerator and immediately stretch a balloon over the opening. What happens to the balloon?

Materials

- dropper
- water
- empty plastic bottle
- plastic disk

Step 1

Step 3

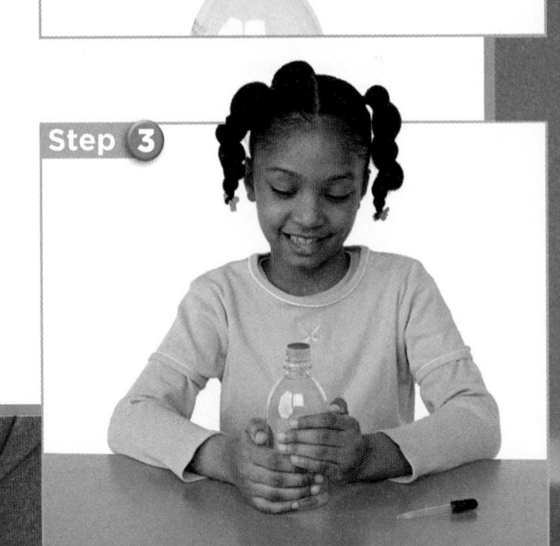

Read and Learn

Essential Question
How can you describe heat?

Vocabulary
heat, p. 480
thermal energy, p. 482
temperature, p. 482
thermometer, p. 483
conductor, p. 484
insulator, p. 484

Reading Skill
Main Idea and Details

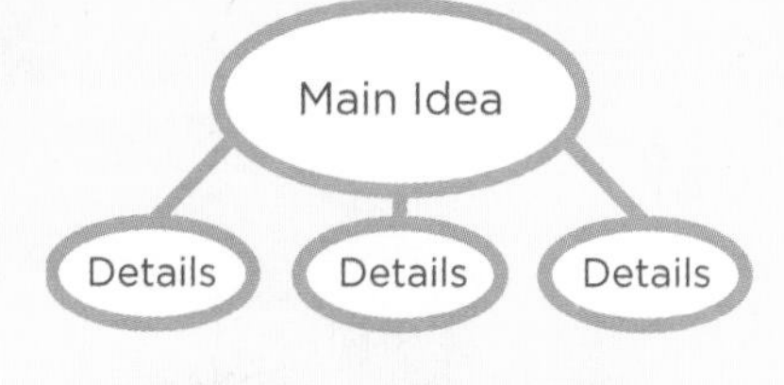

Technology LOG ON
e-Glossary and e-Review online at www.macmillanmh.com

The Sun's energy warms the air, land, and water.

What is heat?

Have you ever put your hands on a bowl of hot soup? What happened to your hands? Your hands got warm. Heat moved from the hot bowl to your cool hands. **Heat** is the flow of energy between objects. Heat can travel through solids, liquids, and gases. It can even travel through space. No matter what it travels through, heat always flows from a warmer object to a cooler one.

Sources of Heat

The Sun is Earth's main source of heat. A *source* is where something comes from. The Sun's heat warms the air, land, and water. Without the Sun's heat, it would be too cold on Earth for most living things to survive.

Quick Lab

Heating Water and Soil

1. **Predict** Which heats up faster, a cup of water or a cup of soil?
2. **Use Variables** Fill one cup with 150 milliliters of water. Fill another cup with 150 mL of soil.
3. **Measure** Put a thermometer in each cup. Measure the temperature of the water and soil. Record the data.
4. **Experiment** Put the cups in a warm place. Record the temperature in each after 15 minutes.
5. **Use Numbers** Find the difference between the first and last readings of each thermometer.
6. **Interpret Data** Which cup warmed up more? How do you know?

Fires, light bulbs, and stoves are some other sources of heat. Fires use chemical changes to produce heat. Light bulbs and some stoves use electricity to produce heat. Rubbing two objects together can also produce heat. This is why your hands get warm when you rub them together.

Heating Objects

Some objects heat up faster than others. For example, at the beach you will find sand and water. Both are warmed by the Sun. The sand gets very hot, but the water stays much cooler.

Quick Check

Main Idea and Details Describe how heat flows.

Critical Thinking What are some ways people use heat?

How does heat affect matter?

Remember that all matter is made of tiny particles. These particles are always moving. The energy that makes them move is called **thermal energy** (THUR•mul E•nur•jee). Heating matter increases how much thermal energy the particles have. A hot object, such as hot soup, has a lot of thermal energy. Its particles move quickly. A cold object, such as an ice cube, has much less thermal energy. Its particles move slowly.

Thermal energy is what makes objects feel hot or cold. In fact, when you measure an object's temperature (TEM•puh•ruh•chur), you are really measuring its thermal energy. **Temperature** is a measure of how hot or cold something is. It describes how much thermal energy an object has. The more thermal energy an object has, the higher its temperature will be.

hot

cold

Measuring Temperature

A thermometer helps you keep the water in this tank at a healthy temperature for the fish.

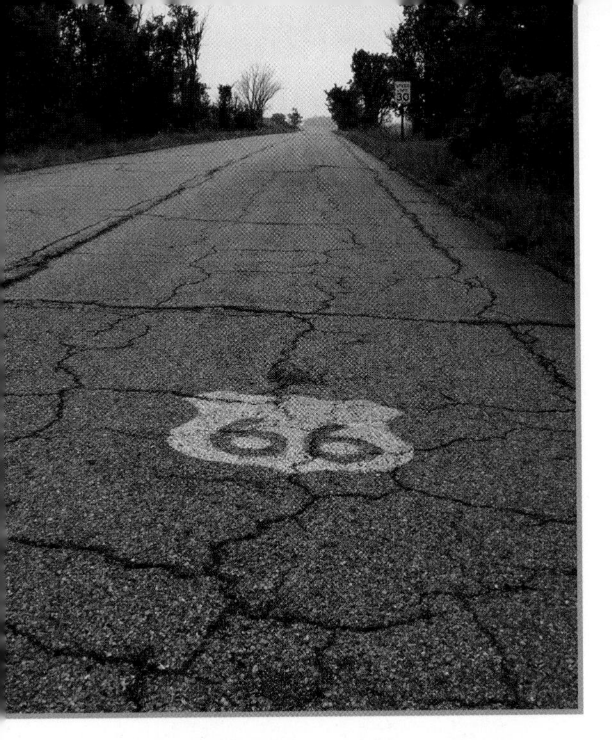

▲ Roads can crack as they expand and contract with changing temperatures.

Expanding and Contracting

When heat flows into an object, the object gains thermal energy. Its temperature increases. Its particles move faster and farther apart. The object gets bigger, or *expands*. When heat flows away from an object, the object loses thermal energy. Its temperature decreases. Its particles move more slowly. The object gets smaller, or *contracts*.

You can see matter expand or contract in a thermometer (thur•MAH•muh•tur). A **thermometer** is a tool used to measure temperature. Some thermometers are made up of a clear tube filled with a liquid. When the temperature of the liquid increases, the liquid expands. It rises and fills more of the tube. When the temperature of the liquid decreases, the liquid contracts. It fills less of the tube.

Changing State

Heat can cause matter to change state. Solids, such as ice cream, can melt when they are heated. Liquids, such as water, can evaporate when they are heated. They can freeze when heat flows from them.

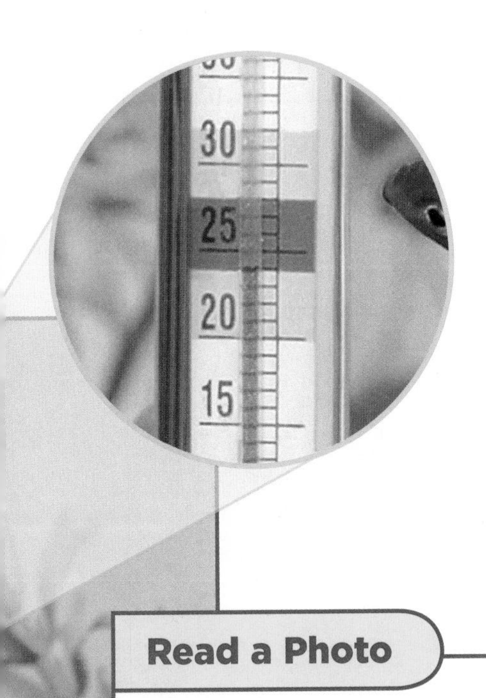

Read a Photo

What is the temperature shown on the thermometer? Give your answer in °C.

Clue: Line up the top of the red liquid with the black markings on the thermometer.

Quick Check

Main Idea and Details List some ways heat affects matter.

Critical Thinking What happens to an ice pop when it is out of the freezer?

How can you control the flow of heat?

Heat moves more easily through some materials than others. That is why pots are often made of metal. Heat moves easily through metals. Heat moves from the stove to the metal pot. The whole pot gets warm. Materials such as metals are good conductors (kun•DUK•turz). A **conductor** is a material that heat moves through easily.

When you are cold, you wrap yourself in a blanket to keep warm. A blanket is an insulator (IN•suh•lay•tur). An **insulator** is a material that heat does not move through easily. Wool, cotton, and fur are examples of insulators.

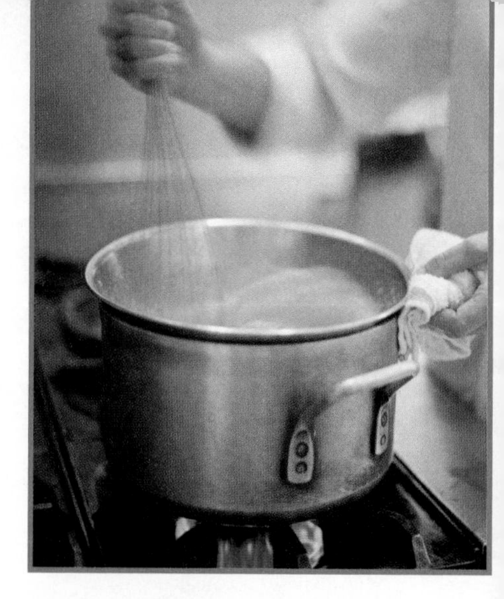

▲ Metal pots are conductors.

Quick Check

Main Idea and Details What is a conductor? What is an insulator? Give an example of each.

Critical Thinking Why do people use foam cups when drinking hot chocolate?

Snow can be an insulator. Heat can not flow easily through the walls of this igloo.

Lesson Review

Visual Summary

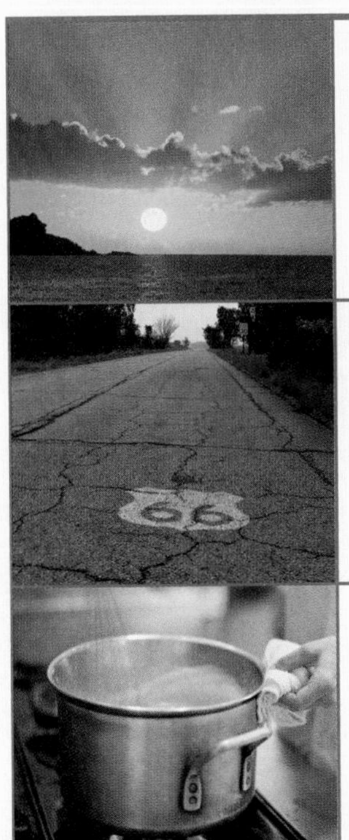

Heat is the flow of energy from a warmer object to a colder object.

Temperature is a measure of how hot or cold something is.

Heat moves easily through **conductors**. Heat does not move easily through **insulators.**

Make a FOLDABLES® Study Guide

Make a three-tab book. Use it to summarize what you learned about heat.

Think, Talk, and Write

1. **Vocabulary** How is temperature different from heat?

2. **Main Idea and Details** How does matter change when heat flows into it?

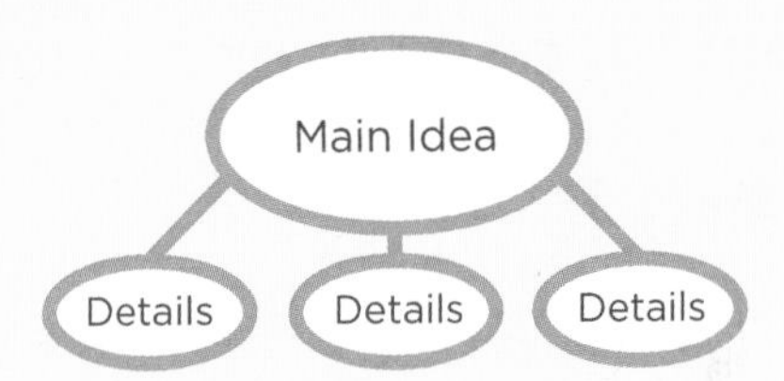

3. **Critical Thinking** Sometimes when people have a fever, they put a cold, wet cloth on their forehead. How does this cool them off?

4. **Test Prep** **Most of Earth's heat comes from**
   - **A** the Sun.
   - **B** water.
   - **C** batteries.
   - **D** electricity.

5. **Essential Question** How can you describe heat?

Math Link

Solve a Problem
Yesterday the temperature outside was 21°C. Overnight the temperature dropped 8 degrees. What is the new temperature?

Health Link

Do Research
Fires are a source of heat. However, fires can be dangerous. Find out ways to prevent fires. Make a pamphlet to share this information.

LOG ON e-Review Summaries and quizzes online at www.macmillanmh.com

Focus on Skills

Inquiry Skill: Experiment

You just learned about heat. You read that an insulator is a material that does not allow heat to pass through it easily. How can you find out whether something is an insulator? You can **experiment** to answer a question like this.

▶ Learn It

When you **experiment**, you perform tests to answer a question. You make observations and collect data. Then you interpret the data to answer a question. When you **experiment**, it is important to test only one thing at a time. This helps you know what caused your results.

▶ Try It

Experiment to find out which is the best insulator: paper, plastic, or foam.

Materials **paper cup, plastic cup, foam cup, 6 ice cubes, plastic wrap, 3 rubber bands**

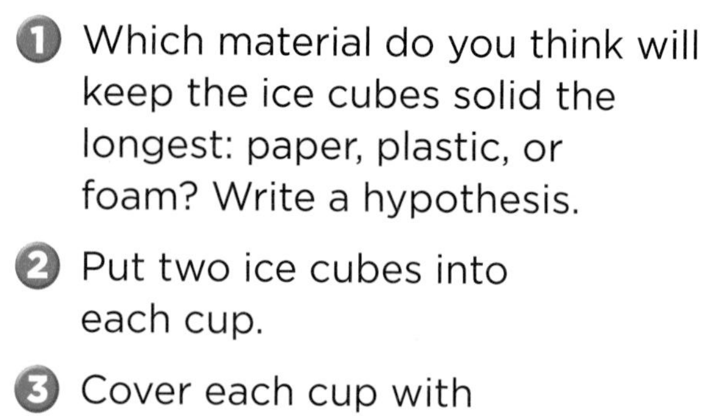

1. Which material do you think will keep the ice cubes solid the longest: paper, plastic, or foam? Write a hypothesis.
2. Put two ice cubes into each cup.
3. Cover each cup with plastic wrap. Use a rubber band to seal the wrap to the cup.

4. Place the cups in a warm place.
5. Observe the ice in the cups every ten minutes for one hour. Record the changes that you observe.

Now use your results to draw conclusions.

- In which cup did the ice cubes melt the slowest?
- Which cup is the best insulator?

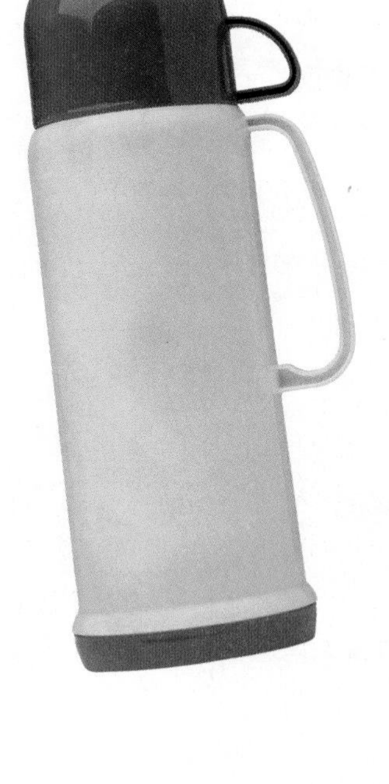

Apply It

Now **experiment** to find out which is the best conductor of heat: aluminum, plastic, or wax paper. Remember that a conductor is a material that lets heat pass through it easily.

Repeat this **experiment** using three different types of wraps and three paper cups. Wrap aluminum foil around one cup, plastic wrap around the second cup, and wax paper around the third cup. Remember to record your observations.

A cooler is an insulator. It keeps your food from getting warm.

Lesson 2

Sound

Look and Wonder

Sounds are all around you. At a parade you hear many different sounds. What makes all of these sounds?

Explore

Inquiry Activity

How can you make sounds?

Make a Prediction

Look at the paper, ruler, and rubber band. What must you do to make a sound with each object? Make a prediction.

Test Your Prediction

⚠ **Be Careful.** Wear goggles.

1. **Observe** Hold a piece of paper by one corner. Wave it around. What happens?
2. **Observe** Place a ruler on a desk. Extend half of it over the edge of the desk. Hold down the end of the ruler on the desk, and tap the other end. What happens?
3. **Observe** Wrap a rubber band around a box. Pluck the rubber band. What happens?

Draw Conclusions

4. What happened when you moved the paper, ruler, and rubber band?
5. **Infer** Can you make a sound using the paper, ruler, or rubber band without making any of them move? Explain your answer.
6. **Infer** How are sounds made?

Explore More

Experiment Test ways to change the sound you made with each object. Try to make the sounds louder or softer, higher or lower. For example, try pulling the rubber band tighter and then plucking it. Record your results and the steps you follow.

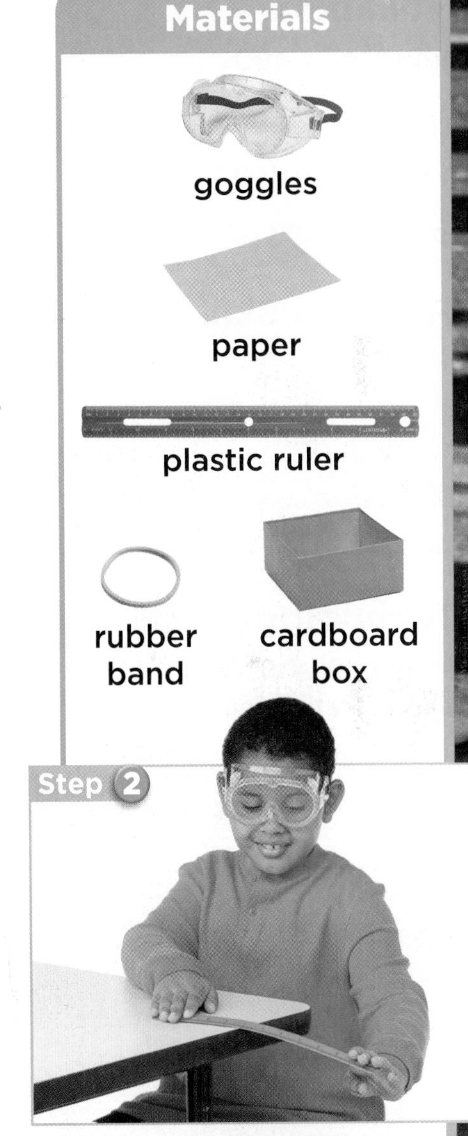

Read and Learn

Essential Question
What is sound?

Vocabulary
vibrate, p. 490
sound, p. 490
volume, p. 492
pitch, p. 493

Reading Skill 
Predict

What I Predict	What Happens

Technology
e-Glossary, e-Review, and animations online at www.macmillanmh.com

What is sound?

Think about the many sounds you hear. Some sounds are musical, such as the notes on a guitar. Other sounds are harsh, such as chickens squawking. All sounds begin when something vibrates (VI•brayts). To **vibrate** is to move back and forth quickly. You can not make a sound without making something move. **Sound** is a form of energy that comes from objects that vibrate.

How Sound Travels

Have you ever dropped a stone into a pond? The stone makes waves in the water that move out in all directions. Sound travels in waves too. When you pluck a guitar string, it vibrates. This creates a sound wave. The sound wave moves through the air. You hear the sound when the sound wave reaches your ear.

You can make sound when you pluck, tap, or blow on an object. ▼

Orca whales use sounds to communicate.

Sound waves can travel through air, a gas. Sound waves can also travel through liquids and solids. Some sea animals communicate by making sounds underwater. You hear a knock on the door because sound travels through the door. Sound waves travel through matter. Sound does not travel in space.

Sound does not travel at the same speed through all materials. Sound travels slowest through a gas. It travels faster through a liquid. It travels fastest through a solid.

Quick Check

Predict You hit a drum with a stick. What happens?

Critical Thinking Do you think that you can hear sounds in outer space?

◀ Tie a piece of string to two cups. Then talk in one end while a friend listens in the other. Why can you hear your friend?

The sounds of jets roaring through the sky are louder than the chirps of a bird.

How are sounds different?

Close your eyes and listen. What sounds do you hear? What makes sounds different?

Volume

Sounds can have different volumes. **Volume** describes how loud a sound is. A plane flying overhead is louder than a bird's song. A plane has a greater volume.

Loud sounds are made by objects that vibrate with a lot of energy. The more energy an object vibrates with, the louder the sound it makes. Tap your foot on the floor. Then stomp your foot. You use more energy to stomp your foot than to tap it. Stomping makes a high-energy vibration, so you hear a loud sound. Tapping makes a low-energy vibration, so you hear a soft sound.

Pitch

Some sounds are high, such as the squeaking of a mouse. Other sounds are low, such as the croaking of a bullfrog. A sound's **pitch** (PICH) is how high or low it is. An object that vibrates quickly has a high pitch. An object that vibrates slowly has a low pitch.

An object's length can affect pitch. Look at the marimba below. When hit, the shorter keys vibrate faster than the longer ones. The shorter a key is, the faster it vibrates and the higher its pitch.

The thickness of an object can also affect pitch. A guitar has thin strings and thick strings. Thin strings vibrate faster than thick strings. Thin strings have higher pitches.

Quick Check

Predict How does stretching a rubber band affect its pitch?

Critical Thinking Compare the sound of a bicycle horn to a car horn.

Quick Lab

Changing Sounds

1. **Predict** How can you change the sound a straw makes?
2. Flatten one end of a straw. Then cut across the tip of this end as shown.
3. **Experiment** Cover your teeth with your lips and then blow hard through the cut end of the straw. Describe what you hear.
   Now blow softer. How does the sound change?
4. **Experiment** Now try using straws of different lengths. Remember to cut the tip before you blow into the straw. Describe what you hear. How does the sound change?

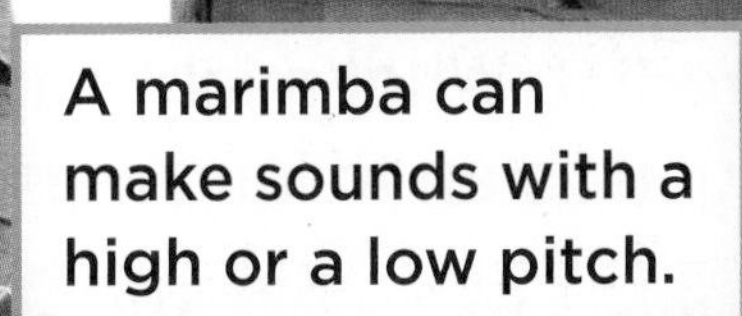
A marimba can make sounds with a high or a low pitch.

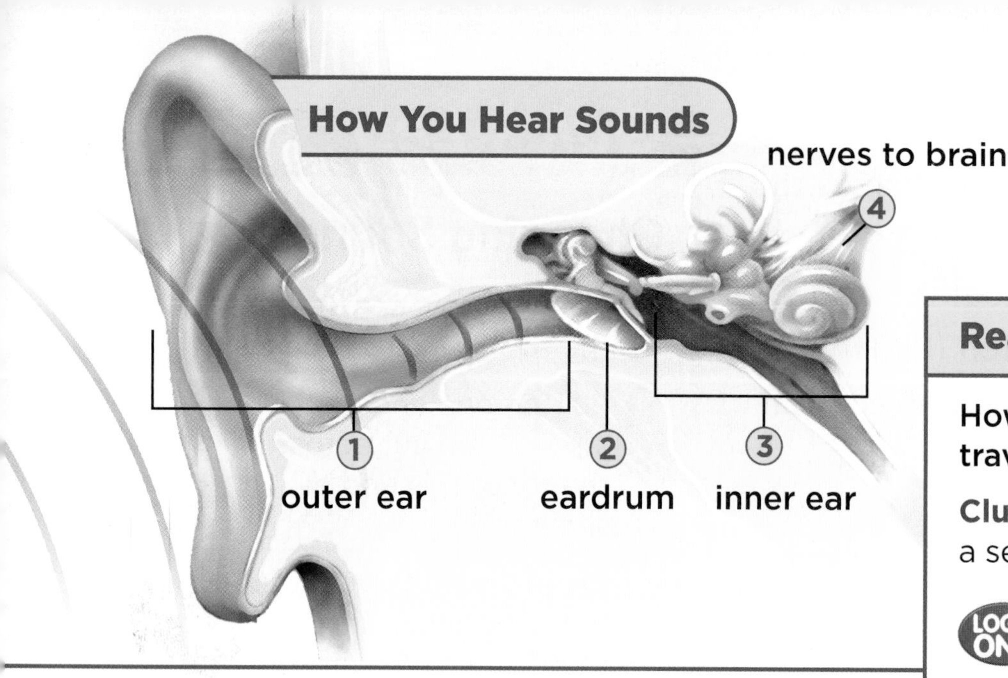

Read a Diagram

How does a sound wave travel through your ear?

Clue: Numbers help show a sequence.

Science in Motion Watch how sound moves at www.macmillanmh.com

How do you hear sounds?

What happens to a sound wave when it reaches your ear? First, the sound wave is collected by the outer ear. Next, the sound wave makes your eardrum vibrate. This causes three tiny bones in your inner ear to vibrate. These vibrations pass through the inner ear to nerves. The nerves send a message to your brain, and you hear a sound.

Protect Your Hearing

It is important to protect your ears. Never put a finger or pencil in your ear. You might hurt the parts inside. Other things can damage your ear as well. Loud sounds have a lot of energy. They can damage the parts inside the ear.

This construction worker must protect his ears.

Quick Check

Predict What might happen to your hearing if you listen to loud music often?

Critical Thinking Which makes your eardrum vibrate faster, a high sound or low sound?

Lesson Review

Visual Summary

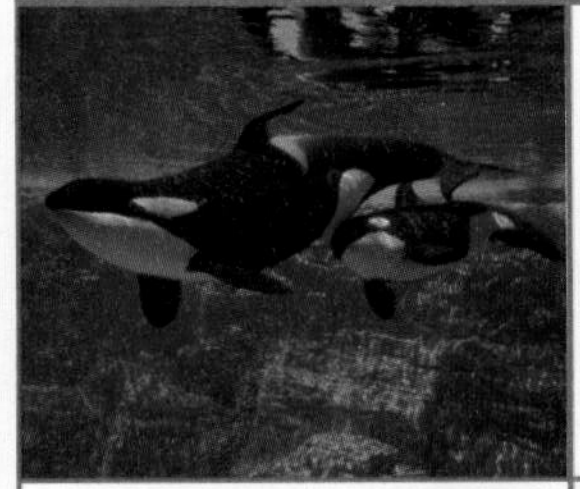

Sound is produced when an object vibrates. Sound can travel through solids, liquids, and gases.

Sounds can be compared by using volume and pitch.

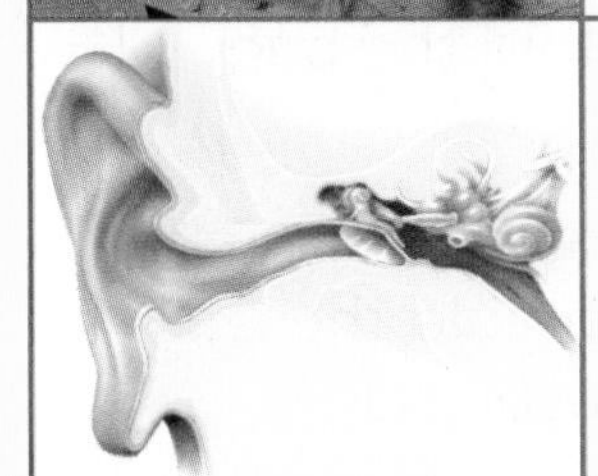

You hear sounds when vibrations travel through your ear.

Make a FOLDABLES Study Guide

Make a trifold book. Use it to summarize what you learned about sound.

Think, Talk, and Write

1. **Vocabulary** What is the difference between pitch and volume?

2. **Predict** How would cymbals sound if you tapped them together lightly? How would they sound if you crashed them together?

What I Predict	What Happens

3. **Critical Thinking** List five different sounds you hear. How are these sounds the same? How are these sounds different?

4. **Test Prep** **Objects that vibrate quickly make sounds with**
   - **A** high volume.
   - **B** low volume.
   - **C** high pitch.
   - **D** low pitch.

5. **Essential Question** What is sound?

Writing Link

Persuasive Writing
It is important to protect yourself from loud sounds. Loud sounds can harm your hearing. Find out how you can protect your hearing. Then write a paragraph telling others what you learned.

Music Link

Make an Instrument
Put on goggles. Stretch rubber bands of different thicknesses around a shoe box. Then use the rubber bands to make sounds. How can you change the pitch? How can you change the volume?

LOG ON e-Review Summaries and quizzes online at www.macmillanmh.com

Be a Scientist

Materials

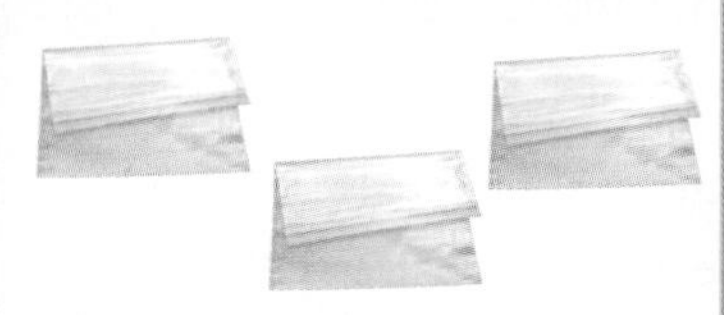

3 plastic bags

tuning fork

water

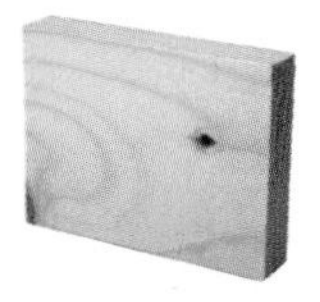

wooden block

Structured Inquiry

How does sound move through different types of matter?

Form a Hypothesis

You just learned that sound travels through solids, liquids, and gases. How does the state of matter affect how sound travels? Write a hypothesis.

Test Your Hypothesis

1. Fill a plastic bag with air and seal it. Hold the bag against your ear.
2. **Experiment** Tap the tines of the tuning fork against the bottom of your shoe. Then hold the base of the tuning fork against the plastic bag. Listen to the sound it makes.
3. Fill a plastic bag with water. Seal it and hold it against your ear.
4. **Experiment** Tap the tuning fork and hold it against the bag. Record any differences you hear.
5. Place a wooden block in a plastic bag. Squeeze out as much air as you can and seal the bag. Hold the bag against your ear.
6. **Experiment** Tap the tuning fork and hold it against the bag. How is the sound different now? Record your observations.

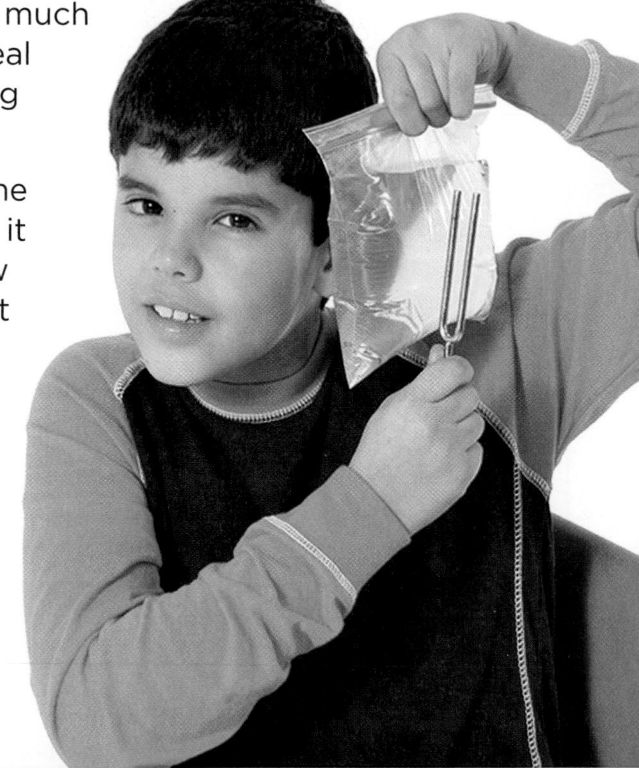

Step 4

Draw Conclusions

7. How did the tuning fork sound different through the different materials?
8. **Interpret Data** Through which material was the sound loudest?
9. **Infer** Does sound travel best through a solid, a liquid, or a gas?

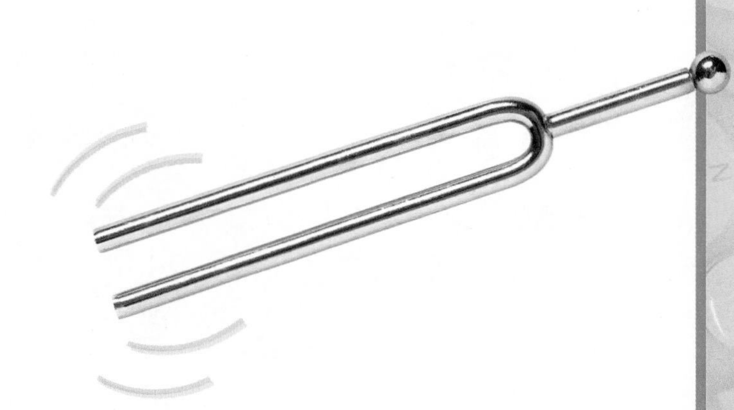

Guided Inquiry

How does sound move through different solids?

Form a Hypothesis

Sound can be stopped, slowed down, or absorbed by different solids. How does sound travel through different solids?

Test Your Hypothesis

Design an experiment to investigate how sound travels through different solids. Decide on the materials you will need. You might want to try plastic, wooden, and metal objects. Write the steps you will follow. Record your results and observations.

Draw Conclusions

Did your results support your hypothesis? Why or why not?

Open Inquiry

What other questions do you have about sound? For example, what objects block sound the best? Design an experiment to find out.

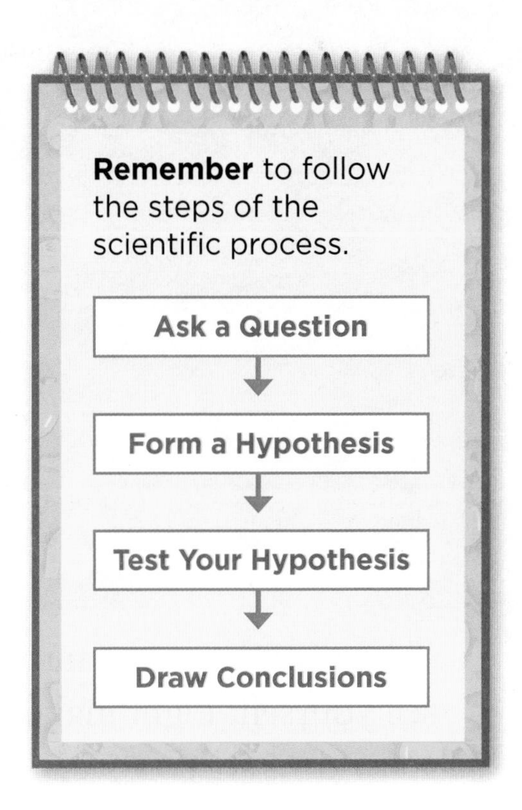

Lesson 3

Light

Millennium Park, Chicago, Illinois

Look and Wonder

When you look at a mirror, you can see yourself. Light makes this possible. How does light move?

Explore

Inquiry Activity

How does light move?

Materials

mirror

flashlight

Make a Prediction

What happens to light when it hits a mirror? Make a prediction.

Test Your Prediction

1. Hold a mirror in front of you. Have a partner shine the flashlight onto the mirror.
2. **Observe** What happens to the flashlight's beam?
3. **Experiment** Pick a spot on the wall. Can you make light bounce off the mirror and shine on that spot? How? Do you have to move the mirror, the flashlight, or both?

Step 1

Draw Conclusions

4. What happened to the beam of light when it hit the mirror? What happened when you moved the mirror? What happened when you moved the flashlight?
5. **Communicate** Make a drawing to show how light moves when it strikes the mirror.

Step 3

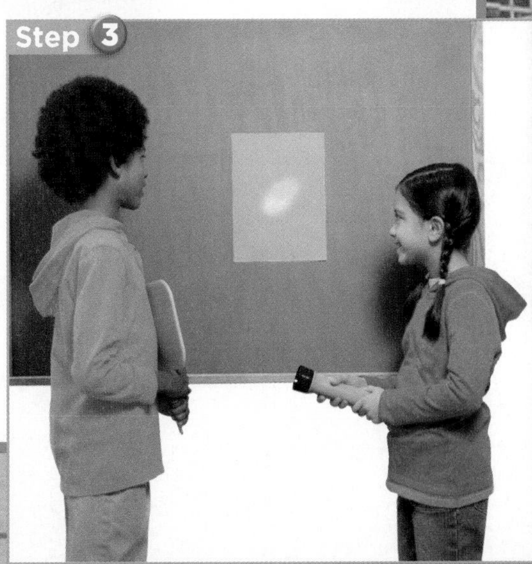

Explore More

Experiment Sit next to your partner. Leave 1 meter of space between you and your partner. Then hold a mirror so you can see your partner. Can your partner see you in the mirror? Can you see yourself and your partner in the mirror at the same time?

Read and Learn

Essential Question
How does light allow you to see objects?

Vocabulary
light, p. 500
absorb, p. 500
reflect, p. 501
opaque, p. 502
shadow, p. 502
transparent, p. 503
translucent, p. 503
refract, p. 503

Reading Skill
Draw Conclusions

Text Clues	Conclusions

Technology 
e-Glossary, e-Review, and animations online at www.macmillanmh.com

What is light?

You use light every day. **Light** is a form of energy. It allows you to see objects. Light comes from many different sources. The Sun is Earth's main source of light. Fires and light bulbs are some other sources of light.

Light travels away from a source in a straight path. When you turn on a flashlight, you can see a straight beam of light. Even light from the Sun travels millions of miles through space in a straight path. Light travels in a straight path until it hits an object.

Absorption

Light can be **absorbed** (ub•ZORBD), or taken in, when it hits an object. Black objects absorb almost all the light that hits them. White objects absorb almost no light.

Light moves away from this lighthouse in a straight path.

◀ When light hits some objects, it reflects off in a different direction.

Reflection

When light hits some objects, it **reflects** (rih•FLEKTS), or bounces off of them. It changes direction and then continues to move in a straight path.

Light bounces off objects in the same way that a ball bounces. When you hit a ball down, it bounces up. When light hits an object, it bounces off in a different direction.

You see an object when light from the object reaches your eyes. Most objects do not make their own light. You see those objects when light reflects off them and goes into your eyes.

Mirrors are very smooth, shiny surfaces. They reflect almost all the light that hits them. ▼

 ***Quick Check***

Draw Conclusions How can a mirror help you see behind you?

Critical Thinking Is it possible for you to see in the dark? Explain.

▲ When the Sun is behind the tree, a shadow forms in front of the tree.

What happens when light hits different objects?

How do you stay dry on a rainy day? You stand under an umbrella. An umbrella blocks raindrops so they do not reach you. Opaque (oh•PAYK) objects act somewhat like an umbrella to light. **Opaque** objects block light from passing through them. A brick wall, a piece of cardboard, and even you are opaque. You can not see through opaque objects.

Opaque objects can cause shadows to form. A **shadow** is a dark space that forms when light is blocked. You have probably seen your shadow on a sunny day. Your body blocked the sunlight. The shadow that formed had a shape similar to your body.

A shadow's size depends on where a light source is. The closer an object is to a light source, the bigger its shadow. Light coming from above creates a short shadow. As the light source gets lower, the shadow gets longer.

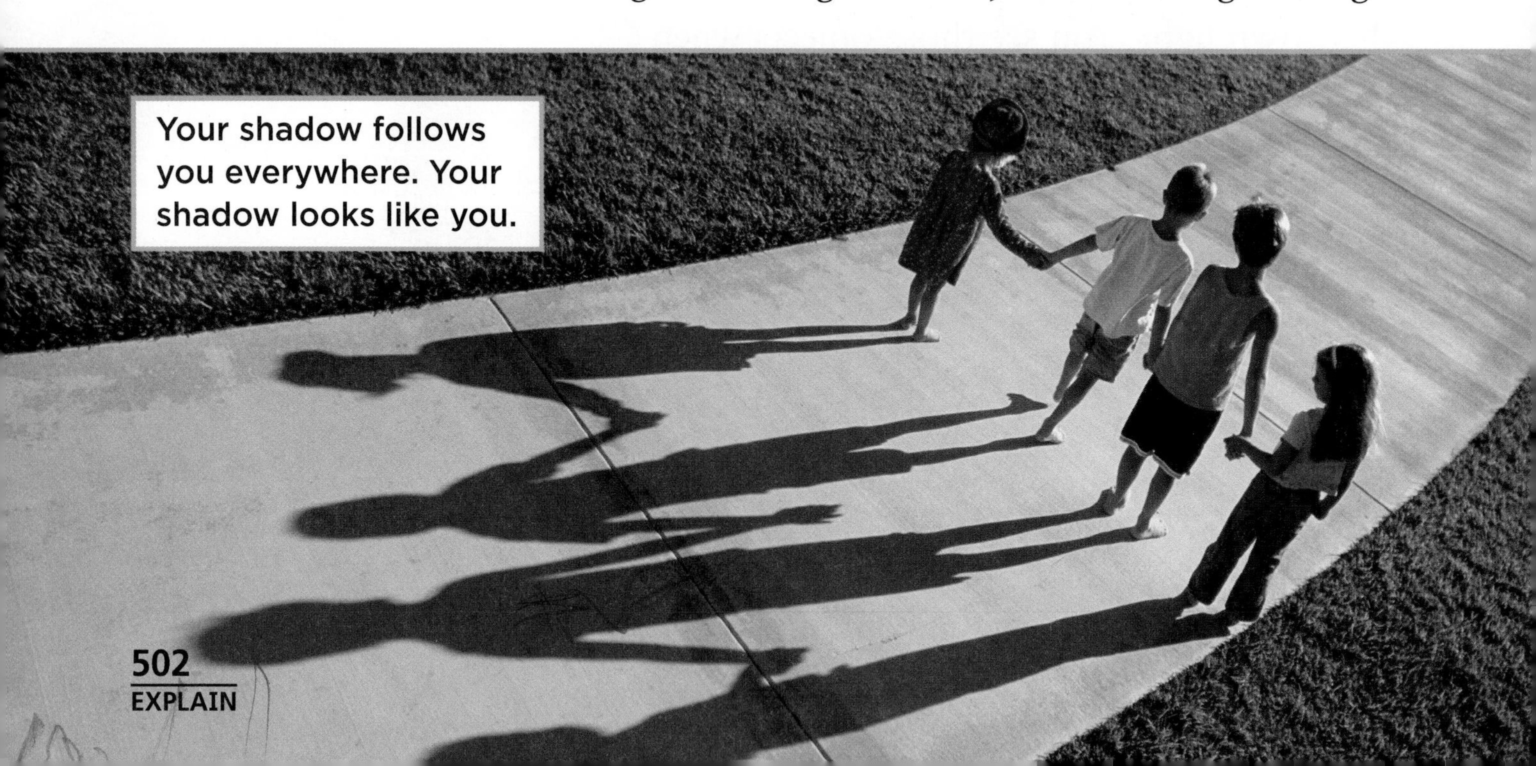

Your shadow follows you everywhere. Your shadow looks like you.

Not all objects are opaque. Light can pass through some objects. **Transparent** (trans•PAYR•unt) objects allow light to pass straight through them. Air, glass, and clear plastic are transparent. You can see clearly through these objects because they allow light to pass through them.

Other objects are translucent (trans•LEW•sunt). **Translucent** objects let light pass through them, but they scatter the light. You can not see clearly through translucent objects for this reason. Wax paper and frosted glass are translucent.

Refraction

Light can refract (rih•FRAKT) when it passes from one material to another. To **refract** means to bend. Look at the photograph of the pencil. Is the pencil broken? No, it just looks that way. Light refracts when it passes from the air to the water. Then it bounces off the pencil. The light refracts again when it moves from the water back to the air. The bending light makes the pencil look broken.

▲ Bending light rays make this pencil look broken.

 ***Quick Check***

Draw Conclusions What three things do you need to make a shadow?

Critical Thinking Why does sunlight pass through a window and not a wall?

▲ When light passes through a prism, it is separated into different colors.

Why can you see colors?

What color is light from the Sun? You might say yellow or white. In fact it is a mixture of many colors. White light, such as sunlight, is made up of every color of light. To show this, you can use a prism (PRIH•zum). A *prism* is a piece of glass that refracts light. Prisms separate white light into all the colors that make it up. They do this by refracting each color of light by a different amount.

When white light strikes a colored object, some colors of light are absorbed. Other colors of light are reflected. The reflected light enters your eyes. You see the object as the color of this reflected light.

Water vapor in the sky can act like tiny prisms. When water vapor refracts sunlight, a rainbow forms.

FACT White light is made up of every color of light.

When white light strikes a green leaf, the leaf absorbs all the colors except for green. Only green light bounces off the leaf. This color is reflected to your eyes. You see the leaf as green. Something different happens when light strikes a red flower. Now the green light is absorbed. All colors are absorbed except for red. Only red light is reflected to your eyes. You see the flower as red. An object that absorbs all light that strikes it looks black. An object that reflects all light that strikes it looks white.

Quick Check

Draw Conclusions What colors make up light from the Sun?

Critical Thinking Why does a banana appear to be yellow?

Quick Lab

Mixing Colors

1. **Predict** Look at the photo below. What happens to the color of the plate when you spin it?
2. Divide a white paper plate into eight equal parts. Color each section of the plate a different color.
3. **Observe** Carefully push a pencil into the center of the plate. Hold the plate away from your body. Spin it. What color do you see when the plate is spinning?

Seeing Colors

Read a Diagram

Why does the leaf look green?

Clue: Look at the color of light that is reflected.

LOG ON *Science in Motion* Watch colors at www.macmillanmh.com

▲ This girl can see the ice-cream cone when reflected light enters her eye.

How do you see?

What happens to light when it reaches your eyes? First, the light is refracted as it goes into the cornea (KOR•nee•uh). Then, light passes through the pupil. The pupil is the black opening in the center of each eye. The pupil controls how much light enters the eye. Next, light travels through the lens. The lens refracts the light so it strikes the back of the eyeball. The optic nerve sends information about the light to the brain. The brain then uses that information to make a picture.

Quick Check

Draw Conclusions How does reflected light allow you to see this page?

Critical Thinking How does the size of a pupil change?

Lesson Review

Visual Summary

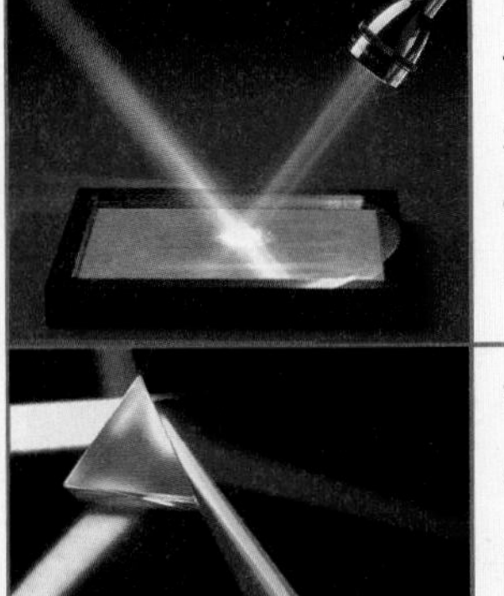

Light travels away from a source in a straight path. Objects can reflect or absorb light.

White light is made of many different colors. You see an object as the color of its reflected light.

When light enters your eye, you see an image.

Make a FOLDABLES Study Guide

Make a trifold book. Use it to summarize what you learned about light.

Think, Talk, and Write

1. **Vocabulary** What happens when light is refracted?

2. **Draw Conclusions** Why does a school bus appear yellow and a fire truck appear red?

Text Clues	Conclusions

3. **Critical Thinking** How could you make the shadow of a marble look like the shadow of a tennis ball?

4. **Test Prep** **A sheet of aluminum foil is an example of what type of material?**
   - **A** translucent
   - **B** shadow
   - **C** transparent
   - **D** opaque

5. **Essential Question** How does light allow you to see objects?

Writing Link

Writing That Informs

Find out how you can protect your skin from absorbing too much sunlight. Find out why wearing white clothes and sunblock can help. Then write about it.

Art Link

Shadow Puppets

Use your hands and a flashlight to make shadows. Try to make different shapes and animals. Move your hand closer and farther away from the light. What happens to the shadow puppet?

LOG ON e-Review Summaries and quizzes online at www.macmillanmh.com

Reading in Science

A BEAM OF LIGHT

A laser creates a narrow beam of light.

Surgeons are doctors who perform operations to fix injuries or treat diseases. They can use scalpels—special tools with sharp blades—to cut through skin, muscles, and organs of the human body. Today, surgeons have another tool they can use to do operations. This tool is a beam of light!

This beam of light is called a *laser.* Lasers are very powerful. They can cut though the human body without causing much bleeding.

Lasers were first used to remove birthmarks on children's skin. Today, surgeons also use lasers to treat injuries to the brain, the heart, and many other parts of the body. Lasers are also used to improve people's eyesight.

Write About It

Summarize Read the article again. List the most important information in a chart. Then use the chart to write a summary of the article.

Summarize

A summary

- gives the important details;
- is brief;
- is told in your own words.

The surgeon is using a laser to perform this operation.

Connect to

at www.macmillanmh.com

Lesson 4

Electricity

Seattle, Washington

Look and Wonder

A lightning bolt flashes through the sky. You turn on a lamp so you can see. What is the same about lightning bolts and light bulbs?

Explore

Inquiry Activity

What makes a bulb light?

Make a Prediction

How can you connect a battery, a wire, and a light bulb to make the bulb light up? Make a prediction.

Test Your Prediction

1. **Experiment** Try to light the bulb using a light bulb, wire, and battery.
2. **Communicate** Draw each setup. Record your results.
3. **Communicate** When your light bulb is lit, compare setups with your classmates. Is there more than one setup that lights the bulb?

Draw Conclusions

4. How many setups could you find that made the bulb light?
5. **Infer** Look at the setups that lit the bulb. What do you think is necessary to make the bulb light up?

Explore More

Experiment How could you light two bulbs using only one battery? Can you think of more than one way? Try it.

Materials

D-cell battery

one 20-centimeter piece of insulated wire

light bulb

Step 1

Step 2

Read and Learn

Essential Question
How do you use electricity?

Vocabulary
electrical charge, p. 512
static electricity, p. 513
electric current, p. 514
circuit, p. 515
switch, p. 515

Reading Skill
Sequence

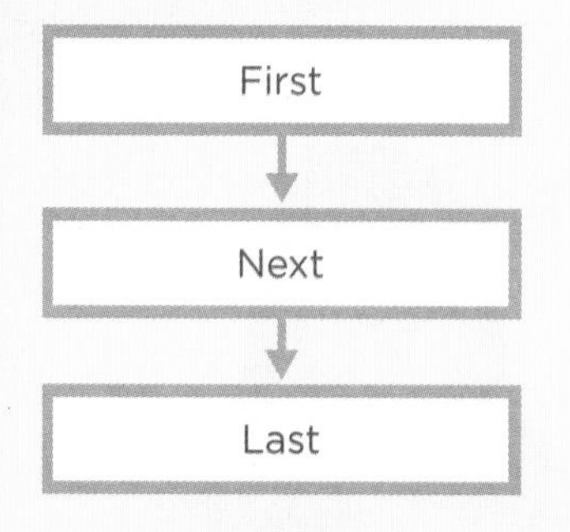

Technology

e-Glossary and e-Review online at www.macmillanmh.com

What is electrical charge?

Have you ever had a shock from touching a doorknob? Why does this happen? The same thing that causes light bulbs to glow and lightning to strike causes this shock. All these things happen because of electricity.

All electricity is the result of electrical charge. Like volume and mass, **electrical charge** is a property of matter. You can not see electrical charge. However, you can understand how objects with different charges interact.

There are two types of electrical charges. One is called positive. The other is called negative. An object with a positive charge pulls on an object with a negative charge. Objects with a positive charge push each other away. Objects with a negative charge also push each other away.

◀ Static electricity can cause you to get a shock when you touch a doorknob.

Static Electricity

All objects are made of charged particles. Most objects have the same number of positive particles as negative particles. The charge is balanced. When two objects touch, however, negative particles can move from one object to the other. Negative particles build up on one object. That object now has a negative charge. A buildup of electrical charge is called **static electricity.**

Rub a balloon on a sweater and hold it near a wall. The balloon sticks to the wall! When you rub the balloon, negative particles move from the sweater to the balloon. The balloon gets a negative charge. It repels the negative particles in the wall and attracts the positive particles. This causes it to stick to the wall.

Static electricity is what sometimes causes you to get a shock when you touch a doorknob. When you walk across the floor, negative particles move from the floor to your body. You get a negative charge. When you touch a doorknob, the negative particles move from you to the knob. You feel this as a shock. When static electricity moves from one object to another, it is called a *discharge.*

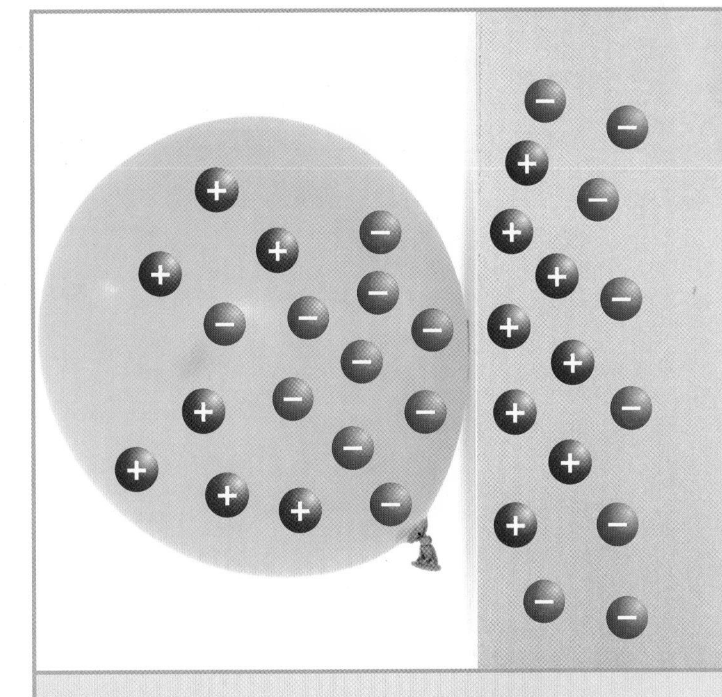

▲ This balloon has a negative charge. It attracts the positive (+) particles in the wall and repels the negative (–) particles. This causes it to stick to the wall.

Quick Check

Sequence What happens when you rub a balloon with a wool cloth?

Critical Thinking Why do clothes stick together when they come out of the dryer?

What is electric current?

Charged particles can build up on objects. They can also be made to flow. A flow of charged particles is called an **electric current** (KUR•unt). You use electric current every day. Electric current provides the energy you need to power lights, radios, computers, hair dryers, and many other products. We use energy from electric current to produce heat, light, sound, and motion.

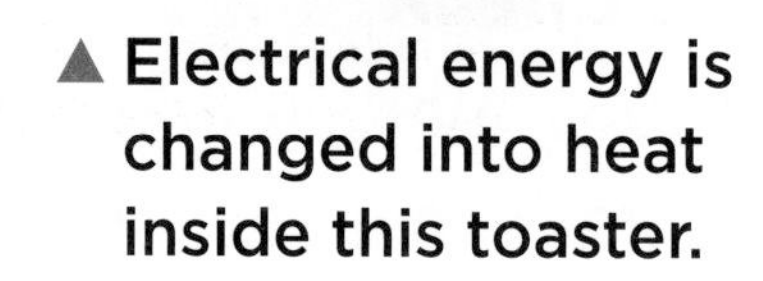

▲ Electrical energy is changed into heat inside this toaster.

These headphones change electrical energy into sound. ▶

Circuits

Electric current needs a path, or circuit (SUR•kut), through which to flow. A **circuit** is a path that is made of parts that work together to allow current to flow. Look at the diagram on this page. Wires connect the bulb to a battery. The battery is the circuit's power source.

To keep an electric current moving, a circuit can not have any breaks. A complete unbroken circuit, like the one shown on top, is called a *closed circuit.* A circuit with breaks or openings is called an *open circuit.*

Switches

You can use a switch (SWICH) to open and close a circuit. A **switch** allows you to control the flow of current. When a switch is in the on position, there is no gap in the path. The circuit is closed, and current can flow. Turn the switch off, and there is a gap in the path. The circuit is open, and current does not flow.

Electric Circuit

closed circuit

When the switch is closed, electric current flows. The bulb lights.

open circuit

When the switch is open, electric current does not flow. The bulb does not light.

Read a Diagram

Why is the second bulb not lit?

Clue: Compare the paths in each diagram.

Quick Check

Sequence What happens when you close the switch on a circuit?

Critical Thinking Where can you find circuits in your home?

FACT Current flows through the light bulb back to the battery.

Quick Lab

Conductors and Insulators

1. Put a battery in a battery holder. Connect a wire to each side of the battery holder.
2. Connect the free end of one of the wires to a socket that has a bulb in it. Then connect a third wire to the socket as shown.

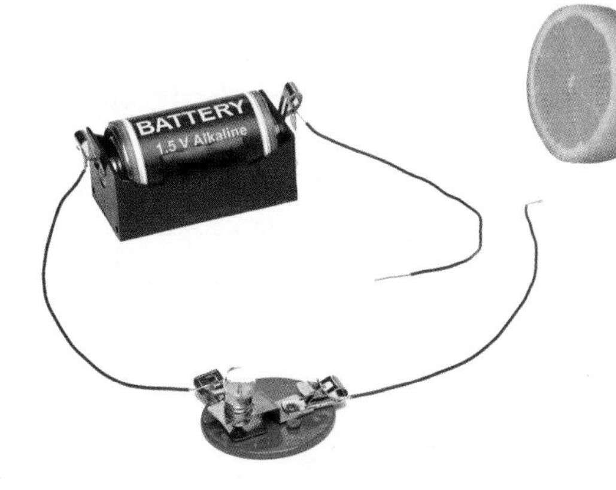

3. **Experiment** Gather objects, such as crayons and paper clips. Touch the free ends of the wires to each object.

4. **Observe** Does the bulb light up with each object? Record what happens.
5. **Infer** Which objects are conductors? Which are insulators?

What are conductors and insulators?

The electric current in your home flows through wires. These wires are usually made of copper and are wrapped in plastic. Copper is a material that allows current to flow through it very easily. Materials that allow current to flow easily are called conductors (kun•DUK•turz). Most metals are conductors.

Plastic is wrapped around the wires in your home because plastic is an insulator (IN•suh•lay•tur). An insulator is a material that does not allow current to flow easily through it. The plastic coating on wires does not let current flow through it. This protects you from getting a shock. Glass, plastic, and rubber are good insulators.

Quick Check

Sequence What happens to current when it reaches an insulator?

Critical Thinking Why are the wires in a circuit often made of copper?

Copper wires are conductors. The plastic around each wire is an insulator. ▶

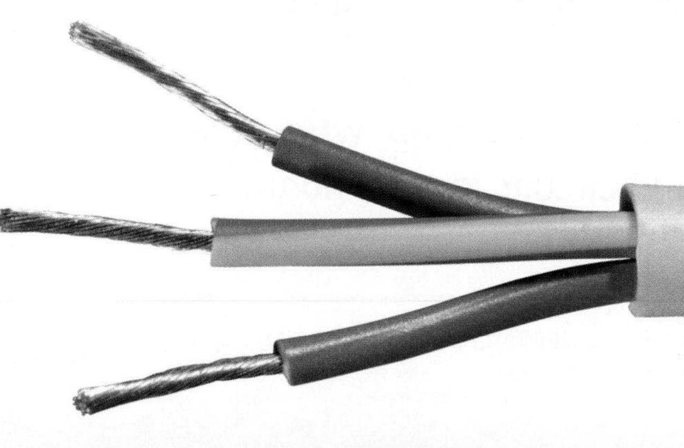

Lesson Review

Visual Summary

	Electricity is made up of charged particles. **Static electricity is** a buildup of charged particles.
	**Electric current travels** in a path called a circuit. A switch can control the flow of current.
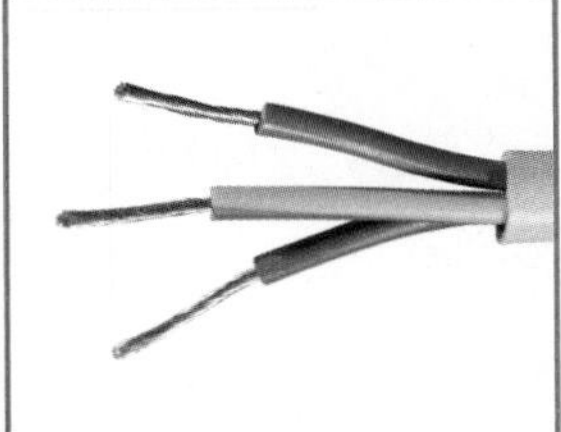	**Conductors** allow electric current to flow through them easily. **Insulators** do not.

Make a FOLDABLES® Study Guide

Make a three-tab book. Use it to summarize what you learned about electricity.

Think, Talk, and Write

1. **Vocabulary** What is a circuit?
2. **Sequence** How do you get a shock from touching a doorknob?

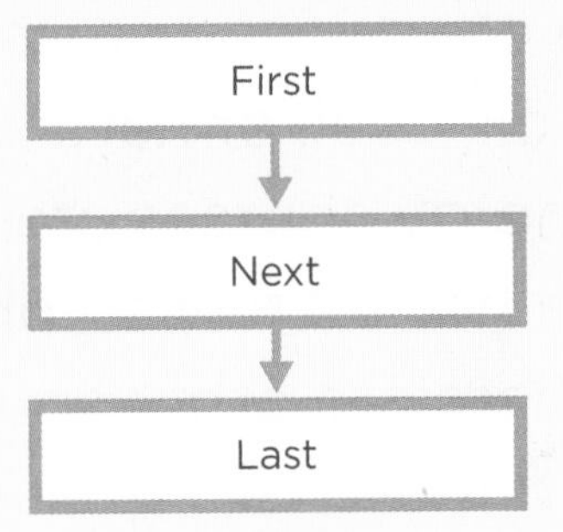

3. **Critical Thinking** You turn the switch on a flashlight. The light does not come on. List things that might be wrong with the flashlight.
4. **Test Prep Which changes electrical energy into motion?**
   - **A** toaster oven
   - **B** paper airplane
   - **C** flashlight
   - **D** electric train
5. **Essential Question** How do you use electricity?

Writing Link

Expository Writing

Make a list of all the ways your life is made easier by electricity. Describe how you would have to do things differently if there were no electricity.

Health Link

Make a Poster

Electrical devices can be dangerous if not used properly. Research how to use electricity safely. Make a poster about it.

Other Energy Sources

Most of the energy we use to produce electricity comes from burning oil, coal, or natural gas. These energy sources are limited. They can not be reused or replaced easily. There are other sources of energy that can be replaced in short periods of time. Wind can turn windmills to make energy. Energy from the Sun can be collected by solar panels. Do you think it is important to find other sources of energy? What are some ways you can encourage people to use other sources of energy?

These windmills use the wind's energy to produce electricity.

Persuasive Writing

A persuasive letter

- **clearly states an opinion;**
- **supports the opinion with reasons and facts;**
- **persuades the reader to agree with that opinion.**

▲ **These solar panels use the Sun's energy to produce electricity.**

Write About It

Persuasive Writing

Write a persuasive letter to a community leader. Tell why you think it is important to find other sources of energy. Be sure to follow the form of a formal letter.

e-Journal Write about it online at www.macmillanmh.com

Math in Science

The Cost of Energy

The chart below shows the amount of energy used by several appliances in a year. The energy used is measured in kilowatt-hours and costs 10 cents per kWh. Multiply to find out how much it costs to use these appliances for one year.

Multiply Decimals

- To multiply a decimal number by a whole number, first multiply the same way you would with two whole numbers.
- Then, count the number of decimal places. Place the decimal point that many places from the right.

86 x $0.10 (2 decimal places)

86 x 10 = 860 (count 2 decimal places 8.60)

86 x $0.10 = $8.60

Solve It

Copy the chart and fill in the missing information. Then solve these problems.

- A DVD player uses 24 kWh of energy in a year. If it cost $0.10 per kWh, what does it cost to use a DVD player for one year?
- If you save $0.50 per week for one year (52 weeks), how much would you save in one year?

Appliances	Yearly Energy Use (kWh)	Cost (per kWh)	Total Yearly Cost
clock radio	86	$0.10	$8.60
refrigerator	2,088	$0.10	$208.80
toaster	50	$0.10	$ ______
telephone answering machine	32	$0.10	$______
color television	238	$0.10	$23.80
clothes washer	108	$0.10	$10.80

CHAPTER 12 Review

Visual Summary

Lesson 1 Heat affects matter in many ways. Heat always moves from warmer objects to cooler objects.

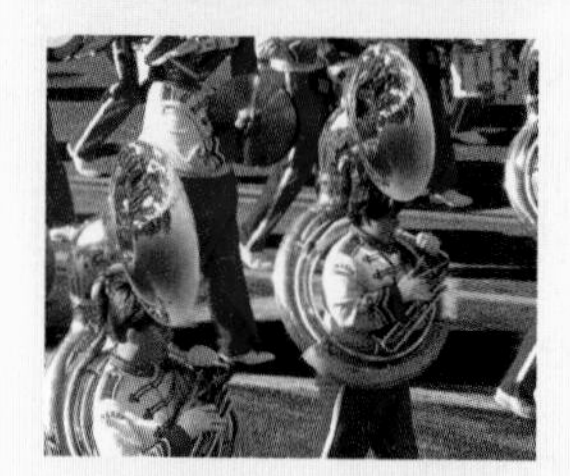

Lesson 2 Sounds are made when objects vibrate. Volume and pitch can be used to compare sounds.

Lesson 3 Light is a form of energy that allows you to see objects. Light moves in straight paths.

Lesson 4 Electricity is made up of charged particles. These charged particles can flow through a circuit.

Make a FOLDABLES® Study Guide

Glue your lesson study guides to a piece of paper as shown. Use your study guide to review what you have learned in this chapter.

Vocabulary

DOK I

Fill each blank with the best term from the list.

circuit, p. 515	**reflect**, p. 501
electric current, p. 514	**shadow**, p. 502
heat, p. 480	**sound**, p. 490
insulator, p. 484	**temperature**, p. 482
light, p. 500	**vibrates**, p. 490

1. When light is blocked by an object, a ________ forms.

2. A path that allows electric current to flow is a ________.

3. The form of energy that allows you to see objects is called ________.

4. When a guitar string vibrates, a ________ is made.

5. When light hits an object, it can bounce, or ________, off the object.

6. Energy that moves from a warm object to a cold object is called ________.

7. When an object moves back and forth very quickly, it ________.

8. A material that heat does not move through easily is an ________.

9. A flow of charged particles is an ________.

10. A thermometer is used to measure ________.

LOG ON e-Glossary Words and definitions online at www.macmillanmh.com

Skills and Concepts

DOK 2–3

Answer each of the following.

11. Summarize What happens when an electrical switch is in the off position? What changes when the switch is turned on?

12. Persuasive Writing What is your favorite kind of music? Write a paragraph explaining why you enjoy it. Include *volume* and *pitch* in your paragraph.

13. Experiment Cover one thermometer with black paper and another with white paper. Put them in a warm place for 15 minutes. Then read each thermometer. Which color heats up faster: black or white? Why?

14. Critical Thinking Suppose you see a picture in a book. Then you see the same picture on a computer screen. Where does the light come from to show you each picture?

15. Critical Thinking A magician says she can bend a wand. She dips the wand into a glass of water. The wand appears to bend. Explain how this trick works.

16. What material is this pot made of? Why?

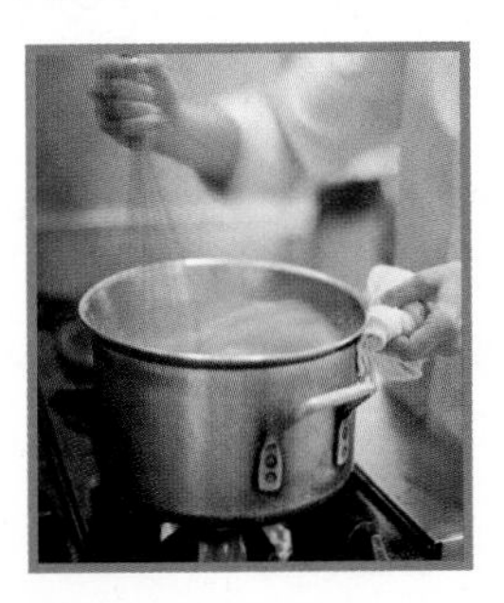

17. True or False *Sound travels fastest through solids, such as metals.* Is this statement true or false? Explain.

18. True or False *Heat always travels from a cooler object to a warmer object.* Is this statement true or false? Explain.

19. How are a lamp and the Sun alike?

A Both give off daylight.

B Both produce electricity.

C Both reflect light.

D Both are sources of heat and light.

The Big Idea

20. What are the main forms of energy, and how are they used?

Performance Assessment

DOK 3

Changing Energy

Many machines work by changing electrical energy. For example, a toaster changes electrical energy into heat. How do other machines change electrical energy?

- ▶ List six machines that change electrical energy.
- ▶ Describe how the energy changes.
- ▶ How can living things change energy from one form into another? Describe two ways that this can happen.

Test Preparation

1 **Look at the picture below.**

Which gives off the most heat?

A the Sun

B the Moon

C the beach

D the ocean

DOK 1

2 **How can you find out whether a substance is a good conductor of heat?**

A Measure how long it takes for it to heat up when it touches a hot object.

B Measure its temperature when it is cold and again when it is hot.

C Heat it and see whether it boils or melts.

D Freeze it and then measure its temperature.

DOK 2

3 **All sounds begin when something**

A spins.

B shifts.

C bends.

D vibrates.

DOK 1

4 **Through which material will sound waves travel slowest?**

A wire

B air

C glass

D water

DOK 1

5 **A rubber mallet is used to strike two different tuning forks with equal force.**

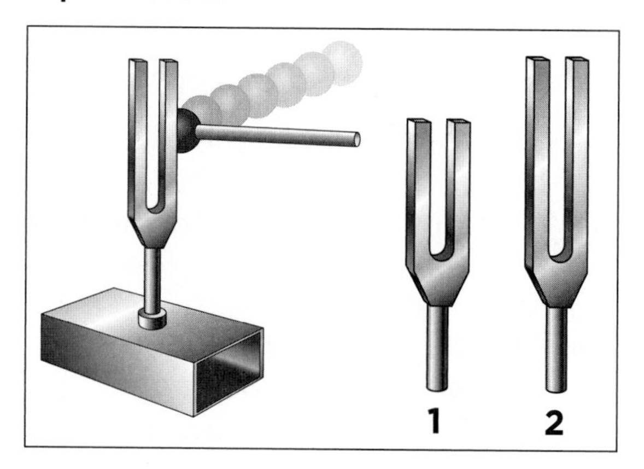

How will the sounds most likely be different?

A They will differ in pitch.

B They will differ in volume.

C They will differ in energy.

D They will differ in loudness.

DOK 2

6 What happens to the beam of a flashlight when it hits a mirror?

A It is refracted.

B It is reflected.

C It is absorbed.

D It is conducted.

DOK 1

7 If you rub two balloons with a wool cloth,

A the balloons will attract each other.

B the balloons will not affect each other.

C the balloons will repel each other.

D the balloons will pop.

DOK 2

8 Jamal made the circuit shown below.

Which item will he need to make the circuit work?

A another bulb

B another battery

C a switch

D another wire

DOK 1

9 Lindsay heated a pot of water. She poured an equal amount of the water into four different cups. After 30 minutes she measured the temperature of the water in each cup.

Cup	Temperature (in °C)
1	22
2	36
3	26
4	33

Which cup is the best insulator?

A cup 1

B cup 2

C cup 3

D cup 4

DOK 2

10 How do rainbows form?

DOK 1

Check Your Understanding

Question	Review	Question	Review
1	pp. 480–481	6	pp. 500–503
2	p. 484	7	pp. 512–513
3	p. 490	8	pp. 511, 515
4	p. 491	9	pp. 484, 486–487
5	pp. 492–493	10	pp. 504–505

Lighting Technician

Have you ever watched a motion-picture awards show? If so, you might have heard actors thank members of the film crew. An important part of the film crew is the chief lighting technician.

The chief lighting technician designs the lighting for the scenes of a movie. The lighting must create a mood that matches the action of the scene. The chief lighting technician uses different combinations of lights for different scenes. The technician also changes the location of the light sources to create different moods.

To become a chief lighting technician, you need to know about light and electrical energy. You also should have some experience in drama or filmmaking. Many chief lighting technicians begin their career as members of a lighting crew.

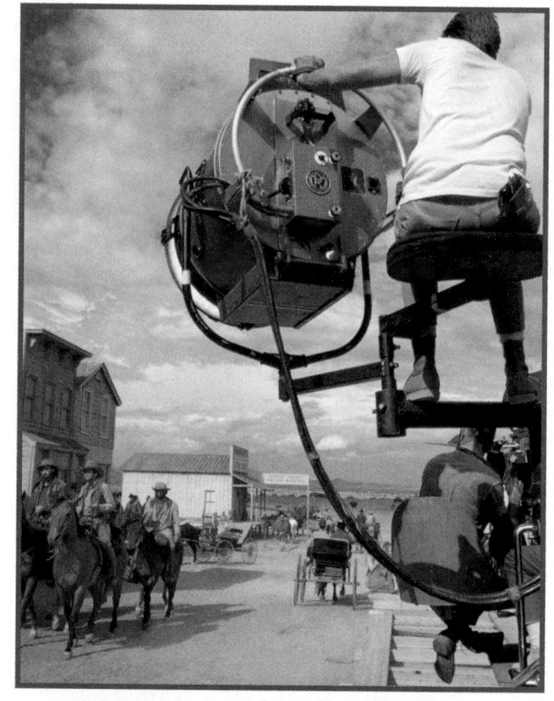

▲ This technician is lighting a set for a motion picture.

▲ A lighting technician knows about light and electrical energy.

Here are some other physical science careers:

- electrician
- engineer
- architect
- car designer

LOG ON e-Careers at www.macmillanmh.com

Reference

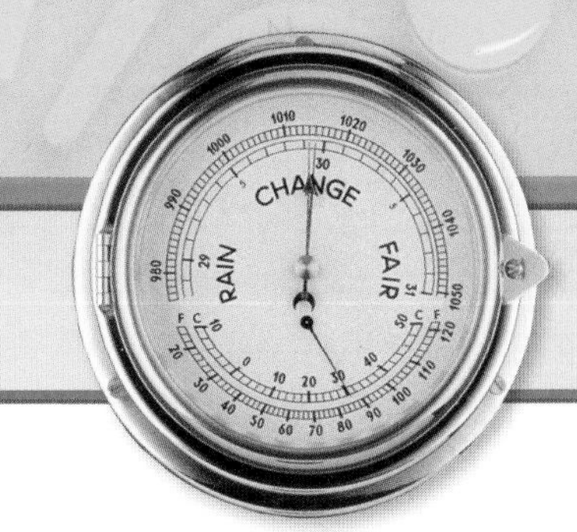

Measurements

Units of Measurement

Temperature

- The temperature on this thermometer reads 86 degrees Fahrenheit. That is the same as 30 degrees Celsius.

Length

- This student is 3 feet plus 9 inches tall. That is the same as 1 meter plus 14 centimeters.

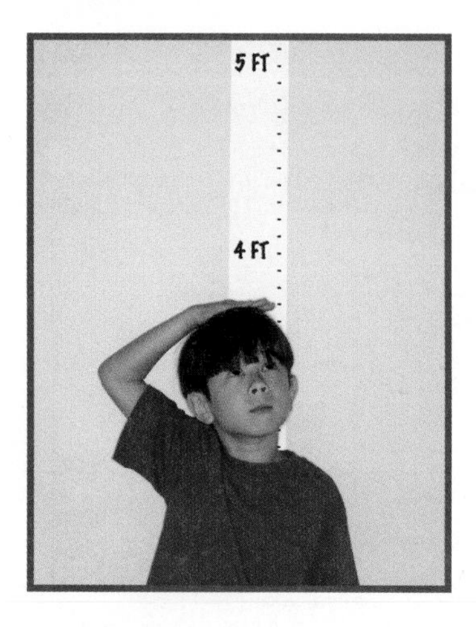

Mass

- You can measure the mass of these rocks in grams.

Volume of Fluids

- This bottle of water has a volume of 2 liters. That is a little more than 2 quarts.

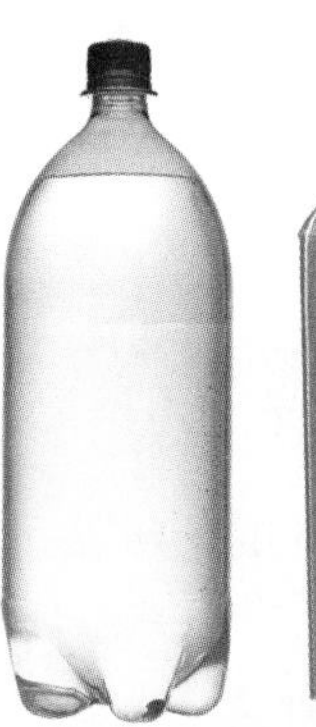

Weight/Force

- This pumpkin weighs about 7 pounds. That means the force of gravity is 31.5 newtons.

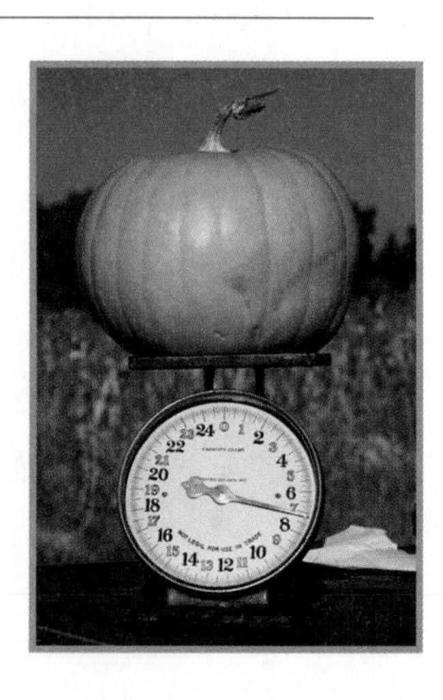

Speed

▶ This student can ride her bike 100 meters in 50 seconds. That means her speed is 2 meters per second.

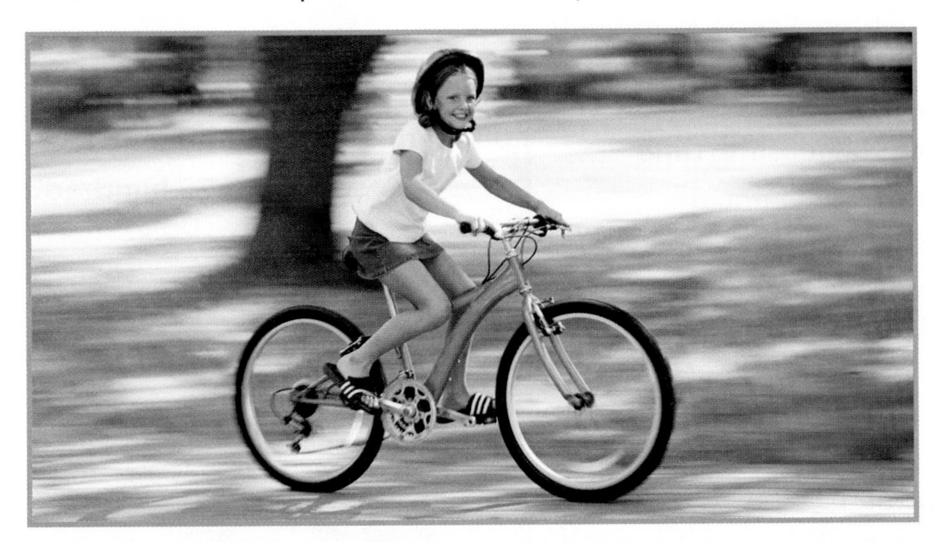

Table of Measures	
SI International Units/Metric Units	**Customary Units**
Temperature Water freezes at 0 degrees Celsius (°C) and boils at 100°C.	**Temperature** Water freezes at 32 degrees Fahrenheit (°F) and boils at 212°F.
Length and Distance 10 millimeters (mm) = 1 centimeter (cm) 100 centimeters = 1 meter (m) 1,000 meters = 1 kilometer (km)	**Length and Distance** 12 inches (in.) = 1 foot (ft) 3 feet = 1 yard (yd) 5,280 feet = 1 mile (mi)
Volume 1 cubic centimeter (cm^3) = 1 milliliter (mL) 1,000 milliliters = 1 liter (L)	**Volume of Fluids** 8 fluid ounces (fl oz) = 1 cup (c) 2 cups = 1 pint (pt) 2 pints = 1 quart (qt) 4 quarts = 1 gallon (gal)
Mass 1,000 milligrams (mg) = 1 gram (g) 1,000 grams = 1 kilogram (kg)	
Area 1 square meter (m^2) = 1 m × 1 m 10,000 square meters (m^2) = 1 hectare	**Area** 1 square foot (ft^2) = 1 ft × 1 ft 43,560 square feet (ft^2) = 1 acre
Speed meters per second (m/s) kilometers per hour (km/h)	**Speed** miles per hour (mph)
Weight/Force 1 newton (N) = 1 kg × 1m/s^2	**Weight/Force** 16 ounces (oz) = 1 pound (lb) 2,000 pounds = 1 ton (T)

Measurements

Measure Time

You measure time to find out how long something takes to happen. Stopwatches and clocks are tools you can use to measure time. Seconds, minutes, hours, days, and years are some units of time.

Try it Use a Stopwatch to Measure Time

1. Get a cup of water and an antacid tablet from your teacher.
2. Tell your partner to place the tablet in the cup of water. Start the stopwatch when the tablet touches the water.
3. Stop the stopwatch when the tablet completely dissolves. Record the time shown on the stopwatch.

0 minutes
25 seconds
75 hundredths (0.75) of a second

▲ **Push the button on the top right of the stopwatch to start timing. Push the button again to stop timing.**

Measure Length

You measure length to find out how long or how far away something is. Rulers, tape measures, and meter sticks are some tools you can use to measure length. You can measure length using units called meters. Smaller units are made from parts of meters. Larger units are made of many meters.

Look at the ruler below. Each number represents 1 centimeter (cm). There are 100 centimeters in 1 meter. In between each number are 10 lines. The distance between each line is equal to 1 millimeter (mm). There are 10 millimeters in 1 centimeter.

Try it Find Length with a Ruler

Place a ruler on your desk. Line up a pencil with the "0" mark on the ruler. Record the length of the pencil in centimeters.

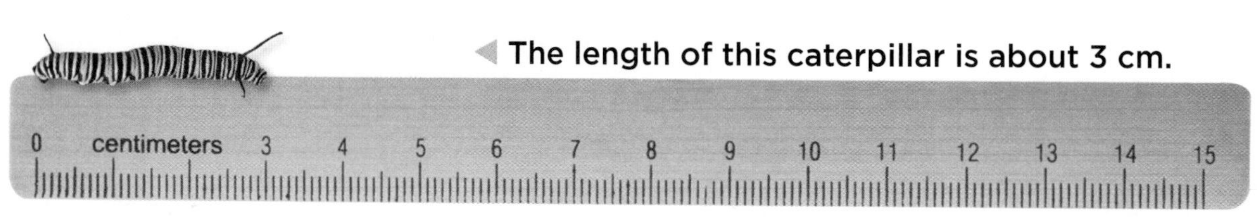

◀ **The length of this caterpillar is about 3 cm.**

Measure Liquid Volume

Volume is the amount of space something takes up. Beakers, measuring cups, and graduated cylinders are tools you can use to measure liquid volume. These containers are marked in units called milliliters.

Try it Measure Liquid Volume

1. Gather a few empty plastic containers of different shapes and sizes.
2. Use a graduated cylinder to find the volume of water each container can hold. To start, fill the graduated cylinder with water, then pour the water into the container. Continue pouring this until the container is full. Keep track of the number of milliliters you add.

10 mL

▲ **This graduated cylinder can measure volumes up to 100 mL. The distance between each number on the cylinder represents 10 mL.**

Measure Mass

Mass is the amount of matter an object has. You use a balance to measure mass. To find the mass of an object, you compare it with objects whose masses you know. Grams are units people use to measure mass.

Try it Measure the Mass of a Box of Crayons

1. Place a box of crayons on one side of a pan balance.
2. Place gram masses to the other side until the two sides of the balance are level.
3. Add together the numbers on the gram masses. This total equals the mass of the box of crayons.

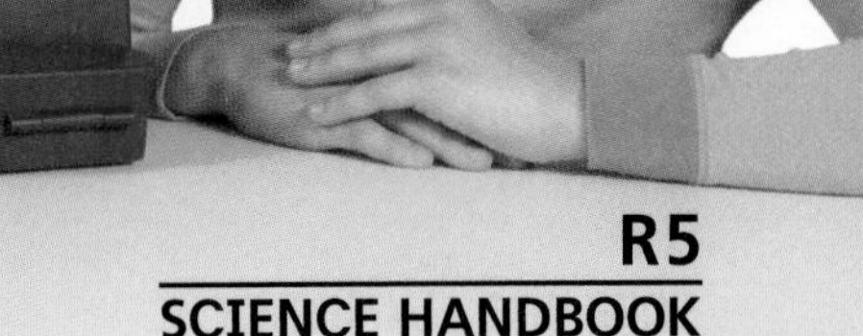

Measurements

Measure Force/Weight

You measure force to find the strength of a push or pull. Force can be measured in units called newtons (N). A spring scale is a tool used to measure force.

Weight is a measure of the force of gravity pulling down on an object. A spring scale measures the pull of gravity. One pound is equal to about 4.5 N.

Try it Measure the Weight of an Object

1. Hold a spring scale by the top loop. Put a small object on the bottom hook.
2. Slowly let go of the object. Wait for the spring to stop moving.
3. Read the number of newtons next to the tab. This is the object's weight.

Measure Temperature

Temperature (TEM•puh•ruh•chur) is how hot or cold something is. You use a tool called a thermometer (thur•MAH•muh•tur) to measure temperature. In the United States, temperature is often measured in degrees Fahrenheit (°F). However, you can also measure temperature in degrees Celsius (°C).

Try it Read a Thermometer

1. Fill a beaker with ice water. Then put a thermometer in the water.
2. Wait several minutes. Read the number next to the top of the red liquid inside the thermometer. This is the temperature.
3. Repeat with warm water.

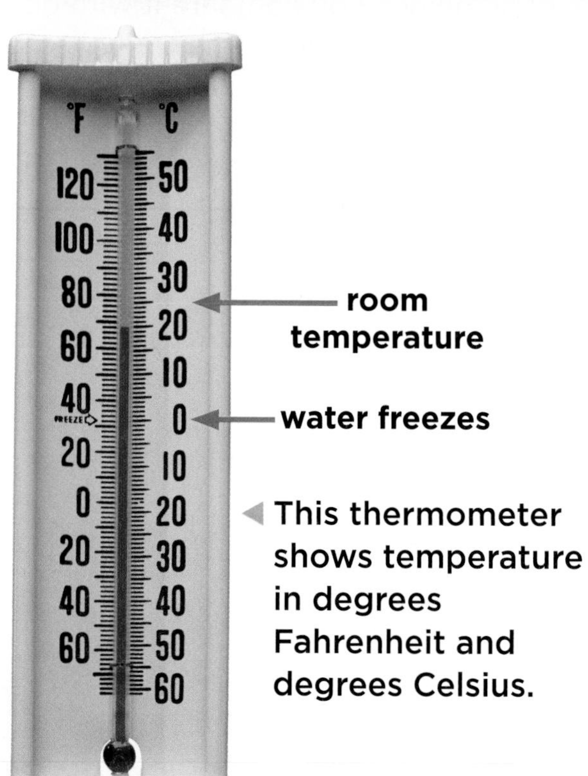

This thermometer shows temperature in degrees Fahrenheit and degrees Celsius.

Tools of Science

Use a Microscope

A microscope (MI•kruh•skohp) is a tool that magnifies objects, or makes them look larger. A microscope can make an object look hundreds or thousands of times larger. Look at the photo to learn the different parts of a microscope.

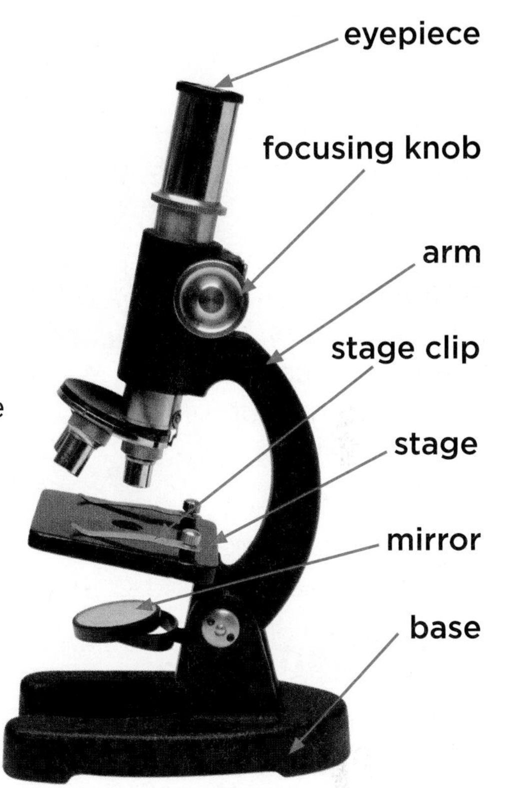

Try it Examine Salt Grains

1. Move the mirror so it reflects light upward toward the stage. △ **Be Careful.** Never point the mirror at bright lights or the Sun. This can cause permanent eye damage.
2. Place a few grains of salt on a slide. Put the slide under the stage clips on the stage. Be sure that the salt grains are over the hole in the stage.
3. Look through the eyepiece. Turn the focusing knob slowly until the salt grains come into focus. Draw a picture of what you see.

Use a Hand Lens

A hand lens is another tool that magnifies objects. It is not as powerful as a microscope. However, a hand lens still allows you to see details of an object that you cannot see with your eyes alone. As you move a hand lens away from an object, you can see more details. If you move a hand lens too far away, the object will look blurry.

Try it Magnify a Rock

1. Look at a rock carefully. Draw a picture of it.
2. Hold a hand lens above the rock so you can see the rock clearly.
3. Fill in any details on your original drawing that you did not see before.

Use a Calculator

Sometimes during an experiment, you have to add, subtract, multiply, or divide numbers. A calculator can help you carry out these operations.

Try it Convert from °F to °C

Water boils at 212°F. Use a calculator to convert 212°F into degrees Celsius.

1. Press the [ON] key. Then, enter the number 212 by pressing [2][1][2].
2. Subtract 32 by pressing [-][3][2].
3. Multiply by 5 by pressing [x][5].
4. Finally, divide by 9 by pressing [÷][9]. Press [=]. This is the temperature in degrees Celsius.

Now, convert 100°F into degrees Celsius.

Use a Camera

During an experiment or nature study, it helps to observe and record changes that happen over time. Sometimes it can be difficult to see these changes if they happen very quickly or very slowly. A camera can help you keep track of visible changes. Studying photos can help you understand what happens over the course of time.

Try it Gather Data from a Photo

The photos below show a panda eight days after birth and then several months later. What differences do you notice? How has the panda changed over those months? Now think of something else that changes over time. With the help of an adult, use a camera to take photos at different times. Compare your photos.

Computers

Use a Computer

A computer has many uses. You can use a computer to get information from compact discs (CDs) and digital video discs (DVDs). You can also use a computer to write reports and to show information.

The Internet connects your computer with computers around the world, so you can collect all kinds of information. When using the Internet, visit only Web sites that are safe and reliable. Your teacher can help you find safe and reliable sites to use. Whenever you are online, never give any information about yourself to others.

Try it **Use a Computer for a Project**

1. Choose an environment to research. Then use the Internet to find out about this environment. Where is the environment located in the world? What is the climate like in the environment? What kinds of plants and animals live there?
2. Use DVDs or other sources from the library to find out more about your chosen environment.
3. Use the computer to write a report about the information you gathered. Then share your report with others.

Make Maps

Locate Places

A map is a drawing that shows an area from above. Many maps have numbers and letters along the top and side. The letters and numbers help you find locations. The Buffalo Zoological Garden, for example, is located at D4 below. To find it, place a finger on the letter D along the left side of the map and another finger on the number 4 at the top. Move your fingers straight across and down the map until they meet. Now find B1. What is there?

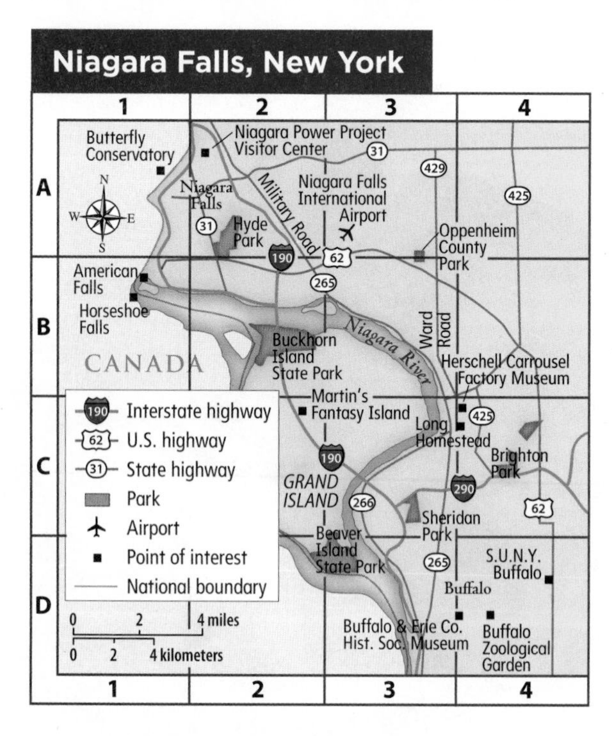

Try it Make a Map

Make a map of an area in your community. It might be a park or the area between your home and school. Include numbers and letters along the top and side. Use a compass to find north, and mark north on your map.

Idea Maps

The Niagara Falls map shows how places are connected to each other. Idea maps, on the other hand, show how ideas are connected to each other. Idea maps help you organize information about a topic.

Look at the idea map below. It connects ideas about water. This map shows that Earth's water can be freshwater or salt water. The map also shows three sources of freshwater. You can see that there is no connection between *rivers* and *salt water* on the map. This can remind you that salt water does not flow in rivers.

Try it Make an Idea Map

Make an idea map about a topic you are learning in science. Your map can include words, phrases, or sentences. Arrange your map in a way that makes sense to you and helps you understand the connection between ideas.

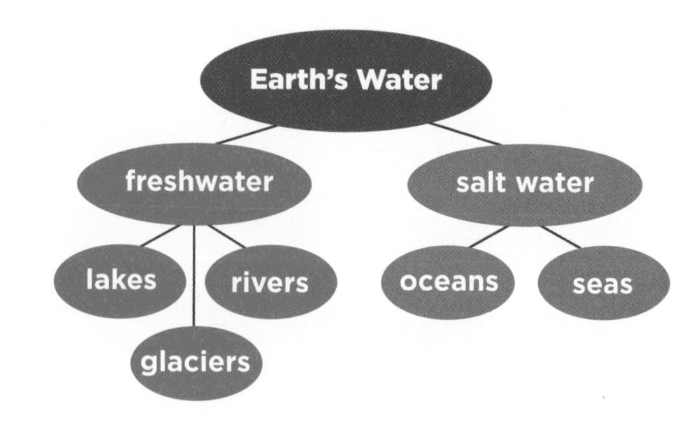

Make Charts

Living	Nonliving
tree	rock
chipmunk	puddle
bird	cloud

Charts are useful for recording information during an experiment and for communicating information. In a chart, only the column or the row has meaning but not both. In this chart, one column lists living things. A second column lists nonliving things.

Try it Organize Data in a Chart

Take a survey of your class. Find out each student's favorite kind of pet. Make a chart to show this information. Remember to show your information in columns or in rows.

Make Tables

Tables can also help organize data, or information. Tables have columns that run up and down and rows that run across. Headings tell you what kind of data that is in each row or column.

The table below shows the properties of some minerals. Which mineral in the table has a white streak? Which mineral is yellow in color?

Try it Organize Data in a Table

Collect a few minerals from your teacher. Observe the properties of each. Make a table like the one shown. Use the same column headings. Record the properties of each mineral.

Mineral Identification Table

	Hardness	Luster	Streak	Color	Other
pyrite	6–6.5	metallic	greenish-black	brassy yellow	called "fool's gold"
quartz	7	nonmetallic	none	colorless, white, rose, smoky, purple, brown	
mica	2–2.5	nonmetallic	none	dark brown, black, or silver-white	flakes when peeled
feldspar	6	nonmetallic	none	colorless, beige, pink	
calcite	3	nonmetallic	white	colorless, white	bubbles when acid is placed on it

Organizing Data

Make Graphs

Graphs also help organize data. Graphs make it easy to notice trends and patterns. There are many kinds of graphs.

Bar Graphs

A bar graph uses bars to show data. What if you want to find the warmest and coldest months for your city? Every month you find the average temperature in the newspaper. You can organize the temperatures in a bar graph so you can easily compare them.

Month	Temperature (°C)
January	6
February	8
March	10
April	13
May	16
June	19
July	22
August	20
September	19
October	14
November	9
December	7

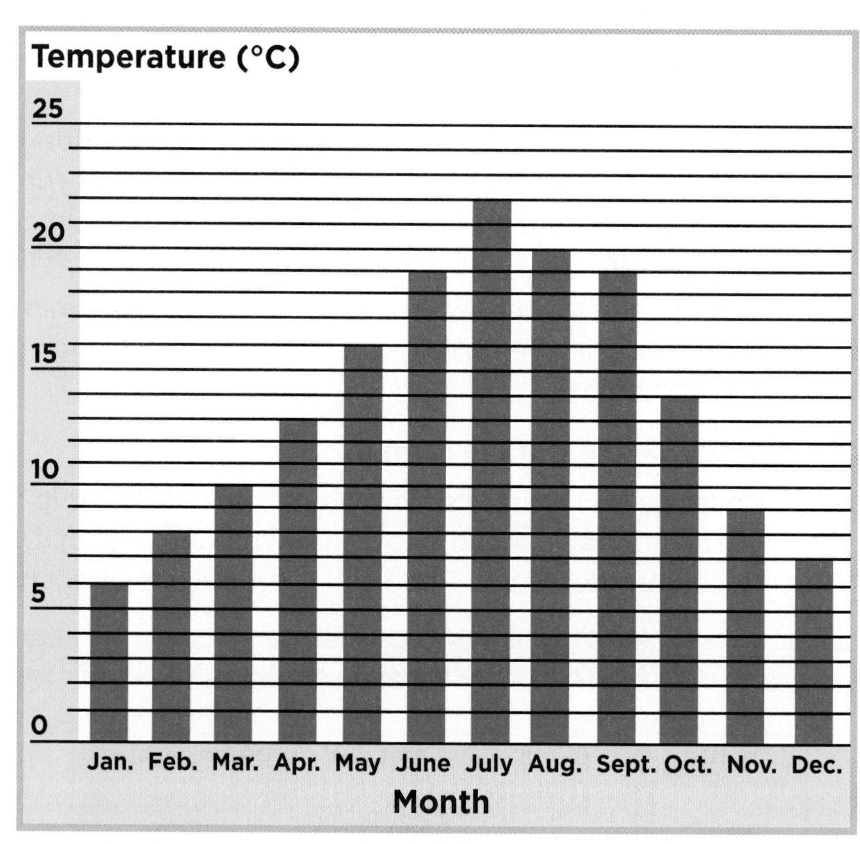

1. Look at the bar for the month of April. Put your finger at the top of the bar. Move your finger straight to the left to find the average temperature for that month.
2. Find the highest bar on the bar graph. This bar represents the month with the highest average temperature. Which month is it? What is the average temperature for that month?
3. Look at the bars on the graph. What pattern do you notice in the temperatures from January to December?

Pictographs

A pictograph uses symbols, or pictures, to show information. What if you collect information about how much water your family uses each day?

Water Used Daily (liters)	
drinking	10
showering	100
bathing	120
brushing teeth	40
washing dishes	80
washing hands	30
washing clothes	160
flushing toilet	50

You can organize this information into a pictograph. In the pictograph below, each bucket means 20 liters (L) of water. A half bucket means half of 20 L, or 10 L, of water.

1. Which activity uses the most water?
2. Which activity uses the least water?

Water Used Daily	
drinking	
showering	
bathing	
brushing teeth	
washing dishes	
washing hands	
washing clothes	
flushing toilet	

= 20 L of water

Line Graphs

A line graph can show how information changes over time. What if you measure the temperature outdoors every hour starting at 6 A.M.?

Time	Temperature (°C)
6 A.M.	10
7 A.M.	12
8 A.M.	14
9 A.M.	16
10 A.M.	18
11 A.M.	20

Now organize your data by making a line graph. Follow these steps.

1. Make a scale along the bottom and side of the graph. Label the scales.
2. Draw a point on the graph for each temperature measured each hour.
3. Connect the points.
4. How do the temperatures and times relate to each other?

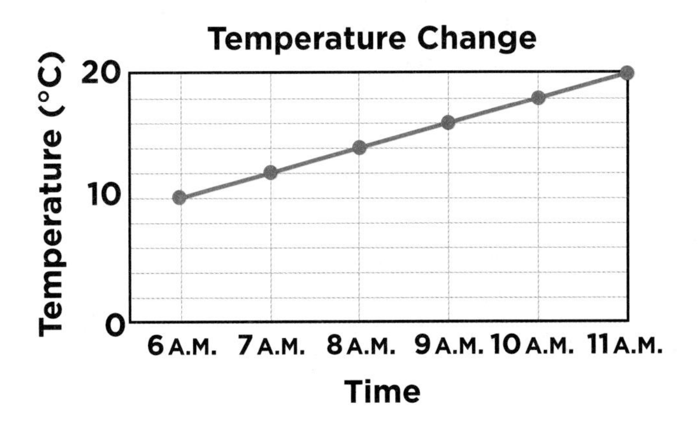

Human Body Systems

The Skeletal System

Feel your elbows, wrists, and fingers. What are those hard parts? Bones! Bones make up the skeletal system. The skeletal system is one of many body systems. A body system is a group of organs that work together to perform a specific job.

The skeletal system is made up of 206 bones. Each bone has a particular job. The long, strong leg bones support the body's weight. The skull protects the brain. The hip bones help you move. Together, bones do important jobs to keep the body active and healthy.

- Bones support the body and give the body its shape.
- Bones protect organs in the body.
- Bones work with muscles to move the body.
- Bones store minerals and produce blood for the body.

Joints

A joint is a place where two or more bones meet. There are three main types of joints.

Immovable joints form where bones fit together too tightly to move. The 29 bones of your skull meet at immovable joints. Partly movable joints are places where bones can move a little. Ribs are connected to the breastbone with these joints. Movable joints, like the knee, are places where bones can move easily. The knee lets the bones of your leg move.

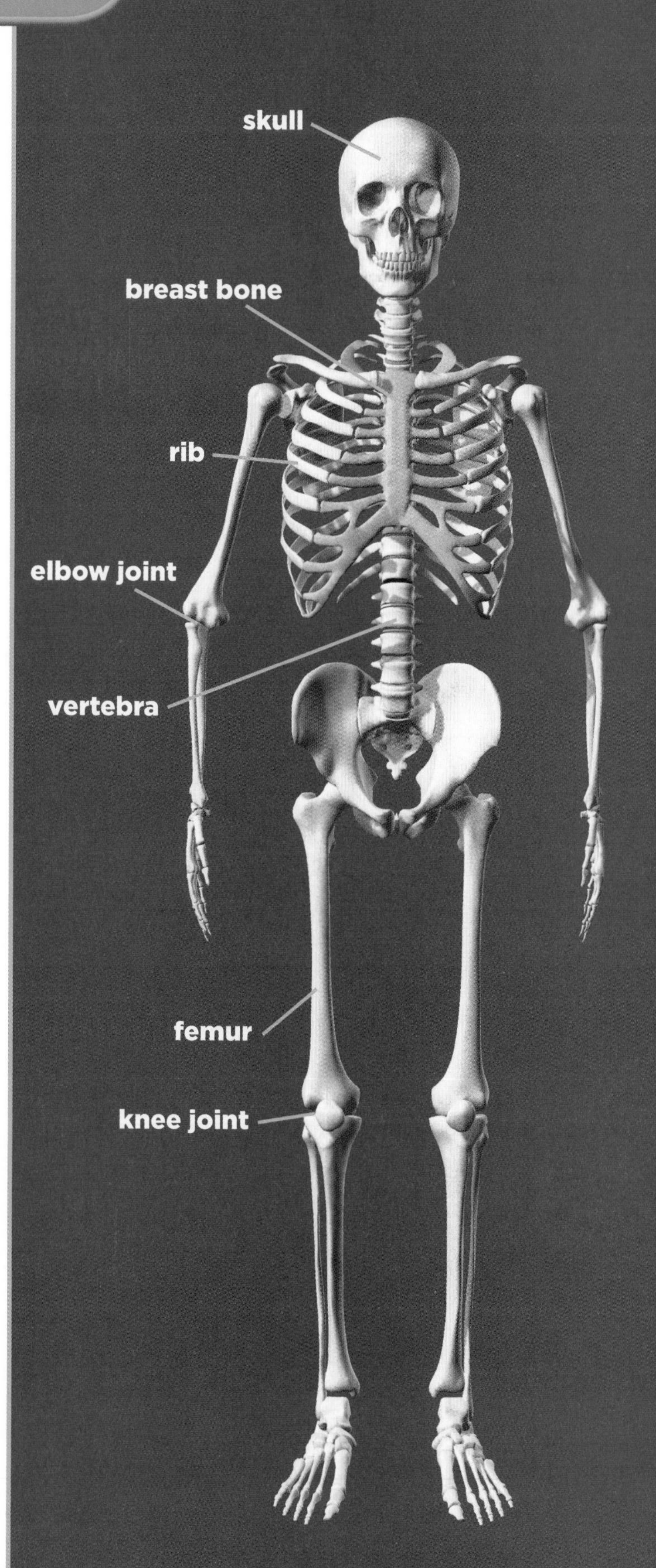

The Muscular System

Together, all the muscles in the body form the muscular system. Muscles allow the body to move. Without muscles, you would not be able to run, smile, breathe, or even blink.

Most muscles are attached to bones and skin. These are called skeletal muscles. To move bones back and forth, skeletal muscles usually work in pairs. Each pulls on a bone in a different direction. When you want to move, your brain sends a message to a pair of skeletal muscles. One muscle contracts, or gets shorter. It pulls on the bone and skin. The other muscle relaxes to let the bone move.

▲ **There are 53 muscles in your face. You use 12 of them whenever you smile.**

biceps

triceps

thigh muscles

calf muscles

◀ **To bend his arm, this boy's biceps contract while his triceps relax.**

Some muscles work without you even thinking about it. The heart is made of muscle. It pumps blood throughout the body even while you sleep. Smooth muscle in the lungs helps you breathe. Smooth muscle in the stomach helps you digest food.

Human Body Systems

The Circulatory System

The body's cells need a constant supply of oxygen and nutrients. The circulatory (SUR•kyuh•luh•tor•ee) system is responsible for sending these things throughout the body. The circulatory system is made up of the heart, blood vessels, and blood.

Blood rich in oxygen travels from the lungs to the heart. The heart is an organ about the size of a fist. It beats about 70 to 90 times each minute, pumping blood through the blood vessels.

Blood vessels are tubes that carry blood. There are two main types of blood vessels. Arteries are blood vessels that carry blood away from the heart. Veins carry blood back to it.

Blood contains plasma, red blood cells, white blood cells, and platelets. Plasma is the liquid part of blood. It carries nutrients and other things the body needs. Red blood cells carry oxygen to all the cells of your body. Red blood cells and plasma also carry wastes, such as carbon dioxide, away from cells. White blood cells work to fight disease. Platelets keep you from bleeding too much when you get a cut.

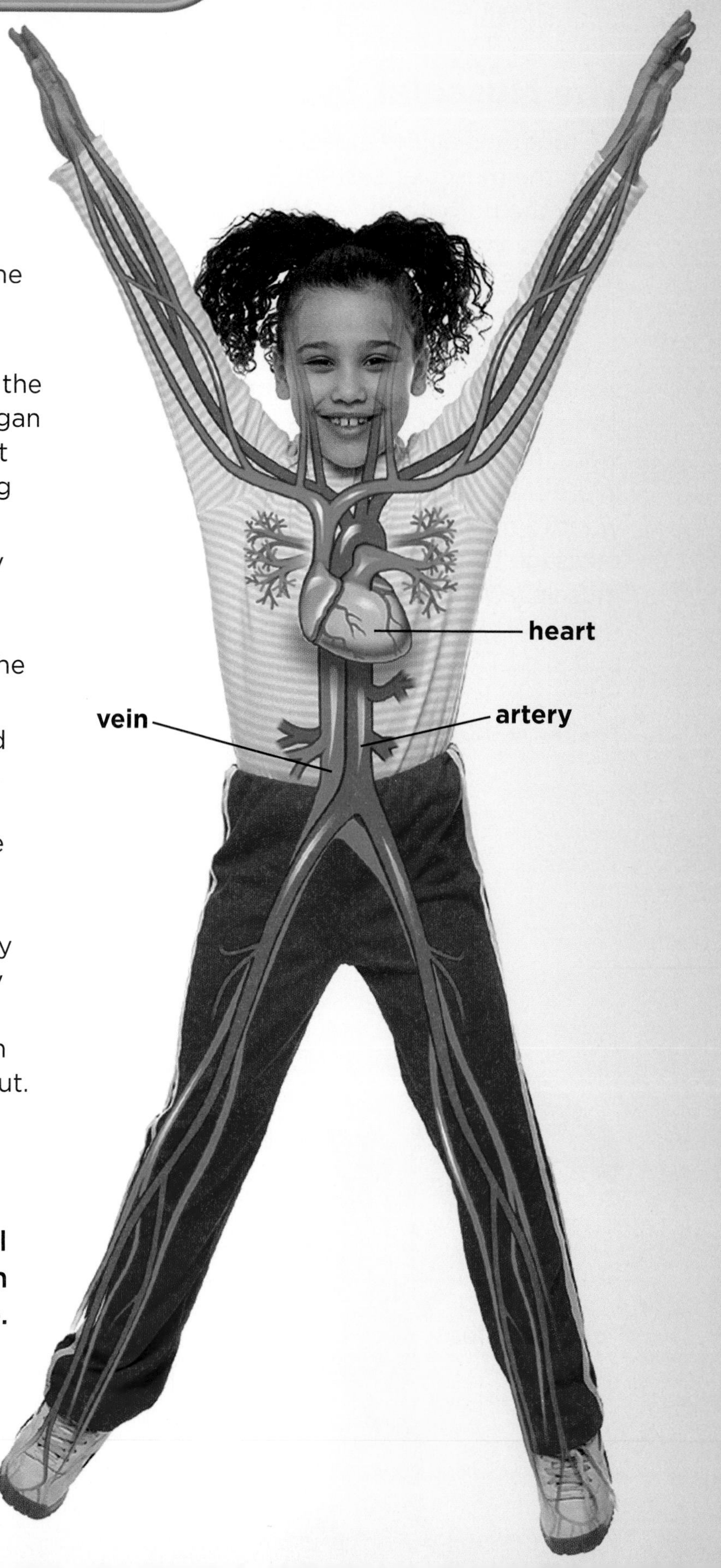

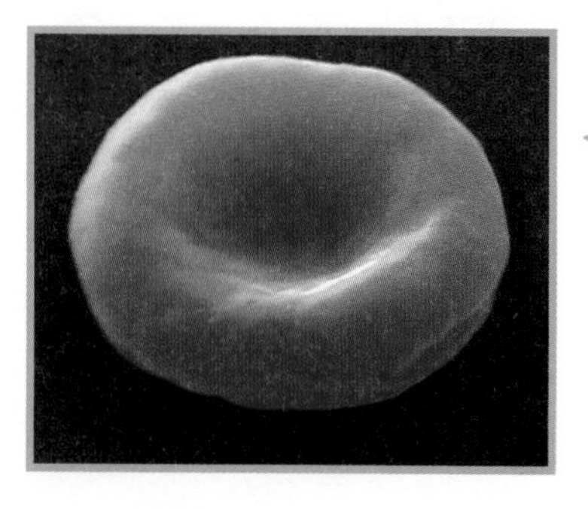

This is how a red blood cell looks through a microscope.

The Respiratory System

The respiratory (RES•pruh•tor•ee) system helps the body take in oxygen and give off carbon dioxide and other waste gases. All the cells in your body require oxygen to work properly. You take in oxygen from the air when you breathe.

Every time you inhale, a muscle called the diaphragm (DI•uh•fram) contracts. This makes room in your lungs for air. Air is taken in through the nose or mouth. This air travels down the throat into the trachea (TRAY•kee•uh).

In the chest, the trachea splits into two bronchial (BRAHN•kee•ul) tubes. Each tube leads to a lung. Inside each lung, the bronchial tube branches off into smaller tubes called bronchioles (BRAHN•kee•olz). At the end of each bronchiole are millions of tiny air sacs. Here, red blood cells release carbon dioxide, a waste gas, and absorb oxygen. When you breathe out, the diaphragm relaxes. This causes the lungs to deflate and push carbon dioxide out of your body through the nose and mouth.

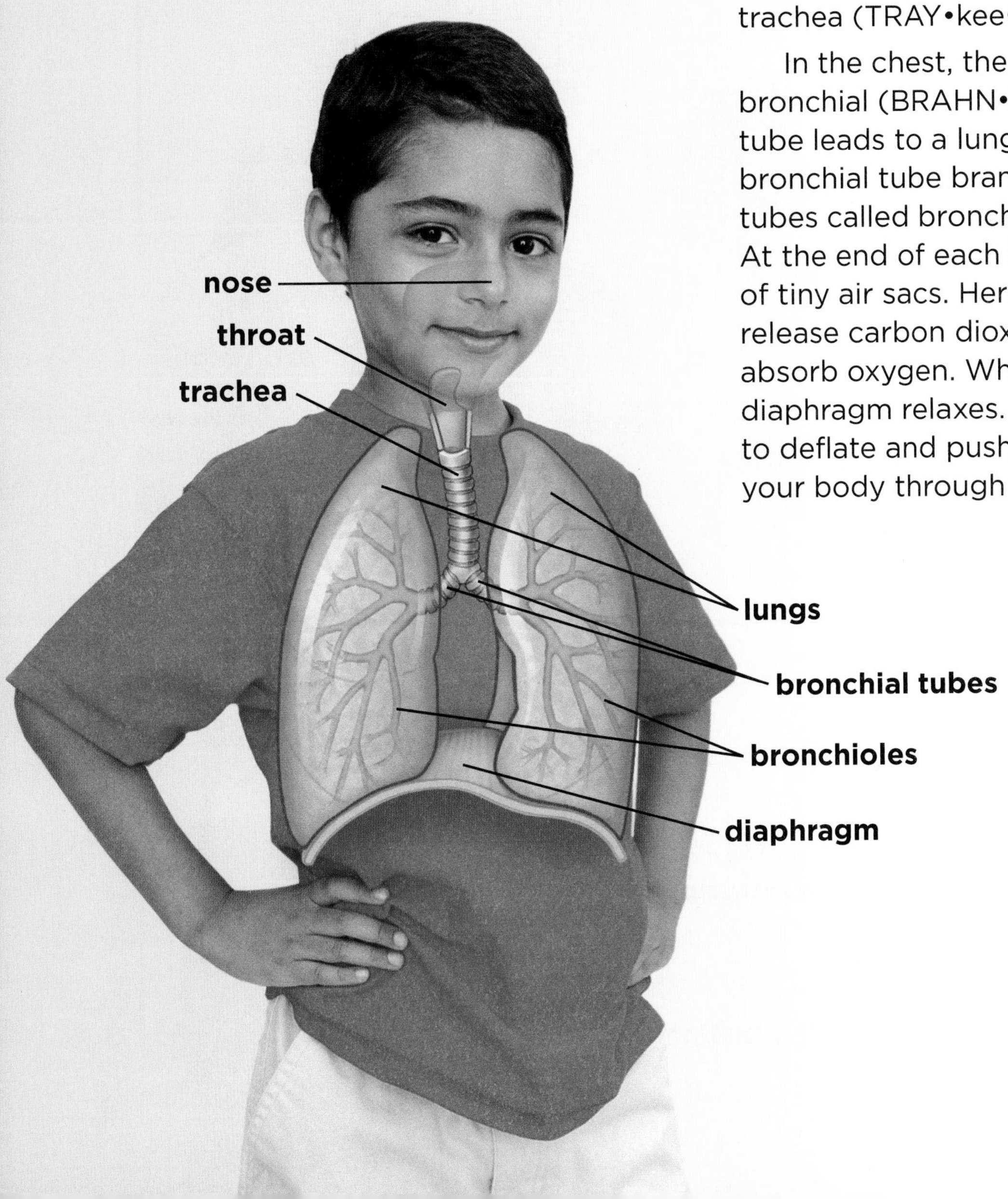

The Digestive System

The digestive (di•JES•tiv) system is responsible for breaking down food into nutrients the body can use. Digestion begins when you chew food. Chewing breaks food into smaller pieces and moistens it with saliva. Saliva helps food travel smoothly when you swallow. The food travels down your esophagus (ih•SAH•fuh•gus) and into your stomach.

Inside the stomach food is mixed with strong, acidic juices. This causes the food to break down further, making it easier for your body to absorb nutrients from the food.

After passing through the stomach, food moves into the small intestine (in•TES•tun). This is where most nutrients are absorbed. The small intestine is a narrow tube about 6 meters (20 feet) long. It is coiled tightly so it fits inside the body. As food passes through the small intestine, digested nutrients are absorbed into the blood. The blood then carries these nutrients to other parts of the body.

After food has passed through the small intestine, it enters the large intestine. The large intestine removes water from the unused food that is left. Then the unused food is removed from the body as waste.

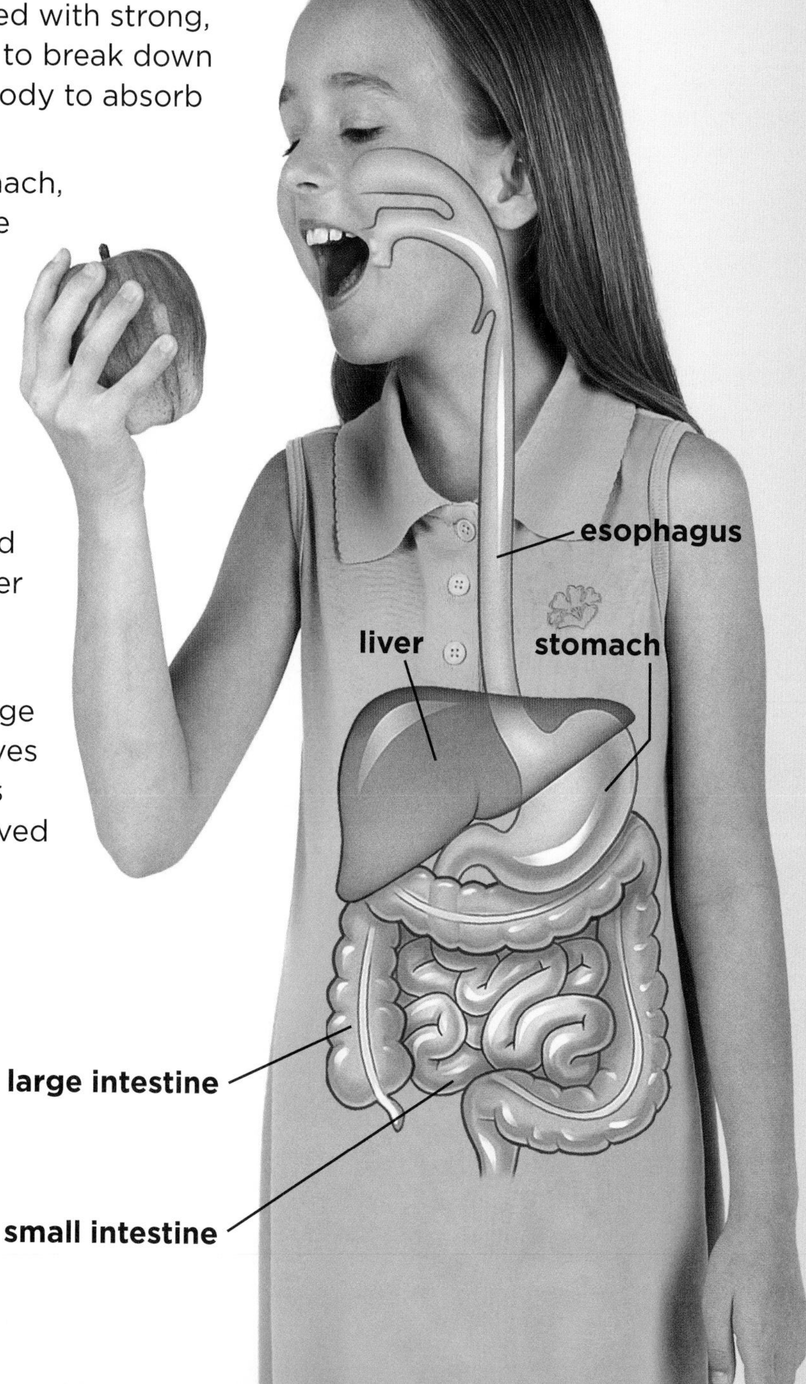

The Excretory System

The excretory (EK•skruh•tor•ee) system gets rid of waste products from your cells. Waste products are materials that the body does not need, such as extra water and salts. The liver, kidneys, bladder, and skin are some organs of the excretory system.

Liver, Kidneys, and Bladder

The liver filters wastes from the blood. It changes wastes into a chemical call urea and sends the urea to the kidneys. Kidneys turn urea into urine. Urine flows from the kidney to the bladder. It is stored in the bladder until it is pushed out of the body through the urethra.

Skin

The skin takes part in excretion when a person sweats. Sweat glands in the inner layer of skin produce sweat. Sweat is made of water and minerals that the body does not need. Sweat is released through the outer layer of the skin. Sweating cools the body and helps it maintain an internal temperature of about 98°F (37°C).

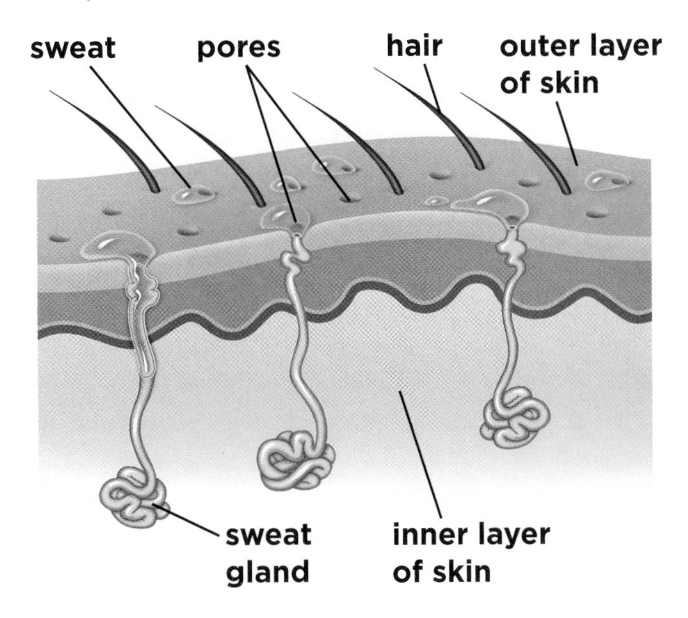

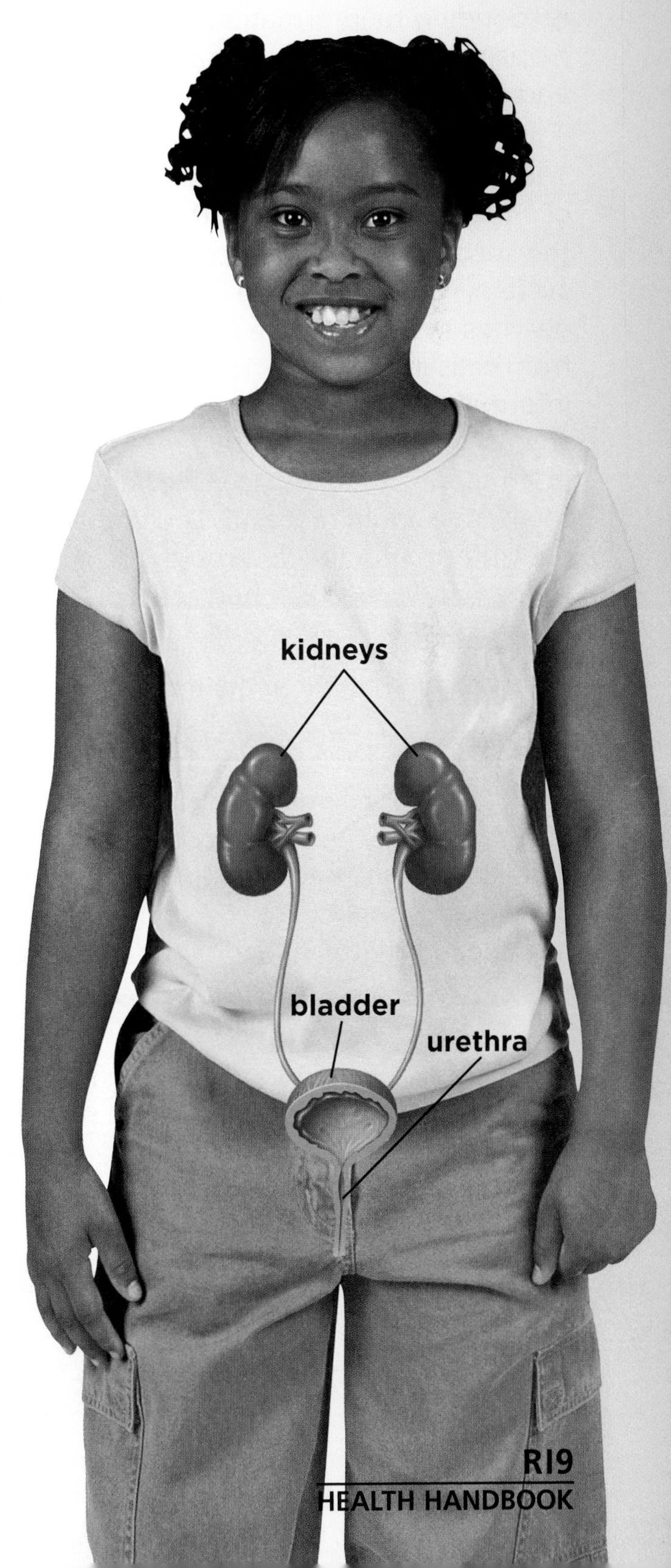

Human Body Systems

The Nervous System

The nervous system is responsible for taking in and responding to information. It controls muscles and helps the body balance. It allows a person to think, feel, and even dream.

The nervous system is made up of two main parts. The first part, the central nervous system, is made up of the brain and spinal cord. All other nerves make up the second part, the peripheral (puh•RIH•frul) nervous system. Nerves from the peripheral nervous system receive sensory information from cells in the body. They pass this information on to the brain through the spinal cord. When the brain receives this information, it makes decisions about how the body should respond. Then it passes this new information back through the spinal cord to the nerves, and the body responds.

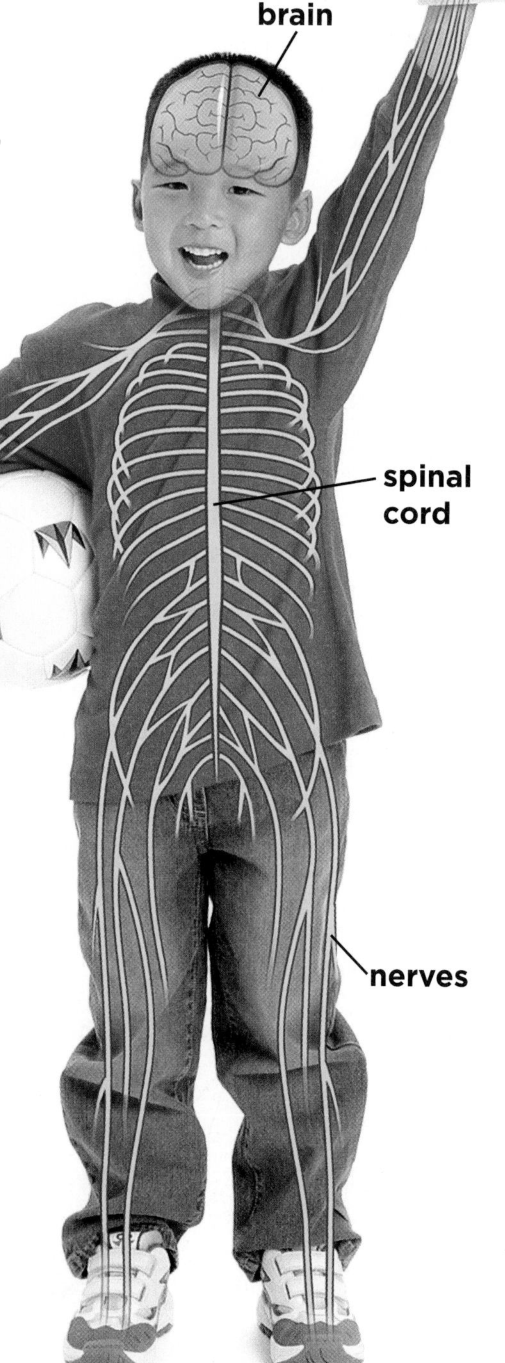

The Brain

The brain has three main parts, the cerebrum (suh•REE•brum), the cerebellum (ser•uh•BE•lum), and the brain stem. The cerebrum is the largest part of the brain. It stores memories and helps control information received by the senses. The cerebellum helps the body keep its balance and directs the skeletal muscles. The brain stem connects to the spinal cord. It controls heartbeat, breathing, and blood pressure.

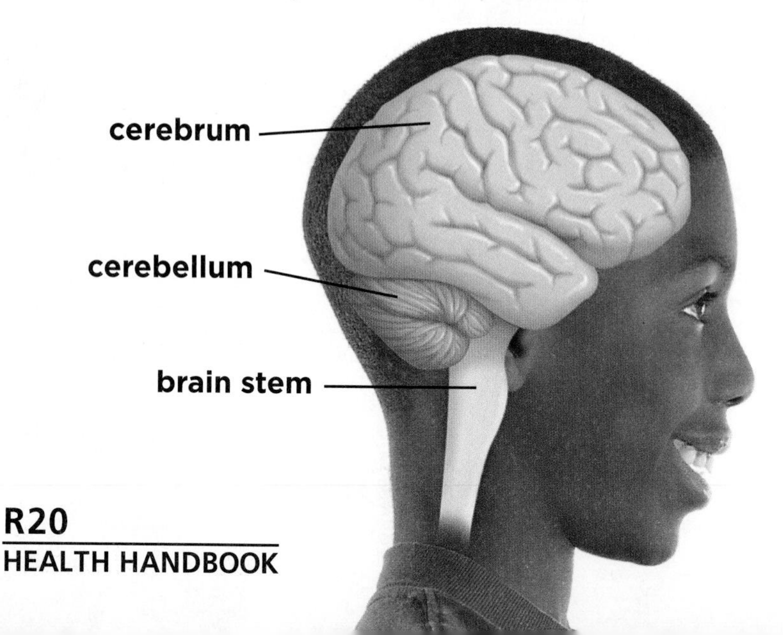

The Senses

Different nerves in the body take in information from the environment. These nerves are responsible for the body's sense of sight, hearing, smell, taste, and touch.

Sight

Light reflects off an object, such as a leaf, and into the eye. The reflected light passes through the pupil in the iris. Cells in the eye change light into electrical signals. The signals travel through the optic nerve to the brain.

Hearing

Sound waves enter the outer ear. They reach the eardrum and cause it to vibrate. Cells in the ear change the sound waves into electrical signals. The signals travel along the auditory nerve to the brain.

Smell

As a person breathes, chemicals in the air mix with mucus in the upper part of the nose. When they reach certain cells in the nose, those cells send information along the olfactory nerve to the brain.

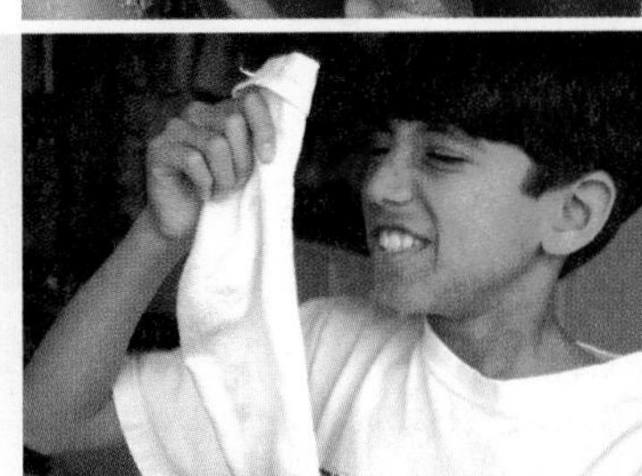

Taste

On the tongue are more than 10,000 tiny bumps, called taste buds. Each taste bud can sense four main tastes—sweet, sour, salty, and bitter. The taste buds send information along a nerve to the brain.

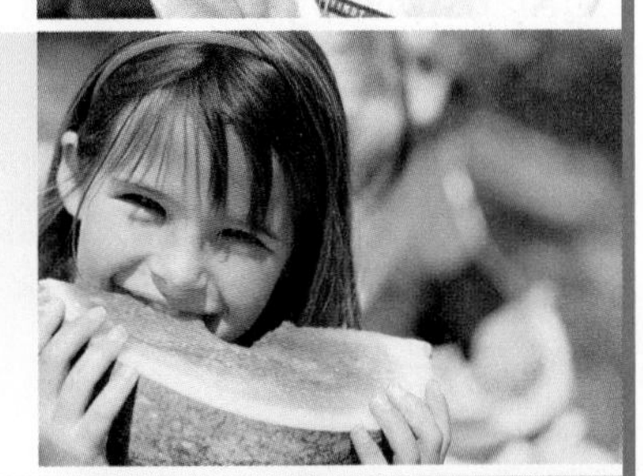

Touch

Different nerve cells in the skin give the body its sense of touch. They help a person tell hot from cold, wet from dry, and hard from soft. Each cell sends information to the spinal cord. The spinal cord then sends the information to the brain.

Immune System

The immune system protects the body from germs. Germs cause disease and infection. Most of the time, the immune system is able to prevent germs from entering the body. Skin, tears, and saliva are parts of the immune system. They work to kill germs and keep them out of the body.

When germs do find a way into your body, white blood cells help find and kill them quickly before you become ill. White blood cells are part of the blood. They travel through blood vessels and lymph (LIMF) vessels. Lymph vessels are similar to blood vessels. However, instead of carrying blood, they carry a fluid called lymph. Many white blood cells are made and live in lymph nodes. Here, they filter out harmful materials from the body.

White blood cells might not always kill germs before the germs start to reproduce in your body. When germs reproduce, they cause illness. Even while you feel ill, the immune system works to kill and remove germs until you are well again.

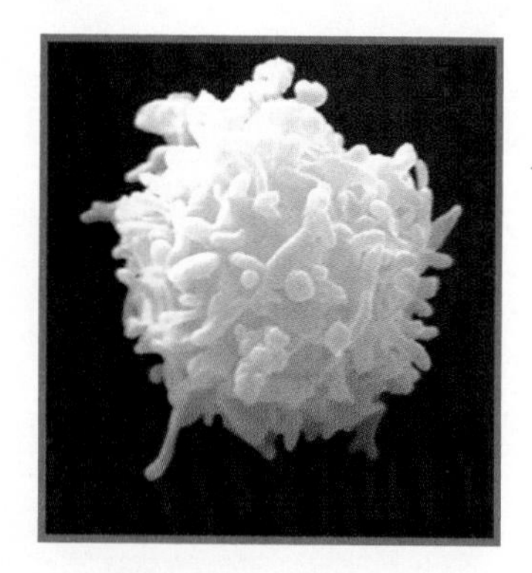

◀ **This is how a white blood cell looks through a microscope.**

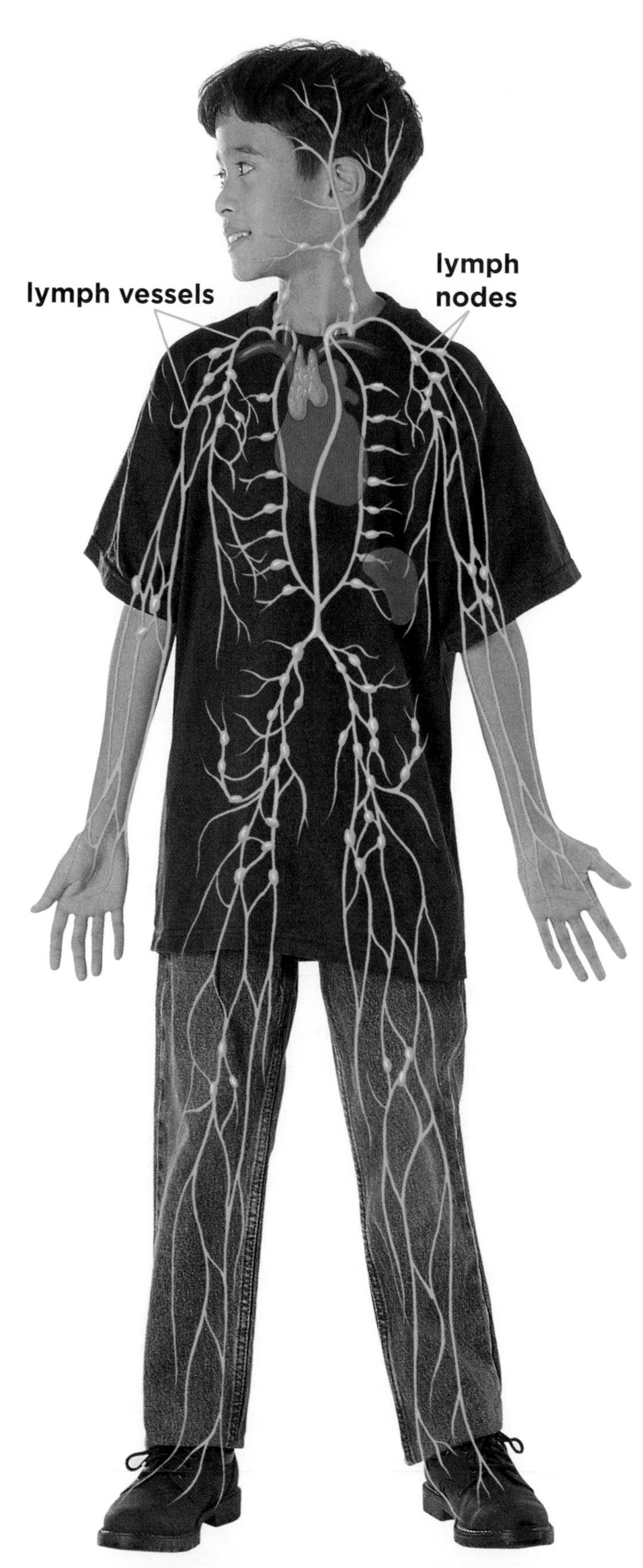

Viruses and Bacteria

One of the main types of germs that makes the body ill are viruses. Illness from a virus like a cold or flu can be a big deal. Yet viruses themselves are very small. In fact, you need a special microscope, an electron microscope, to look at a virus.

Viruses need to be inside living cells, called hosts, to reproduce. As they reproduce, viruses take nutrients and energy from the cell. They can even produce harmful materials that make the body itch or have dangerously high temperatures.

Bacteria is the other main type of germ that can make the body ill. Bacteria are tiny, one-celled organisms. They can live on most surfaces and are able to reproduce outside of cells. Some bacteria can have a harmful effect on the body. Other bacteria, however, are good for the body. Some bacteria in your body, for example, help you digest food.

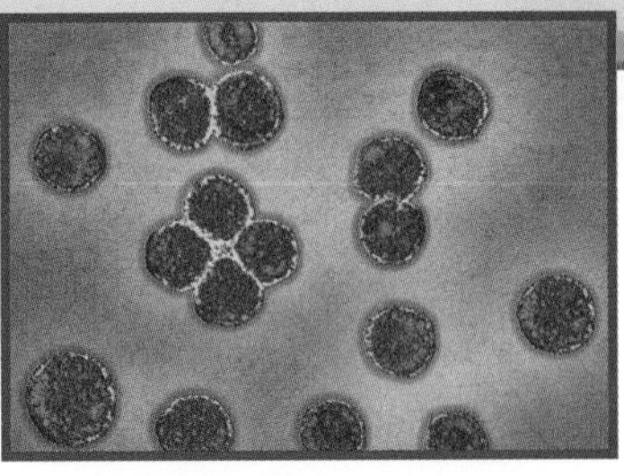

▲ A cold virus as seen through a microscope.

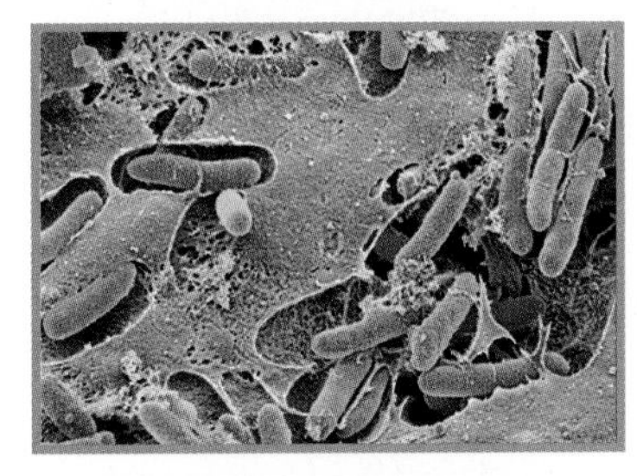

▲ *E. coli* bacteria as seen through a microscope.

You can help your body defend itself against germs. Here is what you can do.

▶ Eat healthful foods. This helps your body get all the nutrients it needs to stay healthy. A healthy body is better able to fight germs.

▶ Be active. Being active makes your body fit. A fit body is better able to fight germs.

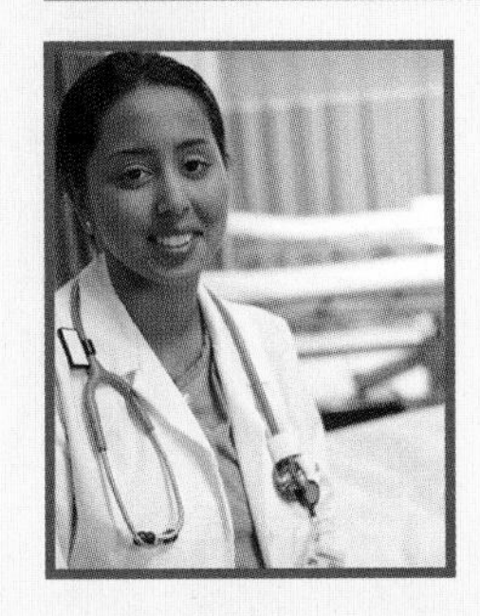

▶ Get a yearly checkup. Make sure you get all of your immunizations. Follow directions when taking medicines given to you by a doctor.

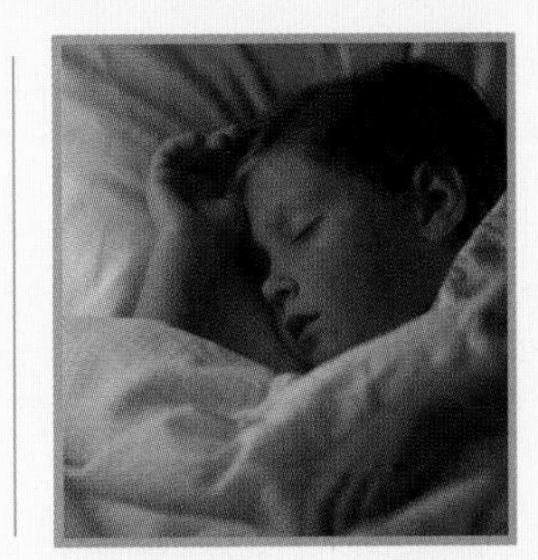

▶ Get plenty of rest. You need about 10 hours of sleep each night. Sleeping helps repair your body. Get extra rest when you are ill.

▶ Do not share cups or utensils with other people. Germs can be on objects you touch. Wash your hands, especially before eating and drinking. By washing your hands, you kill germs and make it harder for harmful things to get into your body.

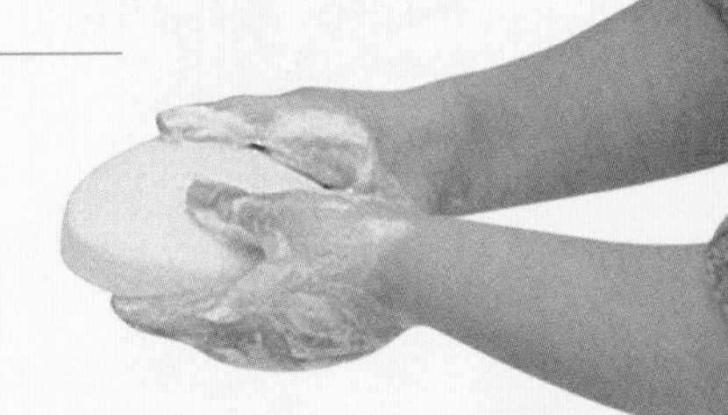

Nutrients

Nutrients are materials in foods that help the body grow, get energy, and stay healthy. By eating a balance of healthful foods, your body gets the nutrients it needs to do all of these things.

There are six kinds of nutrients—carbohydrates, vitamins, minerals, proteins, water, and fats. Each nutrient helps the body in different ways.

carbohydrates

Carbohydrates

Carbohydrates are the main source of energy for the body. Starches and sugars are two types of carbohydrates. Starches come from foods like bread, pasta, and cereal. They provide long-lasting energy. Sugars come from fruits and can be used immediately by the body for energy.

Vitamins

Vitamins help keep the body healthy. They also help to build new cells in the body. The table below shows some vitamins and their sources.

Vitamin	Sources	Benefits
A	milk, fruit, carrots, green vegetables	keeps eyes, teeth, gums, skin, and hair healthy
C	citrus fruits, strawberries, tomatoes	helps heart, cells, and muscles function
D	milk, fish, eggs	helps keep teeth and bones strong

Minerals

Minerals help form new bone and blood cells. They also help your muscles and nervous system work properly. Here are some minerals and their sources.

Mineral	Sources	Benefits
calcium	yogurt, milk, cheese, green vegetables	builds strong teeth and bones
iron	meat, beans, fish, whole grains	helps red blood cells function properly
zinc	meat, fish, eggs	helps your body grow and helps to heal wounds

Fats

Fats help the body use other nutrients and store vitamins. Fats also help the cells of the body work properly. They even help keep the body warm. Fats can be found in foods such as meats, eggs, milk, butter, and nuts. Oils also contain fats. Though some fats help the body, some fats can cause health problems.

Water

Water is one of the most important nutrients. About $\frac{2}{3}$ of the body is made of water! Water makes up most of the body's cells. It helps the body remove waste and protects joints. It also prevents the body from getting too hot.

Proteins

Proteins are a part of every living cell. Proteins help bones and muscles grow. They even help the immune system fight diseases. Foods high in protein are milk, eggs, meats, fish, nuts, and cheese.

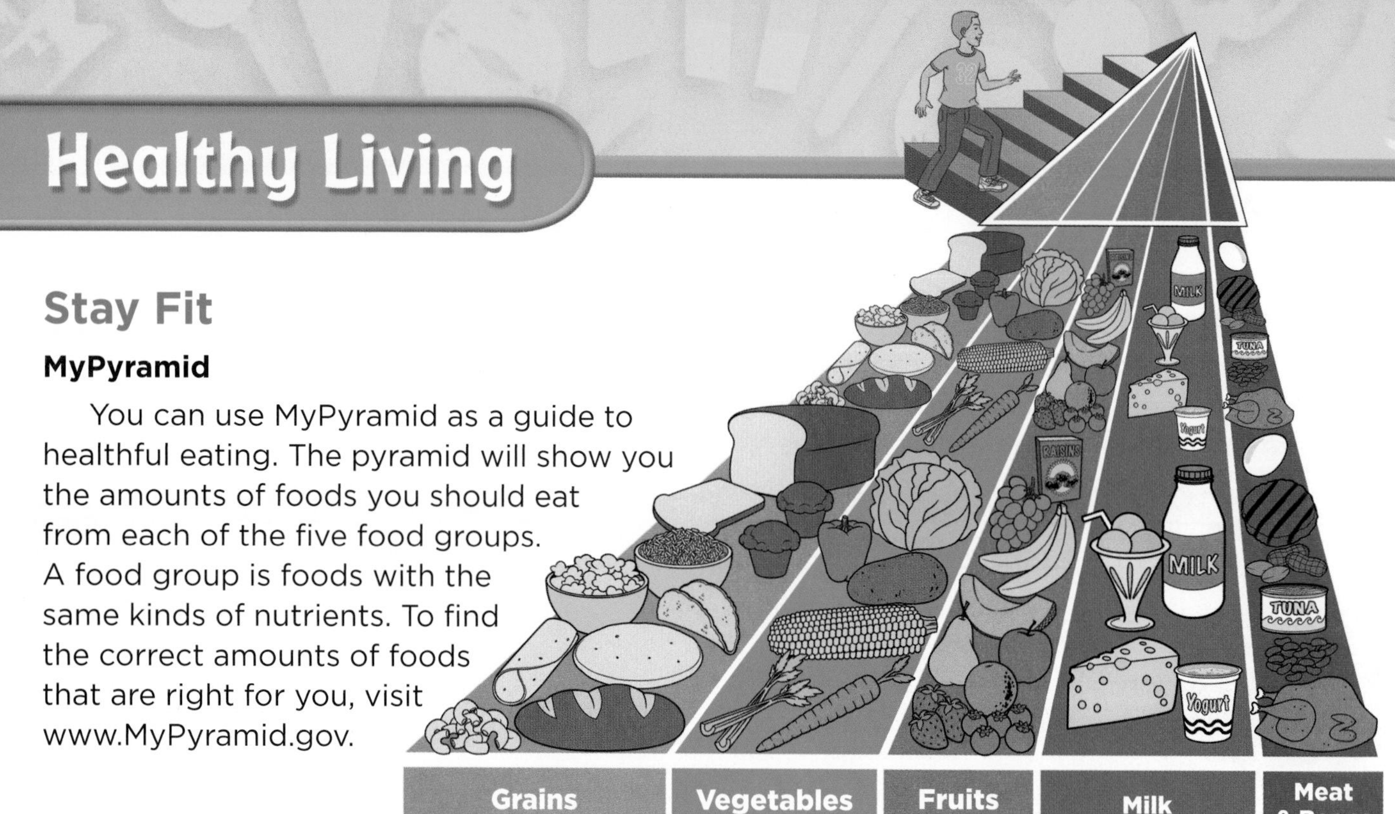

Stay Fit

MyPyramid

You can use MyPyramid as a guide to healthful eating. The pyramid will show you the amounts of foods you should eat from each of the five food groups. A food group is foods with the same kinds of nutrients. To find the correct amounts of foods that are right for you, visit www.MyPyramid.gov.

Be Drug Free

Do not use cigarettes, illegal drugs, or alcohol. These things can harm your body. They can keep you from growing properly and becoming fit.

Be Physically Active

You need to be physically active for at least 60 minutes every day. When you are physically active, you become physically fit. When you are physically fit, your heart, lungs, bones, joints, and muscles stay strong. You keep a healthful weight and lower the risk of disease. You do not have to be on a sports team to be physically active. You just need to move your body. Running, biking, and swimming are just some ways to be physically active.

by Dinah Zike

Folding Instructions

The following pages offer step-by-step instructions about how to make Foldables study guides.

Half-Book

Fold a sheet of paper ($8\frac{1}{2}$" x 11") in half.

1. This book can be folded vertically like a hot dog or . . .
2. . . . it can be folded horizontally like a hamburger

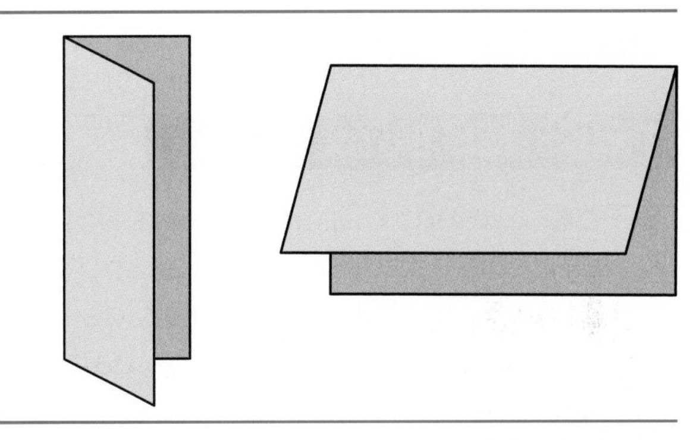

Folded Book

1. Make a half-book.
2. Fold in half again like a hamburger. This creates a ready-made cover and two small pages inside for recording information.

Pocket Book

1. Fold a sheet of paper ($8\frac{1}{2}$" x 11") in half like a hamburger.
2. Open the folded paper. Fold one of the long sides up two inches to form a pocket. Refold along the hamburger fold so the newly formed pockets are on the inside.
3. Glue the outer edges of the two-inch fold with a small amount of glue.

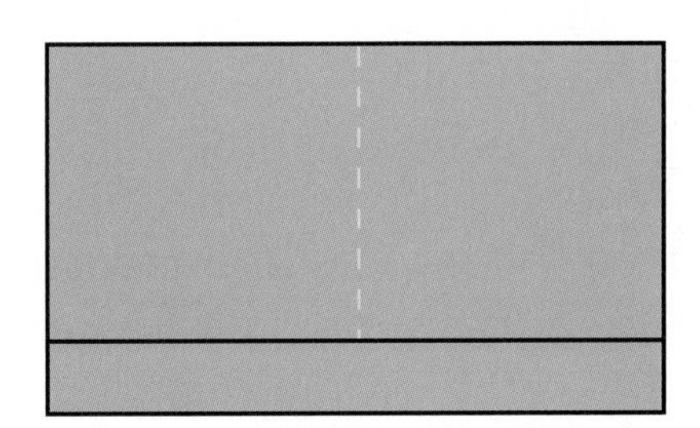

Shutter Fold

1. Begin as if you were going to make a hamburger, but instead of creasing the paper, pinch it to show the midpoint.
2. Fold the outer edges of the paper to meet at the pinch, or midpoint, forming a shutter fold.

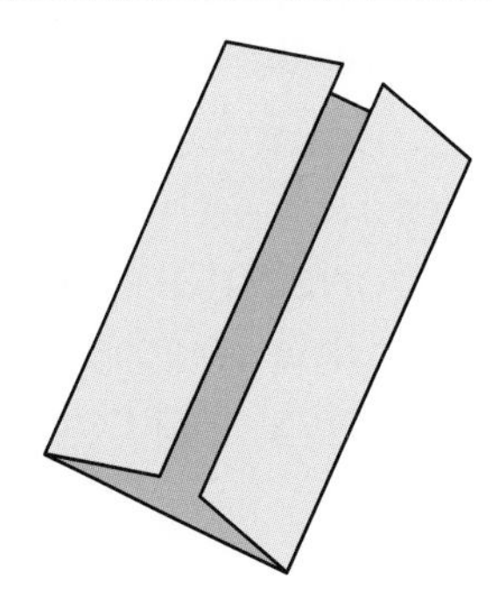

Trifold Book

1. Fold a sheet of paper ($8\frac{1}{2}$" x 11") into thirds.
2. Use this book as is or cut it into shapes.

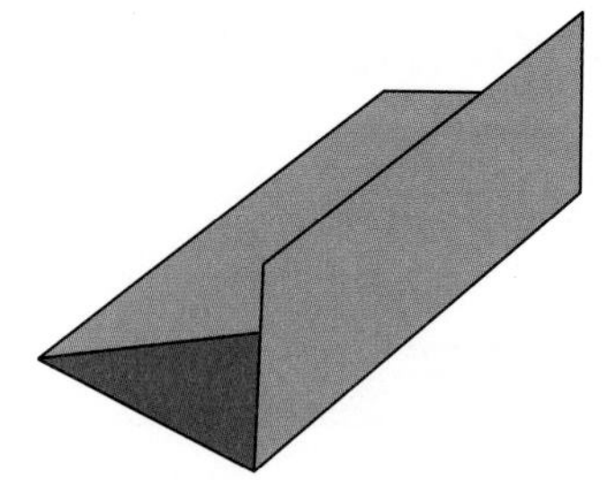

Three-Tab Book

1. Fold a sheet of paper like a hot dog.
2. With the paper horizontal and the fold of the hot dog up, fold the right side toward the center, trying to cover one half of the paper.
3. Fold the left side over the right side to make a book with three folds.
4. Open the folded book. Place one hand between the two thicknesses of paper and cut up the two valleys on one side only. This will create three tabs.

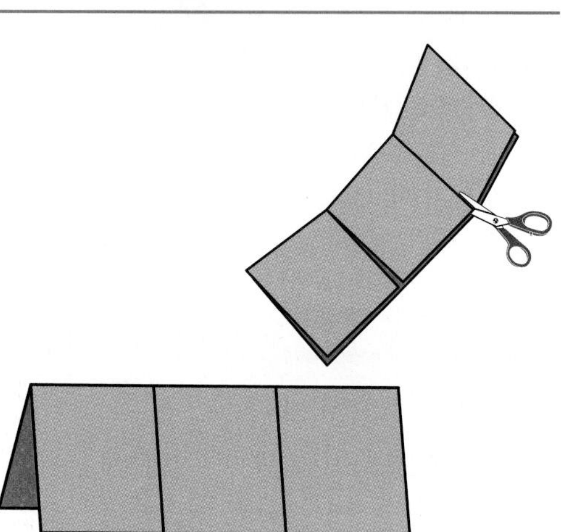

Layered-Look Book

1. Stack two sheets of paper ($8\frac{1}{2}$" x 11") so the back sheet is one inch higher than the front sheet.
2. Bring the bottoms of both sheets upward and align the edges so all the layers or tabs are the same distance apart.
3. When all the tabs are an equal distance apart, fold the papers and crease well.
4. Open the papers and glue them together along the valley, or inner center fold, or staple them along the mountain.

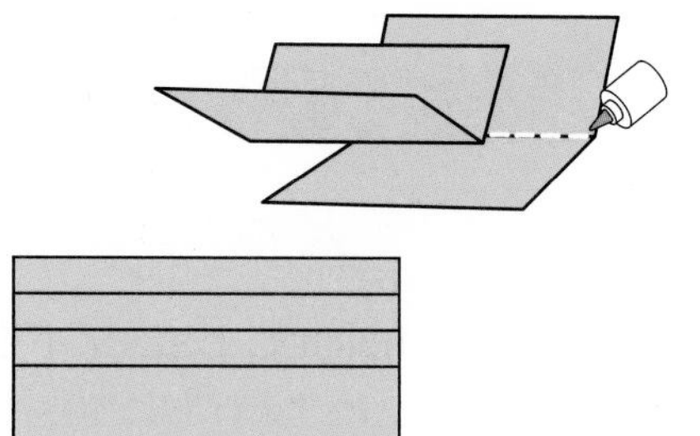

Folded Table or Chart

1. Fold the number of vertical columns needed to make the table or chart.
2. Fold the horizontal rows needed to make the table or chart.
3. Label the rows and columns.

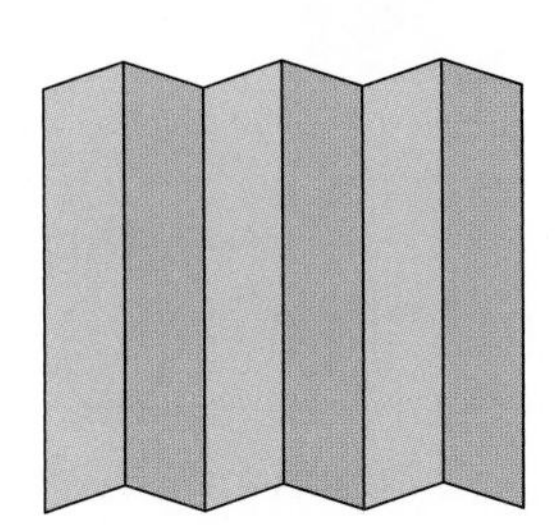

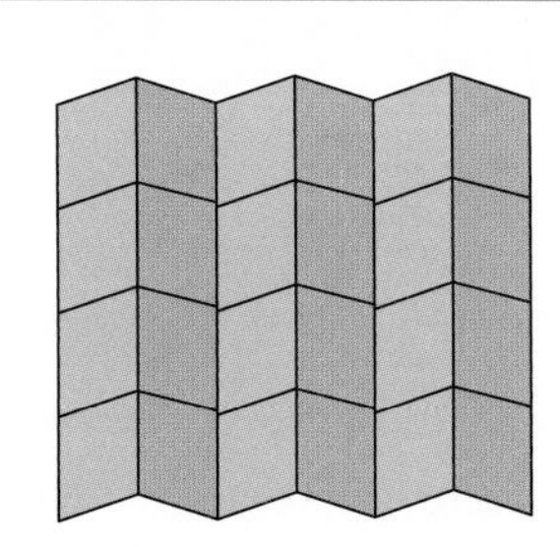

Glossary

Use this glossary to learn how to pronounce and understand the meanings of science words used in this book. The page number at the end of each definition tells you where to find that word in the book.

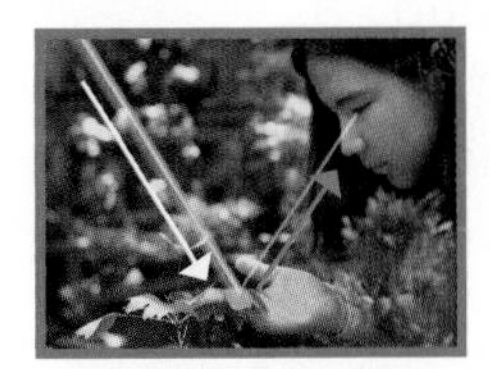

absorb (əb·sôrb′) to take in (p. 500) *A green leaf absorbs all colors of light except for green.*

adaptation (a′dəp·tā′shən) a structure or behavior that helps an organism survive in its environment (p. 134) *Sharp spines are one adaptation that helps a cactus survive.*

air pressure (âr pre′shər) the weight of air pressing down on Earth (p. 283)

amphibian (am·fi′bē·ən) a vertebrate that spends part of its life in water and part of its life on land (p. 59)

atmosphere (at′mə·sfîr) a blanket of gases and tiny bits of dust that surround Earth (p. 280)

axis (ak′səs) a line through the center of a spinning object (pp. 305, 319)

bird (bûrd) a vertebrate that has a beak, feathers, two wings, and two legs and lays eggs (p. 58)

blizzard (bli′zərd) a storm with lots of snow, cold temperatures, and strong winds (p. 297)

Pronunciation Key

The following symbols are used throughout th*is* glossary.

a	**a**t	e	**e**nd	o	s**o**ft	u	**u**p	hw	**wh**ite	ə	**a**bout
ā	**a**pe	ē	m**e**	ō	g**o**	ū	**u**se	ng	so**ng**		tak**e**n
ä	f**a**rther	i	t**i**p	ôr	f**or**k	ü	r**u**le	th	**th**in		penc**i**l
âr	c**ar**e	ī	**i**ce	oi	**oi**l	u̇	p**u**ll	th	**th**is		lem**o**n
ô	l**aw**	îr	f**ear**	ou	**ou**t	ûr	t**ur**n	yü	r**u**le		circ**u**s
								zh	mea**s**ure		

′ = primary accent; shows which syllable takes the main stress, such as **at** in **atmosphere** (at′mə·sfîr′)

′ = secondary accent; shows which syllables take lighter stresses, such as **sphere** in **atmosphere**

boil (boil) the change from a liquid to a gas at a certain temperature (p. 399) *Water will boil if you heat it on the stove.*

camouflage (kam′ə·fläzh′) an adaptation that allows an organism to blend into its surroundings (p. 134)

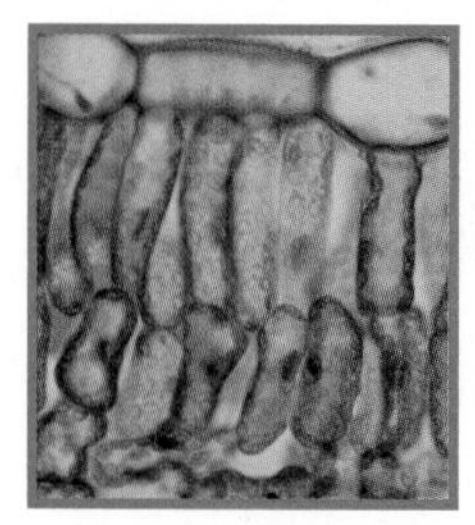

cell (sel) the basic building block of life (p. 26) *You can use a microscope to see that a leaf is made up of many tiny cells.*

chemical change (ke′mi·kəl chānj) a change that causes different kinds of matter to form (p. 418) *Rust forming on this train is a chemical change.*

circuit (sûr′·kət) the path that is made of parts that work together to allow current to flow (p. 515) *This circuit is made of a battery, wires, a light bulb, and a switch.*

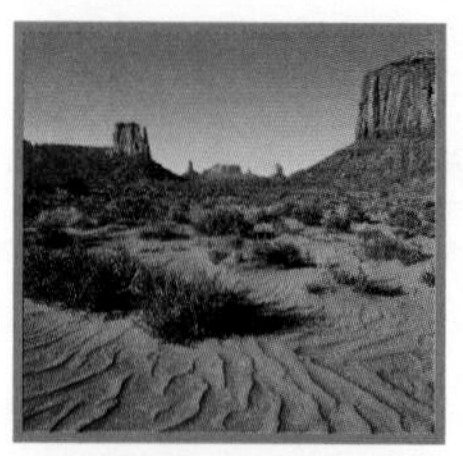

climate (klī′mət) the pattern of weather at a certain place over a long time (pp. 120, 304) *This desert has a hot, dry climate.*

cloud (kloud) a collection of tiny water drops or ice crystals in the air (p. 290)

community (kə·mū′ni·tē) all the living things in one place that interact (p. 166) *All the organisms in this pond make up a community.*

competition (kom′·pə·ti′·shən) the struggle among organisms for water, food, or other needs (p. 153) *There is competition for water between these springbok.*

compound machine (kom′pound mə·shēn′) two or more simple machines put together (p. 470) *Scissors are a compound machine because they are made of levers and wedges.*

condensation (kon′den·sā′shən) the process through which a gas changes into a liquid (p. 293)

LOG ON e-Glossary at www.macmillanmh.com

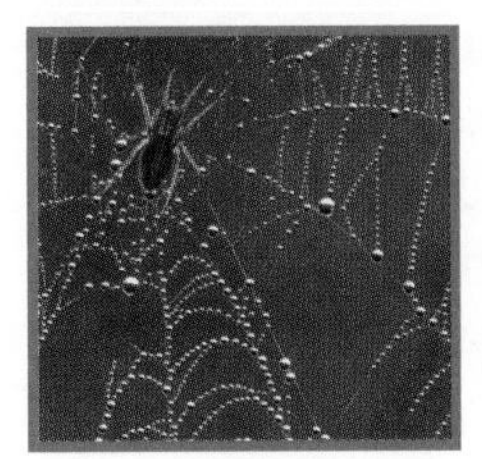

condense (kən·dens′) to change from a gas to a liquid (p. 400) *When water vapor in the air cools and condenses, it can form dew.*

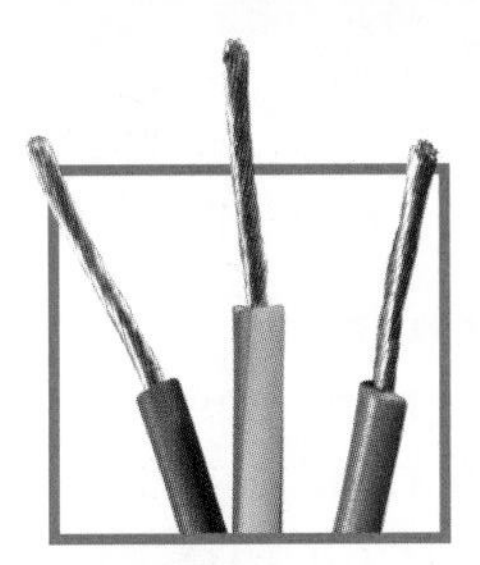

conductor (kən·duk′tər) a material which heat or electric current moves through easily (p. 484) *Copper is a good conductor of heat and electric current.*

cone (kōn) a plant structure where seeds are made in some nonflowering plants (p. 75)

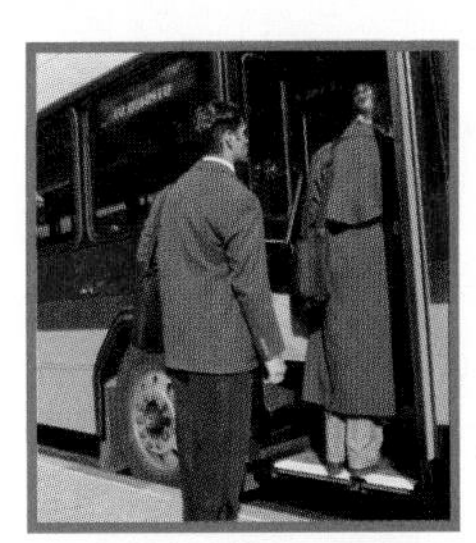

conserve (kən·sûrv′) to use resources wisely (p. 266) *You conserve gasoline when you walk, ride a bicycle, or ride a bus instead of using a car.*

constellation (kon′stə·lā′shən) a group of stars that seem to form a picture (p. 349)

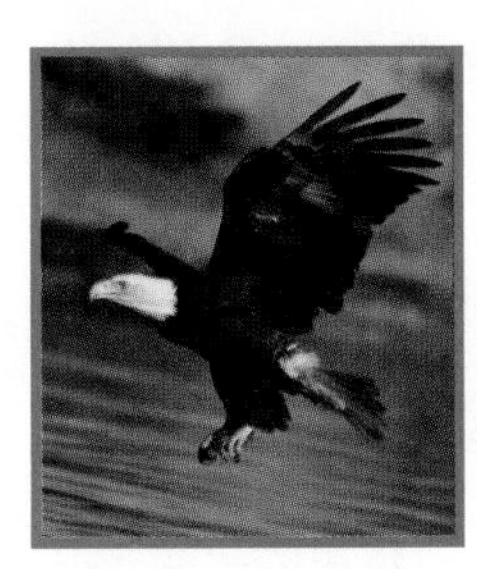

consumer (kən·sü′mər) an organism that eats plants or other animals (p. 110) *Eagles eat fish, snakes, and other small organisms so they are consumers.*

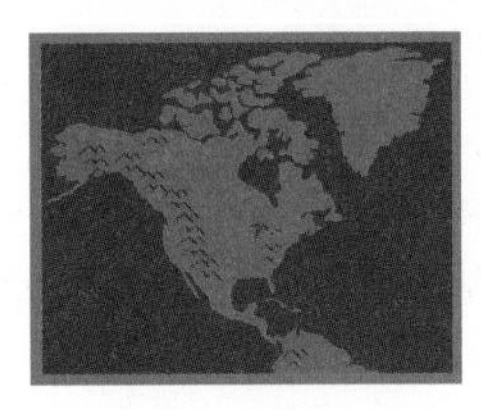

continent (kōn′tə·nənt) a great area of land on Earth (p. 193) *You live on the continent of North America.*

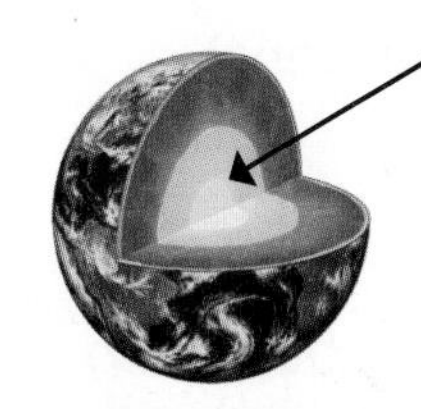

core (kôr) Earth's deepest and hottest layer (p. 198)

crater (krā′tər) a hollow area, or pit, in the ground (p. 332) *There are many craters on the Moon.*

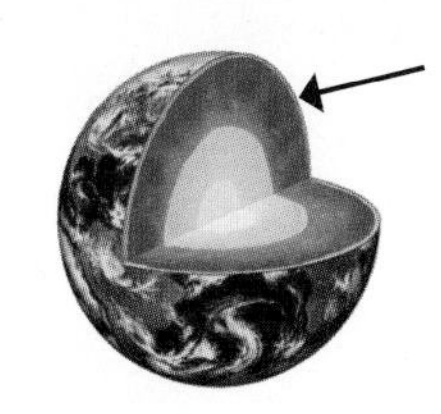

crust (krust) Earth's outermost layer (p. 198)

D

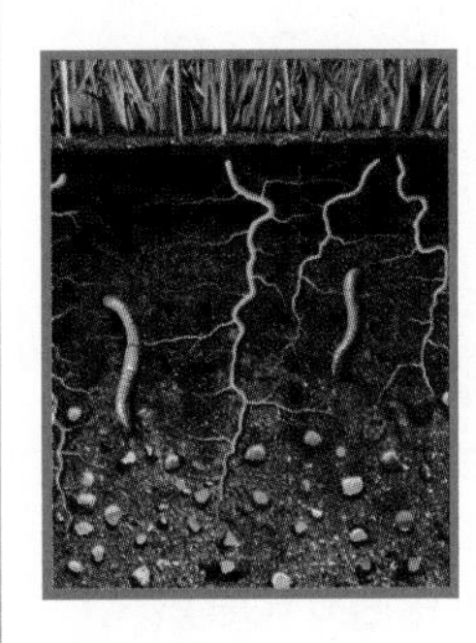

decomposer (dē′kəm·pō′zər) an organism that breaks down dead plant and animal material (p. 111) *Worms are decomposers that eat dead leaves that fall to the ground.*

deposition (de′·pə·zi′shən) the dropping off of weathered rock (p. 216)

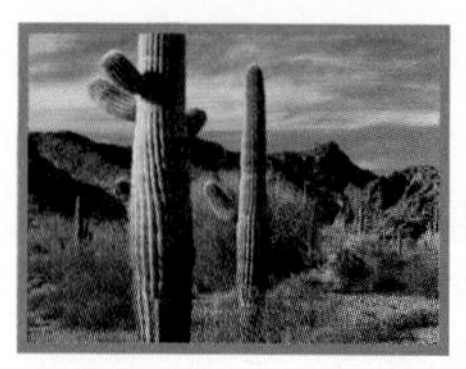

desert (de′zərt) an ecosystem that has a dry climate with little rainfall (p. 122)

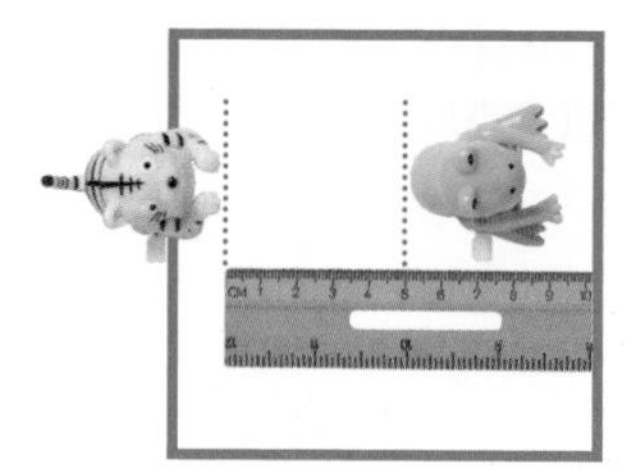

distance (dis′təns) the amount of space between two objects or places (p. 435) *The distance between these two toys is five centimeters.*

drought (drout) when there is no rain in an area for a long period of time (p. 162)

earthquake (ûrth′kwāk) a sudden movement of the rocks that make up Earth's crust (p. 204) *An earthquake caused this road to crack.*

ecosystem (ē′kō·sis′təm) the living and nonliving things that interact in an environment (p. 108)

egg (eg) an animal structure that protects and feeds some very young animals such as birds (p. 83)

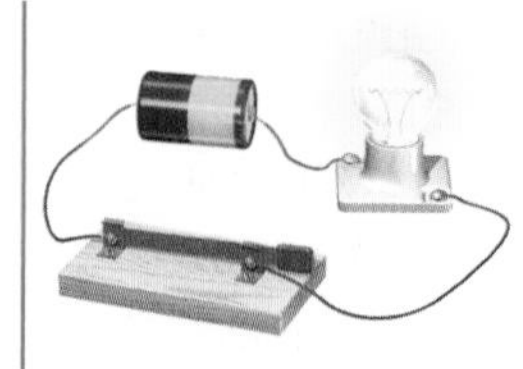

electric current (i·lek′trik kûr′ənt) a flow of charged particles (p. 514) *When electric current flows from a battery to a light bulb, the bulb glows.*

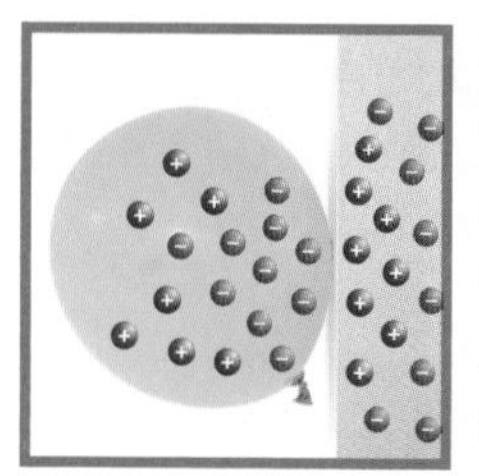

electrical charge (i·lek′tri·kəl chärj) the property of matter that causes electricity (p. 512) *Rubbing a balloon on a sweater gives the balloon a negative electrical charge.*

element (e′lə·mənt) a building block of matter (p. 368) *Gold is an element.*

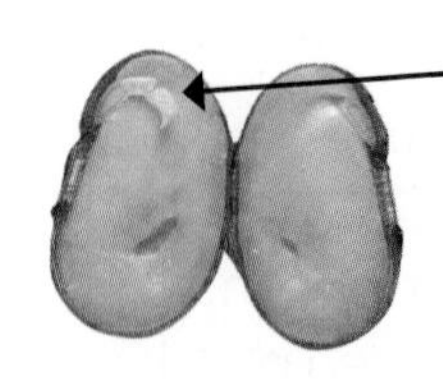

embryo (em′brē·ō) a young plant that is just beginning to grow (p. 70)

endangered (en·dān′jərd) when one kind of organism has very few of its kind left (p. 168) *Bengal tigers are endangered animals because there are very few of them left in the world.*

energy (e′nər·jē) the ability to do work (p. 456) *Energy from this bowling ball causes the pins to fall over.*

environment (en·vī′rən·mənt) all the living and nonliving things that surround an organism (p. 24) *Water, soil, rocks, trees, and zebras are parts of the giraffes' environment.*

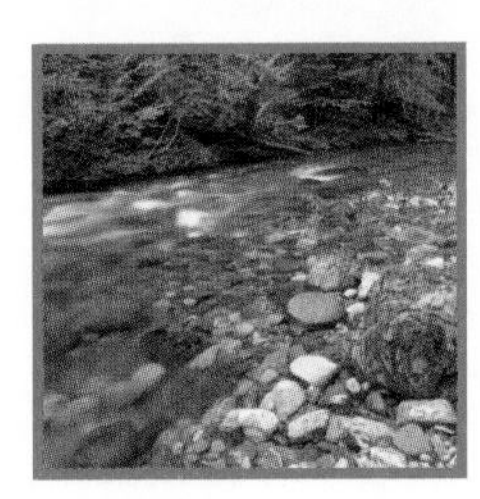

erosion (i·rō′zhən) the weathering away and movement of weathered rock (p. 216) *Erosion happens when water in this stream carries rocks away.*

evaporate (i·vap′ə·rāt′) a change from a liquid to a gas without boiling (p. 399) *These wet clothes will dry when the water in them evaporates.*

evaporation (i·vap′ə·rā′shən) the process through which a liquid changes into a gas without boiling (p. 292)

exoskeleton (ek′sō·skel′i·tən) a hard covering, or shell, that holds up and protects an invertebrate's body (p. 57) *A snail's shell is an exoskeleton.*

extinct (ek·stingkt′) when there are no more of an organism's kind left on Earth (p. 174) *Many dinosaurs became extinct millions of years ago.*

F

fish (fish) a vertebrate that lives in water and breathes oxygen using gills (p. 59)

flood (flud) when dry land becomes covered with water (pp. 162, 208)

flower (flou′ər) a plant structure where seeds are made (p. 72)

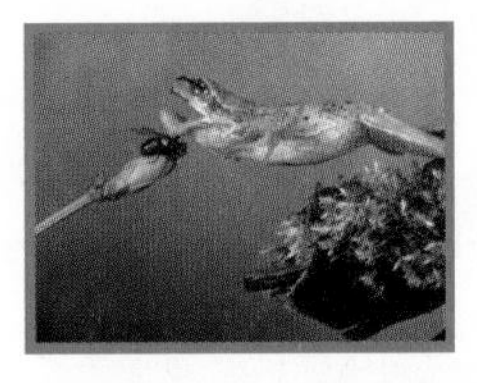

food chain (füd chān) a series of organisms that depend on one another for food (p. 110)

food web (füd web) several food chains that are connected (p. 112)

force (fôrs) a push or a pull (p. 444)

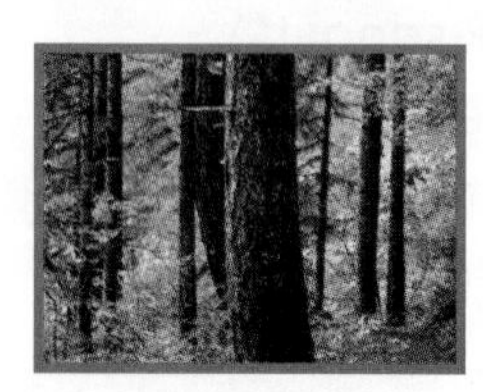

forest (fôr′əst) an ecosystem with many trees (p. 124)

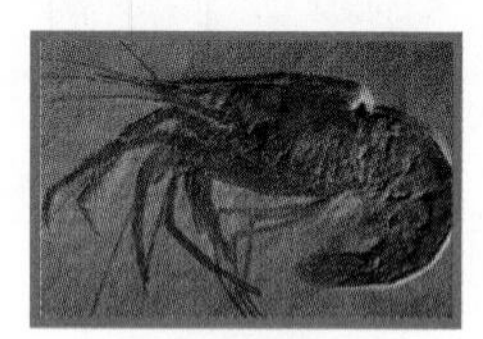

fossil (fä′səl) the trace or remains of something that lived long ago (pp. 174, 250)

freeze (frēz) to change from a liquid to a solid (p. 401) *You can freeze juice to make a juice pop.*

friction (frik′shən) a force that occurs when one object rubs against another (p. 448) *Friction between a break pad and a rim stops a bike.*

fruit (früt) a plant structure that holds seeds (p. 73)

fuel (fū′əl) a material that is burned for energy (p. 252) *Wood and gasoline are examples of fuels.*

gas (gas) matter that has no certain shape or volume (p. 387) *These balloons are filled with a gas called helium.*

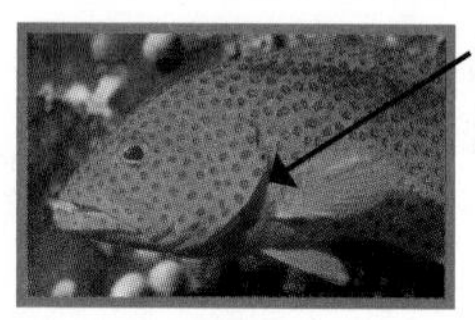

gills (gilz) a structure some animals use to take in oxygen from water (p. 47)

glacier (glā′shər) a large sheet of ice that moves slowly across the land (p. 216)

gravity (gra′və·tē) a pulling force between two objects, such as you and Earth (pp. 378, 447) *Gravity pulls these skydivers toward Earth.*

groundwater (ground′wä′tər) water that is held in rocks below the ground (p. 261)

LOG ON e-Glossary at www.macmillanmh.com

habitat (ha′bə·tat′) the home of a living thing (p. 109) *A coral reef is a habitat for many fish.*

heat (hēt) the flow of energy from a warmer object to a cooler object (p. 480) *The Sun is Earth's main source of heat.*

heredity (hə·re′də·tē) the passing on of traits from parents to offspring (p. 92)

hibernate (hī′bər·nāt) to rest or go into a deep sleep through the cold winter (p. 139)

humus (hū′məs) decayed plant and animal material in soil (p. 240) *Humus makes soil look dark.*

hurricane (hûr′i·kān′) a large storm with strong winds and heavy rain (p. 297)

igneous rock (ig′nē·əs rok) a rock that forms when melted rock cools and hardens (p. 231)

inclined plane (in·klīnd′ plān) a simple machine with a flat slanted surface that is raised at one end (p. 468) *A ramp is an inclined plane.*

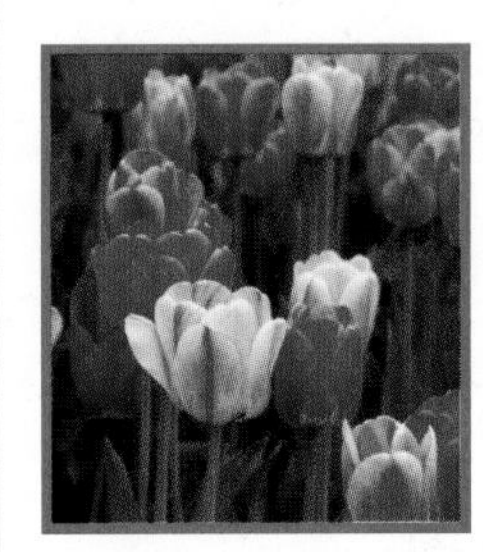

inherited trait (in·her′i·təd trāt) a characteristic that is passed from parents to offspring (p. 92) *A flower's color is an inherited trait.*

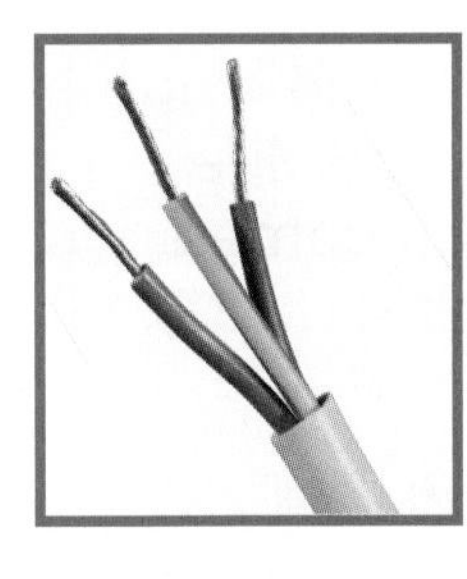

insulator (in′sə·lā′tər) a material that heat or electric current does not move through easily (p. 484) *Plastic is a good insulator of electric current.*

invertebrate (in·vûr′tə·brāt) an animal that does not have a backbone (p. 55)

kinetic energy (kə·ne′tik e′·nər′jē) energy in the form of motion (p. 456)

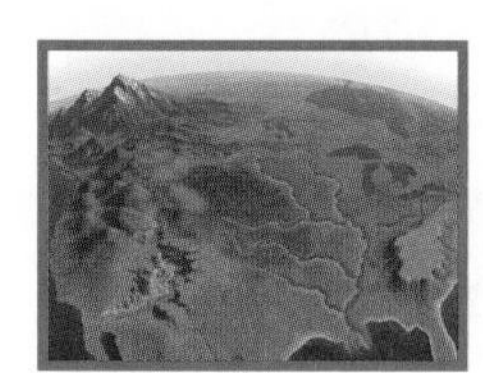

landform (land′fôrm′) a feature of land on Earth's surface (p. 194)

landslide (land′slīd′) the rapid movement of rocks and soil down a hill (p. 208)

larva (lär′və) the stage in some insects' life cycles which comes after hatching (p. 83)

lava (lä′və) melted rock that flows onto land (p. 206)

leaf (lēf) the plant structure where a plant makes food (p. 36)

learned trait (lûrnd trāt) new skills you are taught or learn with experience (p. 94) *Riding a bicycle is a learned trait.*

lever (le′vər) a simple machine that consists of a straight bar that moves on a fixed point, or fulcrum (p. 466)

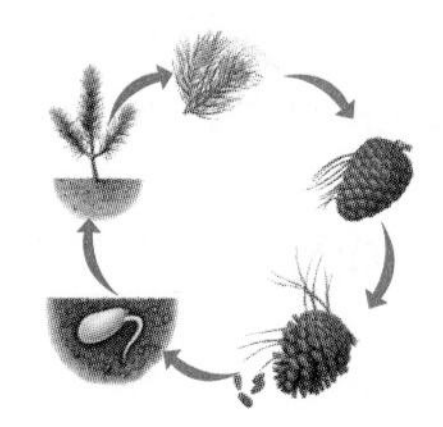

life cycle (līf sī′kəl) how a certain kind of organism grows and reproduces (p. 74)

light (līt) a form of energy that allows you to see objects (p. 500) *Light travels in a straight path from a lighthouse.*

liquid (li′kwəd) matter that has a certain volume but not a certain shape (p. 386)

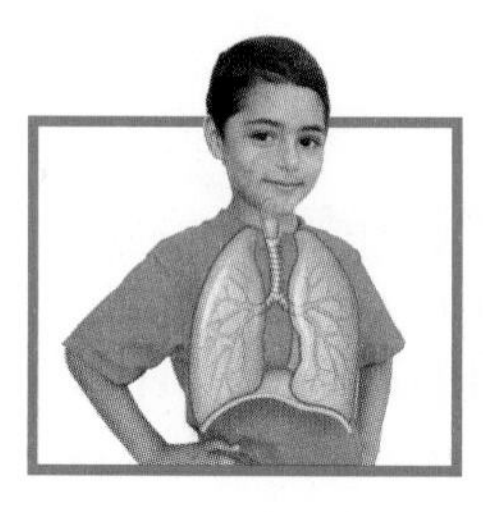

lung (lung) a structure some animals use to take in oxygen from air (p. 47) *Humans breathe oxygen using lungs.*

LOG ON e-Glossary at www.macmillanmh.com

M

magma (mag′mə) melted rock that is below Earth's surface (p. 206)

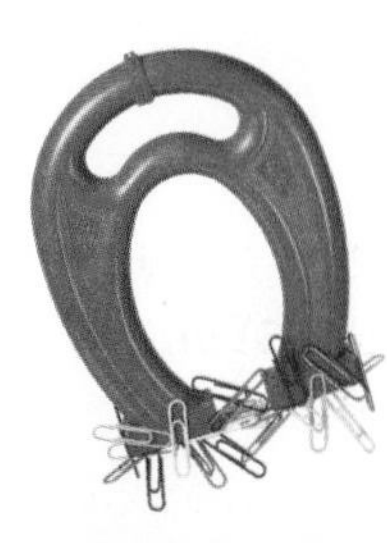

magnet (mag′nət) an object with a magnetic force; magnets can attract or repel certain metals (p. 446)

mammal (ma′məl) a vertebrate that has hair or fur, is born live, and feeds its young with milk (p. 60)

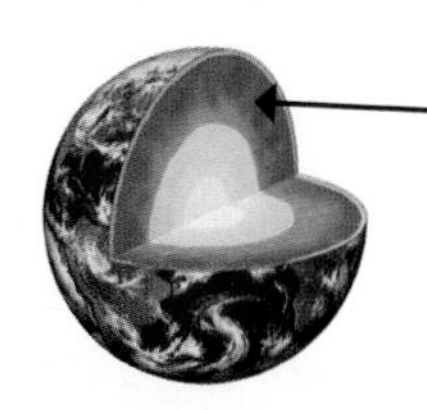

mantle (man′təl) the layer of Earth below the crust (p. 198)

mass (mas) a measure of the amount of matter in an object (p. 365)

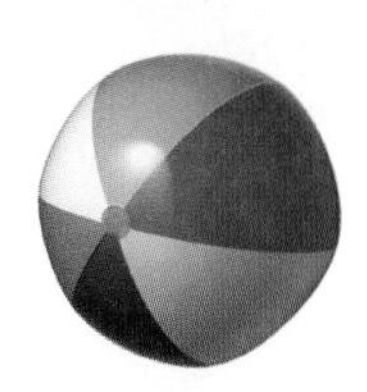

matter (ma′tər) anything that takes up space and has mass (p. 364)

melt (melt) to change from a solid to a liquid (p. 398) *This snowman will melt on a warm day.*

metamorphic rock (me′tə·môr′fik rok) a rock that has been changed by heating and squeezing (p. 233)

metamorphosis (me′tə·môr′fə·səs) a series of changes in which an organism's body changes form (p. 83) *A tadpole becomes a frog through metamorphosis.*

metric system (me′trik sis′təm) a common system of standard units of measurement (p. 374) *A centimeter is a unit in the metric system.*

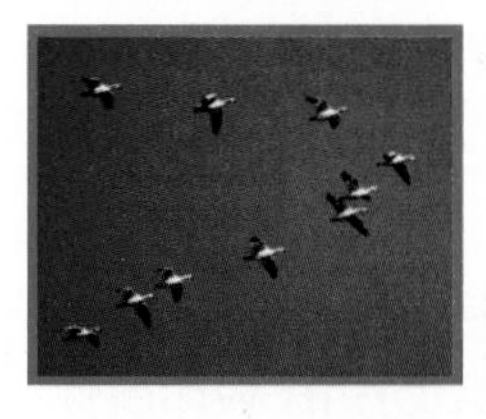

migrate (mī′grāt) to move from one place to another (p. 141) *These geese migrate south when the weather gets cold.*

mimicry (mi′mi·krē) an adaptation in which one kind of organism looks like another kind in color or shape (p. 139)

mineral (mi′nə·rəl) a solid, nonliving substance found in nature (p. 228)

mixture (miks′chər) different kinds of matter mixed together (p. 410)

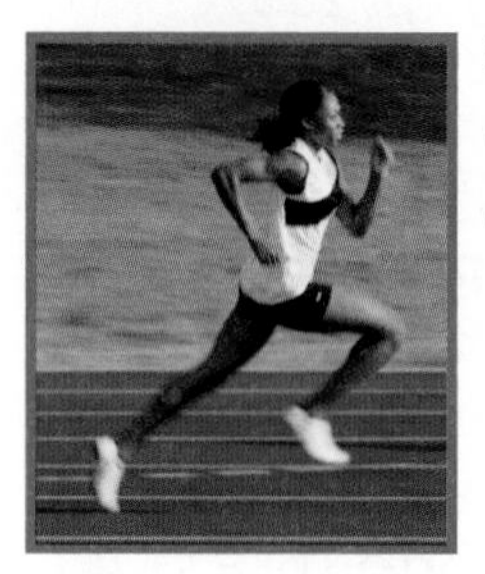

motion (mō′shən) a change in the position of an object (p. 436)

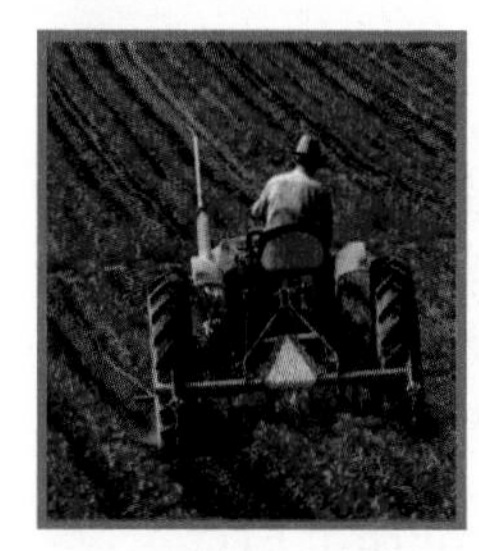

natural resource (na′cha·rəl rē′sôrs′) a material on Earth that is necessary or useful to people (p. 244) *Plants and soil are natural resources.*

nocturnal (nok·tûr′nəl) an adaptation in which an animal is active during the night and asleep during the day (p. 137)

nonrenewable resource (non′ri·nü′ə·bəl rē′sôrs) a resource that can not be replaced or reused easily (p. 253) *Oil is a nonrenewable resource. Once it is used up, it is gone forever.*

nutrient (nü′trē·ənt) a substance that helps living things grow and stay healthy (p. 34) *Spinach has many nutrients that people need.*

ocean (ō′shən) a large body of salt water (pp. 126, 192)

opaque (ō·pāk′) not allowing light to pass through (p. 502) *An ice cream cone is opaque. You can't see through it.*

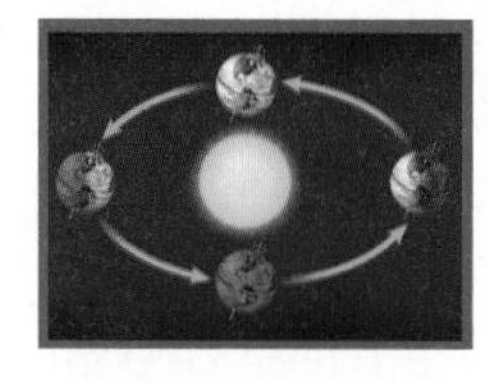

orbit (ôr′bət) the regular path one object travels around another (p. 320) *Earth's orbit around the Sun is nearly round.*

organism (ôr·gə·ni′zəm) a living thing (p. 22) *Koala bears and eucalyptus trees are organisms.*

pan balance (pan ba′ləns) a tool used to measure mass (p. 376)

LOG ON e-Glossary at www.macmillanmh.com

phase (fāz) each shape of the Moon we see (p. 328)

photosynthesis (fō′tə·sin′thə·səs) the process through which plants make food (p. 36)

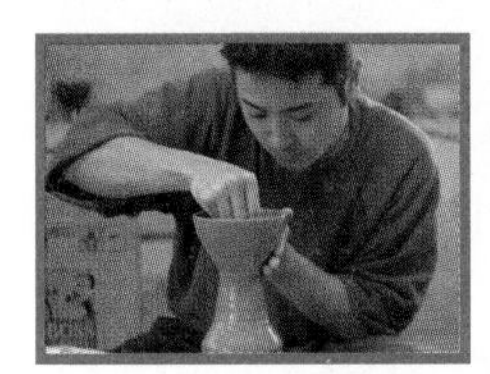

physical change (fi′zi·kəl chānj) a change in the way matter looks (p. 408) *Shaping clay is a physical change.*

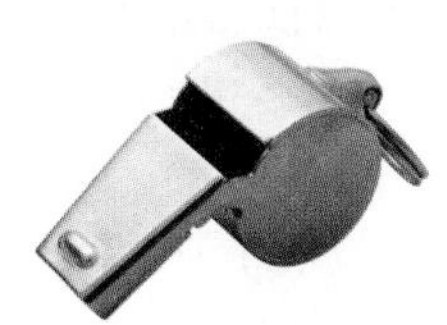

pitch (pich) how high or low a sound is (p. 493) *A whistle has a high pitch.*

planet (pla′nət) a large body of rock or gas with a nearly round shape that revolves around a star (p. 338)

pollination (po′lə·nā′shən) when pollen moves from the male part of a flower to the female part, after which a seed can form (p. 73)

pollution (pə·lü′shən) what happens when harmful materials get into water, air, or land (pp. 154, 264)

population (po·pyə·lā′shən) all the members of one kind type of organism in an ecosystem (p. 166)

position (pə·zi′shən) the location of an object (p. 434)

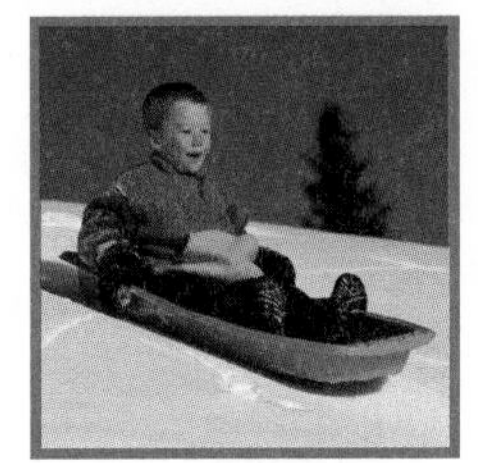

potential energy (pə·ten′shəl e·nər·jē) energy that is stored and ready to be used (p. 456) *A sled at the top of a hill has potential energy.*

precipitation (pri·si′pə·tā′shən) water that falls to the ground from the atmosphere (p. 282)

producer (prə·dü′sər) an organism, such as a plant, that makes its own food (p. 110)

property (pro′pər·tē) any characteristic of matter that you can observe (p. 365) *A sweet taste is a property of this pineapple.*

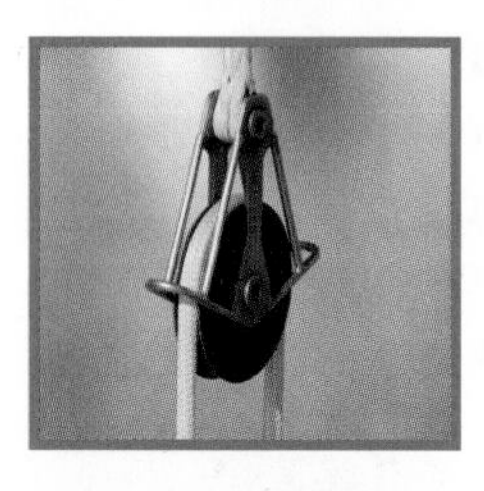

pulley (pu'lē) a simple machine that uses a rope and wheel to move an object (p. 467)

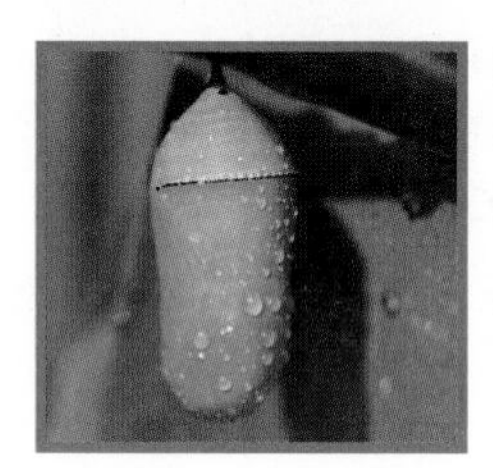

pupa (pū'pə) the stage of some insects' life cycles before becoming an adult (p. 83)

recycle (rē·sī'kəl) to turn old things into new things (p. 156) *Plastic can be recycled to make new bottles and other products.*

reduce (ri·düs') to use less of something (p. 156) *When you fix a leaky faucet, you reduce your use of water.*

reflect (ri·flekt') to bounce off a surface (p. 501) *Light reflects off a mirror.*

refract (ri·frakt') to bend (p. 503) *Light refracts as it moves from the air to the water.*

renewable resource (ri·nü'ə·bəl rē'sôrs) a resource that can be replaced or used again and again (p. 253) *Wind is a renewable resource.*

reproduce (rē'prə·düs') to make more of one's own kind (p. 23)

reptile (rep'tīl) a vertebrate that has scaly, waterproof skin, breathes air with lungs, and lays eggs (p. 58)

resource (rē'sôrs) something in the environment that helps an organism survive (p. 152) *Flowers are a resource for butterflies.*

respond (ri·spond') to react to something (p. 22) *When the weather gets cool in the fall, this tree responds by losing its leaves.*

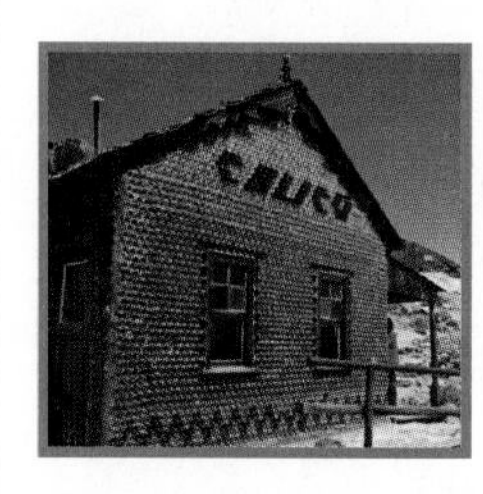

reuse (rē·ūz') to use something again (p. 156) *Old bottles were reused to make this building.*

LOG ON e-Glossary at www.macmillanmh.com

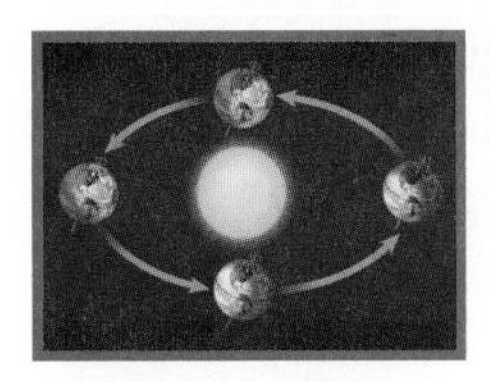

revolve (ri·volv′) to move around another object (p. 320) *Earth revolves around the Sun.*

rock (rok) a nonliving material made of one or more minerals (p. 230)

root (rüt) a plant structure that takes in water and nutrients and holds a plant in place (p. 34)

rotate (rō′tāt) to turn or spin (p. 318) *Earth rotates like a giant top in space.*

S

screw (skrü) a simple machine made up of an inclined plane wrapped into a spiral (p. 468)

season (sē′zən) time of the year with different weather patterns (p. 308)

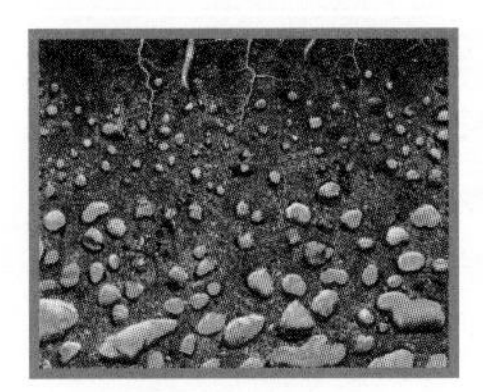

sediment (se′də·mənt) tiny bits of weathered rock or once-living animals and plants (p. 232)

sedimentary rock (se′də·mən′tə·rē rok) a kind of rock that forms from layers of sediment (p. 232)

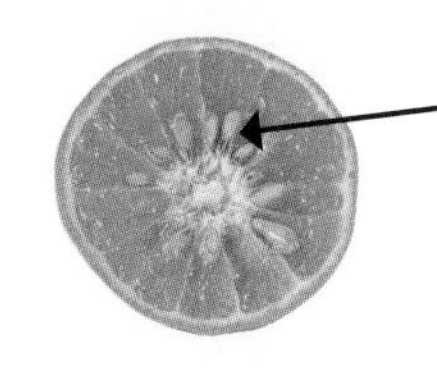

seed (sēd) a structure that can grow into a new plant (p. 70)

shadow (sha′dō) a dark space that forms when light is blocked (p. 502)

shelter (shel′tər) a place in which an animal can stay safe (p. 48) *A nest is a shelter for young birds.*

simple machine (sim′pəl mə·shēn′) a machine with few or no moving parts (p. 465) *A lever is a simple machine.*

soil (soil) a mixture of minerals, weathered rocks, and decayed plant and animal matter (pp. 120, 240)

solar energy (sō′lər e·nər′jē) energy from the Sun (p. 254)

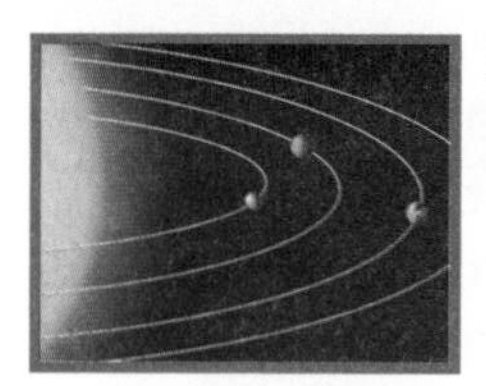

solar system (sō′lər sis′təm) a system made up of a star and the objects that move around it (p. 338)

solid (so′ləd) matter that has a certain shape and volume (p. 384)

solution (sə·lü′shən) a mixture in which one or more kinds of matter are mixed evenly in another kind of matter (p. 411)

sound (sound) a form of energy that comes from objects that vibrate (p. 490)

space probe (spās prōb) a machine that leaves Earth and travels through space (p. 342)

speed (spēd) how fast an object moves over a certain distance (p. 438)

sphere (sfîr) a body that has the shape of a ball (p. 305) *A globe is a sphere.*

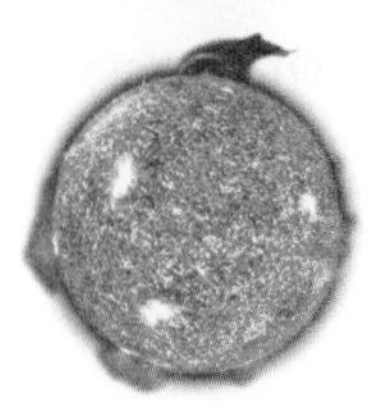

star (stär) a ball of hot, glowing gases (p. 322) *Our Sun is a star.*

state of matter (stāt uv ma′tər) a form of matter, such as solid, liquid, and gas (p. 384)

static electricity (sta′tik i·lek′tri′sə·tē) the build up of an electrical charge on a material (p. 513) *Static electricity can cause you to get a shock from touching a doorknob.*

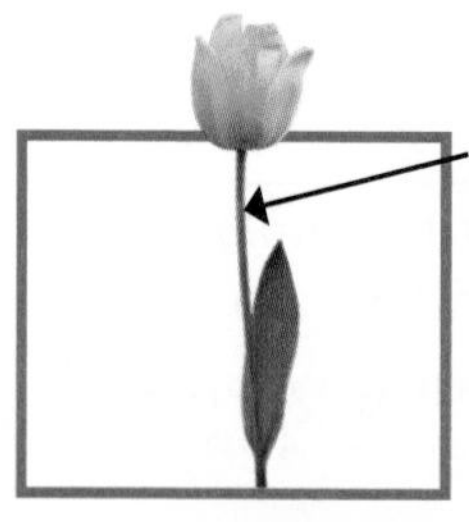

stem (stem) a plant structure that holds a plant up and helps leaves reach sunlight (p. 35)

structure (struk′chər) a part of an organism (p. 33) *Horns are structures that help wild sheep keep safe.*

LOG ON e-Glossary at www.macmillanmh.com

switch (swich) a device that allows you to control the flow of electric current through a circuit (p. 515)

T

telescope (te′lə·skōp′) a tool used to make far away objects appear closer and larger (p. 342)

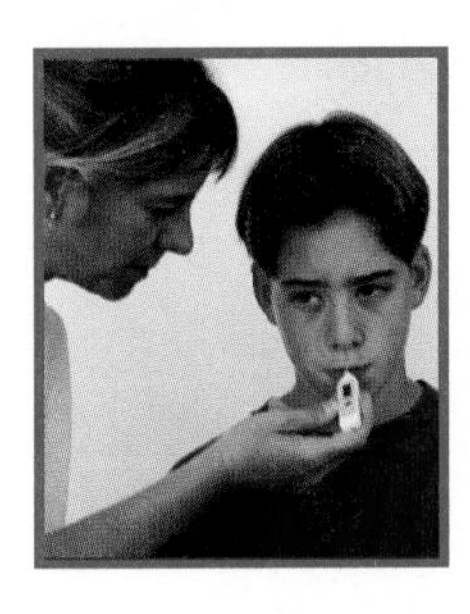

temperature (tem′pə·rə′chər) a measure of how hot or cold something is (pp. 280, 482) *When you visit the doctor, she measures your body's temperature.*

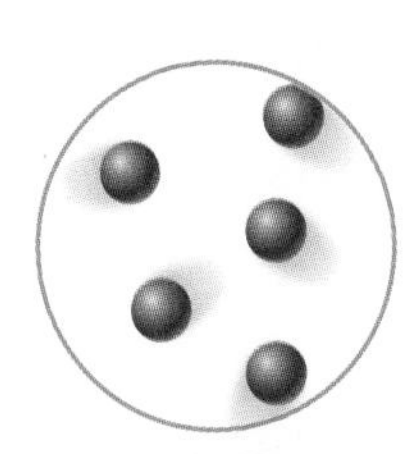

thermal energy (thûr′məl e·nər′jē) the energy of moving particles of matter (p. 482)

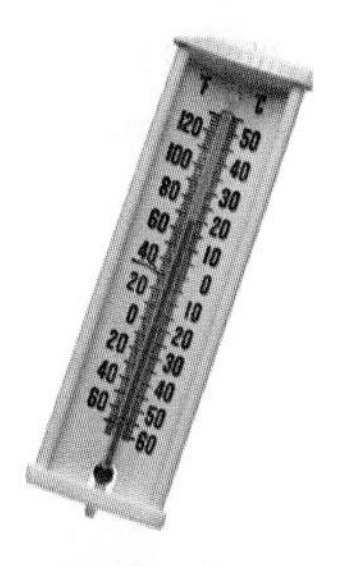

thermometer (thûr′mo′mə·tər) a tool that is used to measure temperature (p. 483)

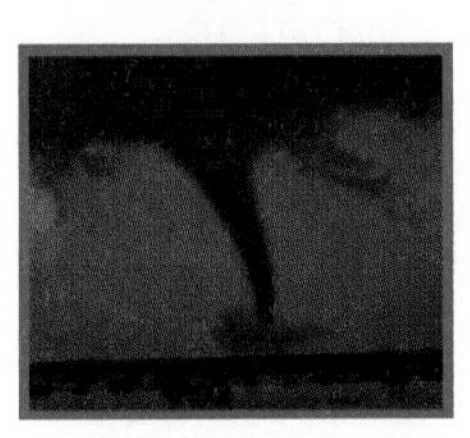

tornado (tôr·nā′dō) a powerful storm with rotating winds that forms over land (p. 296)

trait (trāt) a feature of a living thing (p. 92) *Spots are a trait of Dalmatians.*

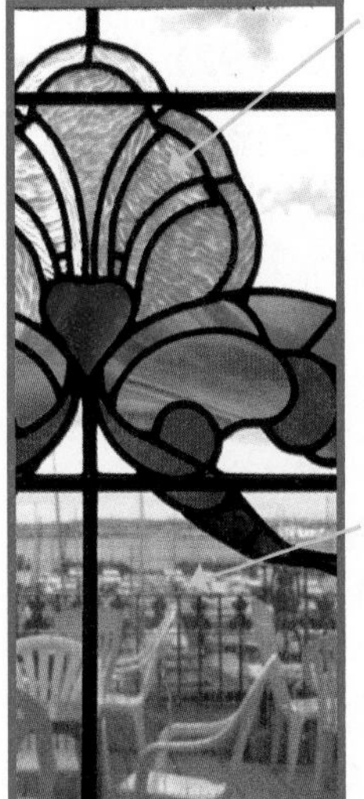

translucent (trans·lü′sənt) scattering light that passes through, so that objects on the other side appear blurry (p. 503) *Frosted glass is translucent.*

transparent (trans·pâr′ənt) letting all light through, so that objects on the other side can be seen clearly (p. 503) *Clear glass is transparent.*

V

vertebrate (vûr′tə·brāt′) an animal with a backbone (p. 54)

vibrate (vī′brāt) to move back and forth quickly (p. 490) *A guitar string vibrates after you pluck it.*

volcano (vol′kā·nō) a mountain that builds up around an opening in Earth's crust (p. 206)

volume (vol′ūm)
1. a measure of how much space an object takes up (p. 365)

2. how loud or soft a sound is (p. 492) *A whisper has a low volume.*

water cycle (wä′tər sī′kəl) describes how water moves between Earth's surface and the atmosphere (p. 294)

water vapor (wä′tər vā′pər) water in its gas form (p. 292) *Water vapor is an invisible gas.*

weather (we′thər) what the air is like at a certain time and place (p. 280) *The weather shown here is warm and sunny.*

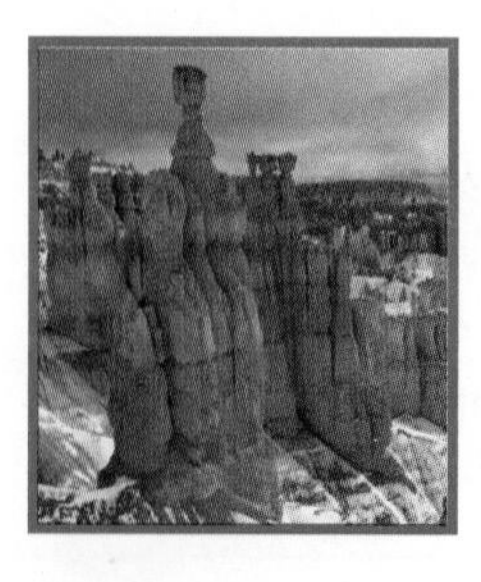

weathering (we′thə·ring) the breaking down of rocks into smaller pieces (p. 214) *Weathering caused these interesting rock shapes to form.*

wedge (wej) a simple machine that uses force to split objects apart (p. 469)

weight (wāt) a measure of the pull of gravity on an object (pp. 378, 447)

wetland (wet′land) an ecosystem where water covers the soil for most of the year (p. 128)

wheel and axle (wēl and ak′səl) a simple machine that consists of a wheel that moves around a post; the post is called an axle (p. 467)

wind (wind) moving air (p. 283)

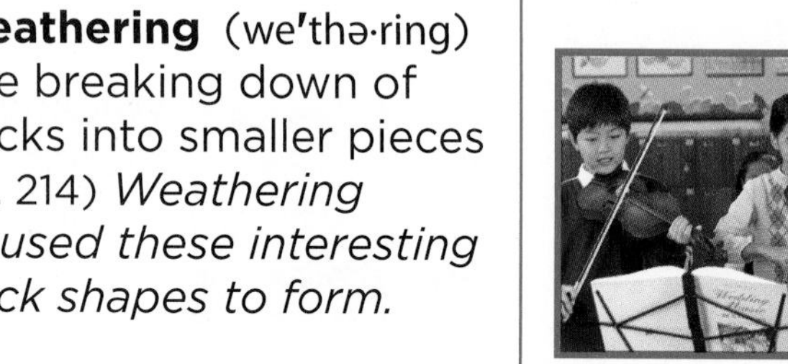

work (wûrk) what is done when a force changes an object's motion (p. 454) *You do work when you move a bow to play the violin.*

LOG ON e-Glossary at www.macmillanmh.com

Index

Note: Page references followed by an asterisk indicate activities.

C

F

G

N

O

Y

Z

Credits

Abbreviation key: MMH=Macmillan/McGraw-Hill

Cover Photography Credits: Front Cover: Gary Bell/oceanwideimages.com; **Spine:** Andre Seale/Alamy; **Back Cover:** (t)Andre Seale/Alamy, (b)Chris Newbert/Minden Pictures, (bkgd)Gary Bell/oceanwideimages.com.

Literature Credits: "Monarch Butterfly" by Marilyn Singer from *Fireflies at Midnight* by Marilyn Singer. Text copyright © 2003 by Marilyn Singer. Reprinted with permission by Atheneum Books for Young Readers, an imprint of Simon & Schuster Children's Publishing Division. "One Cool Adventure" from *TIME for Kids*, Vol. 6, Issue 18. Copyright © 2006 by *TIME for Kids.* All rights reserved. "The Good Ship Popsicle Stick" from *TIME for Kids,* Vol. 11, Issue 1. Copyright © 2006 by *TIME for Kids.* All rights reserved. "Jump Rope" from *In the Spin of Things* by Rebecca Kai Dotlich. Text copyright © 2003 by Rebecca Kai Dotlich. Published by Boyds Mills Press, Inc.

Illustration Credits: All illustrations are by Macmillan/McGraw-Hill (MMH).

Photography Credits: TitlePg Chris Newbert/Minden Pictures; **a–b** Gary Bell/oceanwideimages.com; **b** Chris Newbert/Minden Pictures; **i** Gary Bell/oceanwideimages.com; **ii–iii** Getty Images; **iii** Stockbyte; **iv** Ken Karp/MMH; **ix** (t)PunchStock, (b)Getty Images; **vi** (t)Susan Cruz/MMH, (c)Alamy, (b)Siede Preis/Getty Images; **vii** (t)Alamy, (b)Russ Merne/Alamy; **viii** William Manning/CORBIS; **x** James Sparshatt/CORBIS; **xi** (t, bl)Ken Cavanagh/MMH, (bc)Michael Scott/MMH, (br)Richard T. Nowitz/CORBIS; **xii** (t)Roger Phillips/DK Images, (bl)SuperStock, (bc)CORBIS, (br)Getty Images; **xiii** (t)Mark A. Schneider/Photo Researchers, Inc., (bl)Ken Cavanagh/MMH, (br)PunchStock; **xiv** Jayme Thornton/Getty Images; **1** Frans Lanting/Minden Pictures; **2–3** Michael Melford/NGS Images; **3** Courtesy American Museum of Natural History; **4–5** Mike Powles/Jupiterimages; **6–7** Courtesy American Museum of Natural History; **8** Getty Images; **8–9** Fabrice Bettex/Alamy; **9** (l)Nick Garbutt/Getty Images, (r)Ken Cavanagh/MMH; **10** (tl)Piotr Naskrecki/Minden Pictures, (tr)Mitsuhiko Imamori/Minden Pictures, (b)Chris Hellier/CORBIS; **11** (t)Piotr Naskrecki/Minden Pictures, (bl)Wolfgang Kaehler/CORBIS, (br)Pete Oxford/Minden Pictures; **12** (r)CORBIS, (others)PunchStock; **13** (tl)Getty Images, (tr)Siede Preis/Getty Images, (bl)Alamy, (tc, br)PunchStock; **14** (tl)Jacques Cornell/MMH, (tr)Ken Cavanagh/MMH, (b)Stockbyte; **15** Gary Bell/Oceanwide Images; **16–17** Getty Images; **17** (t)Stockbyte, (others)T. Kitchin & V. Hurst/NHPA; **18–19** Zigmun Leszczysnki/Animals Animals; **19** (t to b)Kike Calvo/Alamy, Joseph Van Os/Getty Images, Lester V. Bergman/CORBIS, Craig K. Lorenz/Photo Researchers, Inc., Amos Nachoum/CORBIS; **20–21** Joseph Van Os/Getty Images; **21** CORBIS; **22** Alamy; **23** (t)Zigmund Leszczynski/Animals Animals, (b)Gary Vestal/Getty Images; **24** (t)Bob Jensen/Bruce Coleman USA, (b)David Cavagnaro/Visuals Unlimited; **25** (t)Kike Calvo/Alamy, (b)Alamy; **26** (t)Image100 Ltd., (bl)Getty Images, (br)Lester V. Bergman/CORBIS; **27** (t to b)Zigmund Leszczynski/Animals Animals, David Cavagnaro/Visuals Unlimited, Lester V. Bergman/CORBIS, Jacques Cornell/MMH; **28–29** Brian Sytnyk/Masterfile; **29** (l)Ken Graham/Accent Alaska, (r)Dr. Dennis Kunkel/Visuals Unlimited; **30–31** Mark Bolton/CORBIS; **31** (cr, br)Michael Scott/MMH, (others)MMH; **32–33** DK Images; **34** (t)PunchStock, (c)Wally Eberhart/Visuals Unlimited, (b)CORBIS; **35** (t)Ken Cavanagh/MMH, (b)Rosemary Calvert/Alamy; **36** (t)Siede Preis/Getty Images, (b)Matthew Ward/DK Images; **37** (t)Pete Oxford/Minden Pictures, (b)Dr. Gerald Van Dyke/Getty Images; **38** (tl)Karl-Heinz Haenel/CORBIS, (tr, cl)Alamy, (cr)Mark Bretherton/Alamy, (bl)Betsy Dupuis/iStockphoto, (br)Stefan Auth/Alamy; **39** (t)DK Images, (c)Betsy Dupuis/iStockphoto, (b)Jacques Cornell/MMH; **40** (b)Ken Karp/MMH, (others)MMH; **41** Ken Karp/MMH; **42–43** Purestock; **43** MMH; **44** (t)Pete Oxford/naturepl.com, (b)Alamy; **45** (t)Michael Fogden/Animals Animals, (b)Tim Fitzharris/Minden Pictures; **46** (t)Alamy, (b)Richard Du Toit/naturepl.com; **47** (t)PunchStock, (b)Jack Jackson/Alamy; **48** (t)PunchStock, (b)Owen Newman/naturepl.com; **49** (t to b)Michael Fogden/Animals Animals, Richard Du Toit/naturepl.com, PunchStock, Jacques Cornell/MMH; **50** Getty Images; **51** (l to r, t to b)PunchStock, Getty Images, PunchStock, Alan & Sandy Carey/Getty Images, PunchStock, PunchStock, Alan & Sandy Carey/Getty Images, CORBIS; **52–53** Tim Lama/Getty Images; **53** MMH; **54** Craig K. Lorenz/Photo Researchers, Inc.; **55** (t)MMH; (b)Amos Nachoum/CORBIS; **56** (tl)CORBIS, (tr)Mark Moffett/Minden Pictures, (bl)Georgette Douwma/Getty Images, (br)PictureQuest; **57** (tl)Andrew J. Martinez/Photo Researchers, Inc., (tr)Alamy, (bl)Reinhard Dirscherf/Alamy, (br)G.K. & Vikki Hart/Getty Images; **58** Alamy, (b)Frans Lemmins/CORBIS; **59** (t)Tim Flach/Getty Images, (c)PunchStock, (b)Alamy; **60** (t)Mark Newman/Alamy, (b)Paul A. Souders/CORBIS; **61** (t to b)Craig K. Lorenz/Photo Researchers, Inc., Georgette Douwma/Getty Images, Getty Images, Jacques Cornell/MMH; **62** (l)SuperStock, (r)Anthony Mercieca/SuperStock; **63** (l)PunchStock, (r)Getty Images; **64** (t to b)Joseph Van Os/Getty Images, CORBIS, PunchStock, Tim Lama/Getty Images, Jacques Cornell/MMH; **65** (l)DK Images, (r)MMH; **66–67** Frank Krahmer/PictureQuest; **67** (t to b)Roger Phillips/DK Images, Hans Pfletschinger/Peter Arnold Inc., Frans Lanting/Minden Pictures, Jane Burton/naturepl.com, David Stoecklein/CORBIS; **68–69** Keith Neale/Masterfile; **69** (br)Michael Scott/MMH, (others)MMH; **70** (l)Roger Phillips/DK Images, (c, r)J. Brown/Animals Animals; **71** (t)Nigel Cattlin/Alamy, (bl, bc)J. Brown/Animals Animals, (br)Roger Standen/Science Photo Library; **72** (t)Michael P. Gadomski/SuperStock, (b)Hans Pfletschinger/Peter Arnold Inc.; **73** (t to b)PunchStock, Getty Images, Punchstock, Alamy; **74** SuperStock; **75** Susan Cruz/MMH; **76** Nigel Cattlin/Alamy; **76–77** DK Images; **77** (t)J. Brown/Animals Animals, (b)Jacques Cornell/MMH; **78** Roger Phillips/DK Images; **79** (t)Michael Scott/MMH, (b)MMH; **80–81** Frans Lanting/Minden Pictures; **81** (t to b)MMH, Michael Scott/MMH, Ken Cavanagh/MMH; **82** (cw from top)Philippe Clement/naturepl.com, Stephen Dalton/Photo Researchers, Inc., George Bernard/NHPA, Stephen Dalton/Minden Pictures; **83** (cw from top)Perennou Nuridsany/Photo Researchers, Inc., Troy Bartlett/Alamy, Bill Beatty/Visuals Unlimited, Ted Kinsman/Photo Researchers, Inc.; **84** (cw from top)Gerard Lacz/Animals Animals, Kevin Schafer/CORBIS, James Watt/Animals Animals, D. Fleetham/Animals Animals; **85** (tr)Siede Preis/Getty Images, (cl)Jane Burton/naturepl.com, (tl, cr)Getty Images, (bl)Norbert Wu/Minden Pictures, (br)LUTRA/NHPA; **86** (cw from top)Alamy, Mitsuaki Iwago/Minden Pictures, Anup Shah/Alamy; **87** (t to b)Stephen Dalton/Photo Researchers, Inc., James Watt/Animals Animals, Alamy, Jacques Cornell/MMH; **88** Rene Morris/Getty Images; **89** (l)Lee Karney/US Fish & Wildlife Service, (r)MMH; **90–91** David Stoecklein/CORBIS; **91** Michael Scott/MMH; **92** (t)Bonnie Sue Rauch/Photo Researchers, Inc., (others)Susan Cruz/MMH; **93** Alamy; **94** (t)Tom & Pat Leeson/Photo Researchers, Inc., (b)Sharon Bailey/Alamy; **95** (t to b)Bonnie Sue Rauch/Photo Researchers, Inc., Tom & Pat Leeson/Photo Researchers, Inc., Sharon Bailey/Alamy, Jacques Cornell/MMH; **96** (t)Denis Finnin/American Museum of Natural History, (b)Daniel Heuclin/NHPA; **97** Daniel Heuclin/NHPA; **98** (t to b)Keith Neale/Masterfile, Frans Lanting/Minden Pictures, David Stoecklein/CORBIS, Jacques Cornell/MMH; **99** (t)Frans Lemmins/CORBIS, (tr)Lee Karney/US Fish & Wildlife Service, (b)J. Brown/Animals Animals; **100** (t)Flip Nicklin/Minden Pictures, (b)Richard Sobol/Animals Animals; **101** Tim Fitzharris/Minden Pictures; **102–103** Alamy; **103** Joe McDonald/Visuals Unlimited; **104–105** Ian West/Jupiterimages; **105** (t to b)Georgette Douwma/Getty Images, Jim Allan/Alamy, Jim Brandenburg/Minden Pictures, Steve Maslowski/Photo Researchers, Inc., Bill Beatty/Visuals Unlimited, Peter David/Getty Images; **106–107** Yva Momatiuk/John Eastcott/Minden Pictures; **107** (b)Michael Scott/MMH, (others)MMH; **110** (tl)Jim Allan/Alamy, (tr)James Carmichael Jr./NHPA, (c)CORBIS, (bl)C. Milkens/Animals Animals, (br)Stephen Dalton/Minden Pictures; **111** (tl)Michael Gadomski/Animals Animals, (tr)Alan & Sandy Carey/Getty Images, (bl)Animals Animals, (br)Steve Maslowski/Photo Researchers, Inc.; **114** (t)Joe Polillio/MMH, (b)Gary Braasch/CORBIS, (inset)Photo Researchers, Inc.; **115** (t)Stephen Dalton/Minden Pictures, (c)Gary Braasch/CORBIS, (b)Jacques Cornell/MMH; **116–117** James Worrell/Getty Images; **117** (tl, tc)CORBIS, (tr)Getty Images, (bl)SuperStock, (br)Konrad Wothe/Minden Pictures; **118–119** Georgette Douwma/Getty Images; **119** (br)Michael Scott/MMH, (others)MMH; **120** Jim Brandenburg/Minden Pictures; **121** (t)Annie Griffiths Belt/Getty Images, (b)Andre Seale/Alamy; **122** John Cancalosi/Peter Arnold, Inc.; **122–123** (b)David L. Brown/PictureQuest; **123** David Kjaer/naturepl.com; **124** (inset)Stuart Westmorland/Getty Images, (bkgd)Mike Dobel/Masterfile; **125** (inset)David Whitaker/Alamy, (bkgd)Russell Illig/Getty Images; **126** Siede Preis/Getty Images; **126–127** PunchStock; **127** (inset)Ken Cavanagh/MMH; **128** (t)Steven David Miller/naturepl.com, (b)Momatiuk Eastcott/The Image Works; **129** (t to b)Jim Brandenburg/Minden Pictures, Russell Illig/Getty Images,

PunchStock, Jacques Cornell/MMH; **130** (l)Franz Lanting/Minden Pictures, (r)MMH, (inset)A. L. Porzecanski and J. W. McDonough; **130–131** Gabriel Rojo/naturepl.com; **131** Franz Lanting/Minden Pictures; **132–133** Paul Nicklen/Getty Images; **133** MMH; **134** Fritz Rauschenbach/CORBIS; **135** (t)Gregory G. Dimijian/Photo Researchers, Inc., (b)Bryan & Cherry Alexander/Alamy; **137** (tl)Dr. Merlin D. Tuttle/Photo Researchers, Inc., (tr)MMH, (b)John Cancalosi/naturepl.com; **138** (t)Jacques Jangoux/Photo Researchers, Inc., (b)Jonathan Need/Alamy; **139** (t)Bill Beatty/Visuals Unlimited, (c)Paul Sterry/Alamy, (b)Getty Images; **140** (t)Ralph A. Clevenger/CORBIS, (b)Marty Snyderman/Alamy; **141** (t)Doug Perrine/naturepl.com, (b)Peter David/Getty Images; **142** (t)M. Timothy O'Keefe/Bruce Coleman USA, (b)Alamy; **143** (t to b)Gregory G. Dimijian/Photo Researchers, Inc., Bryan & Cherry Alexander/Alamy, Doug Perrine/naturepl.com, Jacques Cornell/MMH; **144** (t to b)MMH, Ken Cavanagh/MMH; **145** Ken Cavanagh/MMH; **146** (t to b)Yva Momatiuk/John Eastcott/Minden Pictures, Georgette Douwma/Getty Images, Paul Nicklen/Getty Images, Jacques Cornell/MMH; **147** (l)Jim Brandenburg/Minden Pictures, (tl)Annie Griffiths Belt/Getty Images, (tc)Russell Illig/Getty Images, (tr)PunchStock; **148–149** Jim Brandenburg/Minden Pictures; **149** (t to b)David Cavagnaro/Visuals Unlimited, Nigel J. Dennis/Photo Researchers, Inc., T. O'Keefe/Getty Images, Alamy, Annie Griffiths Belt/CORBIS, Wardene Weisser/Bruce Coleman USA; **150–151** Claude Ponthieux/Alamy; **151** MMH; **154** (t)T. O'Keefe/Getty Images, (b)Chip Porter/Getty Images; **155** (t)Robert W. Ginn/PhotoEdit, (b)Michael Scott/MMH; **156** (t)Stephen Saks/Lonely Planet Images, (b)Getty Images, (inset)Scott Mitchell/Getty Images; **157** (t)T. O'Keefe/Getty Images, (c)Getty Images, (b)Jacques Cornell/MMH; **158** MMH; **159** Kari Marttila/Alamy; **160–161** Curt Maas/AgStockUSA; **161** MMH; **162** Peter Arnold Inc.; **163** (t)John McColgan/Bureau of Land Management Alaska Fire Service, (c)Hal Beral/CORBIS, (b)Mark Turner/PictureQuest; **164** Nigel J. Dennis/Photo Researchers, Inc.; **164–165** Steve Bloom/Alamy; **165** Rod Patterson/Animals Animals; **168** Alamy; **169** (t)Peter Arnold Inc., (c)Rod Patterson/Animals Animals, (b)Jacques Cornell/MMH; **170** John Cancalosi/Peter Arnold, Inc.; **171** Lynn M. Stone/Bruce Coleman USA; **172–173** Annie Griffiths Belt/CORBIS; **173** MMH; **174** Russ Merne/Alamy; **175** (t)The Image Works, (b)Dr. Rebecca Cairns-Wicks/ARKive; **176** (t)The Natural History Museum, London, (c)Albert Copley/Visuals Unlimited, (b)Chris Howes/Alamy; **177** The Natural History Museum, London; **178** (t)Bruce Coleman USA, (b)Wardene Weisser/Bruce Coleman USA; **179** (t to b)Russ Merne/Alamy, The Natural History Museum, London, Wardene Weisser/Bruce Coleman USA, Jacques Cornell/MMH; **180** (l)Jupiterimages, (r)CORBIS, (inset)The Natural History Museum, London; **180–181** John Sibbick/Natural History Museum Picture Library; **181** Stan Honda/Getty Images; **182** (t to b)Claude Ponthieux/Alamy, Curt Maas/AgStockUSA, Annie Griffiths Belt/CORBIS, Jacques Cornell/MMH; **183** (l)Chris Howes/Alamy, (r)Getty Images; **184** (t)Penny Tweedie/CORBIS, (b)Paul Nicklen/Getty Images; **185** Pawel Kumelowski/Omni-Photo Communications; **186** yourexpedition.com; **186–187** SuperStock; **187** yourexpedition.com; **188–189** Jupiterimages; **189** (t to b)Chad Ehlers/Alamy, Lloyd Cluff/CORBIS, Greg Vaughn/Alamy, Yva Momatiuk/John Eastcott/Minden Pictures, Tom & Pat Leeson/Photo Researchers, Inc.; **190–191** Chad Ehlers/Alamy; **191** (t)MMH, (b)Michael Scott/MMH; **193** PunchStock; **195** Michael Scott/MMH; **200** (b)Fred Hirschmann/Getty Images, (others)Michael Scott/MMH; **201** Michael Scott/MMH; **202–203** Lloyd Cluff/CORBIS; **203** (b)Michael Scott/MMH, (others)MMH; **204** Daniel Sambraus/Photo Researchers, Inc.; **207** (t)Michael Scott/MMH, (b)Greg Vaughn/Alamy; **208** (t)CORBIS, (b)Josh Ritchie/Getty Images; **209** (t to b)Daniel Sambraus/Photo Researchers, Inc., Greg Vaughn/Alamy, CORBIS, Jacques Cornell/MMH; **210** (t)Craig Lovell/CORBIS, (b)Mark Edwards/Peter Arnold, Inc.; **210–211** Vince Streano/CORBIS; **212–213** Tom Bean/CORBIS; **213** (bl, br)Michael Scott/MMH, (others)MMH; **214** Scott Darsney/Lonely Planet Images; **215** (t)Yva Momatiuk/John Eastcott/Minden Pictures, (b)Louis Psihoyos/IPN; **216** (t)Tom & Pat Leeson/Photo Researchers, Inc., (b)CORBIS; **217** (t)Michael Scott/MMH, (b)Bernhard Edmaier/Photo Researchers, Inc.; **218** (l)CORBIS, (r)Danny Lehman/CORBIS; **219** (t to b)Scott Darsney/Lonely Planet Images, Bernhard Edmaier/Photo Researchers, Inc., CORBIS, Jacques Cornell/MMH; **220** Getty Images; **221** Danita Delimont/Alamy; **222** (t to b)Chad Ehlers/Alamy, Lloyd Cluff/CORBIS, Tom Bean/CORBIS, Jacques Cornell/MMH; **223** (l)Tom & Pat Leeson/Photo Researchers, Inc., (r)MMH; **224–225** Don Smith/Alamy; **225** (t to b)Ken Cavanagh/MMH, Jerome Wyckoff/Animals Animals, Ian Francis/Alamy, Toyohiro Yamada/Getty Images, Layne Kennedy/CORBIS, Owaki-Kulla/CORBIS; **226–227** Alamy; **227** (br)Michael Scott/MMH, (others)MMH; **228** (l)Albert Copley/Visuals Unlimited, (c)Roger Weller/Cochise College, (r)Mark A. Schneider/Photo Researchers, Inc.; **229** (t)Roger Weller/Cochise College, (c)Ken Cavanagh/MMH, (b)ER Degginger/Photo Researchers, Inc.; **231** (t)Michael Scott/MMH, (c)Photo Researchers, Inc., (b)Ken Cavanagh/MMH; **232** (t)Wally Eberhart/Visuals Unlimited, (b)Jonathan Blair/CORBIS; **232–233** William Manning/CORBIS; **233** (t)Jerome Wyckoff/Animals Animals, (c)Dr. Parvinder Sethi, (b)Roger Weller/Cochise College; **234** (t)Roger Weller/Cochise College, (c)Richard Cummins/SuperStock, (b)Alamy; **235** (t to b)Roger Weller/Cochise College, William Manning/CORBIS, Richard Cummins/SuperStock, Jacques Cornell/MMH; **236** (l)Lester Lefkowitz/Getty Images, (r)Richard Nowitz/National Geographic Image Collection; **237** Jacques Cornell/MMH; **238–239** Joseph Van Os/Getty Images; **239** (l, b)Michael Scott/MMH, (others)MMH; **240** (b)Photo Researchers, Inc., (inset)Kevin Schafer/Peter Arnold Inc.; **241** Jacques Cornell/MMH; **242** (l)Franck Jeannin/Alamy, (r)Ian Francis/Alamy; **243** (t)Michael Scott/MMH, (others)Jacques Cornell/MMH; **244** (t)Richard Hamilton Smith/CORBIS, (b)David Frazier/The Image Works; **245** (t to b)Photo Researchers, Inc., Ian Francis/Alamy, David Frazier/The Image Works, Jacques Cornell/MMH; **246–247** Lee White/CORBIS; **247** (t)Michael Scott/MMH, (b)PunchStock; **248–249** Layne Kennedy/CORBIS; **249** (br)Michael Scott/MMH, (others)MMH; **250** (t)Tom Bean/CORBIS, (b)Michael & Patricia Fogden/CORBIS; **251** (t)Michael Scott/MMH, (b)Charles R. Belinky/Photo Researchers, Inc.; **252** Andrew J. Martinez/Photo Researchers, Inc.; **252–253** Owaki-Kulla/CORBIS; **253** Getty Images; **254** (t)Toshiyuki Aizawa/CORBIS, (c)Kimimasa Mayama/CORBIS, (b)Jeremy Woodhouse/Masterfile; **255** (t to b)Tom Bean/CORBIS, Owaki-Kulla/CORBIS, Jeremy Woodhouse/Masterfile, Jacques Cornell/MMH; **256** (l)Wisconsin Historical Society, (r)Klaus Guldbrandsen/Photo Researchers, Inc.; **256–257** Cristina Pedrazzini/Photo Researchers, Inc.; **257** (l)Tommaso Guicciardini/Photo Researchers, Inc., (r)Warren Gretz/Science Photo Library; **258–259** Toyohiro Yamada/Getty Images; **259** (l to r, t to b)MMH, Michael Scott/MMH, Michael Scott/MMH; **260** PunchStock; **260–261** Hans Strand/CORBIS; **262** Mira/Alamy; **264** Vincent Munier/naturepl.com; **264–265** PunchStock; **265** Getty Images; **266** CORBIS; **267** (t to b)Getty Images, CORBIS, Jacques Cornell/MMH; **268** (t to b)Ken Karp/MMH, MMH; **269** MMH; **270** (t to b)Alamy, Joseph Van Os/Getty Images, Layne Kennedy/CORBIS, Toyohiro Yamada/Getty Images, Jacques Cornell/MMH; **271** (t)Owaki-Kulla/CORBIS, (b)Roger Weller/Cochise College; **272** (t)Tony West/CORBIS, (b)David Hiller/Getty Images; **273** Alamy; **274–275** Frank Whitney/Getty Images; **275** (t)Alamy, (b)Gary W. Carter/CORBIS; **276–277** Tim Fitzharris/Masterfile; **277** (t to b)Erwan Balanca/PictureQuest, Alamy, David R. Frazier/Photo Researchers, Inc., Richard Thom/Visuals Unlimited, Layne Kennedy/CORBIS; **278–279** Rolf Bruderer/CORBIS, (bkgd)Bruce Heinemann/Getty Images; **279** (b)Ken Karp/MMH, (others)MMH; **280–281** Getty Images; **281** Jupiterimages; **282** (t)Eric Nguyen/CORBIS, (bl)Tony Freeman/PhotoEdit, (br)Alamy; **283** (l)Matthias Engelien/Alamy, (r)Leonard Lessin/Photo Researchers, Inc.; **284** David Hay Jones/Science Photo Library; **285** (t)Rolf Bruderer/CORBIS, (c)Leonard Lessin/Photo Researchers, Inc., (b)Jacques Cornell/MMH; **286** MMH; **287** (t)MMH, (b)Michael Scott/MMH; **288–289** Erwan Balanca/PictureQuest; **289** (l to r, t to b)Michael Scott/MMH, MMH; **290** (t)John Da Silva/Getty Images, (b)Getty Images; **291** (t)Naoki Okamoto/Getty Images, (b)David R. Frazier/Photo Researchers, Inc.; **292** Getty Images; **292–293** Darrell Gulin/CORBIS; **293** (t)Michael Scott/MMH, (b)Michael Newman/PhotoEdit; **296** Jim Zuckerman/CORBIS; **297** (t)CORBIS, (tl)Daniel Aguilar/CORBIS, (b)Alamy; **298–299** PictureQuest; **299** (t to b)David R. Frazier/Photo Researchers, Inc., Darrell Gulin/CORBIS, PictureQuest, Jacques Cornell/MMH; **300–301** Aaron Horowitz/CORBIS; **301** (t)NOAA, (b)Howard Bluestein/Photo Researchers, Inc.; **302–303** Neil McAllister/Alamy; **304** (tl)Richard Thom/Visuals Unlimited, (tr)Mark E. Gibson/CORBIS, (bl)Alamy, (br)Silvestre Machado/Getty Images; **306** (l)CORBIS, (r)Bob Winsett/CORBIS; **307** (l)Alamy, (r)Michael Scott/MMH; **308** (t)Philip & Karen Smith/Lonely Planet Images, (b)Layne Kennedy/CORBIS; **309** (t to b)Silvestre Machado/Getty Images, Bob Winsett/CORBIS, Layne Kennedy/

CORBIS, Jacques Cornell/MMH; **310** Konrad Wothe/Minden Pictures; **311** Bob Krist/eStock Photo; **312** (t to b)Rolf Bruderer/CORBIS, Erwan Balanca/PictureQuest, Neil McAllister/Alamy, Jacques Cornell/MMH; **313** (l)David R. Frazier/Photo Researchers, Inc., (r)PunchStock; **314–315** Getty Images; **315** (c)Eckhard Slawik/Photo Researchers, Inc., (b)Getty Images; **316–317** Getty Images; **317** MMH; **319** MMH; **321** Herman Eisenbeiss/Photo Researchers, Inc.; **322** PunchStock; **323** (t)PunchStock, (b)MMH; **324** (l)CORBIS, (r)D. Hurst/Alamy; **325** Masterfile; **326–327** James Blank/Getty Images; **327** (bkgd)Getty Images, (inset)J. Beckman/MMH; **328** Eckhard Slawik/Photo Researchers, Inc.; **329** (t)CORBIS, (others)Eckhard Slawik/Photo Researchers, Inc.; **330** Eckhard Slawik/Photo Researchers, Inc.; **331** MMH; **332** (t)CORBIS, (b)PunchStock; **333** (t)Eckhard Slawik/Photo Researchers, Inc., (c)PunchStock, (b)MMH; **334** MMH; **335** Getty Images; **336–337** Alamy; **337** (t to b)MMH, Janet Beckman/MMH; **339** MMH; **340** (tl)Getty Images, (tr)Photo Researchers, Inc., (b)Photo Researchers, Inc.; **341** (tl)PunchStock, (tr)Getty Images, (bl)NASA-JPL, (br)Elvele Images/Alamy; **342** (t)Charles C. Place/PunchStock, (b)Cornell University; **343** (t)Charles C. Place/PunchStock, (c)Photo Researchers, Inc., (b)MMH; **344** Alamy; **345** (l to r, t to b)Elvele Images/Alamy, Photo Researchers, Inc., Getty Images, Photo Researchers, Inc., CORBIS, PunchStock, Getty Images, Photo Researchers, Inc.; **346** Daryl Pederson/Alaska Stock; **346–347** Shigemi Numazawa/Photo Researchers, Inc.; **347** (l to r, t to b)MMH, Janet Beckman/MMH; **349** (t)Janet Beckman/MMH, (b)Gerard Lodriguss/Photo Researchers, Inc.; **351** (t)Gerard Lodriguss/Photo Researchers, Inc., (b)MMH; **352** Denis Finnin/American Museum of Natural History; **352–353** Photo Researchers, Inc.; **354** (t to b)Getty Images, James Blank/Getty Images, Alamy, Shigemi Numazawa/Photo Researchers, Inc., MMH; **355** (t)Cornell University, (b)Eckhard Slawik/Photo Researchers, Inc.; **356** (t)Joe Raedle/Getty Images, (b)Luiz C. Marigo/Peter Arnold, Inc.; **357** Alaska Stock; **358–359** Peter Dejong/AP Images; **359** (cr)Courtesy Sea Heart Foundation, (others)Ken Karp/MMH; **360–361** Alamy; **361** (t to b)Paul Barton/CORBIS, Matthias Stolt/Alamy, Ariel Skelley/CORBIS, Alamy, age fotostock, Janet Beckman/MMH; **362–363** CORBIS; **363** MMH; **364** Paul Barton/CORBIS; **365** (tl)PunchStock, (tr)Alamy Images, (b)PunchStock; **366** (t)Matthias Stolt/Alamy, (b)Burnett-Palmer/Bruce Coleman USA; **367** (l)Ken Cavanagh/MMH, (r)MMH; **368** (l to r, t to b)Getty Images, Alamy, Alamy, Ken Cavanagh/MMH, Lawrence Lawry/Photo Researchers, Inc., CORBIS; **369** (t to b)Paul Barton/CORBIS, Burnett-Palmer/Bruce Coleman USA, CORBIS, Jacques Cornell/MMH; **370** Denis Finnin/American Museum of Natural History; **370–371** Photo Researchers, Inc.; **372–373** Alexis Rosenfeld/Photo Researchers, Inc.; **373** Michael Scott/MMH; **374** (t)MMH, (bl)Jupiterimages, (b)Frank Gaglione/Getty Images; **375** (l to r)Ken Karp/MMH, MMH, Ken Cavanagh/MMH; **376** (t)Alamy, (b)Michael Scott/MMH; **377** (bl)Janet Beckman/MMH, (others)MMH; **378** (t)MMH, (b)NASA; **379** (t to b)Ken Cavanagh/MMH, Michael Scott/MMH, NASA, Jacques Cornell/MMH; **380, 381** MMH; **382–383** Danita Delimont; **383** (b)Michael Scott/MMH, (others)MMH; **384** Ariel Skelley/CORBIS; **385** (t)Alamy, (bl)CORBIS, (br)Gary S. Chapman/Getty Images; **386** (t)MMH, (b)age fotostock; **386–387** Janet Beckman/MMH; **387** (t)MMH, (b)Janet Beckman/MMH; **388** (t)Richard Hutchings/PhotoEdit, (c)David Young-Wolff/PhotoEdit, (b)Tom Stewart/CORBIS; **389** (t to b)Alamy, MMH, Janet Beckman/MMH, Jacques Cornell/MMH; **390** Getty Images; **391** MMH; **392** (t to b)CORBIS, Alexis Rosenfeld/Photo Researchers, Inc., Danita Delimont, Jacques Cornell/MMH; **393** (t)Jack Hollingsworth/Getty Images, (c)Alamy, (b)Ken Cavanagh/MMH; **394–395** James Allan Brown/Visuals Unlimited; **395** (t to b)John A. Rizzon/Getty Images, MMH/MMH, Jose Luis Pelaez, Inc./CORBIS, Getty Images, Alamy, James Sparshatt/CORBIS; **396–397** Ei Katsumata/Alamy; **397** (b)Michael Scott/MMH, (others)MMH; **398** Masterfile; **399** (tl, tr)Alamy, (others)MMH/MMH; **400** (t)Viesti Associates, (b)Alamy; **401** (t)MMH, (b)Michael Newman/PhotoEdit; **402** CORBIS; **403** (t to b)Masterfile, Michael Newman/PhotoEdit, CORBIS, Jacques Cornell/MMH; **404** Michael DeYoung/CORBIS; **405** MMH; **406–407** Ulana Switucha/Alamy; **407** (b)Michael Scott/MMH, (others)MMH; **408** Jose Luis Pelaez, Inc./CORBIS; **409** (t to b)Charles E. Rotkin/CORBIS, James Schnepf/Getty Images, Alamy, Dan Holmberg/Getty Images; **410** (t)Getty Images, (b)Alamy; **411** Getty Images; **412** (t)Ken Karp/MMH, (bl)CORBIS, (bc)Jacqui Hurst/CORBIS, (br)David R. Frazier/The Image Works; **413** (t to b)Dan Holmberg/Getty Images, Getty Images, CORBIS, Jacques Cornell/MMH; **414** (t)Steven Puetzer/Getty Images, (b)Brian Hagiwara/Getty Images; **414–415** CORBIS; **415** Jan Suttle/Alamy; **416–417** (t)PunchStock, (b)Richard Ransier/Alamy; **417** MMH; **418** (l)John A. Rizzo/Getty Images, (c)PunchStock, (r)Alamy; **419** (t)MMH, (b)James Sparshatt/CORBIS; **420** (t)age fotostock, (b)Ken Cavanagh/MMH; **420–421** Jose Fuste Raga/CORBIS; **421** (t to b)James Sparshatt/CORBIS, Alamy, age fotostock, Jacques Cornell/MMH; **422, 423** MMH; **424** (t to b)Ei Katsumata/Alamy, Ulana Switucha/Alamy, PunchStock, Jacques Cornell/MMH, Richard Ransier/Alamy; **425** CORBIS; **426** (t)Colin Cuthbert/Photo Researchers, Inc., (b)Natalie Fobes/CORBIS; **427** Richard Cummins/CORBIS; **428–429** The Image Works; **430–431** CORBIS; **431** (t to b)Kwame Zikomo/SuperStock, Franco & Bonnard/CORBIS, Michael Newman/PhotoEdit, Jeff Greenberg/Alamy, CORBIS, Steve Cole/Getty Images; **432–433** Alamy; **433** (t to b)MMH, Michael Scott/MMH; **434** Kwame Zikomo/SuperStock; **435** (t)Ken Cavanagh/MMH, (b)Ken Karp/MMH; **436** (t)Michael Prince/CORBIS, (others)Franco & Bonnard/CORBIS; **437** (t)CORBIS, (bl)Mark Junak/Getty Images, (br)Jason Molyneaux/Masterfile; **438** (tl)MMH, (cl)Michael Scott/MMH, (cr)Getty Images, (b)Masterfile; **439** (t to b)Kwame Zikomo/SuperStock, Mark Junak/Getty Images, Getty Images, Masterfile, Jacques Cornell/MMH; **440** The Image Works; **441** (l)CORBIS, (r)Robert Mullan/Alamy; **442–443** Tsuneo Nakamura/Alamy; **443** (b)Michael Scott/MMH, (others)MMH; **444** Michael Newman/PhotoEdit; **445** (t)Tim Pannell/CORBIS, (c)Jim Cummins/CORBIS, (b)Tim McGuire/CORBIS; **446** (t)Alamy, (bl)Richard T. Nowitz/CORBIS, (bc)Ken Cavanagh/MMH, (br)Michael Scott/MMH; **447** (b)Andi Duff/Alamy, (others)Ken Karp/MMH; **448** (t)CORBIS, (b)Alamy; **449** (t to b)Michael Newman/PhotoEdit, Andi Duff/Alamy, Alamy, Jacques Cornell/MMH; **450** (b)Ken Karp/MMH, (others)MMH; **452–453** Marc Steinmetz/The Image Works; **453** (b)Michael Scott/MMH, (others)MMH; **454** PhotoEdit; **455** (tl)David Young-Wolff/PhotoEdit, (tr)Alamy, (cl)Bob Daemmrich/PhotoEdit, (cr)Alan & Sandy Carey/Getty Images, (bl)Jeff Greenberg/Peter Arnold, Inc./Alamy, (br)Jonathan Nourak/PhotoEdit; **456–457** CORBIS; **458** Alan Thornton/Getty Images; **459** (t to b)Jeff Greenberg/Alamy, CORBIS, Alan Thornton/Getty Images, Jacques Cornell/MMH; **460** Neil McAllister/Alamy; **461** MMH; **462–463** Jim Corwin/Index Stock Imagery; **463** (b)Michael Scott/MMH, (others)MMH; **464** David Sailors/CORBIS; **465** (tl)Michael Scott/MMH, (tc)Bartomeu Amengual/SuperStock, (tr)CORBIS, (bl)Rob Rayworth/Alamy, (bc)Tony Freeman/PhotoEdit, (br)Juan Silva/PunchStock; **467** (t)Gary Rhijnsburger/Masterfile, (b)Neil Beer/Getty Images, (inset)Sebastien Baussai/Alamy; **468** (t)Steve Cole/Getty Images, (b)Wally Bauman/Alamy; **469** (t)Michael Scott/MMH, (b)Tony Freeman/PhotoEdit; **470** (l)Getty Images, (r)CORBIS; **471** (t to b)Juan Silva/PunchStock, Neil Beer/Getty Images, Getty Images, Jacques Cornell/MMH; **472** Alamy; **473** (l)PunchStock, (r)Ken Cavanagh/MMH; **474** (t to b)Alamy, Tsuneo Nakamura/Alamy, Marc Steinmetz/The Image Works, Jim Corwin/Index Stock Imagery, Jacques Cornell/MMH; **475** (tl)Alamy, (tc)Alan & Sandy Carey/Getty Images, (tr)CORBIS, (bl)Tony Freeman/PhotoEdit, (bc)Wally Bauman/Alamy, (br)Getty Images; **476–477** Alamy; **477** (t to b)Peter Gridley/Alamy, Alamy, Yoav Levy/PhotoTake, David Young-Wolff/Alamy, Chad Ehlers/Alamy; **478–479** Peter Gridley/Alamy; **479** (b)Ken Karp/MMH, (others)MMH; **480–481** Heather Perry/Getty Images; **481** Ken Karp/MMH; **482** (t)John A. Rizzo/Getty Images, (c)John A. Rizzo/Getty Images, (b)Alamy; **483** (t)CORBIS, (b)Alamy Images; **484** (t)Lisa Romerein/Getty Images, (b)Alamy; **485** (t to b)Heather Perry/Getty Images, CORBIS, Lisa Romerein/Getty Images, MMH; **486** (t)PunchStock, (b)Ken Karp/MMH; **487** (t)PunchStock, (b)Getty Images; **488–489** David Young-Wolff/Alamy; **489** (b)Ken Karp/MMH, (others)MMH; **490** (b)David Gregs/Alamy, (inset)Yoav Levy/PhotoTake; **491** (t)Alamy, (c)PunchStock, (b)Jayme Thornton/Getty Images; **492** (t)Joe Raedle/Getty Images, (b)Alan Williams/NHPA; **493** P. Narayan/age fotostock; **494** Michael Newman/PhotoEdit; **495** (t to b)Alamy, P. Narayan/age fotostock, MMH; **496** (bl)Materials - Water/MMH, (others)MMH; **497** MMH; **498–499** Ben Hays/Alamy; **499** MMH; **500** Chad Ehlers/Alamy; **501** (t)Richard Megna/Fundamental Photographs, NYC, (b)Richard Hutchings/PhotoEdit; **502** (t)David Keaton/CORBIS, (b)Liane Cary/age fotostock; **503** (t)Philippa Lewis/CORBIS, (b)Richard Megna/Fundamental Photographs, NYC; **504** (t)Alfred Pasieka/Photo Researchers, Inc., (b)David Olsen/Alamy; **505** (t)MMH, (b)David Fischer/Getty Images; **506** PunchStock; **507** (t to b)Richard Megna/Fundamental Photographs, NYC, Alfred Pasieka/Photo Researchers, Inc., PunchStock, MMH; **508** Rich LaSalle/Getty Images; **508–509** Kurt Coste/Getty Images; **510–511** R. Morely/Getty Images; **511** (t to b)MMH, MMH, Joe Polillio/

MMH; **512** (l)Ken Cavanagh/MMH, (r)Joe Polillio/MMH; **513** Joe Polillio/MMH; **514** (t)Ken Cavanagh/MMH, (cl)Dynamic/Alamy, (cr)Brian Hagiwara/Jupiterimages, (b)PunchStock; **516** (t to b)CORBIS, Ken Karp/MMH, MMH, David Hebden/Alamy; **517** (t to b)Joe Polillio/MMH, David Hebden/Alamy, MMH; **518** (t)Alfred Buellesbach/ The Image Works, (b)Alamy; **519** CORBIS; **520** (t to b)Peter Gridley/Alamy, David Young-Wolff/Alamy, Ben Hays/Alamy, R. Morely/Getty Images, MMH; **521** (l)Lisa Romerein/Getty Images, (r)Alamy; **522** (t)Adam Woolfitt/CORBIS, (b)Michael Newman/PhotoEdit; **R01** (t)Leonard Lessin/Photo Researchers, Inc., (c)Michael Scott/MMH, (b)Jonathan Nourak/PhotoEdit; **R02** (tl)David Young-Wolff/ PhotoEdit, (tr)Ken Cavanagh/MMH, (cl)Ken Karp/MMH, (cr)Alamy, (bl)Randy Faris/ CORBIS, (br)Getty Images; **R03** PunchStock; **R04** (t)MMH, (c)Ken Cavanagh/MMH, (b)Amos Morgan/Getty Images; **R05** (tl)Ken Cavanagh/MMH, (tc, tr)MMH, (b)Michael Scott/MMH; **R06** (t)Frans Lanting/Minden Pictures, (b)Jupiterimages; **R07** (t)Getty Images, (b)Joe Polillio/MMH; **R08** (t)PhotoEdit, (b)Gerry Ellis/Minden Pictures, (inset)Katherine Feng/Minden Pictures; **R09** PunchStock; **R14** Roger Harris/ Photo Researchers, Inc.; **R15** (t)Michael Newman/PhotoEdit, (b)Richard Hutchings/ MMH; **R16** (l)Photo Researchers, Inc., (r)MMH; **R17** Dan Bigelow/Getty Images; **R18** Alamy; **R19** Amos Morgan/Getty Images; **R20** (t)Index Stock Imagery, (b)Richard Hutchings/MMH; **R21** (t to b)CORBIS, Hoby Finn/Getty Images, Cindy Charles/PhotoEdit, PunchStock, PunchStock; **R22** (l)Photo Researchers, Inc., (r)MMH; **R23** (l to r, t to b)Dr. Gopal Murti/Visuals Unlimited, Dr. Gary D. Gaulger/ PhotoTake, Steven Mark Needham/Jupiterimages, Ken Karp/MMH, Ryan McVay/ Getty Images, PunchStock, MMH; **R24** (cw from top)Alamy, Getty Images, PunchStock, Getty Images, PunchStock, Getty Images, PunchStock; **R25** (tl)Getty Images, (tc)Ken Cavanagh/MMH, (tr)Jacques Cornell/MMH, (cl)Jonell Weaver/Getty Images, (cr)Pornchai Mittongtare/Jupiterimages, (b)Steven Mark Needham/Jupiterimages; **R26** (t)MMH, (b)Ariel Skelley/CORBIS; **R29** (tl)David Fischer/Getty Images, (tr)Darrell Gulin/Getty Images, (c)Lee Karney/US Fish & Wildlife Service, (bl)Tim Flach/ Getty Images, (br)Alamy, (inset)Getty Images; **R30** (l to r, t to b)MMH, PunchStock, Gregory G. Dimijian/Photo Researchers, Inc., CORBIS, Lester V. Bergman/CORBIS, Nigel J. Dennis/Photo Researchers, Inc., James Sparshatt/CORBIS, Getty Images, Michael Newman/PhotoEdit; **R31** (l to r, t to b)Viesti Associates, David Hebden/ Alamy, CORBIS, Siede Preis/Getty Images, John A. Rizzo/Getty Images, Gerard Lodriguss/ Photo Researchers, Inc., Alan & Sandy Carey/Getty Images, CORBIS; **R32** (l to r, t to b)David L. Brown/PictureQuest, Ken Karp/MMH, Ken Cavanagh/ MMH, Joe Polillio/MMH, Peter Arnold Inc., Alamy, Lloyd Cluff/CORBIS, Nigel Cattlin/ Alamy, Siede Preis/Getty Images, Alan & Sandy Carey/Getty Images; **R33** (l to r, t to b)Alan Thornton/Getty Images, Russ Merne/Alamy, Joseph Van Os/ Getty Images, PunchStock, Tom & Pat Leeson/Photo Researchers, Inc., Josh Ritchie/ Getty Images, Alamy, Susan Cruz/MMH, Fritz Rauschenbach/CORBIS, G.K. & Vikki Hart/ Getty Images, Michael Newman/PhotoEdit; **R34** (l to r, t to b)Russell Illig/Getty Images, Janet Beckman/MMH, Jonathan Blair/CORBIS, Jack Jackson/Alamy, Michael Newman/ PhotoEdit, Bernhard Edmaier/Photo Researchers, Inc., Alamy, Andi Duff/Alamy, Getty Images, age fotostock; **R35** (l to r, t to b)PunchStock, Jacques Cornell/MMH, Heather Perry/Getty Images, Steve Cole/Getty Images, Alamy, Bonnie Sue Rauch/ Photo Researchers, Inc., Paul Sterry/Alamy, PunchStock, David Hebden/Alamy, Daniel Aguilar/CORBIS, PictureQuest; **R36** (l to r, t to b)CORBIS, Siede Preis/Getty Images, PunchStock, CORBIS, Chad Ehlers/Alamy, Troy Bartlett/Alamy, MMH, Masterfile, Dan Bigelow/Getty Images; **R37** (l to r, t to b)Dr. Parvinder Sethi, Ken Cavanagh/ MMH, Stephen Dalton/Photo Researchers, Inc., PunchStock, Ken Cavanagh/MMH, CORBIS, Alamy, PunchStock, Bill Beatty/Visuals Unlimited, Ei Katsumata/Alamy, Siede Preis/Getty Images; **R38** (l to r, t to b)Getty Images, Getty Images, CORBIS, PunchStock, Dr. Merlin D. Tuttle/Photo Researchers, Inc., PunchStock, Owaki-Kulla/ CORBIS, MMH; **R39** (l to r, t to b)Eckhard Slawik/Photo Researchers, Inc., Annie Griffiths Belt/Getty Images, Kwame Zikomo/SuperStock, Ulana Switucha/ Alamy, CORBIS, Siede Preis/Getty Images, Getty Images, CORBIS, Hans Pfletschinger/ Peter Arnold Inc., CORBIS, Brian Sytnyk/Masterfile, PunchStock; **R40** (l to r, t to b)Sebastien Baussai/Alamy, Glen Allison/Getty Images, PunchStock, Zigmund Leszczynski/Animals Animals, Getty Images, PunchStock, Getty Images, PunchStock, Richard Megna/Fundamental Photographs, NYC, Gary Vestal/Getty Images, Richard Megna/Fundamental Photographs, NYC, Stephen Saks/Lonely Planet Images; **R41** (l to r, t to b)Jerome Wyckoff/Animals Animals, Jacques Cornell/MMH, MMH, Ken Cavanagh/MMH, Liane Cary/age fotostock, PunchStock, Tony Freeman/PhotoEdit, Michael Scott/MMH, PunchStock, Jacques Cornell/MMH; **R42** (l to r, t to b)CORBIS, Jacques Cornell/MMH, PunchStock, Alamy, Paul Barton/CORBIS, John A. Rizzo/Getty Images, Ken Cavanagh/MMH, Joe Polillio/MMH, David Gregs/Alamy, Susan Cruz/MMH, Cornell University, CORBIS, Alan & Sandy Carey/Getty Images; **R43** (l to r, t to b)Getty Images, Philippa Lewis/CORBIS, Russell Illig/Getty Images, Craig K. Lorenz/Photo Researchers, Inc., Burke/Jupiterimages, Yoav Levy/PhotoTake, CORBIS; **R44** (l to r, t to b)Ken Cavanagh/MMH, Juan Silva/PunchStock, PunchStock, MMH, CORBIS, Neil Beer/Getty Images, Yva Momatiuk/John Eastcott/Minden Pictures, Alamy; **R66–67** Gary Bell/oceanwideimages.com; **R67** Chris Newbert/Minden Pictures.